DORDT INFORMATION SERVIC

3 6520 0102643 W

D0903594

DATE DUE

Methods of Experimental Physics

VOLUME 12

ASTROPHYSICS

PART A: Optical and Infrared

METHODS OF EXPERIMENTAL PHYSICS:

L. Marton, *Editor-in-Chief*

Claire Marton, *Assistant Editor*

DORDT COLLEGE LIBRARY
Sioux Center, Iowa 51250

Volume 12

Astrophysics

PART A: Optical and Infrared

Edited by

N. CARLETON

Smithsonian Astrophysical Observatory
and Harvard University
Cambridge, Massachusetts

1974

ACADEMIC PRESS · New York and London

A Subsidiary of Harcourt Brace Jovanovich, Publishers

60238
60238

YRARBIL 3G3LLOC TGROD
0ặ210 awol ,iełneQ xuoiS

COPYRIGHT © 1974, BY ACADEMIC PRESS, INC.
ALL RIGHTS RESERVED.
NO PART OF THIS PUBLICATION MAY BE REPRODUCED OR
TRANSMITTED IN ANY FORM OR BY ANY MEANS, ELECTRONIC
OR MECHANICAL, INCLUDING PHOTOCOPY, RECORDING, OR ANY
INFORMATION STORAGE AND RETRIEVAL SYSTEM, WITHOUT
PERMISSION IN WRITING FROM THE PUBLISHER.

ACADEMIC PRESS, INC.
111 Fifth Avenue, New York, New York 10003

United Kingdom Edition published by
ACADEMIC PRESS, INC. (LONDON) LTD.
24/28 Oval Road, London NW1

Library of Congress Cataloging in Publication Data
Main entry under title:

Optical and infrared astronomy.

    (Methods of experimental physics, v. 12: Astrophysics,
pt. A)
    Includes bibliographical references.
    1.   Astronomical spectroscopy.   2.   Photometry,
Astronomical.   3.   Spectrum, Infra-red.   I.   Carleton,
Nathaniel, Date   ed.   II.   Series: Methods of
experimental physics, v. 12, pt. A.   III.   Series:
Astrophysics, pt. A.
QB465.067   522'.6   73-17150
ISBN 0−12−474912−1

PRINTED IN THE UNITED STATES OF AMERICA

CONTENTS

1. Photomultipliers: Their Cause and Cure

by ANDREW T. YOUNG

2. Other Components in Photometric Systems

by ANDREW T. YOUNG

8. Polarization Techniques

by K. SERKOWSKI

11. Fourier Spectrometers

by HERBERT W. SCHNOPPER and RODGER I. THOMPSON

12. Fabry-Perot Instruments for Astronomy

by F. L. ROESLER

CONTRIBUTORS

Numbers in parentheses indicate the pages on which the authors' contributions begin.

G. G. FAZIO, *Smithsonian Astrophysical Observatory, Cambridge, Massachusetts* (315)

DONALD M. HUNTEN, *Kitt Peak National Observatory, Tucson, Arizona* (193)

GERALD E. KRON, *US Naval Observatory, Flagstaff Station, Flagstaff, Arizona* (252)

D. W. LATHAM, *Smithsonian Astrophysical Observatory, Cambridge, Massachusetts* (221)

F. J. LOW, *Lunar and Planetary Laboratory, University of Arizona, Tucson, Arizona* (415)

JOHN L. LOWRANCE, *Princeton University Observatory, Princeton University, Princeton, New Jersey* (277)

G. H. RIEKE, *Lunar and Planetary Observatory, University of Arizona, Tucson Arizona* (415)

F. L. ROESLER, *Department of Physics, University of Wisconsin, Madison, Wisconsin* (531)

HERBERT W. SCHNOPPER, *Department of Physics, Massachusetts Institute of Technology, Cambridge Massachusetts* (491)

DANIEL J. SCHROEDER, *Thompson Observatory, Beloit College, Beloit, Wisconsin* (463)

K. SERKOWSKI, *Lunar and Planetary Laboratory, University of Arizona, Tucson, Arizona* (361)

RODGER I. THOMPSON, *Steward Observatory, University of Arizona, Tucson, Arizona* (491)

E. J. WAMPLER, *Lick Observatory, Board of Studies in Astronomy and Astrophysics, University of California, Santa Cruz, California* (237)

ANDREW T. YOUNG, *Department of Physics, Texas A & M University, College Station, Texas* (1)

PAUL ZUCCHINO, *Princeton University Observatory, Princeton University, Princeton, New Jersey* (277)

FOREWORD

In planning possible expansions of our series, and with it a better coverage of physics, we became aware of the need to present the methods used in astrophysics. After discussing with friends the possible scope of such a volume it became advisable to split the subject into two logical subdivisions: one covering optical and infrared astrophysics and a second covering radio astrophysics. The present volume, edited by Professor N. Carleton, represents the first part; and Dr. L. Meeks has edited the second.

At this time it is my pleasure to announce two more expected additions to this series: a volume on Fluid Dynamics, to be edited by Professor R. Emrich, and a volume on Polymer Physics, edited by Dr. A. Peterlin.

I wish to express my heartfelt thanks to Dr. Carleton and to all the authors for their efforts. I hope this volume will be as well received as its predecessors.

L. Marton

PREFACE

Each contributor to this volume was asked simply to provide an exposition of the methods of working in his field that would be useful particularly to graduate students and to others entering the field. The guiding idea has been to discuss methods of observation in terms of the basic physics involved, with as little reference as possible to specific apparatus. The length and degree of detail in the various articles, as planned out with the authors, are functions of the amount and quality of our knowledge in the area, of its apparent stability, and of the availability of already existing articles. Variations in these parameters have suggested large variations in the style and length of the contributions.

I wish to thank the authors for their thoughtfulness in working within this framework, and hope that the readers of the volume will indeed find that our judgments have generated a useful reference work.

NATHANIEL P. CARLETON

1. PHOTOMULTIPLIERS: THEIR CAUSE AND CURE*†

1.1. Introduction

Photoelectric photometry is a fundamental technique in observational astronomy. It is done at one time or another by most observational astronomers. However, in spite of its importance, it is often inadequately taught and badly practiced, and there is no modern text or reference book on the subject. Observers rarely give details of their technique, and a multitude of sins may be concealed with the phrase "the observations were made and reduced in the usual way"—the "usual way" being different for each worker. Some even seem to regard their techniques as trade secrets.

In fact, photometry comparable to good modern work was done by Prager and Guthnick[1] in 1914, using gas-filled photodiodes. Although they reached a precision of about 1% (0.01 mag), they could only measure the brightest stars in the sky. Because the equipment was delicate, and the observing techniques rather tedious, photoelectric photometry was generally regarded as difficult or impractical (although very accurate) for the next 30 years. A good review of the difficulties of the older photometric techniques was published at the end of this period by Weaver.[2] When photomultipliers were introduced after World War II, many persons felt the millennium had arrived, and that anyone could do good photometry without effort.

The principles of photoelectric photometry appear simple. The photoemissive detector produces a current supposedly proportional to the light intensity falling on it, thus reducing photometry to a simple electrical

---

[1] P. Guthnick and R. Prager, *Veröff. Kgl. Sternwarte Berlin Babelsberg.* Band I, Heft 1 (1914); Band II, Heft 3 (1918).

[2] Harold F. Weaver, *Popular Astron.* 54, 211, 287, 339, 389, 451, 504 (1946).

---

† See also Volume 2, Chapter 11.1.

* Part 1 is by Andrew T. Young.

measurement that can be made very accurately. However, systematic errors on the order of 0.1 mag have been discovered in published photometric data that were nominally accurate to 0.01 mag or better. Although fainter stars have been reached with modern photomultipliers, the general accuracy of photoelectric photometry[2a] has not improved in 50 years. Thus, although it would be nice to offer "photometry without tears," it appears that good photometry only comes with tears. However, it is better to shed them while doing the work than after the results are published, when someone else shows they are wrong.

The thesis of the first three parts of this volume is that *accurate* photometry is still a difficult undertaking, but that it can be done if one is willing to take the necessary pains. Although it is written primarily for astronomers, it may be useful to the many physicists, chemists, and biologists who must also use photomultipliers without really understanding how they work, or why they often fail to do what is expected of them. Such workers may study fluorescence, flash photolysis, Raman scattering, or more ordinary photometric problems, using such laboratory apparatus as spectrophotometers, colorimeters, polarimeters, and microdensitometers. The first two parts of this volume deal with the principles of photoelectric photometry, and are directed to these workers as well as to astronomers. The third part is more specifically astronomical, although the discussions of atmospheric extinction and transformation of data to a standard system are relevant to disciplines outside astronomy. The emphasis throughout is on physical principles rather than on specific hardware and software, not only because the latter are rapidly outdated by new photometric techniques, but also because the user needs to *understand* what he is doing, rather than to be given a book of recipes.

If sufficient care is given to *each* of the photometric problems described below, substantial improvements should be possible. There is no reason why astronomical photometry cannot be done to a real accuracy of 0.002 or 0.003 mag, rather than the 0.01 or 0.02 usually achieved[2a] today. We do not need better photomultipliers or skies, or bigger telescopes and computers to reach this goal. We do need better design and more careful use of photometers, standard systems, and observation and reduction techniques. Better results will help solve problems throughout astrophysics, galactic structure, and extragalactic astronomy. So let us get on with it!

[2a] M. P. FitzGerald, *Astron. Astrophys. Suppl.* **9**, 297 (1973).

1.2. An Idealized Photomultiplier

The "textbook" photomultiplier is fairly simple. Light, striking a photocathode enclosed in a vacuum tube, releases photoelectrons. These are attracted to a positively charged electrode, called a *dynode*, where (owing to the now considerable energy of the initial electrons, which have been accelerated by about 100 V) more electrons are "splashed out." These secondary electrons are in turn accelerated to a still more positive dynode, and the process is repeated. The gain at each dynode is typically a factor of four; after perhaps ten such multiplications, the original photocurrent has been increased by the electron multiplier roughly a millionfold, and is easily measured. Even the individual electron avalanches due to single photoelectrons are readily detected as pulses at the anode, the final collecting electrode. Since the probability of photoelectron emission is independent of the light intensity, the average anode current (or the pulse-counting rate) is directly proportional to the light intensity over a great dynamic range.

Unfortunately, this "textbook" model is extremely oversimplified and misleading. It is an unfortunate fact of life that photomultipliers are influenced by almost every conceivable environmental variable: temperature, humidity, electric and magnetic fields, time (aging and fatigue effects), and even radioactivity and cosmic rays, in addition to an extreme sensitivity to changes in operating potentials. Consequently, these devices are susceptible to an appalling range of interactions with associated equipment, and a procedure intended to cure one trouble may introduce or intensify another.

To understand the properties of real photomultipliers, we must understand the physical processes that occur in them. We begin with a close look at photoemission.

1.3. Basic Physics of Photomultipliers

1.3.1. Photoemission

1.3.1.1. Introduction. If light of sufficiently short wavelength falls on a solid, electrons are emitted. This process (*photoemission*, or the external *photoelectric effect*) was discovered by Hertz[3] in 1887 during his experiments on the generation of electromagnetic waves.

[3] H. Hertz, *Ann. Phys.* **31** (3), 983 (1887); and *Wied. Ann.* **31**, 383 (1887).

The effect was explained in Einstein's classic papers[4] of 1905–1906, using Planck's idea that matter absorbs radiation of wavelength λ only in *quanta* whose energy is

$$E = hc/\lambda, \tag{1.3.1}$$

where $h = 6.63 \times 10^{-27}$ erg sec is Planck's constant and $c = 3 \times 10^{10}$ cm/sec is the speed of light. Einstein proposed that electrons are bound to the material by an energy W, called the *work function*. This explained the absence of photoemission for light of wavelength longer than the *photoelectric threshold* wavelength λ_c. If $\lambda < \lambda_c$, the maximum energy of the emitted photoelectrons is

$$E_{\max} = hc/\lambda - W. \tag{1.3.2}$$

Notice that $E_{\max}$ is not the average or typical energy of the emitted electrons, but their maximum energy. Most of them will have much less energy; many are insufficiently energetic to escape at all. Thus, the probability of emission, sometimes called the *quantum efficiency* or *quantum yield*, is always less than unity. For practical work, the quantum yield per incident photon is the important parameter; but for theoretical studies of photoemission, only the quantum yield per absorbed photon (i.e., the yield corrected for reflection and transmission losses) is important.

Since $E_{\max}$ is measured electrically (as the retarding potential required to suppress the photocurrent), it is convenient to express energies in electron volts (eV). Equation (1.3.1) then becomes

$$E = 12{,}398/\lambda \tag{1.3.3}$$

if λ is measured in angstroms. Thus, the energies of visible quanta are of the order of 2 or 3 eV. The work functions of pure metals range from about 2 V for alkali metals (Cs, Rb) to about 5 V for noble metals (Pt, Au).

At the threshold wavelength λ_c, the energy of the photoelectrons should be zero, so we have

$$\lambda_c = hc/W. \tag{1.3.4}$$

Experimental determination of λ_c thus gives the work function $W = hc/\lambda_c$, which generally agrees well with values of W obtained from thermionic emission studies of clean metallic surfaces.

Since the number of quanta or *photons* in a beam of monochromatic light is proportional to its intensity, and the probability of photoemission

[4] A. Einstein, *Ann. Phys.* **17** (4), 132 (1905); **20**, 199 (1906).

is the same for each photon, the average number of photoelectrons emitted per second should be exactly proportional to the intensity of the light. This basic linearity of photoemission is very important astronomically, since the brightest stars are about 10^{10} times brighter than the faintest, and the sun is about 10^{10} times brighter still. The determination of astronomical distances depends heavily on accurate measurements of such large intensity ratios.

For reasons which will appear shortly, pure metals are not practical photocathode materials. In order to understand such materials, it is necessary to introduce some elementary solid-state physics in the next few pages. The picture that will be introduced is oversimplified, but will suffice to give the user some appreciation of what goes on in his instrument. Readers who want a better understanding of solid-state physics should consult Kittel's textbook.[5]

1.3.1.2. Photoemission from Metals. The simple "textbook" picture of photoemission given above does not adequately describe the observed emission from metals. As λ increases, the photocurrent falls off exponentially near λ_c, instead of cutting off sharply. Since the material is not at absolute zero, the electrons have some thermal energy which helps them overcome the potential barrier W.

A detailed explanation can be given by considering the distribution of electron energies. Because of their high density, the electrons in a cold metal (like those in a white dwarf star) form a degenerate gas, filling all energy states up to an energy E_F, the Fermi level (Fig. 1). A continuum of states is available above the Fermi level, but these are all empty at 0 K. The difference in energy between the vacuum level E_v and the Fermi level is the work function W. These relations are usually shown in an energy-level diagram analogous to the Grotrian diagram of atomic energy levels (Fig. 2).

At temperatures above absolute zero, some electrons are thermally excited into states above the Fermi level. These excited electrons can then be removed by photons with somewhat less energy than W. (This process is analogous to photoionization of an atom from an excited state.) Because of the continuum of excited states in a metal, the experimental photoelectric threshold is smeared out. A detailed theory for photoelectric emission from metals was first given by Fowler,[6] based on the above picture.

[5] C. Kittel, "Introduction to Solid State Physics," 3rd ed. Wiley, New York, 1966.
[6] R. H. Fowler, *Phys. Rev.* **38**, 45 (1931).

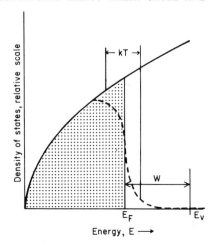

FIG. 1. The density of states available to electrons in a metal, as a function of energy E. The dashed curve represents the density distribution of filled states at temperature T. The shaded area represents the filled states at absolute zero. E_F is the Fermi level and E_V is the vacuum level. The work function $W = E_V - E_F$.

The continuous distribution of energy levels in a metal allows an electron near the Fermi level to move up into a nearly empty state by gaining a very small amount of energy. This has two important consequences. First, by gaining a small energy from a weak external electric field, an electron can easily move through the metal—that is, metals are good electrical conductors. Second, a moving electron in a metal can readily share its energy with its neighbors, so that the mean free path is only a few angstroms.

However, the mean free path of photons in a metal is about 100 Å. Thus only a few percent of the absorbed photons liberate electrons near enough to the surface to escape while they still have more energy than

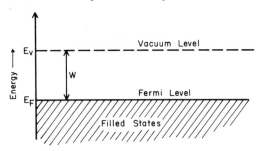

FIG. 2. Energy-level diagram for a cold metal. The shaded area represents filled states at absolute zero (see Fig. 1).

the work function. Furthermore, because the many free electrons make metals good reflectors, the fraction of incident photons that produces photoelectrons (that is, the *quantum efficiency* or probability of photo-emission) is still smaller, generally a few tenths of a percent or less.

Since the electronic structure that gives metals their characteristic properties of high electrical and thermal conductivity, optical reflectivity and absorption, and also the mechanical properties of ductility and malleability—in short, the electronic structure that makes metals "metallic"—also makes them poor photoemitters, we must look to other materials for efficient photoemissive detectors. The first practical photoemitters were in fact the hydrides of alkali metals.

1.3.1.3. Photoemission from Nonmetals

1.3.1.3.1. ELECTRONIC STRUCTURE. Clearly, the recipe for an efficient photoemitter calls for strong optical absorption, combined with comparatively long mean free paths for electrons. The latter could be achieved if there were few energy levels available near the Fermi level, so that excited electrons could not readily share their energy with their neighbors. This situation is found in certain semiconductors, which are, in fact, the practical photoemissive cathodes.

The electrons in crystals are restricted to discrete broad bands of energy, just as the electrons in an isolated atom are restricted to discrete narrow energy levels. In metals, the Fermi level falls within a band, so that there is a high density of states available near the Fermi level.

In semiconductors and insulators, on the other hand, the Fermi level falls between bands, in a region of extremely low state density (Fig. 3).

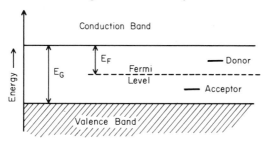

FIG. 3. Energy-level diagram of a semiconductor. The band-gap energy E_G and Fermi level (E_F below the conduction band) are shown. Localized donor and acceptor states in the band gap are also indicated. At absolute zero, all such states below the Fermi level are filled, and all states above are empty. Excitation of an electron from the valence band to an acceptor leaves a hole in the valence band; similarly, donors can ionize to contribute electrons to the conduction band.

The highest filled band is called the *valence band*; the empty band above it is the *conduction band,* and the energy difference E_G between the top of the valence band and the bottom of the conduction band is the *band-gap* energy. Since the Fermi level falls in the band gap, electrons cannot contribute to the electrical conductivity unless they acquire a large energy E_F and enter the conduction band. (Thus, relatively strong external electric fields are required to produce an appreciable current flow, and these materials are poor electrical conductors.) Such *conduction electrons* have long mean free paths, because they cannot share their energy with valence electrons: there are no intermediate states (in the band gap) for valence electrons to move up into, or for conduction electrons to move down into.

The conductivity of metals depends primarily on the length of electron mean free paths, but the conductivity of semiconductors and insulators depends primarily on the availability of charge carriers. At an absolute temperature T, the number of electrons thermally excited into the conduction band will be proportional to $\exp(-E_F/kT)$, where Boltzmann's constant $k = 1.38 \times 10^{-16}$ erg/K or 8.62×10^{-5} eV/K. If E_F is very large compared to kT, there will be practically no conduction electrons and the material will be called an *insulator*. If E_F is not too large, there will be a little conduction, and the material will be called a *semiconductor*. At room temperature, $T = 300$ K, $kT = 0.026$ eV, and the dividing line is at about $E_F = 0.75$ V or roughly 30 times kT. At higher temperatures, many room-temperature insulators show considerable conductivity; an example is the ceramic element of the Nernst glower.

1.3.1.3.2. OPTICAL PROPERTIES. Since the smallest energy that can be given to a valence electron is E_G, a semiconductor is essentially nonabsorbing (a transparent dielectric) for radiation of energy $hc/\lambda < E_G$. However, for $hc/\lambda > E_G$, the material absorbs strongly, like a metal. Thus, at

$$\lambda_a = hc/E_G, \qquad (1.3.5)$$

there is an *absorption edge*. On the short-wavelength side of the absorption edge, both the real and the imaginary parts of the complex refractive index are large (typically ~3). Absorbed photons with $\lambda < \lambda_a$ excite electrons into the conduction band, increasing the number of carriers and hence the conductivity. This *photoconductivity* (or internal photoelectric effect) is useful for radiation detection in the infrared. However, it is not very important for $\lambda < 1$ μm because photoemissive detectors in this region have higher detectivity.

Unsensitized (blue-sensitive) photographic emulsions show a visible absorption edge. In this case, the photosensitive semiconductor is usually AgBr, with a band gap of about 2.8 eV. The absorption edge lies in the blue at about 4500 Å, which accounts both for the yellowish color of the unexposed emulsion and for its rapid drop in sensitivity between 4000 and 5000 Å. In photography, photoelectrons excited into the conduction band move through the crystal and are trapped at "sensitivity specks" (acceptor impurities; see below). When several electrons have been trapped at the same site, the silver halide crystal becomes chemically developable. Thus, the primary process in photography—the excitation of a photoelectron into the conduction band of a semiconductor—is the same as in photoemissive and photoconductive detection.

The insulators have absorption edges in the ultraviolet, corresponding to their large band gaps. Thus, they are transparent to visible light. For example, quartz absorbs strongly below 1600 Å, corresponding to a band gap of 7.75 eV. On the other hand, semiconductors like silicon and germanium, with band gaps of 1.1 and 0.7 eV, respectively, can only be used as transparent (dielectric) optical elements in the infrared beyond their absorption edges at 1.1 and 1.8 μm.

1.3.1.3.3. HOLES, DONORS, AND ACCEPTORS. An unoccupied state in the valence band is called a *hole*. In many ways, a hole behaves like a positively charged electron. Holes in the valence band can move through the crystal and conduct an electric current, just as electrons in the conduction band do. An electron–hole *pair* is normally produced by exciting an electron from the valence band to the conduction band.

A semiconductor in which holes and conduction electrons are about equally numerous is called an *intrinsic* semiconductor. However, an excess of holes (or electrons) can be produced by defects in the crystal lattice; the material is then classified as *p-type* (or *n-type*), according to the sign of the majority carriers. The p- and n-type materials are called *extrinsic* semiconductors.

Defects can be interstitial atoms, vacancies, foreign (impurity) atoms substituted in normal lattice positions, structural dislocations, etc. They produce localized energy levels in the band gap. An electron in such a state near the top of the band gap can be thermally excited into the conduction band; such a state is called a *donor* (see Fig. 3). Excess donors make the material n-type.

On the other hand, a defect can be an electron *acceptor*. When an acceptor traps a thermally excited electron from the valence band, a hole is created in the valence band. An excess of acceptors makes the material p-type.

Donors generally have low ionization energies, i.e., they lie near the lower edge of the conduction band. Acceptors generally lie near the valence band. Donors (acceptors) are neutral in their ground state, and are ionized into an electron (hole) and a positive ion (negative ion) when excited.

1.3.1.3.4. PHONONS. We have described the motion of the electrons as though the atoms of the crystal were fixed. In fact, they are always oscillating about their equilibrium positions. Because each atom is coupled to its neighbors, mechanical vibrations (sound waves) can propagate through the crystal. These vibrations can be analyzed into normal modes, just as the electromagnetic vibrations in a box are analyzed into normal modes in deriving the Planck and Rayleigh–Jeans formulas for temperature radiation. Energy can be added to or subtracted from each mode only in quantized amounts; by analogy with the photons in a black box, these elementary excitations are called *phonons*.

Phonons behave very much like ordinary particles. For example, the phonons in a semiconductor are primarily responsible for thermal conduction, just as, in metals, the more numerous free electrons contribute the conductivity. Interactions between electrons (or holes) and the crystal lattice are conveniently treated as collisions with phonons. Electrons exchange energy and momentum with phonons, just as with other "particles." From the wave point of view, the electron wave can be scattered by irregularities in the periodic potential field of the crystal lattice. Such lattice scattering is an important energy-loss mechanism for slow-moving electrons. Because the lost energy adds to the lattice vibrations, this is often called *phonon production*. Because the lattice atoms are very massive compared to the electrons, these collisions are almost elastic; the direction of the electron's momentum is drastically changed in each collision, but the energy loss is small. In the alkali antimonides, the energy loss is 0.01 eV or less, and the mean free path between collisions is 30 or 40 Å. Conduction electrons thus perform a random walk as they diffuse through the crystal, slowly losing energy and producing phonons. (The process is analogous to the diffusion of photons in a scattering atmosphere.)

If the temperature is lowered, the phonon number density decreases (i.e., the lattice becomes more regular) and the electrons' mean free path between scatterings increases. However, some scattering is produced by permanent defects (such as impurities) that do not disappear on cooling.

1.3.1.3.5. OTHER ENERGY-LOSS MECHANISMS. Conduction electrons can lose energy in other ways. They may be trapped by ionized donors or neutral acceptors, or they may recombine with holes in the valence band. The electron energy may be carried off by phonons or photons. If radiative deexcitation is important, the material is called a *phosphor*.

A very important energy-loss mechanism for energetic electrons is *pair production*: a fast electron collides with a valence electron, raising it into the conduction band and leaving a hole in the valence band. Since the minimum energy required to produce the electron–hole pair is E_G, the original electron must have had at least this much energy above the bottom of the conduction band (total energy above the valence band greater than $2E_G$). In fact, the threshold photon energy E_T for pair production (collisional ionization) is about $3E_G$ in typical photoemitters. Above this energy, the electron mean free path is short (10–20 Å) and independent of temperature.

We can now say that the large value of E_G (and hence of E_T) is what makes possible long mean free paths for photoelectrons, and hence, efficient photoemitters.

1.3.1.3.6. ELECTRON AFFINITY AND WORK FUNCTIONS. However, the excited photoelectron must not only diffuse to the surface of the material, but must also be able to escape. In general, the crystal–vacuum interface presents a potential barrier to electrons. The height of this barrier above the bottom of the conduction band is called the *electron affinity E_A*. Thus the photoelectric work function, the minimum energy required to extract an electron from the valence band, is (see Fig. 4)

$$W_{ph} = E_G + E_A , \qquad (1.3.6)$$

since the few filled states in the band gap make only negligible contributions to the optical absorption (and hence to the photoemission).

Thus an additional requirement for an efficient photoemitter is $E_G > E_A$, so that the photoelectric work function ($E_G + E_A$) is less than the threshold energy for pair production ($E_T \approx 2E_G$). If $E_G < E_A$, conduction electrons energetic enough to escape will rapidly be degraded below the escape energy by pair-producing collisions with valence electrons.

On the other hand, the thermionic work function is

$$W_{th} = E_F + E_A , \qquad (1.3.7)$$

since electrons excited thermally from near the Fermi level dominate the

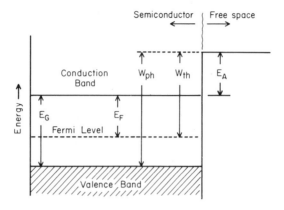

FIG. 4. Energy levels at the semiconductor–vacuum boundary. To escape, a conduction electron must have enough energy to surmount the potential barrier of height E_A (the electron affinity) at the interface.

thermionic emission; valence electrons contribute negligibly, owing to the very small value of $\exp(-W_{ph}/kT)$. Since $E_F < E_G$, $W_{th} < W_{ph}$, and the thermionic emission from a semiconductor may be larger than one would expect from its photoelectric work function. The photoelectric work function measures the energy (below vacuum) of the top of a region of densely filled states, but the thermionic work function measures the position of the Fermi level. (However, this difference is slight for practical photoemitters, which are strongly p-type, so that the two work functions are nearly equal.) In metals, the Fermi level is the top of the densely filled region, so the two work functions are exactly equal.

The angular distribution of the emitted electrons follows approximately a cosine law, as we might expect for diffusely emitted, multiply scattered particles.[†] That is, the photoelectrons have no "memory" of the direction of the incident photons; in fact, photoemission is always a three-body process, in which part of the photon's momentum is transferred to the crystal, so that even the initial velocity vector of the photoelectron is different from that of the photon.

The energy distribution of the photoelectrons is limited according to Einstein's relation [Eq. (1.3.2)]; but, owing to pair production, most of them have energies less than E_T, even for more energetic photons (see Fig. 5). For still more energetic photons, one or more of the electrons released in pair production may escape, as well as (or instead of) the

---

[†] Cf. the quasi-Lambertian limb darkening of a scattering planetary atmosphere.

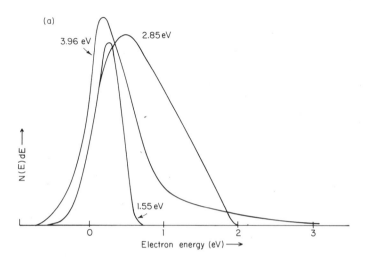

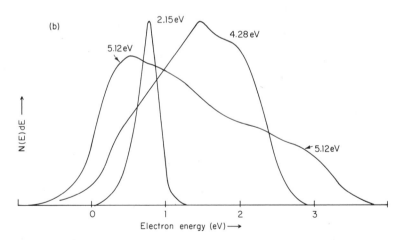

Fig. 5. Photoelectron energy distributions, for two common cathode materials and several different wavelengths of light. Each curve is labeled with the incident photon energy in electron volts. (a) S-1 cathode; data taken from K. S. Neil and C. H. B. Mee [*Phys. Status Solidi* **2**, 43 (1970)]. (b) K_2CsSb cathode; data from R. Nathan and C. H.B. Mee [*Phys. Status Solidi* **2**, 67 (1970)]. In both cases a decrease in mean electron energy, due to pair production, appears at high photon energies.

original photoelectron; thus, in the far ultraviolet, several photoelectrons can be produced by a single photon.

The electron affinity is strongly affected by impurities, dislocations, etc., at the surface of the material. Thus, different samples of nominally identical materials may have very different E_A's and consequently different photoelectric thresholds, although the band-gap energy E_G may be very similar for all of them.

The Fermi level, and consequently the thermionic work function, is also strongly influenced by defects, being pushed upward by donors (n-type material) and downward by acceptors (p-type). For example, if enough donors are added, the Fermi level moves up to within kT of the conduction band, and the material becomes a nearly metallic conductor.

The profound effect of small impurity concentrations on both the electron affinity and the Fermi level—and hence on the photoelectric threshold and thermionic emission—are a considerable nuisance to the user of photoemissive devices, for no two detectors ever have the same spectral response or dark emission. In fact, these properties even vary across the sensitive area of a single detector. However, although these adventitious variations cause practical difficulties, they should not mislead the user into a belief in black magic, or to despair of obtaining a sound understanding of the basic processes involved.

1.3.2. Secondary Emission

Although photoemission is an efficient process in modern cathode materials, it would not play such an important part in astronomical photometry today if the emitted electrons had to be detected directly. The most sensitive electrometers require hundreds of electrons to produce a detectable signal, and are too delicate to be used outside the laboratory. Although a stable gain of about ten is achieved in gas-filled phototubes, the critical advance in photoelectric technology was the invention of the electron-multiplier phototube, or *photomultiplier*.

The electron multiplier uses the phenomenon of pair production described earlier. If an electron is accelerated through about 100 V and then strikes a semiconductor, it will produce many electron–hole pairs near the surface of the material. A few of the electrons will diffuse to the surface and escape; if these can be collected and accelerated, the process can be repeated over and over, until a very large gain is achieved. The escape of these secondary electrons is called *secondary emission*.

Secondary emission is very similar to photoemission, the chief difference being that the escaping electrons are initially produced by an energetic electron rather than by an energetic photon. The factors that facilitate the escape of a photoelectron (e.g., long mean free path and low electron affinity) work equally well to allow the escape of secondary electrons. Thus it is hardly surprising that materials useful for photo-cathodes are also useful for secondary emission. However, because the incident (primary) electrons may be made very energetic, and because the optical absorption is irrelevant for secondary emission, various wide-band-gap materials such as MgO and KCl are also useful secondary emitters.

At first sight it might appear that the secondary yield (average number of secondary electrons emitted per incident primary) could be increased indefinitely merely by increasing the primaries' energy. However, a faster primary has less time to interact with electrons in the material, just as (in the theory of stellar dynamics[7]) a fast star interacts less in each encounter with field stars than does a slow one. Thus the number of electron–hole pairs produced per unit length of path by the incident primary actually decreases with increasing energy. Consequently, when the penetration depth of the primary increases beyond the mean escape depth for the secondaries, the secondary yield decreases with increasing primary energy; there is a maximum yield available from a given material. This maximum is near unity for metals, which are not practical secondary emitters; but it may be as large as 200 or more for some semiconductors. The primary energy required to reach maximum yield is typically a few hundred to a few thousand volts, but much lower energies are ordinarily used. An excellent quantitative discussion of secondary emission is given by Simon and Williams.[8]

The energy spectrum of electrons leaving a secondary-emitting surface is rather complex (see Fig. 6). Some of the incident primary electrons are reflected (or rather, diffracted by the crystal lattice) with little change in energy. Some penetrate the crystal and are scattered back, with some loss of energy; likewise, some energetic secondaries produced near the surface may escape before being reduced to the threshold energy E_T. However, most of the secondaries will quickly be reduced to energies below E_T by additional pair-producing collisions. The energy spectrum of the secondaries alone would thus resemble the energy spectrum of photo-

[7] See, for example, D. Mihalas and P. M. Routly, "Galactic Astronomy," Chapter 11. Freeman, San Francisco, California, 1968.

[8] R. E. Simon and B. F. Williams, *IEEE Trans. Nucl. Sci.* **NS-15**, no. 3, June, 167 (1968).

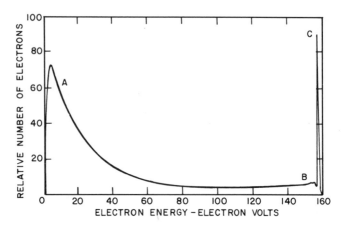

FIG. 6. Typical energy distribution of electrons leaving a secondary emitting surface. A—true secondary electrons; B—inelastically reflected primary electrons; C—elastically reflected primaries.

electrons produced by 100-V X rays. These low-energy electrons are easily focused from one stage of a photomultiplier to the next, but the energetic component can be troublesome.

1.4. Real Photomultipliers

1.4.1. Materials and Construction

We now leave the idealized world of physics and enter the messy world of reality, where we find photomultipliers composed of nonideal materials, contaminated by impurities, and used under varying environmental conditions. We begin at the beginning, for although the user may never have to make a photomultiplier, he should know how a typical photomultiplier is made; many troubles are built in at the factory (often for very good reasons). An excellent account of phototube technology is given by Sommer.[9]

1.4.1.1. General. To begin with, we must have a photocathode, a dynode system, a (partially!) evacuated envelope to keep air out, and some electrical connections that penetrate this envelope. There must also be some mechanical supporting structure to keep all the parts inside the envelope fixed in the proper relative positions.

[9] A. H. Sommer, "Photoemissive Materials." Wiley, New York, 1968.

Right away, we have problems. The metallic electrical connections have a large coefficient of thermal expansion, so the envelope is usually made of a "soft" (soda–lime) glass of similar expansion; otherwise, thermal stresses might break the glass or cause the metal to pull away from it, breaking the seal and admitting air. (Sometimes this happens still, especially under extreme cooling, as with liquid nitrogen.) However, this glass is opaque below 3400 Å or so; if we need to work in the ultraviolet, special windows such as quartz or sapphire must be used. Quartz is attached through a series of graded seals to glasses of intermediate expansion; even so, thermal strains are produced, and occasionally a tube comes apart. Sapphire and other special windows are usually cemented or bonded to the glass envelope; here again is a possible point of weakness. Just to preserve its mechanical integrity, a tube should not be subjected to either rapid or extreme heating or cooling. One or two degrees per minute is usually a safe rate; the manufacturer should specify the allowable temperature range, and should be *asked*, if he fails to specify it on data sheets.

A particular problem of PMT manufacture is the chemical reactiveness of the materials (especially alkali metals) used in cathodes and dynodes. Most of these are rapidly attacked by oxygen and water vapor (i.e., they cannot be exposed to air). This means that the cathode (and often also the dynode) material must be chemically produced in the same chamber—usually the tube envelope itself—in which the tube is evacuated and sealed off. Two of the commonest ingredients, cesium and antimony, have high vapor pressures and are readily deposited by evaporation. Unfortunately, they also tend to migrate in the vapor phase, and deposit on everything inside the tube, including "insulating" supports (which thereby become unreliable resistors).

A closely related problem is that most photocathode materials decompose at relatively low temperatures (80–200°C); and this, together with the volatility mentioned above, prevents the high-temperature ($\sim$600°C) bake-out that is usual in vacuum technology. As a result, photomultipliers end up with a greater burden of adsorbed and residual gas than one would like, which produces a variety of undesirable effects that increase noise and decrease stability and dynode gain. Residual gas pressures are typically 10^{-6} Torr.

Another dilemma is that one wants a photocathode to be uniform across its useful area; but one also wants a close cathode-to-first-dynode geometry, both for good photoelectron collection efficiency and (in pulse counting) for a small spread in electron transit times and hence a short pulse width. The close spacing may preclude uniform deposition of evaporated materials across the cathode; and a multicomponent cathode may require so many evaporators as to prevent close spacing.

Finally, one must bear in mind possible incompatibilities among tube materials. Mechanical substrates for cathodes and dynodes must not react with the active materials deposited on them, nor may cathode and dynode materials degrade each other. For example, although the silver–cesium–oxide (S-1) photocathode was one of the first commercially used in photodiodes, it was one of the last to be used successfully in a photomultiplier. The first multipliers used cesium antimonide dynodes, and it turns out that S-1 cathodes are fatally poisoned by traces of antimony. Alkali metals react with most glasses above 200°C, and even attack materials as inert as quartz and gold.

Now, to see how these (and other!) problems affect actual tube manufacture, we consider the production of a simple PMT with cesium antimonide cathode and dynodes. The mechanical parts are formed and assembled in air; electrical connections are spot-welded to a base-pin feed-through assembly; and the glass envelope is slipped on over the internal "works" and sealed to the edges of this assembly. The assembled tube is then sealed (usually via a tubulation in the base assembly) to a vacuum system, pumped down, and baked at a relatively low temperature (300–400°C). This removes most, but not all, of the adsorbed air and water vapor. Now, depending on the dynode geometry, the antimony for the cathode and dynode surfaces may be evaporated, or the antimony may have been deposited on these surfaces before assembly. (In either case, the volatile Sb has prevented a high-temperature bake-out.) Generally, "fast," focused dynode systems have exposed insulators that prohibit antimony evaporation after assembly; these antimonated surfaces have become somewhat contaminated by exposure to air. On the other hand, some "slow" (e.g., venetian-blind) systems can be antimonated after assembly and bake-out, producing cleaner, more uniform surfaces. Thus, "fast" tubes tend to be noisier than some "slow" tubes; so high counting rates and low dark noise tend to be mutually exclusive.

To activate the tube, cesium vapor is introduced, either from an external (or sometimes a removable) source, or from a chemically reacting "bead" assembled in the tube. The cesium is allowed to react with the antimony at about 150°C; when the reaction is complete (usually as measured by the cathode's photoemission), excess cesium is pumped out, and the tube is sealed off. One must, of course, try not to slop excess cesium all over the inside of the tube envelope and insulating supports;

an external source produces a cleaner, quieter tube, but is awkward and costly.

Clearly, tube manufacture involves trade-offs and compromises. Although some of these problems are technically soluble, the solution may raise the price beyond the means of prospective users. Astronomers should be aware that they are an economically negligible market, and must use tubes developed for large markets such as scintillation counters and laboratory spectrophotometers. The tubes first used by astronomers (the RCA 931-A) were mass-produced in World War II for use as wide-band noise generators in military equipment; they are still excellent noise generators, but that is not what the astronomer wants!

1.4.1.2. Photocathodes

1.4.1.2.1. CESIUM ANTIMONIDE. The oldest photomultiplier cathode used in astronomy is still one of the commonest. Its properties are well understood,[9] and its role in photoelectric photometry is like that of unsensitized plates in photographic photometry, i.e., it is the basis of the internationally adopted standard system.

In the older literature these are sometimes called "alloy" or "Sb–Cs" photocathodes, but they are now known to be a well-defined semiconducting compound Cs_3Sb. Actually, the best photoelectric yield is obtained with an appreciable cesium deficiency, at about the composition $Cs_{2.9}Sb$. This deficiency makes the material strongly p-type, so the Fermi level lies near the top of the valence band. It also produces a high density of acceptor states, which are populated by thermally excited electrons at room temperature; these give the spectral response curves a long red tail beyond the expected photoelectric threshold near 6000 Å. This tail, which extends well beyond 7000 Å, is a nuisance in filter photometry, because most blue and ultraviolet filters are quite transparent beyond about 7000 Å; with some cool stars, this "red leak" may contribute some tens of percent of the observed signal.[10]

The long red tails of Cs_3Sb cathodes hinder the determination of the energy-level diagram, as no sharp threshold exists. However, Spicer[11] developed a simple theory for the threshold region, from which he was able to deduce the band gap and electron affinity for Cs_3Sb and other alkali antimonides.[12] Spicer attributes the remaining exponential red tail

---

[10] C.-Y. Shao and A. T. Young, *Astron. J.* **70**, 726 (1965).

[11] W. E. Spicer, *Phys. Rev.* **112**, 114 (1958).

[12] W. E. Spicer, *J. Appl. Phys.* **31**, 2077 (1960).

to photoemission from thermally populated acceptors, which are very numerous owing to the cesium deficiency in the strongly p-type material. This model predicts the disappearance of the tail as $T \to 0$, just as is observed.[11–14] Thus the logarithm of quantum yield is a nearly linear function of photon energy in the tail region, whose slope is proportional to the reciprocal of the absolute temperature.

The extent of the red tail depends both on the acceptor density and on the electron affinity of the cathode material. These are primarily influenced by the degree of cesiation the tube received in manufacture, and to some extent by the residual gas, and the thermal history of the tube. Consequently, the tail varies tremendously from tube to tube;[14,15] even at 6100 Å, the response (relative to that for blue light) varies by a factor of at least 16, and an extrapolation to 7000 Å suggests a factor exceeding a thousand. Thus the red tail (or leak) may be negligible with one tube, and disturbingly large in another of the same type. Because a lower electron affinity enhances blue response (and thermionic emission) as well as prolonging the tail, more sensitive tubes tend to have longer, higher tails. EMI makes a special ("S") version to reduce both dark current and tail response, at a cost of about 20% in blue response. (However, the potassium–cesium "bialkali" antimonide cathode provides still better red rejection, with superior blue response.) Finally, the tail strongly influences the cathode's apparent response to tungsten (incandescent) light—that infamous "microamperes per lumen" rating that is so deceptive, and so ubiquitous on data sheets.

Cesium antimonide is used both as opaque (metal-backed) and "semi-transparent" (glass-backed) cathodes. In the first type, the light enters at the vacuum side of the material; in the second, the light enters from the substrate (window) side. The two types differ slightly[9,14] in spectral response, because the light has been somewhat filtered by passing through the "transparent" cathode by the time it reaches the exit (vacuum) face. Since photoelectrons produced near the vacuum face are more likely to escape than those produced deeper in the cathode material, and since the optical absorption increases rapidly at shorter wavelengths, the transparent cathodes are, on the average, more red-sensitive than opaque cathodes; thicker cathodes have more red and less blue response. Consequently, the two types have different "S" (spectral response) numbers:

[13] R. B. Murray and J. J. Manning, *IRE Trans. Nucl. Sci.* **NS-7**, 80, June-Sept. (1960).

[14] A. T. Young, *Appl. Opt.* **2**, 51 (1963).

[15] A. H. Mikesell, *Astron. J.* **54**, 191 (1949).

opaque cesium antimonide cathodes are designated S-4, and transparent ones S-11. The corresponding response curves contain the ultraviolet absorption of the soft glass envelopes; if these are replaced by uv-transmitting glass, the curves are called S-5 and S-21, and quartz windows give S-19 and S-13, respectively. All these types have nearly the same quantum efficiency ($\sim$15%) near 4000 Å, however.

The response, especially around 5000 Å, can be nearly doubled by placing a semitransparent cathode on a reflecting substrate; the resulting spectral response is called S-17. A similar effect can be produced by multipassing end-window photocathodes by total- or multiple-reflection techniques.[16] In either case, most of the improvement is in the red tail, where optical absorption of Cs_3Sb is relatively weak.

The apparent spectral response and quantum efficiency, in all cases, depend strongly on the amount of light absorbed in the cathode material, as most of the photoelectrons produced within it will escape (except near the threshold wavelength). Much of the incident light is usually reflected, because of the high refractive index ($n \approx 3$) and absorption ($k \approx \frac{1}{2}$) of the material. For example, about one third of the light normally incident on an opaque cathode from vacuum is Fresnel-reflected; but only about half as much is reflected at the glass–photocathode interface of a semitransparent cathode, owing to the lower relative index. Reflection losses are further increased for opaque photocathodes because they are usually used at oblique incidence, and because the light suffers an extra 4% Fresnel reflection at the inside surface of the window. Thus, about 20% more light actually gets into a typical end-on cathode than into a typical opaque cathode, even if no special enhancement technique is used. In the blue, most of this light is absorbed; but red light is strongly transmitted through an "end-on" cathode, or absorbed at depths exceeding the electron escape depth in an opaque one. The various enhancement schemes[16,17] reduce the reflection and transmission losses. Thus, they generally less than double the blue response, but can increase red response three or more times.

A detailed treatment of these optical effects must include interference phenomena, taking account of the considerable phase changes due to the high absorption (imaginary part of the refractive index). A typical cathode is about 300 Å thick (about a sixth of a wavelength, allowing for the high

[16] W. D. Gunter, Jr., G. R. Grant and S. A. Shaw, *Appl. Opt.* **9**, 251 (1970).

[17] R. Gelber and P. Baumeister, *Appl. Opt.* **9**, 863 (1970). See also J. A. Love, III and John R. Sizelove, *ibid.* **7**, 11 (1968).

refractive index), which is about the electron escape depth. Thus, a very small change in cathode thickness can grossly affect a tube's spectral response. Similarly, even a monolayer of oxygen can reduce the electron affinity by several tenths of a volt; but further oxidation gradually destroys the Cs_3Sb. It is hardly surprising that no two tubes have the same spectral response!

1.4.1.2.2. OTHER ALKALI ANTIMONIDES, AND RELATED COMPOUNDS. In general, apart from somewhat different threshold wavelengths and peak quantum efficiencies, all the alkali antimonides resemble Cs_3Sb. A very thorough comparison is given in Sommer's book,[9] so we mention only a few of practical importance here (see Table I). All the efficient photoemitters are p-type, cubic crystals, containing an excess of antimony compared to the stoichiometric ratio (M_3Sb).

TABLE I. Properties of Some Cathode Materials

Cathode material	Band gap E_G (V)	Electron affinity E_A (V)	Pair-production threshold (V)	Electrical conductivity
Cs_3Sb	1.6	0.45	4.0	Moderate
$(Cs)Na_2KSb$	1.0	0.55	3.0	High
K_2CsSb	1.0	1.1	—	Low
Cs_2Te	3.3	0.25	—	Very low
CsI	6.3	0.1	—	Very low
CuI	3	3	—	Moderate
GaAs(Cs)	1.4	~ 0	—	—

By far the most important at present is $(Cs)Na_2KSb$, the so-called "tri-alkali" or S-20 cathode. It is primarily Na_2KSb, (which is sometimes used without cesiation, and called S-24), but a very small cesium content seems to allow a greater Sb excess and a lower electron affinity than for pure Na_2KSb. It is the most efficient photoemitter known for blue light; when optically enhanced to minimize reflection and transmission losses, it has shown a quantum yield[16,18] near 60%. (Yields over 50% are possible, so long as conduction electrons are not absorbed at the substrate interface. If it were not for inelastic lattice scattering, all photoelectrons might

[18] R. J. Jennings, W. D. Gunter, Jr., and G. R. Grant, *J. Appl. Phys.* **41**, 2266 (1970).

diffuse to the vacuum interface and escape, just as all the photons eventually diffuse out of a semiinfinite conservative scattering atmosphere.)

Because of the great electron escape depth, it is practical to use relatively thick layers of this material for transparent cathodes. The result, as discussed above, is an increase in red response, at the expense of blue response; these are called "extended-red multi-alkali" (ERMA) or S-25 cathodes. As might be expected from their greater thicknesses, they are somewhat less susceptible to improvement by multipassing[19,20] than are regular S-20 cathodes.[21]

The large Sb excess in S-20 cathodes gives them a much higher electrical conductivity than Cs_3Sb. They can thus be used as semitransparent layers, without a conductive (e.g., "NESA" or tin-oxide coated glass) undercoating, down to considerably lower temperature ($\sim$liquid nitrogen) than can S-11 cathodes, which have inadequate conductivity [13,22] below dry-ice temperature ($-78°C$). The conductivity seems to be due, in both cases, to the acceptor states near the Fermi level which also produce the long red tail in the spectral response.[9]

Another alkali antimonide that is becoming important is K_2CsSb, which has a spectral response resembling that of Cs_3Sb, but with higher blue sensitivity and a sharper red cutoff. The reduced red tail appears to result from a lower density of defect states (closer approach to stoichiometry), and is accompanied by much higher electrical resistance. Thus these "bi-alkali" cathodes cannot be used much below room temperature without a conducting substrate; however, their good uniformity and reproducibility and low noise make them attractive for room-temperature applications.

For many applications in the vacuum ultraviolet, such as rocket and satellite astronomy, response to visible light constitutes an unacceptable "red tail." Thus there is a need for so-called "solar blind" cathodes. A number of semiconducting compounds have been used for such work, including a variety of alkali halides (CsI, KBr, etc.), but the most important at present seems to be cesium telluride (Cs_2Te).

Pure Cs_2Te shows a fairly sharp photoelectric threshold near 3.5 eV ($\sim$3500 Å). However, it readily takes up a slight excess of metallic

[19] C. D. Hollish and K. R. Crowe, *Appl. Opt.* **8**, 1750 (1969). See also W. Robinson, J. Williams, and T. Lewis, *ibid.* **10**, 2560 (1971).

[20] T. Hirschfeld, *Appl. Opt.* **7**, 443 (1968).

[21] G. R. Grant, W. D. Gunter, Jr., and E. F. Erickson, *Rev. Sci. Instrum.* **36**, 1511 (1965).

[22] J. P. Keene, *Rev. Sci. Instrum.* **34**, 1220 (1963).

cesium, which produces an unwanted "tail" extending through the visible region, as well as increased ultraviolet response. The electrical conductivity is very low, so that a conductive undercoating (usually metallic) is required. The properties of tubes with Cs_2Te cathodes have been studied extensively[23] and comparisons with other uv cathodes (CsI and CuI) have been made by Heath and McElaney.[24] These and other uv cathodes are discussed in Chapter 12 of Sommer's book.[9]

1.4.1.2.3. THE S-1 OR Ag-O-Cs CATHODE. Although this infrared-sensitive cathode has been used since 1930, its properties are still not well understood.[9] Its main components seem to be metallic silver, dispersed as small single crystals a few hundred angstroms across, and normal cesium oxide (Cs_2O), in nearly equal amounts. For some unknown reason, the size and degree of aggregation of the silver particles is very important.

In some ways, the S-1 cathode resembles a cesium-rich Cs_2Te cathode. Below 3000 Å, the photoemission is large and reproducible, and is clearly due to Cs_2O, which (as might be expected) is quite similar to the homologous chalcogenide Cs_2Te. At longer wavelengths, there is a long red tail, due to the free metal; for example, the spectral response always shows a narrow minimum at 3200 Å, where thin silver films are relatively transparent.

However, the detailed behavior of the (infra-) red tail depends strongly on the amount of silver (which apparently affects its state of aggregation). Cathodes with lower than normal silver content have almost constant response through the visible, with a very long infrared tail. As more silver is added, the tail becomes steeper, and a pronounced hump develops in the 6000–8000-Å region, leaving a minimum of response in the 4500–5500-Å region. Thus, the response around 7000 Å grows, while that beyond 1 μm decreases. Unlike the Cs_3Sb cathode, where an excess of cesium can be baked out during processing, an S-1 cathode cannot be corrected if too much of any ingredient is added. Because of the non-reproducible spectral response, all Ag–O–Cs cathodes are called "S-1," regardless of window or substrate material.

Much of the behavior seems explainable in terms of optical absorption. A smooth, continuous silver layer has low sensitivity because it reflects most of the incident light. A finely divided aggregate (usually produced

[23] W. N. Charman, *J. Sci. Instrum.* (*J. Phys. E*) **2** (2), no. 2, 157 (1969). See also J. W. Campbell, *Appl. Opt.* **10**, 1232 (1971); G. B. Fisher, W. E. Spicer, P. C. McKernan, V. F. Pereskok, and S. J. Wanner, *ibid.* **12**, 799 (1973).

[24] D. F. Heath and J. H. McElaney, *Appl. Opt.* **7**, 2049 (1968).

by reducing silver oxide with cesium vapor) is strongly absorbing, just like "gold black," "platinum black," or indeed the silver grains in a developed photographic emulsion, which are also a few hundred angstroms across.[25] The change in properties with increasing silver content (which probably means increasing grain size) seems similar to the change in spectral absorption of other fine metallic suspensions, as in gold sols and certain colored glass filters. If the size distribution of silver grains determines the optical absorption, and hence the spectral response, the wide variations from one cathode to another are easier to understand.

However, pure silver has no photoemission in the visible or ir, because of its high work function. It seems that the role of the Cs_2O is to lower the work function of the silver enough to allow emission; probably electron tunneling occurs at the $Ag–Cs_2O$ interface. (Possibly other cesium oxides, or even metallic cesium, may play a part in this as well.[9]) Thus, S-1 is a two-phase material, not a simple semiconductor.

S-1 cathodes are known to have considerable inhomogeneity, particularly at the longer wavelengths. Occasional "hot spots" respond to wavelengths beyond 1.5 μm, and may (because of their low work function) be the major source of thermionic emission. The thermionic emission is generally quite large at room temperature, and is extremely variable (10^{-11}–10^{-14} A/cm²).

Cathodes with longer infrared tails generally tend to have higher thermionic dark current, so one manufacturer is now deliberately processing tubes to maximize the dark current! In any case, the thermionic emission is usually unacceptable at room temperature, so S-1 photomultipliers are nearly always refrigerated. Unlike tubes with alkali antimonide cathodes, which usually give minimum dark noise at $-40°C$ or above, the S-1 tubes continue to improve even when cooled to liquid nitrogen temperature ($-170°C$). The metallic silver in the cathode provides good electrical conductivity, even at these low temperatures.

Finally, the great drawback of S-1 cathodes is their inherently low quantum efficiency, typically a few tenths of 1% in the red and infrared.

1.4.1.2.4. GALLIUM ARSENIDE AND RELATED COMPOUNDS. The thorough understanding of photoemission from semiconductors obtained by the mid-1960s allowed solid state physicists to formulate a prescription for an efficient photocathode. It should be a p-type semiconductor (to keep the Fermi level near the valence band, thus giving good electrical conduction with low thermionic emission); and the electron affinity

---

[25] J. F. Hamilton, *Appl. Opt.* **11**, 13 (1972).

should be smaller than the band gap (to prevent losses by electron–hole pair production, thus assuring a large escape depth), and as small as possible (to give high escape probability and good red response). Furthermore, it was known that an adsorbed electronegative layer (such as cesium) could produce an n-type region near the surface of p-type bulk material; the valence and conduction bands are "bent" by this p–n junction as shown in Fig. 7. The Fermi level is fixed by the impurity or defect states, which lie near the valence band in the p-type bulk but near the conduction band in the n-type surface region. Because of the band bending, photoelectrons produced in the p-type region can escape if they have only a little more energy than E_G, provided that $E_A = E_G$, and that the doping is strong enough to produce maximum bending. Thus, the apparent electron affinity E_A' can be made nearly zero.

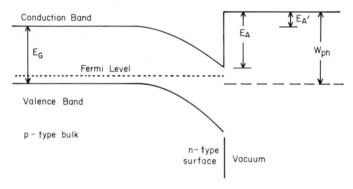

FIG. 7. Bent-band structure near the surface of gallium arsenide coated with cesium. The Fermi level is fixed by donors in the n-type surface, and by acceptors in the p-type bulk material. Because of band bending, the *apparent* electron affinity E_A' is very small, so the photoelectric work function W_{ph} is nearly the same as the band gap E_G of the bulk material.

Now it was known that adsorbed cesium layers produce $E_A \approx 1.4$ eV on a variety of bulk materials; so an efficient photoemitter should result from a cesium coating on a p-type semiconductor with $E_G = 1.4$ eV. Gallium arsenide can be made p-type by suitable doping (e.g., with Zn), and has the required 1.4-V band gap. Thus it was possible to predict that this combination would produce a very efficient photoemitter. The prediction was borne out in practice, and GaAs(Cs) is becoming an important photocathode for $\lambda \leq 1$ µm. This was the first efficient photocathode to be designed from physical principles, rather than discovered empirically.

A further advance occurred when it was realized that Cs_2O surface layers gave even lower E_A than Cs alone. (For example, it is responsible for the 1-eV threshold of the S-1 cathode.) This has allowed the use of related compounds with still lower band gaps, such as indium–gallium arsenide, and arsenide mixtures with phosphides or antimonides, coated with Cs_2O. Further study of the junction between the Cs_2O surface and the III–V bulk material has also produced composite materials with a negative effective electron affinity E_A'. (The physics is complicated by an intervening layer of Cs metal.) A good review of these new materials is given by Bell and Spicer.[26]

Because of the negative apparent electron affinity, any valence electron excited into the conduction band of the p-type material has a high probability of escape. Thus, these cathodes can show over 50% quantum yield per absorbed photon.[9] They also show a sharp red cutoff corresponding to the band gap E_G, instead of the long red tail characteristic of other photocathodes.

Because these materials have only recently become available in photomultipliers, and because several new ones will soon be introduced, it is not possible to give a detailed discussion of their properties here. However, they are already beginning to displace the S-1 photocathode in some applications, because of their much higher quantum efficiency (typically $\sim 10\%$ except very near the threshold).

On the other hand, the commercial cathodes have all been opaque, because of difficulties associated with semitransparent cathodes of this type. These difficulties are primarily of two types: (a) the substrate (window) tends to absorb photoelectrons from the III–V compound; and (b) because of the rapid variation in optical absorption near the bandgap energy (which is also the photoelectric threshold in negative-affinity cathodes), the thickness can be optimized for optical absorption only in a narrow wavelength range, so that broad-band response would require high-order optical multipassing of a thin cathode (in which substrate absorption is troublesome). This makes these materials unattractive for image-tube cathodes. For photomultipliers, opaque cathodes avoid these problems in a satisfactory and simple way.

1.4.1.3. Dynodes

1.4.1.3.1. CONFIGURATIONS. A successful electron multiplier must have a configuration that accelerates secondary electrons and collects

[26] R. L. Bell and W. E. Spicer, *Proc. IEEE* **58**, 1788 (1970). See also W. E. Spicer and R. L. Bell, *Pub. Astron. Soc. Pacific* **84**, 110 (1972).

them efficiently. The basic problem at each dynode is to collect accelerated electrons from the previous electrode, while providing an electric field to accelerate new secondaries away from the surface, toward the next dynode. The solution is to have a weak field at each dynode that retards the incoming ($\sim$100-V) electrons only slightly, but that also directs the low-energy (1–2-eV) secondaries into a stronger field which accelerates them toward the following dynode. This has been done with a variety of geometries; in some cases, magnetic fields have been used to assist in focusing the secondary electrons on the following dynode.

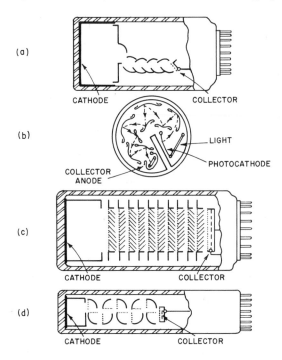

FIG. 8. Typical dynode configurations: (a) linear focused; (b) squirrel-cage; (c) venetian-blind; (d) box-and-grid. In each case, the anode (collector) is a wire grid surrounded by the last dynode.

Dynode systems are classified as *focused* or *unfocused*, although the difference is more of degree than of kind. The *box-and-grid* (BG) system (see Fig. 8) is a typical unfocused structure. Each dynode is a rectangular metal box with two adjacent sides missing. One of these sides is covered with a fine wire mesh or grid; the other is open. The inside surface is coated with a good secondary emitter. Accelerated electrons enter through the grid; secondaries leave through the adjacent open side, which faces the grid of the next dynode. Since each grid forms an equipotential surface, the accelerating field between boxes "reaches in" slightly through the open side of the previous box, providing a weak acceleration to collect the secondaries produced within the box, which is coated on the inside with the secondary-emitting material.

The *venetian-blind* (VB) system is a modification of the box-and-grid design. A wire grid covers the entrance face of each dynode, as before; but the dynode is a row of metal slats like a venetian blind. The space between the slats of a dynode is analogous to the inside of a box in the BG system. In effect, a VB dynode is an array of elongated "boxes" laid side-by-side. The secondary emitter is placed on the front (upper) surfaces of the slats.

The VB system has the advantage of being more compact than the BG system for a given area of secondary emitter. Thus the multiplier itself can be made smaller, or the current density can be made smaller for a fixed physical size. The VB system can also be made more opaque to escaping electrons and to positive ions; these advantages will be explained later. On the other hand, a BG system may be easier to fabricate, depending on the type of secondary emitter used.

Both BG and VB systems are described as "unfocused," which means that secondary electrons produced at one dynode are not mapped into a small area of the next dynode, but tend to spread out. The reason for this is the weakness of the accelerating field; the initial velocities of the secondaries thus have a large influence on their trajectories, and some may not even be captured by the next dynode. A second consequence of the low accelerating field is that the transit time of electrons between dynodes is both long and variable, depending markedly on the initial velocities and positions.

On the other hand, dynodes can be designed to produce large accelerating fields, and these are the "strongly focused" types. The *linear focused* system shown in Fig. 8 is typical. The dynodes may be cylinders with the cross section shown, or they may be curved in three dimensions and nearly hemispherical.

An important modification of the linear focused structure is obtained by bending the entire system around into a circle, producing the so-called "squirrel-cage" structure, which is very compact. In these focused structures, each dynode "sees" some fringing field from the (physically) adjacent dynode, which is actually the next-plus-one electrically. This, plus the openess of the geometry, provides high initial acceleration, short transit time, and short time spread. Such "fast" tubes are often designed to provide minimum spread in transit times, for use in coincidence counting and other applications where timing is important. In some cases, the linear focused structures is modified by the insertion of additional focusing and accelerating electrodes near the central line.

One disadvantage of the focused systems is that each area of the photocathode is mapped onto a small area of the first dynode (D1), and so on down the line, so that variations of efficiency across the face of each dynode appear as variations in anode response when a spot of light is moved across the cathode. That is, the variation in apparent sensitivity across the face of the cathode is partly due to variations across the dynodes as well. Furthermore, the focused tubes tend to collect electrons with variable efficiency depending on their place of origin on the preceding electrode, which further enhances the nonuniformity of response.

All the above types depend solely on electric fields to put the electrons in the proper place, and indeed are adversely affected by magnetic fields of even a few gauss. However, an additional type of multiplier that depends on a magnetic field has become popular in some applications (though not, so far, in photomultiplier tubes). This is the resistance-strip or "crossed-field" multiplier, shown in Fig. 9a. The magnetic field is directed into the paper. Two parallel strips of resistive material, one

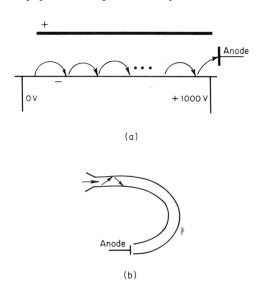

(a)

(b)

FIG. 9. Two types of resistive-film multipliers: (a) resistance-strip; (b) channel.

DORDT COLLEGE LIBRARY
Sioux Center, Iowa 51250

coated with secondary emitter, are subjected to a large potential gradient (say, from left to right). The upper strip is made more positive than the lower strip. An electron produced at the lower strip is accelerated toward the upper strip by the electric field between strips, but cannot reach it because the magnetic field bends the electron's path back to the first strip. However, since the place where the electron reenters the first strip is 50–100 V more positive than the place it left, the electron has enough energy to produce secondaries. These proceed along the strip in a series of hops, being multiplied at each hop.

This system is sometimes erroneously described as a multiplier with an infinite number of stages, although it actually has a small (say $\sim$20) but indefinite number of stages, equal to the number of hops. Its main advantage is simplicity of construction. A closely related type, the channel multiplier, does away with the need for a magnetic field. The electrons are again accelerated by a large gradient between the input and output ends of the resistive dynode, which is bent into a hollow tube. To achieve a large number of bounces, the tube must have a large length/diameter ratio. If the tube is bent around into a helix, ion feedback is reduced, and the whole multiplier can be very compact.

Various solid state multipliers have also been used. These employ an avalanche effect in a semiconductor with a high applied field to produce rather modest gains. Usually, a back-biased junction is used, as in solid-state charged-particle detectors.[26a] They are very compact, but do not provide the high gain and signal/noise ratio of dynode-type multipliers.

1.4.1.3.2. DYNODE MATERIALS.   Almost any good photoemitter can be used as a secondary emitter, because the low electron affinity that allows efficient photoemission also allows efficient escape of secondary electrons. Thus cesium antimonide is widely used, especially in tubes with alkali antimonide cathodes.

However, because the electrons incident on a dynode are much more energetic than visible photons, materials with larger band gaps can be used for secondary emission. For example, MgO and other metallic oxides, and KCl and other alkali halides, are sometimes used.

Recently, cesiated gallium phosphide [analogous to the GaAs(Cs) cathode] has been used. This material has a larger band gap ($E_G \approx 2.24$ eV) than gallium arsenide, so it has not attracted attention as a photo-

---

[26a] See, for example, E. A. Beaver and C. E. McIlwain, *Rev. Sci. Instrum.* **42**, 1321 (1971).

60238

YAARBU 3031100 TOQOG
0251 gwol ,181089 xun12

cathode material. The properties of GaP(Cs) dynodes have been discussed by Simon and Williams[8] and by Krall et al.[27] A much larger gain per stage (over 200) is possible with this material than with conventional dynodes, because of the great electron escape depth associated with negative electron affinity. However, very large accelerating potentials are required to achieve very high stage gain; for a given potential difference, more gain can be obtained from several conventional stages than from a single gallium phosphide dynode. Thus its primary advantages are in very fast tubes, where high voltages and few stages are desirable anyway, and in pulse-height work such as nuclear scintillation counting, where very high first-dynode gain is needed. A disadvantage of this material is that about $\frac{1}{3}$ of the incident electrons are reflected without multiplication.[27a]

1.4.1.3.3. GENERAL BEHAVIOR.   Both theory[8] and experience show that the gain at each dynode is roughly proportional to the energy of the impinging electrons. Usually a slightly weaker law applies, so that we can represent the stage gain g as a power law in the accelerating voltage v for that stage:

$$g = Av^p, \qquad (1.4.1)$$

where $p \leq 1$. The apparent gain g is not just the secondary-emission yield, however, because of such complications as "lost" electrons (imperfect collection of electrons emitted by the previous electrode), dynode "skipping" by elastically reflected primaries, and nonuniformities in dynode surfaces. However, Eq. (1.4.1) represents normal tube behavior reasonably well, with $p \approx 1$ for many dynode types (see Table II). Poor collection efficiency at low voltages can give $p > 1.0$. In general, p decreases slowly with increasing voltage, as the secondary-emission process saturates.[8]

Since the current through a photomultiplier consists of a succession of pulses, any detailed description should consider the step-by-step voltage changes that occur as the electron avalanche proceeds down the tube. Fortunately, in many cases it is possible to assume that each individual current pulse produces negligible voltage changes, so that only the average effect of many pulses need be considered. We shall call this the dc approximation.

[27] H. R. Krall, F. A. Helvy, and D. E. Persyk, IEEE Trans. Nucl. Sci. NS-17, no. 3, 71 (1970).
[27a] L. Birenbaum and D. B. Scarl, Appl. Opt. 12, 519 (1973).

TABLE II. Typical Values of p for Different Dynode Types and Interstage Voltages

Manufacturer	Geometry	Surface	Interstage voltage[a] (V)			
			50	100	150	200
EMI	Venetian blind	CsSb	0.75	0.72	0.65	0.54
EMI	Venetian blind	AgMg	—	0.93	0.82	0.71
EMI	Box and grid	CsSb	0.75	0.62	0.47	(0.4)
EMI	Box and grid	AgMg	—	1.48	1.12	0.94
RCA	Squirrel cage (1P21)	CsSb	0.76	0.76	0.76	—
RCA	Linear focused (C31000)	BeO	—	0.82	0.82	—
ITT	Box and grid	AgMg	(0.55)	0.82	—	—

[a] Parentheses indicate uncertain data.

In a tube with n identical stages of secondary multiplication, the overall tube voltage is

$$V = (n + 1)v, \qquad (1.4.2)$$

assuming the anode collection potential is the same as the interstage voltage. Then the overall gain of the multiplier is

$$G = g^n = (Av^p)^n = A^n[V/(n + 1)]^{np}. \qquad (1.4.3)$$

We do not generally observe p directly. However, the tube manufacturer usually supplies curves showing G or anode sensitivity as a function of V, the overall voltage applied to the voltage divider. In any case, a curve of anode sensitivity versus V is easily constructed if a variable high-voltage supply is available. If we then represent this relation near the operating point by

$$G = BV^q, \qquad (1.4.4)$$

comparison with Eq. (1.4.3) shows that $q = np$ or $p = q/n$. The parameter q is readily obtained from the empirical gain–voltage curve, since

$$q = d(\log G)/d(\log V). \qquad (1.4.5)$$

Hence, p can be found.

It is well known that the photomultiplier gain is sensitively dependent on the power supply voltage. In fact, taking the logarithm of Eq. (1.4.3) and differentiating gives the fractional gain change

$$dG/G = q\,dV/V = np\,dV/V. \tag{1.4.6}$$

Since $p \approx 1$ and $n \approx 10$ for typical tubes, the voltage regulation must be about an order of magnitude better than the gain stability required from the photomultiplier. For example, if we want 0.1% stability in the photomultiplier, we need 0.01% stability in the high-voltage supply. Such stability is commercially available.

One may ask why so many stages are necessary. Would not a tube with fewer dynodes and higher interstage potentials provide the same gain with less sensitivity to voltage variations? If we hold the overall voltage V fixed, we have

$$G = A^n[V/(n+1)]^{np} \tag{1.4.7}$$

or

$$\log G = n \log A + np[\log V - \log(n+1)]. \tag{1.4.8}$$

To find how G varies with the number of stages n we differentiate Eq. (1.4.8):

$$d(\log G)/dn = \log A + p \log[V/(n+1)] - np/(n+1)$$
$$= \log g - np/(n+1). \tag{1.4.9}$$

The term $np/(n+1)$ is near unity, so $\log G$ (and hence G) increases with the number of stages if $\log g \geq 1$, or $g \geq e$. Thus, overall gain is largest, for a fixed overall voltage, if the number of stages is increased until the gain per stage falls to $e \approx 2.7$. As will be explained below, a somewhat higher stage gain is beneficial; under typical operating conditions ($v \approx 100$–150 V), one finds $g \approx 4$, so that $n = 10$–14 gives $G = 10^6$–10^8, and $q \approx 10$, with V in the 1–2-kV range. Thus, real tubes represent a reasonable compromise between maximum overall gain and maximum gain stability.

Let us now look briefly at the individual charge pulses or electron avalanches that appear at the anode. For the present, we assume that each pulse is caused by a single electron entering the dynode system, although other events (cosmic rays, gas ions, etc.) can also produce anode pulses.

In the zero-order approximation, the initial electron strikes the first dynode and causes g electrons to be released; each of these is multiplied g times at the second dynode, giving g^2 electrons; and after n dynodes, we have g^n electrons. However, this description is meaningless if g is not an integer, because there are no fractional electrons. Clearly, we must treat secondary emission as a *statistical* process. Just as the quantum efficiency of a photocathode is a statistical average, the "gain" of a dynode is a mean value.

If we regard the secondary electrons as independent particles, emitted at random with some definite probability, we have a first-order statistical description of secondary emission. In a classical paper,[28] Lombard and Martin assumed that each stage of dynode multiplication is a Poisson process. Thus, if the mean gain is g, the actual number of secondary electrons may be any integer k which occurs with probability

$$P_g(k) = (g^k/k!)e^{-g}. \tag{1.4.10}$$

The standard deviation [root mean square value of $(k - g)$] of this distribution is $g^{1/2}$. Thus, if the mean gain is $g = 4$, its standard deviation is 2. Another important property of the Poisson distribution is the probability of observing zero secondaries (i.e., losing a pulse at the first dynode): $P_g(0) = e^{-g}$. For $g = 2$, $P_2(0) = 0.14$; for $g = 3$, the loss is 5%; for $g = 4$, it is 2%. Clearly, a substantial gain is necessary to avoid losing significant numbers of photoelectrons.

Lombard and Martin[28] found that, for moderate values of g, the result of the compound Poisson process is a distribution resembling the Poisson distribution at D1, but with somewhat higher tails and a broader peak. This is obvious if one considers that the fractional standard deviation contributed at each stage is inversely proportional to the number of electrons in the avalanche at that stage. Thus, the later stages contribute only negligible smearing to the pulse-height distribution passed on by earlier stages; the first two or three dynodes should essentially determine the whole result, no matter how many stages there are.

In comparing such a theory with observed data, one must carefully distinguish between the mean gain g (which includes lost electrons), and the mean height of all nonzero pulses observed at the anode.[29] However, even when such corrections are made, the pulse-height distributions of *real* tubes are always broader than expected; in many cases they do

[28] F. J. Lombard and F. Martin, *Rev. Sci. Instrum.* **32**, 200 (1961).
[29] J. R. Prescott, *Nucl. Instrum. Methods* **39**, 173 (1966).

not even show a peak, but are nearly exponential in shape. An obvious explanation is that real dynodes are not uniform, but have a variable gain from place to place. This additional variance serves to smear out the expected distribution still further.

Thus, Prescott[29] provided a second-order theory, in which a more general distribution function is used, to allow for a variation of gain across each dynode. Prescott's theory includes a parameter b related to the degree of dynode nonuniformity; $b = 0$ generates the classical compound-Poisson distribution of anode pulses, and $b = 1$ generates an exponential distribution. The observed distributions for real tubes can be fairly well matched by theoretical distributions with b typically in the range 0.05–0.20. However, there is not much physical basis for Prescott's mathematical assumptions, and one wonders if the adjustable parameter really has physical significance, or merely provides an interesting exercise in curve fitting. It appears that quasi-exponential distributions are associated uniquely with venetian-blind dynodes,[30,31] and in this case a variable collection efficiency from the front to the back edge of each slat seems likely.[31a]

Even if they are regarded as merely empirical curves, Prescott's distributions allow one to estimate the total fraction of pulses (photoelectrons) lost at all dynodes. In the worst cases, this fraction is about 25%, which agrees reasonably well with independent estimates[32-35] that are not based on a theoretical pulse-height distribution.

Finally, one may remark that all theories based on Poisson statistics are physically in error, as the underlying assumption of a small probability of success (escape) in a very large population (conduction-band secondaries) is not met. Thus, the Poisson law predicts a finite (but small) probability of very large gain; but the total number of electron–hole pairs is limited by the energy of the incident primary electron. For example, a 100-V primary cannot produce more than 50 electron–hole pairs in a typical secondary emitter with a band gap of 2 eV. The number of secondaries actually generated with enough energy to escape is even

[30] A. T. Young, *Appl. Opt.* **8**, 2431 (1969).

[31] A. T. Young, *Appl. Opt.* **10**, 1681 (1971).

[31a] A. F. J. van Raan, J. van der Weg, and J. van Eck, *J. Phys. E* (*Sci. Instrum.*) **5**, 964 (1972).

[32] M. Gadsden, *Appl. Opt.* **4**, 1446 (1965).

[33] A. T. Young and R. E. Schild, *Appl. Opt.* **10**, 1668 (1971).

[34] R. S. Lakes and S. K. Poultney, *Appl. Opt.* **10**, 797 (1971).

[35] Robert Evrard and Claude Gazier, *J. Appl. Phys.* **26**, 37A (1965).

smaller, because they must also overcome the electron affinity, and because some primary energy goes into phonon production and other losses instead of pair production. The upper limit on possible secondaries suggests that the binomial law would be a better parent distribution for secondaries than the Poisson law; unfortunately, this would reduce the tails of the calculated distribution, and require greater dynode non-uniformities to match the observations.

For the present, one must conclude that we have only a qualitative understanding of dynode statistics.

1.4.2. Undesirable Properties of Photomultipliers

We now scrutinize the properties of real tubes in detail. Here we find a host of "small" effects, which are a real problem if we want to reach an accuracy of 1% (0.01 mag). In many cases, these effects seem individually negligible, or nearly so; but in concert, they can have appreciable effects.

If we regard any unwanted influence on our experimental results as *noise*, we can distinguish two main types: additive noise and multiplicative noise. In addition, there are fluctuations associated with the photon nature of light and the discrete charge on the electron. We shall treat this "quantum noise" separately; some people call the statistical photon fluctuations "noise in signal."

We first deal with additive or "dark" noise; then with multiplicative noise, i.e., phenomena that alter the photomultiplier output by a constant fraction, when the light input remains constant.

1.4.2.1. Additive (Dark) Noise

1.4.2.1.1. THERMAL NOISE. It is well known that photomultipliers have an appreciable anode current even in total darkness. This "dark current" is very temperature-dependent; it decreases rapidly on cooling below room temperature. The dark current itself is only a nuisance, as its mean value is simply a zero offset; but the large *fluctuations* in the dark current are a source of additive noise that limits the detection and measurement of low light intensities. Therefore, tubes are always cooled below room temperature in low-level work.

In general, the dark current is an exponential function of absolute temperature, at least above $0°C$, so the logarithm of the dark current is nearly a linear function of T. The most obvious cause of such behavior is cathode thermionic emission, and indeed the dark current in tubes with S-1 cathodes appears to be largely thermionic.

However, in tubes with alkali antimonide photocathodes, the apparent "thermionic work function" derived from the temperature dependence is usually 0.7–1.4 eV, much less than the expected values near the photo-electric work functions (1.6–2.1 eV) for these p-type materials. A further anomaly is that S-20 cathodes give lower dark current than Cs_3Sb, although the latter has half a volt higher work function. Finally, the pulse-height distribution of dark pulses usually looks quite different from the height distribution of single cathode photoelectron pulses, especially in tubes with Cs_3Sb cathodes.[30] That is, while the light-pulse distribution usually shows a bump or maximum, followed by a steep decline at large pulse heights, the dark pulses show a distribution $N(h)$ that is roughly proportional to $1/h^2$, up to four or more times the height of a typical photoelectron pulse. These observations justify the conclusion[36] that thermionic emission is *not* an important mechanism in these photocathodes; the term "thermal dark current" seems preferable to "thermionic."

A study[30] of the dark pulse-height distribution, including both thermal and temporal variations, concludes that residual gas ions are primarily responsible for the dark current. This is further supported by the observation[37] that a 1P21 showed five times lower dark current when electrode voltages were pulsed for a time less than the ion transit time. Ion-caused pulses are clearly observed in many tubes as "afterpulses," which tend to occur roughly a microsecond after a normal photoelectron-initiated avalanche.[35,38–40] The ion afterpulses, like the thermal dark pulses, extend to heights several times a typical light-pulse height. It seems reasonable that if a 100-eV electron can eject 4 or more secondary electrons from a low-work-function dynode, a 100-eV positive ion can eject a similar number of electrons when it strikes a photocathode after being accelerated from the neighborhood of D1.

Both H^+ and H_2^+ ions are important, but heavier ions (with the oc-

[36] W. E. Spicer and F. Wooten, *Proc. IEEE* **51**, 1119 (1963).

[37] M. L. Bhaumik, G. L. Clark, J. Snell, and L. Ferder, *Rev. Sci. Instrum.* **36**, 37 (1965).

[38] R. M. Matheson and F. A. Helvy, *IEEE Trans. Nucl. Sci.* **NS-15**, no. 3, 195 (1968).

[39] J. C. Barton, C. F. Barnaby, and B. M. Jasani, *J. Sci. Instrum.* **41**, 599 (1964).

[40] G. A. Morton, H. M. Smith, and R. Wasserman, *IEEE Trans. Nucl. Sci.* **NS-14**, no. 1, 443 (1967).

casional exception of nitrogen) are not; in particular, cesium or other alkali-metal ions are not observed.[†] Hydrogen in photomultipliers comes from reduction of water vapor (adsorbed on electrodes during exposure to air in assembly and not fully baked out) by alkali metals. However, one might expect it to occur as alkali hydrides such as CsH. It is remarkable that the binding energies of solid alkali hydrides are near 1 eV, much like the apparent activation energy determined from the temperature dependence of the thermal dark current. Also, the lighter alkali metals form more stable hydrides than do the heavier ones; and tubes with a higher content of potassium have lower dark currents than those with pure cesium antimonide cathodes: S-20 [(Cs)Na_2KSb] tubes have lower dark current than S-11 [Cs_3Sb] tubes, and bialkali [K_2CsSb] tubes are lower still.

These facts suggest that ionized hydrogen gas, in equilibrium with alkali hydrides, is the major source of thermal dark current. At higher temperatures, hydrogen is given off, and at lower temperatures it recombines, or at least is adsorbed onto the chemically reactive cathode and dynode surfaces.

The most likely cause of ionization in the dark is high potential gradients near the edge of electrodes; hydrogen gas has a rather low dielectric strength and breaks down readily. Afterpulses, of course, result from electron-impact ionization. Notice that these collisions must occur in the gas phase, not at dynodes, because the field direction at dynode surfaces prevents the escape of positive ions. However, electron impacts at dynodes may promote the release of adsorbed neutral gas; an elevated dark current, lasting from a few seconds to several hours, commonly follows the drawing of large average anode currents. This is particularly noticeable in tubes with small dynode area (hence, high current densities), such as the ITT box-and-grid tubes (FW118, etc.).

1.4.2.1.2. BACKGROUND RADIATION. When a tube is cooled sufficiently to suppress the thermal component of dark current, a fairly constant component, containing some very large pulses (>100 times an average photoelectron pulse), sets the noise limit. This nonthermal component

---

[†] The absence of cesium ions supports the "low" room-temperature vapor pressure of Cs_3Sb, near 10^{-14} Torr; "high" values near 10^{-10} Torr are probably due to poor experimental technique. However, cesium ions *are* important in high-voltage devices such as image tubes; they appear to be produced by Schottky emission (field-induced desorption) at a rate proportional to $V^{1/2}/T$ [M. Oliver, *Advances in Electronics & Electron Physics* **33A**, 27 (1972)].

of dark noise is due to several causes, including cosmic rays and radio-active atoms in the tube.

Cosmic-ray (CR) noise consists primarily of very large pulses due to Cerenkov radiation from relativistic mu-mesons (and a few CR electrons) traversing the tube window. In tubes with transparent cathodes, these light flashes in the window are strongly coupled to the photocathode, producing 100 or more photoelectrons simultaneously. (This "optical enhancement" effect[16] was not taken into account in early studies,[41] which thereby underestimated the size of Cerenkov pulses by a factor of 2 or 3.) The average rate is about 1 to 2 events/cm²/min at sea level, increasing rapidly with altitude;[42] at 2000 m it is doubled, and at balloon altitudes the rate is up another factor of 10. Thus in a tube with low ion noise, such as the EMI 6256, CR noise can be the major component of anode-current noise,[41] even at temperatures as warm as 0°C. On the other hand, in tubes with inefficient (S-1) or opaque cathodes (such as S-4), CR noise is minor.

The size, but not the rate, of Cerenkov pulses depends on the spectral response of the window-photocathode combination.[43] The light is nearly white (in the sense of constant power per unit optical bandwidth), so the number of photons produced is proportional to $(1/\lambda_1 - 1/\lambda_2)$, where λ_1 and λ_2 are the short and long cutoff wavelengths of the window and cathode, respectively. Ultraviolet-transmitting windows (quartz or sapphire) give very big pulses, partly because of the large bandwidth, and partly because of multiple-photoelectron production from photons beyond the pair-production threshold (about 3000 Å for Cs_3Sb cathodes[35]).

For each giant CR pulse, there may be 10–30 smaller pulses,[43] which seem to be due to phosphorescence in the window. Faceplate phosphorescence is known[44] to result from exposure to uv light, which is abundant in Cerenkov radiation. Much of this phosphorescent emission appears to be in the red part of the spectrum, as tubes with red-sensitive (S-20) cathodes show more response to gamma rays (which produce fast Compton electrons,[45] and thus Cerenkov flashes) than do blue (Cs_2Te or Cs_3Sb) tubes. We would expect the statistical fluctuations of these "daughter"

[41] A. T. Young, *Rev. Sci. Instrum.* **37**, 1472 (1967).

[42] G. Chodil, D. Hearn, R. C. Jopson, H. Mark, C. D. Swift, and K. A. Anderson, *Rev. Sci. Instrum.* **36**, 394 (1965).

[43] R. L. Jerde, L. E. Peterson, and W. Stein, *Rev. Sci. Instrum.* **38**, 1387 (1967).

[44] H. R. Krall, *IEEE Trans. Nucl. Sci.* **NS-14**, no. 1, 455 (1967).

[45] F. A. Johnson, *Nucl. Instrum. Methods* **87**, 215 (1970).

pulses to depend on the smaller number of "parent" events, and this is observed: the number of counts per statistically independent event is 5 for S-20, 2 for S-11, and 1 for Cs_2Te cathodes on glass windows; the numbers are 10–20 times higher for sapphire windows.[46] Coates[46a] finds that the number of daughter pulses is proportional to the height of the parent pulse, and that most of the daughters follow within 50 or 100 μsec.

Photomultipliers are efficient detectors of cosmic and gamma rays,[30,43–46] and other (α- and β-) ionizing radiation.[43] Thus they respond to radioactive decays in the tube envelope.[47] In ordinary glasses, the principal radioactive contaminants[44] are radium (a trace impurity in sand) and K-40 (a naturally occurring isotope); K_2O makes up about 5% of most glasses. Special glasses made from purified sand and with very low potassium content are used to reduce these problems, but even in the best cases, several radioactive decays occur in a tube each minute.[47] Additional decays in the local environment (tube housing, building walls, etc.) contribute appreciable noise: Jerde et al.[43] found that large dark pulses in a refrigerated ($-40°C$) RCA 7265 fell from 3/sec to 1/sec when the tube was raised to the 750-mbar level ($\sim$2 km) by a balloon. Clearly, radioactivity and cosmic rays set the noise limit for cooled tubes.

The numerous "daughter" pulses seen by red-sensitive cathodes are particularly troublesome, because they occur in clumps; thus, the dark noise is much higher[48] (typically by a factor of 2 or 3) than the expected value (the square root of the total count rate). Because the dark pulses are correlated, they do not obey Poisson statistics, and the nonthermal dark noise is sometimes called "nonstatistical." This component becomes worse at lower temperatures,[48,49] which favor phosphorescent (i.e., radiative rather than phonon) decays of metastable states excited by cosmic or background radiation. Thus, there is usually an optimum temperature for a given tube, usually -20 to $-40°C$, that minimizes the total dark noise. The disadvantages of excessive cooling are summarized elsewhere.[50]

1.4.2.1.3. MISCELLANEOUS DARK NOISE.   Unfavorable operating conditions may also contribute to dark noise. Among these are electrical

[46] K. Dressler and L. Spitzer, Jr., Rev. Sci. Instrum. 38, 436 (1967).

[46a] P. B. Coates, J. Phys. E (Sci. Instrum.) 4, 201 (1971).

[47] J. Sharpe, IRE Trans. Nucl. Sci. NS-9, no. 3, 54 (1962).

[48] J. P. Rodman and H. J. Smith, Appl. Opt. 2, 181 (1963).

[49] M. Gadsden, Appl. Opt. 4, 1446 (1965).

[50] A. T. Young, Rev. Sci. Instrum. 38, 1336 (1967).

leakage; corona, electroluminescence, and other sources of spurious light; and noise associated with various electronic components used with the photomultiplier.[46a]

As tube voltages are usually 1–2 kV, and the dark current of a refrigerated tube may be 10^{-11} A or less, leakage resistances as high as 10^{14} Ω (100,000,000 MΩ!) between electrodes can be significant. Thus it is not surprising that leakage is hard to avoid.

Leakage may occur either inside or outside the tube. Internal leakage is usually due to cathode material, either on the inside of the envelope between electrode feedthrough connections, or on insulating supports between electrodes. It can get slopped on during tube manufacture, or it can be distilled onto insulating surfaces by high temperatures or thermal gradients.

External leakage is often due to water vapor absorption on sockets, bases, and wiring. Condensation is a particular problem if tubes are refrigerated. Many plastics, like polystyrene and bakelite, are surprisingly leaky at high humidities, especially if contaminated (fingerprints, in particular, are offensive because of their salt content). An instrument may perform normally in an air-conditioned laboratory, or at a desert location, but be unacceptably noisy on a humid night at the telescope. The best treatment is scrupulous cleanliness, and adherence to low-leakage sockets, bases, and connectors (glass and Teflon are quite satisfactory if kept clean). Humidity can be decreased with drying agents, or by keeping critical parts above ambient temperature. However, silica gel is a rather poor dessicant; and it gives up water to cooler surfaces like a wet sponge, so it should be kept at the coldest region of the tube housing (hence, not near heat-dissipating resistors). The heat dissipated by dynode resistors may keep a socket dry, but may also cause cesium migration within the tube itself.

Pulse-counting and ac (chopped) systems are considerably less sensitive to leakage than dc systems, because leakage noise tends to have a $1/f$ character; but good practice dictates avoiding leakage as much as possible, in any case. Leakage is often detectable as a dark current directly proportional to high voltage (i.e., ohmic leakage), while cathode dark current rises with a large power (~ 10) of the tube voltage. Thus, internal leakage is more noticeable at low overall voltages. Leakage in sockets and connectors can be measured with the tube removed from its socket. Leakage between anode and ground can shunt a large load resistance, producing calibration problems (apparent gain variations).

Spurious sources of light include corona, envelope luminescence, scintillations, and light leaks. Corona can be avoided by keeping high-voltage components well separated and free of sharp corners or edges. If necessary, the tube may have to be pressurized: corona can grow to complete breakdown and gas discharge at balloon and rocket altitudes. Even at aircraft and high-altitude observatory pressures, corona may develop in apparatus that works well at sea level.

Some gases (e.g., hydrogen and helium) break down more easily than air; such atmospheres must be avoided. Hydrogen–air mixtures may explode if ignited by electric discharge. Helium is especially insidious, as it diffuses through glass and quartz very readily, and will raise ion noise in minutes, or even destroy a tube after a few hours. On the other hand, an electronegative gas such as SF_6 can be used to increase the dielectric strength of air and reduce corona, although it also decreases thermal conductivity and can alter the heat balance of the tube.

Envelope luminescence is produced not only by background radiation, but also by electrons or ions striking the glass, and even by strong electric fields.[44] Since electrons that escape from the dynode assembly are much more numerous than positive ions, the best policy is to repel the electrons with an external tube shield, near cathode potential,[51] and live with the ions. If the electrons are attracted by a positive (anode-potential) external object, the resulting glow is easily visible to the eye.[52]

Electrode glow at the last few dynodes is also visible, and can be photographed.[44] Feedback of this radiation (which may include recombination radiation from ionized gas) from the last stages to the cathode generally limits the maximum gain at which a tube can be reliably operated, although ion feedback from anode to cathode may also play a part.[37,52] Additional luminescence may come from insulating supports (headers) within the tube.

There is a tendency, particularly in end-on tubes, for the envelope to act as a light-pipe, returning envelope or anode glow directly to the cathode. In some cases the tube design prevents this (e.g., by interposing a metal–glass seal that interrupts the optical path). In most cases, extraneous light can be absorbed by optically contacting a black coating, such as paint, aquadag, or black tape, to all parts of the glass except the portion of the tube window actually needed to admit light. However, such coatings may become brittle in refrigerated chambers, and flake off.

[51] Zs. Náray and P. Varga, *Brit. J. Appl. Phys.* **8**, 377 (1957). See also W. E. R. Davies, *Rev. Sci. Instrum.* **43**, 556 (1972).

[52] R. Gerharz, *J. Electron.* **2**, 409 (1957).

The cathode-potential shield need not touch the tube envelope to be effective in keeping secondary electrons confined to the dynode structure. It can therefore be separate from the optically absorbing coating. In any case, shielding should not extend much closer to the tube base than the end of the dynode structure, or it may cause high fields near the anode pin that can produce electroluminescence of the envelope.

Additional light may be generated in cold-box windows, and in prisms or other elements attached to the tube face for optical enhancement of cathode response.[16] The amount of light caused by cosmic rays is roughly proportional to the volume of glass in front of the cathode; but optical-glass prisms will raise the dark noise still further because of their high K-40 content. Quartz or low-potash glass should be used, and such devices should be kept as small as possible. Some abrasives used in optical grinding are sufficiently radioactive to contaminate optical parts appreciably, too. Even silica gel, used as a drying agent, can give off light, and should be kept away from the PMT cathode.

Finally, it would hardly seem necessary to point out light leaks in apparatus as a source of spurious dark noise; but experience shows them to be remarkably common. Any well-designed piece of equipment should be operable in full daylight without a noticeable effect on dark noise. Unfortunately, most astronomical instruments are not well-designed; they tend to be "designed" by astronomers instead of by competent engineers. The number of scientists who expect a simple butt joint to be light-tight is simply amazing. Therefore, let it be said that each and every joint in a photoelectric instrument should be designed as a light trap; that all flat surfaces are specular reflectors (including "black" anodized aluminum); that one or even two 90° corners may provide insufficient attenuation; and that, if any of this is news to you, you had better not try designing your own instrument. Of course, there are subtler problems than these (such as the infrared transparency of many "black" materials), but the above list encompasses the commoner faults.

A good test for light-tightness is to go over an instrument with a flash-light, shining it straight into every joint. The simpler expedient of merely looking for an increase in dark level when the room lights are turned on may fail, if a crack happens to be looking at the floor, or into a dark corner of the room. Sometimes one sees a rise in dark level when a person wearing a white shirt passes in front of some part of the instrument, or (conversely) a fall when a hand shadows some sensitive location. Some leaks only occur at a particular knob setting, such as a filter position. Such phenomena may be difficult to detect, and can easily ruin an experiment. Black tape is the standard cure; but it should not be used as a substitute for good practice in the first place.

Dark noise can also come from amplifiers, power supplies, and other apparatus used in photoelectric systems. Such noise can be found by operating all the equipment with the high voltage off; or, with the high voltage on, but the PMT removed from its socket. Very peculiar behavior can come from dynamical impedance mismatches that occur with some unit unplugged, however. We have seen a pulse-counting system that recorded more "dark" counts with the high-voltage supply turned off than with 500 V on the photomultiplier tube. Apparently the high-voltage lead was picking up transient noise and coupling it (through the socket wiring) to the anode lead; but this pickup was shorted out by the low output impedance of the supply when it was operating. Similarly, removing the tube from its socket, or (worse!) disconnecting the pre-amplifier from the PMT assembly, can alter the capacitative loading of the preamplifier input and give an incorrect noise reading. The whole system's behavior is thus *not* equivalent to the sum of the disconnected parts, so one must be careful.

One must be particularly careful in an observatory dome, where the large motors used to slew the dome and telescope can generate large transients and high-frequency noise. Even the relays used to control the slow motions have been known to introduce spurious pulses in a counting system. In the laboratory, elevator motors cause similar troubles. Such noise problems usually require the services of a good electrical engineer; the astronomer's job is to beware of such problems and detect them if they exist, not to fix them.

Although pulse-counting systems are vulnerable to transient noise, dc systems are equally susceptible to other noise sources. Leakage has already been mentioned. Cable noise is also a problem: at the low currents measured in dc photometry, the charge generated by friction inside or-dinary cables when touched or flexed is quite large. "Low-noise" cables reduce, but do not fully eliminate, this problem.

Another difficulty occurs in anode-circuit connectors: an oxide film builds up on silver contacts, which the tiny anode current (usually microamperes or less) cannot break down. Eventually, erratic behavior results. This "dry-circuit" problem can be alleviated by using gold-plated connectors, or by simply unplugging and reconnecting the anode connector a few times, to break the oxide layer mechanically. This problem occurs mainly with installations that are more or less permanent; apparatus that is disconnected and reconnected every few weeks never has time to accumulate enough oxide to cause trouble. The usual symptom is erratic performance, such as complete on–off modulation of the apparent anode current. This is really multiplicative rather than additive noise, however.

1.4.2.2. Multiplicative Noise; Stability Problems

1.4.2.2.1. TEMPERATURE EFFECTS. Temperature changes affect both cathodes and dynodes; they also produce changes in associated electronic, mechanical, and optical components. At the 1% level, *everything* is temperature-dependent.

Temperature effects in photomultipliers were discovered in the 1950s; the early literature is summarized, and extensive newer data are presented, in two reviews.[13,14] As tubes are usually cooled to reduce additive noise, we shall describe the effects of cooling rather than heating. The effects in alkali-antimonide tubes are mainly (a) a rather uniform increase of blue response, usually on the order of 1%/°C; and (b) a large, rapid decrease of response in the tail region, which can be several percent per degree at very long wavelengths. Spicer's observations on photodiodes[11,12,36] suggest that some attempts to separate cathode and dynode effects in multiplier tubes[53] were only partly successful, and that both cathode blue response and individual dynode gain are about equally affected, with coefficients of -0.05 to $-0.10\%/°C$. This appears to be due to an increase in electron escape depth, due to a reduction in lattice disorder on cooling. Another way to say this is that the lower phonon density at low temperatures reduces the electrons' energy losses on their way to the vacuum interface. Oxidized silver–magnesium dynodes seem to have less temperature dependence than Cs_3Sb dynodes,[14,53,54] possibly because scattering by permanent defects dominates phonon scattering.

The drop in red-tail response is attributed to the freezing out of electrons from the thermally filled acceptor states which are believed to be the source of tail response. Tubes with longer tails lose less of this response on cooling,[14] as Spicer's theory[11] predicts. In fact, the temperature dependence in the tail can be fairly well estimated from this theory; for the quantum yield $S(\lambda, T)$ should be approximately

$$S(\lambda, T) = S(\lambda_c) \exp\left[-\frac{A}{T}\left(\frac{1}{\lambda_c} - \frac{1}{\lambda}\right)\right]. \qquad (1.4.11)$$

[53] M. Lontie-Bailliez and A. Meessen, *Ann. Soc. Sci. Bruxelles* **73**, 390 (1959).

[54] G. M. de'Munari, G. Mambriani, and F. Giusiano, *Rev. Sci. Instrum.* **38**, 551 (1967).

for $\lambda > \lambda_c$, where the parameter A, which represents the size of the tail, should depend on the density of acceptors, and λ_c is the photoelectric threshold derived by Spicer.[11,12] We readily find the temperature coefficient

$$d \ln S(\lambda, T)/dT = -(1/T)\{\ln[S(\lambda, T)/S(\lambda_c)]\}. \qquad (1.4.12)$$

That is, the tail-size parameter A has been eliminated. Furthermore, the quantum yield at Spicer's threshold $S(\lambda_c)$ is typically about 1/10 of the peak quantum yield in the blue S_{max}; so we can write

$$d \ln S(\lambda, T)/dT \approx -(1/T)\{\ln[S(\lambda, T)/S_{max}] + 2.3\}. \qquad (1.4.13)$$

This last approximation is independent of both A (i.e., tail size) and λ_c, the cutoff wavelength. Thus, it may be regarded as a universal formula for temperature coefficients in the red-tail region of alkali–antimonide cathodes. For example, typical values for Cs_3Sb cathodes at room temperature $(T = 300 \text{ K})$ are $S_{max} \approx 0.15$; $\lambda_c = 6000 \text{ Å}$; $S(\lambda_c) = 0.015$; and, at $\lambda = 7000 \text{ Å}$, $S(\lambda) = 3 \times 10^{-4}$ and $d \ln S/dT = +1\%/\text{deg}$, or $10^{-2}/\text{K}$. Then either Eq. (1.4.12) or Eq. (1.4.13) predicts a temperature coefficient of $+1.3\%/\text{deg}$. Similar coefficients occur in S-20 cathodes at about 8000 Å, if one allows for the longer λ_c and higher S_{max} of the trialkali cathodes. The model also explains why $(d \ln S/dT)$ is much more positive at lower temperatures[53] for a fixed wavelength: both the smaller value of T in the denominator, and the lower value of $S(\lambda, T)$ contribute.

On extreme cooling, tubes with transparent cathodes lose response at all wavelengths, unless a conductive undercoating is used. The temperature at which this occurs depends on cathode current;[13] it can occur at room temperature in diodes.[21] It is clearly due to a loss of cathode conductivity, which allows a potential gradient to develop across the cathode, thereby decreasing the effective K-D1 accelerating potential, and also reducing collection efficiency by altering the electric field configuration in the K-D1 region. This effect, which is most severe for low-conductivity cathodes (see Table I), may look like nonlinearity or an anomalous spectral response, depending on the type of measurement.

All the above phenomena are reversible. However, irreversible changes have been seen; Johnson[55] found large, long-lasting changes in spectral response after cooling a 1P21 (Cs_3Sb) cathode with dry ice. The tube gradually recovered over a period of several years. Such changes are

[55] H. L. Johnson, *Astrophys. J.* **135**, 975 (1962).

probably due to cesium migration; they do not occur in good tubes if temperature changes are gradual and uniform. Temperature gradients should be avoided if stable operation is required.

Unfortunately, some people have tried cooling just the cathode, instead of the whole tube; this is just inviting trouble. The dynode voltage–divider resistors, which dissipate several watts and heat the tube socket and base, can cause similar problems. Large temperature gradients always tend to distill alkali metals from the warmer parts onto the cooler ones.

Temperature extremes are also harmful. Leaving a tube near a radiator, or in direct sunlight, may cause trouble; some cathodes can be permanently injured by temperatures above 50°C. At the other extreme, tubes may break or connections may open up, at very low temperatures. Cooling also invites condensation of moisture on windows and electrical connections.

Other peculiar effects have been reported in transparent cathodes at low temperatures. Hora and his co-workers[56,57] have reported non-linearities, discontinuities, and hysteresis effects in S-20 diode photocells at temperatures at, or a little above, liquid nitrogen (77 K). These effects have not been duplicated by others, and may be due to experimental difficulties rather than photocathode physics. For example, the spectral-response anomalies[56] may be due to cathode resistivity effects, and the discontinuities and hysteresis effects[57] may be due to contact irregularities caused by thermal contraction. Somewhat similar hysteresis effects[58] reported in Cs_3Sb may be due to window icing.[14] However, the point is that *problems* can occur at low temperatures, whether these are intrinsic or accidental; the cautious worker will not use more cooling than necessary.[50]

Finally, one must allow sufficient time for a tube to come to thermal equilibrium and stabilize. Even an elementary study[14] of the heat balance of a tube shows that thermal conductances are very small, on the order of 1 mW/°C. For comparison, the radiative heat exchange between two blackbodies near room temperature is 0.6 mW/cm² for each degree of temperature difference. As tube electrodes usually have areas of several square centimeters, thermal equilibrium within a tube depends largely

[56] H. Hora, R. Kantlehner, N. Riehl, and P. Thoma, Z. Naturforsch. 21a, 324 (1966).

[57] H. Hora, R. Kantlehner, and N. Riehl, Z. Naturforsch. 20a, 1591 (1965). See also Z. Phys. 190, 286 (1966).

[58] G. M. de'Munari, G. Mambriani, and F. Giusiano, Rev. Sci. Instrum. 38, 1158 (1967).

on radiation, not conduction. For a 1-gm metal electrode with 5 cm² of area, the thermal time constant is 2 min at room temperature, and longer when cooled. The low thermal emissivity of metal surfaces considerably extends this time.

Once the heat reaches the tube envelope, it must be passed on to the walls of the container. The conductivity of a 2-mm-thick layer of air is about equal to the radiative heat transfer at 300 K; so for wider spacing between walls, radiation is again the major contributor. Even for perfectly black walls, this gives a thermal time constant of 12 min for a glass tube window 2 mm thick, not counting the time required for heat to diffuse through the glass.

If we are cooling a tube 100°C (e.g., from room temperature to dry ice), we must let 5 thermal time constants elapse before it is within one degree of its final temperature. With the optimistic assumptions above (perfectly black surfaces at 300 K, and neglecting heat transfer from electrodes to envelope), this time is 1 hr. Allowing for imperfect emissivity, lower temperatures, and internal heat transfer would prolong this estimate by a factor of 3 to 4; this is in agreement with the time (several hours) actually required for real tubes to cool down.[14] Even a much smaller ΔT requires nearly as long, as the time to reach 1° stability is proportional to the logarithm of the temperature change; a 20° change requires 3 instead of 5 time constants, or over half as long as a 100° change.

A thick, internal cold-box window can be helpful[14] in lowering cathode temperature, but it increases cosmic-ray noise and so may not reduce the total dark noise.

1.4.2.2.2. FATIGUE AND NONLINEARITY. These effects are considered together, because they both tend to occur at high anode currents. Thus, fatigue effects may appear as nonlinearity in experimental results. A good general reference on both topics is Keene's paper.[22]

Fatigue is something of a misnomer, as tubes are observed to increase their responsivity almost as often as they decrease. It is frequently not monotonic;[59] a rapid change of one sign (seconds or minutes) followed by a slower drift (hours) back in the other direction often suggests that two or more mechanisms are involved. The short-term effect[60] is probably due to charging of insulating supports, especially in tubes with cylindrical (e.g., squirrel-cage, and some linear focused) or box-and-grid dynodes,

[59] W. Michaelis, H. Schmidt, and C. Weitkamp, *Nucl. Instrum. Methods* **21**, 65 (1963).
[60] O. Youngbluth, Jr., *Appl. Opt.* **9**, 321 (1970).

where the electrons can "see" a large solid angle of dynode supports. This effect is least in venetian-blind tubes, and focused tubes with hemispherical dynodes. The longer-term effects seem to involve changes in dynode gain, caused by cesium migration, which alters the electron affinity of dynode surfaces and thus stage gain.

These effects typically occur for average anode currents on the order of a microampere. They appear at lower currents in tubes with small dynode area (e.g., around 0.1 μA in the ITT box-and-grid tubes). At much higher anode currents ($\sim 10^{-3}$ A), enough power is dissipated by electron bombardment to heat the dynodes;[22] the resulting loss of gain is about what one would expect from typical dynode temperature coefficients (-0.05 to -0.10% per stage per degree) and thermal conductances (~ 1 mW/deg).

All the above effects are reversible; a tube usually recovers its former response if the anode current is reduced (or cut off), either by reducing the incident light[59] or by reducing tube voltage (and hence, overall gain). These phenomena, together with the lack of change in cathode areal[60] and spectral[61] response, show that only dynode effects are involved.

Slow changes in dynode gain (~ 0.1–1.0%/hr)[27,61] also occur at low anode currents. Such drifts are not usually considered "fatigue," but are attributed to "aging" of the tube. Most likely, gradual cesium migration is the mechanism. This is supported by Latham's observation[61] that larger dynode gain drifts occurred in a cooled tube than in the same tube at room temperature; for temperature gradients are likelier to occur under refrigeration, partly because of localized heat leaks,[14] and partly because thermal radiation is less effective in reducing gradients in a refrigerated tube, owing to the T^4 factor in the Stefan–Boltzmann law.

Aging effects in photocathodes have been studied in vacuum photodiodes.[62,63] Generally, these appear to be random changes during many months; daily changes are a few tenths of a percent. More constant conditions (temperature control, and constant application of supply voltages) improve stability.[63]

However, large cathode currents produce a permanent loss of cathode response.[62,64,65] At current densities exceeding a few microamperes per square centimeter, loss of red response is rapid (several percent per hour)

[61] D. W. Latham, SAO Spec. Rep. No. 321, p. 2–27 (1970).

[62] L. Foitzik, *Optik* **15**, 628 (1958). See also A. H. Sommer, *Appl. Opt.* **12**, 90 (1973).

[63] J. G. Edwards, *J. Phys. E* (*Sci. Instrum.*) **3**, 567 (1970).

[64] L. Lavoie, *Rev. Sci. Instrum.* **38**, 833 (1967).

[65] F. W. Schenkel, *IEEE Trans. Electron Devices* **ED-15**, 40 (1968).

in both alkali-antimonide and S-1 cathodes. The loss is faster at high temperatures, or if a potential gradient exists between the cathode and the outside of the envelope.[64] Evidently, migration of alkali-metal ions is an important mechanism; either loss of alkali from the cathode, or poisoning of the cathode by ions from the glass envelope, may be involved. Such electrolysis does not normally occur at the low cathode currents used in multiplier phototubes, especially if the external shield is at cathode potential, and if the tube is kept cool. It is fortunate that these precautions to reduce additive noise also reduce the likelihood of cathode deterioration.

Short-term, reversible "fatigue" effects can look like nonlinearity,[65a] especially if (as often happens in astronomical observations) a single measurement can take a minute or more, with a comparable interval for recovery between measurements. However, photomultipliers also have intrinsic nonlinearities.

Perhaps the most obvious tube nonlinearity is the one common to all vacuum-tube devices; space–charge limitation of anode current. This effect increases with anode current (or, light level). A typical figure[22] is 10% or less at 1 mA. Of course, this is far above the anode currents (less than or equal to 1 μA) normally used in accurate work. Furthermore, the space–charge nonlinearity is a steep function of anode current, varying almost with the square of the current; so we can normally ignore it if the anode current is kept low enough to avoid fatigue. However, in some low-duty-cycle pulsed applications,[37] one may safely draw instantaneous peak currents in the milliampere range. As the space–charge effect (unlike fatigue) depends on the instantaneous rather than the average current, it can be important in these applications. As in other electron tubes, the only treatment is to go to higher collecting voltages;[22,37,65a] doubling the voltage roughly halves the nonlinearity, but the detailed effects will depend on the pulse width, dynode configuration, and other factors, and should be measured in each individual case. An accurate method of measuring nonlinearity has been published by Bennett;[66] his accurate attenuator is now commercially available. A more sensitive method, measuring the beat frequency in the anode current of a tube illuminated with light beams chopped at two different frequencies, has been developed by Jung[66a] and applied by Sauerbrey.[65a]

[65a] G. Sauerbrey, *Appl. Opt.* **11**, 2576 (1972).

[66] H. E. Bennett, *Appl. Opt.* **5**, 1265 (1966). See also K. D. Mielenz and K. L. Eckerle, *ibid.* **11**, 594 (1972).

[66a] H. J. Jung, *Z. Angew. Phys.* **30**, 338 (1971).

A second intrinsic nonlinearity is due to cathode resistivity.[13,22] In cooled tubes with alkali-antimonide cathodes, this can be an important effect even at microampere anode currents[13] (picoampere cathode currents); in photodiodes, it is important at room temperature at nanoampere cathode currents.[22] The effect can be reduced by using less cooling, or by using cathodes with conductive (tin oxide, or metal) substrates.

Other nonlinearities occur because of "loading" effects in the associated circuitry. That is, the currents drawn by PMT electrodes cause changes in electrode potentials.

The simplest of these, anode loading, is the decrease of last-dynode-to-anode collecting voltage by the voltage drop across the anode load resistor (R_L in Fig. 10). The PMT anode does not present a pure current source to this load, but has some (large) dynamic resistance. Thus it is an oversimplification to say that tubes are operated so as to collect all electrons, or in the (supposedly) constant-current or saturation portion of the plate characteristic. However, as long as the D_n–A collecting potential exceeds

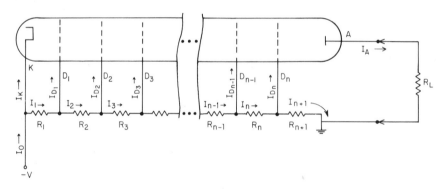

FIG. 10. A typical PMT voltage-divider string. Standard tube-circuit notation is used :K is the photocathode; D_1, D_2, ... are dynodes; A is the anode, R_L is the anode load resistance (often the input resistance of the electrometer used to read anode current). The resistors R_1, R_2, ... usually are on the order of 10^4 to 10^6 Ω. The electron flow in each branch of the circuit is denoted by I (with an appropriate subscript).

about 75 V, the characteristic curves are reasonably flat, and one can expect that nonlinearities will be considerably smaller than the fractional change in anode voltage. Typically, a 10-V change in collecting potential will produce a few tenths of a percent nonlinearity. This may be adequate in some cases; for the highest accuracy, the anode load should not be allowed to develop more than 1 V. Anode nonlinearity should be measured[66] if high linearity is needed; it depends somewhat on the dynode design, as at least the last two dynodes "see" the anode voltage changes in most tubes.

Loading of the dynode-supply chain is more complicated; an analysis and a good discussion are given by Lush,[67] and the experimental effects have been investigated by Sauerbrey[65a] and by Engstrom and Fischer.[68] Refer again to Fig. 10; the voltage divider draws a current

$$I_0 = \frac{V}{(n+1)R} \qquad (1.4.14)$$

if the tube currents can be neglected. However, the currents drawn by the tube load the voltage divider and disturb the accelerating potentials, and, hence, the overall gain of the tube.

A simple heuristic argument shows roughly the effect of loading by the dynode currents. For conceptual simplicity, we consider the electron currents, rather than the "conventional" current which flows in the opposite direction. We suppose that the gain per stage is very high, so that only the last dynode draws an appreciable current. This must be the anode current I_A minus the much smaller current supplied by the previous dynode, which we neglect. Since there are many dynodes, the combination of the voltage V in series with the n resistors of value R before the last dynode acts nearly as a constant-current source; thus, we suppose that the electron current I_0 is divided at D_n into two parts, one I_A flowing to the anode, and the other $I_0 - I_A$ flowing to ground through the resistor R. Evidently the effect of the anode current is to decrease the voltage across the last resistor by an amount $I_A R$. Then the anode collection voltage drops by $I_A R$, but the voltage across all the dynodes increases by the same amount. Since this is equivalent[†] to raising the overall voltage by

$$dV = \frac{n+1}{n} \cdot I_A R, \qquad (1.4.15)$$

we have

$$\frac{dV}{V} = \frac{n+1}{n} \cdot \frac{I_A R}{V} = \frac{1}{n} \frac{I_A}{I_0} \qquad (1.4.16)$$

[67] H. J. Lush, *J. Sci. Instrum.* **42**, 597 (1965). See also A. W. Sloman, *Rev. Sci. Instrum.* **43**, 356 (1972).

[68] R. W. Engstrom and E. Fischer, *Rev. Sci. Instrum.* **28**, 525 (1957).

[†] We assume that anode collection efficiency is not significantly affected by the change in collection voltage.

from Eq. (1.4.14); hence, applying Eq. (1.4.6),

$$\frac{dG}{G} = np\,\frac{dV}{V} = p\,\frac{I_A}{I_0}. \tag{1.4.17}$$

Although this result is approximate, we shall see below that it differs from the exact answer by only a very small correction term.

In order to calculate exactly the effects of tube loading of the voltage divider, we label the currents as in Fig. 10. The current in the ith resistor of the dynode chain, counted from the cathode, is I_i. The current flowing to the ith dynode from the bleeder chain is I_{Di}. The input, cathode, and anode currents are, respectively, I_0, I_K, and I_A. Then

$$\begin{aligned}
I_0 &= I_1 + I_K,\\
I_1 &= I_2 + I_{D1},\\
I_2 &= I_3 + I_{D2},\\
&\ \ \vdots\\
I_n &= I_{n+1} + I_{Dn},
\end{aligned} \tag{1.4.18}$$

and

$$I_0 = I_A + I_{n+1}. \tag{1.4.19}$$

The current within the tube incident on each dynode is g times greater than that incident on the previous dynode; the anode current is g times larger than the current incident on the last dynode. However, each dynode must supply a secondary-emission current, g times larger than the current incident on it, to the next dynode. Then the current supplied to the last dynode is

$$I_{Dn} = I_A - I_A/g = I_A(1 - 1/g) = I_A(1 - r) \tag{1.4.20}$$

where $r = 1/g$ is the ratio of the currents in successive stages.

Similarly,

$$I_{Di} = I_A r^{(n-i)}(1 - r) \tag{1.4.21}$$

and

$$I_K = I_A r^n = I_A/G. \tag{1.4.22}$$

Substituting Eqs. (1.4.20)–(1.4.22) into Eqs. (1.4.18) and (1.4.19),

$$I_n = I_{n+1} + I_{Dn} = I_{n+1} + I_A(1 - r), \tag{1.4.23a}$$

$$\begin{aligned}
I_{n-1} &= I_n + I_{Dn-1} = [I_{n+1} + I_A(1 - r)] + I_A r(1 - r)\\
&= I_{n+1} + I_A(1 - r)(1 + r) = I_{n+1} + I_A(1 - r^2),
\end{aligned} \tag{1.4.23b}$$

and, in general,

$$I_i = I_{n+1} + I_A(1 - r^{n-i+1}) \tag{1.4.23c}$$

and also, from Eqs. (1.4.18) and (1.4.19),

$$I_0 = I_1 + I_K = I_A + I_{n+1}, \tag{1.4.24}$$

so that

$$I_{n+1} = I_1 - I_A + I_K. \tag{1.4.25a}$$

Since $I_1 > I_A = GI_K$, where $G \approx 10^5-10^8$ in most cases, it is a very good approximation to neglect I_K:

$$I_{n+1} \approx I_1 - I_A. \tag{1.4.25b}$$

Since the voltage v_i on the ith stage is I_iR, the total voltage

$$V = \sum_{i=1}^{n+1} v_i = \sum_{i=1}^{n+1} I_iR = R \sum_{i=1}^{n+1} I_i, \tag{1.4.26}$$

or, using Eqs. (1.4.23) and (1.4.25),

$$V/R = (n + 1)I_{n+1} + I_A\left[n - \sum_{i=1}^{n} r^{n-i+1}\right]. \tag{1.4.27}$$

The summation in the last term is

$$\sum_{i=1}^{n} r^{n-i+1} = \left[\frac{1 - r^{n+1}}{1 - r}\right] - 1 \approx \left(\frac{1}{1 - r}\right) - 1$$

$$= \frac{r}{1 - r} = \frac{1}{g - 1}, \tag{1.4.28}$$

using the same approximation as before, that r^{n+1} is negligibly small. This gives

$$\frac{V}{R} = (n + 1)I_{n+1} + I_A\left(n - \frac{1}{g - 1}\right) = I_0(n + 1), \tag{1.4.29}$$

from Eq. (1.4.14). Solving for I_{n+1}, we have

$$I_{n+1} = I_0 - I_A\left(\frac{1}{n + 1}\right)\left(n - \frac{1}{g - 1}\right) = I_0 - I_A\left[\frac{n - 1/(g - 1)}{n + 1}\right]$$

$$= I_0 - I_A\left[\frac{n(g - 1) - 1}{(g - 1)(n + 1)}\right]. \tag{1.4.30}$$

We can substitute this value for I_{n+1} in the expression for I_i and find the current I through the ith resistor is

$$I_i = I_0 + I_A \left[\frac{g}{(g-1)(n+1)} - r^{n+1-i} \right]. \qquad (1.4.31)$$

If g_i is the gain of the ith stage,

$$\frac{dg_i}{g_i} = p \frac{dv_i}{v_i} = p \frac{dI_i}{I_i} = p \frac{(I_0 - I_i)}{I_0}$$

$$= p \frac{I_A}{I_0} \left[\frac{g}{(g-1)(n+1)} - r^{n+1-i} \right]. \qquad (1.4.32)$$

However,

$$\frac{dG}{G} = \sum_{i=1}^{n} \frac{dg_i}{g_i} = p \frac{I_A}{I_0} \sum_{i=1}^{n} \left[\frac{g}{(g-1)(n+1)} - r^{n+1-i} \right]$$

$$= p \frac{I_A}{I_0} \left[\frac{ng}{(n+1)(g-1)} - \sum_{i=1}^{n} r^{n+1-i} \right]; \qquad (1.4.33)$$

the last summation was evaluated in Eq. (1.4.28), so we have

$$\frac{dG}{G} = p \frac{I_A}{I_0} \left[\frac{ng}{(n+1)(g-1)} - \frac{1}{g-1} \right]$$

$$= p \frac{I_A}{I_0} \left[1 - \frac{g}{(n+1)(g-1)} \right]. \qquad (1.4.34)$$

The first term of Eq. (1.4.34) is the approximate result of Eq. (1.4.17); for typical cases, the second term is less than $1/7$ as large as the first.

In some cases, larger resistors are used at the anode end of the voltage divider to reduce space–charge effects. The simple argument given earlier indicates that nonlinearity due to voltage divider loading should be nearly proportional to the last resistor in the chain, i.e., the one that establishes the voltage between last dynode and anode. As the resistor may be two or even three times as large as the earlier ones, we can expect the loading effects to be correspondingly larger than indicated by Eq. (1.4.17). In fact, as long as the overall voltage is much larger than the voltage across the last resistor, the approximation

$$dG/G \approx qI_A R_{n+1}/V \qquad (1.4.35)$$

should provide a reasonable estimate of the increase in gain due to loading effects, if the parameter q is measured with the same voltage-divider chain.

Because p is near unity (see Table II), and equal resistors are commonly used in divider strings, Eq. (1.4.17) provides the basis for the rule of thumb that the string current should be 100 times the maximum anode current if linearity within 1% is required. If better linearity is needed, the string current should be correspondingly larger. For example, a 1-mA string current will give 1% nonlinearity at an anode current of 10 μA, and 0.1% nonlinearity at 1 μA. At these anode currents, fatigue effects are likely to be comparable to the nonlinearity, so string currents of 1 to 10 mA are usually satisfactory.

This nonlinearity could be reduced[65a] by using Zener diodes to stabilize the last few interstage potential differences. This would permit the use of lower string currents; however, Zeners become unstable and noisy at low current levels. [Zener diodes are often used to maintain a high K–D1 voltage, to produce good photoelectron collection efficiency, and high first-dynode gain (and hence a narrow pulse-height distribution, and good signal noise ratio).] A lower string current reduces the heat dissipated in the string, and hence the heat input and thermal gradients at the tube base. As Lush[67] points out, each additional regulated stage at the anode end reduces the nonlinearity by a factor of g^2. However, Sauerbrey[65a] finds that other nonlinear effects are dominant when the last stage is regulated.

Finally, the situation is somewhat different in focused dynode systems than in unfocused ones, because the stage gain g also depends on the voltage ratios in adjacent stages in focused multipliers. Thus the increase in gain due to string loading is partly offset by the decrease that unequal interstage voltages produce; for unregulated (purely resistive) strings, this improves the linearity considerably.[67,68] In general, Zener diodes are not very useful with focused tubes, particularly if the overall voltage is varied to control gain.

All the above analysis is based on the dc approximation. If we look at the behavior of individual electron avalanches (pulses), we have to consider the charging of interstage capacitances, and also the spread in pulse sizes. In scintillation counting or pulsed applications, pulses corresponding to many photoelectrons produced simultaneously must be considered; large Cerenkov pulses also occur in normal dark noise. Loading effects and nonlinearity will be greater for the larger pulses, so that the pulse-height distribution is distorted. Furthermore, if capacitors are used across the last elements of the divider chain, the divider will have a "memory" of recent pulses because of the time required to recharge a capacitor after drawing a pulse from it. Thus, a pulse that immediately follows another will, on the average, receive a different amplification than an isolated pulse would receive. In extreme cases, this could degrade the resolution of scintillation equipment; the resolution becomes worse at higher counting rates because of the reduced spacing of pulses.

We can imagine several situations. If the capacitors can store enough charge to supply a pulse without substantial voltage drop, and the resistors are small enough to recharge the capacitors before the next pulse occurs, there should be no nonlinearity. Since the average charge per anode pulse is mGe, where m is the number of photoelectrons in the pulse at the cathode and e is the electronic charge, a voltage change

$$dV = -mGe/C \qquad (1.4.36)$$

will occur across the last resistor when the pulse passes; C is the capacitance across the last stage. Combining this with Eq. (1.4.6) gives

$$dG/G = q \, dV/V = qmGe/CV. \qquad (1.4.37)$$

In order to keep the gain change per pulse less than a specified fraction f,

$$C > qmGe/Vf. \qquad (1.4.38)$$

In practice, the capacitors should be several times larger because of the spread in pulse heights.

For a typical tube counting single-photoelectron pulses ($m = 1$) with a gain of 10^6 and $q \approx 10$, we have $C \gtrsim 2 \times 10^{-13}$ F (0.2 pF) for $f = 0.01$. In most cases, the distributed capacitances amount to several picofarads, so that even without added capacitors Eq. (1.4.38) is met. However, at $G = 10^7$, the distributed capacitances are marginal, and the gain for large pulses ($m \gg 1$) may differ by several percent from that for small ones. This shows up as a distortion of the measured anode pulse-height distribution. In scintillation counting, where one wants to measure pulse heights accurately up to at least $m = 10^2$ or 10^3, it is obvious that capacitors will have to be added to the later stages.

Now consider the recharging problem. If the pulses occur at an average rate of P pulses per second, then the time constant RC of the voltage divider network should be small compared to P^{-1} if the capacitors are to recharge between pulses. That is,

$$R \ll 1/PC < Vf/PqmGe. \qquad (1.4.39)$$

If we assume $C = 5\,\text{pF}$ for tube and wiring capacitances, a typical divider string ($R \approx 10^5\,\Omega$, giving a current of 1 mA for 100-V interstage potential) requires $P \ll 2 \times 10^6$ for recharging to occur between pulses. At a gain $G = 10^6$, $P = 2 \times 10^6$ gives an anode current of approximately 3×10^{-7} A, and $G = 10^7$ gives microampere anode currents for single photoelectrons ($m = 1$). At these and higher counting rates (anode currents), recharging does not occur: thus, the dc approximation is valid for the range of interest. If capacitors are added to the later stages, the dc approximation is valid for even lower anode currents. Thus the nonlinearity analysis is generally satisfactory.

The last paragraph also demonstrates that nonlinear behavior readily occurs in fast pulse-counting systems. Because of the limited gain-bandwidth product of amplifier components, fast discriminators require the use of large values of G (10^7–10^8). This gets one into microampere anode currents (and hence, gain changes) at counting rates of $P = 10^5$–10^6 per second, however. Thus, gain-shift nonlinearity may occur before pulse-overlap nonlinearity with modern fast ($\sim$10-nsec resolution) counting systems. (It will be shown below that, although counting systems can be made less sensitive to gain changes than dc systems, the advantage is not large if optimum signal/noise ratio is required.)

The problem of nonlinearity due to pulse overlap (finite time resolution) in counting systems is much more complex[69,70] than is usually supposed. If one takes an idealized model of identical pulses of equal width, the classical result

$$f = P \cdot \tau \tag{1.4.40}$$

(where τ is the dead-time of the counter) is easily derived. If the probability of a pulse arriving during the time τ is 0.01, then 1% of the pulses arriving at random will occur while the counter is paralyzed by an earlier pulse, and the fraction of pulses missed will be $f = 0.01$. Thus the rule of thumb is that there will be 1% nonlinearity in counting a 1-MHz signal with a 100-MHz counter. Unfortunately, real pulses come in all shapes and sizes; and real counters may be paralyzed longer by big ones than by little ones.[69] Furthermore, two small pulses may look like one big pulse when they overlap, so the effective pulse-height distribution varies with the count rate.[70] While analytical and numerical methods exist for making the necessary corrections, they require such

[69] Brano Souček, *Rev. Sci. Instrum.* **36**, 1582 (1965).
[70] J. H. Williamson, *Rev. Sci. Instrum.* **37**, 736 (1966).

detailed knowledge of the equipment that a direct measurement of the nonlinearity[66,71] is usually much simpler, and more accurate.

The linearity problem for laboratory photometers has been reviewed by Hawes.[71] His article (and its references) should be studied carefully by anyone concerned with accurate measurements. Astronomers may be surprised to learn that commercial laboratory instruments can achieve a degree of linearity ($\sim 10^{-3}$) an order of magnitude better than the accuracy of most astronomical photometry.

Actually, the linearity problem is slightly more complex in astronomical photometry than in laboratory work, because of atmospheric scintillation. With a small telescope, the intensity modulation due to scintillation can approach 100%, and (owing to the fact that the logarithm of the intensity is normally distributed) instantaneous intensities several times the average occur fairly frequently. Thus, the nonlinearity for a twinkling star can be considerably worse than for a steady light source; and the amount depends on air mass. Hence, corrections for nonlinearity derived from steady light sources may not apply to stars, and systematic errors can result. This problem can be serious (several percent) in absolute-energy comparisons against a (steady) standard lamp, because small telescope apertures are used. Furthermore, Sauerbrey[65a] finds that nonlinearity can vary with the position of the illuminated spot on the cathode, and with wavelength. One should try to avoid nonlinear operation altogether, rather than use a nonlinear instrument and try to correct for it.

1.4.2.2.3. ELECTROMAGNETIC EFFECTS AND NONUNIFORMITY. The influence of electrostatic fields, and the need for proper electrical shielding, was discussed in Section 1.4.2.1.3. The nonlinear effects discussed above (1.4.2.2.2) that depend on electrode potentials may also be regarded as electrostatic effects, as may be fatigue effects which are due to charging of internal insulators.

An additional voltage-dependent effect may be called the "voltage-color" effect: the apparent spectral response varies with applied voltage (in particular, with K–D1 voltage). Unfortunately, there are only a few reports of this effect in the literature.[72,73] Usually, the effect of raising tube voltage is to make the spectral response relatively bluer. If we try to represent the effect by a single parameter, the percent change in relative response at two wavelengths, resulting from a fixed percent

[71] R. C. Hawes, *Appl. Opt.* **10**, 1246 (1971).

[72] F. Rufener, *Pub. Obs. Genève* **15**, 189 (1962).

[73] W. M. Irvine, C. Pikoos, J. Charon, and G. Lecomte, *Astrophys. J.* **140**, 1629 (1964).

change in tube voltage, provides a suitable measure. Rufener[72] found an 8% change in the ($\lambda 3451/\lambda 5850$) response ratio of a Cs_3Sb cathode, on increasing tube voltage from 650 to 1400 V. This corresponds to a 0.024-magnitude zero-point shift in (b–y) color, from 900 to 1400 V; or a voltage-color coefficient of $-0.04\%/1000\text{Å}$ for 1% change in voltage. Irvine et al.[73] found less than a $0.03\%/1000$ Å/% ΔV effect in an EMI 6256, an S-13 tube with very close K–D1 spacing. In an S-1 tube (FW-118), they found a large effect, rather parabolic in wavelength, on the order of $0.1\%/1000$ Å/% ΔV. These observations are probably best explained by the improved collection of fast photoelectrons at higher K–D1 voltage; as blue light produces more energetic photoelectrons than red light (cf., Fig. 5), the wavelength dependence is understandable.

However, an opposite effect has been observed in the red-tail region.[19,74–76] Here, an increased K–D1 voltage appears to reduce the effective electron affinity, thus extending the tail and enhancing red response. In specially constructed tubes with a close-spaced mesh[19,75] or accelerating electrode,[76] fields up to about 5000 V/cm can be used reliably (breakdown occurs at higher fields). Typically, such fields reduce the electron affinity by several hundredths of a volt, producing a large apparent gain ($\sim$3–5 times) in the far tail region, but little effect ($<10\%$) at shorter wavelengths. The effect is not a linear function of voltage, particularly in the case of the rather lumpy S-1 cathode;[19] however, a rough estimate for typical K–D1 fields (at most, a few hundred volts per centimeter) in ordinary tubes suggests changes in E_A on the order of a few millivolts, with some tens of percent change in relative tail response over the normal range of tube voltages. The effect should be least in tubes like the ITT FW-118/129/130 and EMI 9558,[32] where the whole front end of the tube is at cathode potential and the accelerating field at the cathode surface is relatively weak; the effect should be largest in tubes with close K–D1 spacing (e.g., EMI 9502/6256) or those with fast, focused dynodes (e.g., Amperex 56AVP, etc.) where high fields occur at the cathode surface.

Finally, some voltage-color effect, particularly in focused tubes, must come from interaction between the nonuniform spectral response across the cathode[77,78] and the variation of local collection efficiency, or effective

[74] K. R. Crowe and J. L. Gumnick, *Appl. Phys. Lett.* **11**, 249 (1967).
[75] W. D. Gunter, Jr., R. J. Jennings, and G. R. Grant, *Appl. Opt.* **7**, 2143 (1968).
[76] E. G. Burroughs, *Appl. Opt.* **8**, 261 (1969).
[77] G. S. Birth and D. P. DeWitt, *Appl. Opt.* **10**, 687 (1971).
[78] E. A. Ballik, *Appl. Opt.* **10**, 689 (1971).

cathode area, with voltage.[60,79] As a general rule, the nonuniformity of cathode response is greater in the red-tail region[9,77] (presumably, owing to local variations in E_A and/or cathode thickness), just as is the variation from one tube to another.[14] The influence of voltage on the apparent "shading map" is known to be higher in focused structures (e.g., RCA 931-A/1P21, and Amperex tubes) than in unfocused tubes;[60,79] part of this may be due to the contribution of nonuniform dynodes[80] to the apparent shading in focused tubes.

Apparent nonuniformity in response to white or monochromatic light has been widely observed[23,48,79-82] The commonest pattern seen in responsivity maps of end-on tubes is roughly symmetrical, with either a maximum or a local minimum near the center. Such a pattern is largely due to nonuniform deposition of ingredients during cathode manufacture. Gross variations are seldom less than 15%, and can exceed a factor of 2. Small-scale structure[82] can be factors of 3 or 4. Small cathodes tend to be more nonuniform than large ones; the region of good response can be only a small part of the nominal cathode area.[79,81] Thus, although it is common practice with 1P21's to illuminate only a small part of the cathode, and to move the tube in two coordinates to find and use the area of maximum response, this should be done with end-on tubes as well. In tubes with a focus electrode, maximum response and maximum uniformity occur at nearly the same focus voltage,[60] which seems also to give the narrowest pulse-height distribution.[35,83] Finally, the sensitivity map varies with the angle of incidence of the light[77] as well as with tube voltages and wavelength; and, from the above discussion, we can expect it to vary with temperature as well. Thus, one must emphasize Ballik's conclusion[78] that the variations for other cathodes "are as alarming as those for the S-1 photocathode. Therefore, it is important that the area of the photocathode illuminated during measurements, and the wavelength of the radiation [as well as all other factors], are the same as those used during calibration."

As might be expected, photomultipliers are just as sensitive to magnetic fields as to electric fields. Even the earth's field is enough to cause variations of 5 to 10% in the response of most tubes. Some manufacturers provide quantitative information of the effects of fields with various

[79] H. C. Ingrao and J. M. Pasachoff, *Rev. Sci. Instrum.* **32**, 866 (1961).

[80] S. J. Roth, *IRE Trans. Nucl. Sci.* **NS-7**, 57 (1960).

[81] I. Reif, R. N. Kniseley, and V. A. Fassel, *Appl. Opt.* **9**, 2398 (1970).

[82] L. E. Wood, T. K. Grady, and M. C. Thompson, Jr., *Appl. Opt.* **8**, 2143 (1969).

[83] M. Brault and C. Gazier, *J. Phys.* **24**, 345 (1963).

orientations; in general, fields at right angles to the electron trajectories have the most severe effects. Of course, fields from nearby transformers, motors, and other current-carrying conductors can be more severe than the geomagnetic field; so magnetic shielding should always be used.

On the other hand, magnetic effects can be used to improve signal/noise ratio in end-on tubes with large cathodes, as was first shown by Farkas and Varga.[84] An axial magnetic field on the order of 100 G prevents electrons emitted from the side walls from entering the dynode system. Thus, if only a small spot of light falls on the center of the cathode, the dark noise can be drastically reduced with very little loss of signal; the signal/noise ratio can be improved by an order of magnitude. This technique is especially applicable to tubes with multialkali cathodes,[85,86] which require several bulky evaporators and consequently cannot be made small in area. The magnetic field artificially modifies the responsivity map, reducing the effective area to a small area.[84] A particular consequence of the reduced sensitive area is a reduction in the number of photoelectrons produced simultaneously in energetic events (CR; radioactive decays), so the relative number of many-electron pulses in the dark noise is greatly reduced.[85] (This greatly improves the S/N ratio in dc detection, as will be explained below; the improvement is not so big in pulse-counting.[86])

Both electromagnets[84-86] and permanent magnets[85] have been used. The former have the disadvantage of heating the photocathode;[85,86] the latter may contain radioactive contamination. In either case, there is a danger that the magnet may saturate the tube's magnetic shielding, or that the shielding may deform the magnet's field in an undesirable way.[84] Permanent magnets offer lower fringing fields than a simple solenoid; but the latter seems to provide better performance.[85] Finally, one should be aware that permanent magnets vary in strength with temperature, and may lose strength if cycled in temperature. These complications suggest that cooling may generally be preferable to magnets for reducing dark noise, especially in pulse-counting work.

1.4.2.2.4. POLARIZATION EFFECTS. Photomultipliers are also sensitive to the state of polarization of the incident light, and the polarization effect varies strongly with wavelength and angle of incidence. There are only a

[84] G. Farkas and P. Varga, *J. Sci. Instrum.* **41**, 704 (1964).
[85] W. Knight, Y. Kohanzadeh, and G. Lengyel, *Appl. Opt.* **7**, 1115 (1967).
[86] J. A. Topp, H. W. Schrötter, H. Hacker, and J. Brandmüller, *Rev. Sci. Instrum.* **40**, 1164 (1969).

few reports of polarization sensitivity in the literature, mostly for the
RCA 931-A/1P21 (whose obliquely illuminated cathode could be expected to show considerable polarization[87,88]). These tubes are up to
30% more sensitive to light whose electric vector is perpendicular to the
tube's axis, than to the orthogonal polarization; the effect is generally
greater at longer wavelengths. However, even normally illuminated end-on
tubes also show a polarization of response,[77] typically in the range 1–10%.
Part of this may be due to light transmitted by the cathode and incident
at an oblique angle on the first dynode, or to light reflected from the
dynode and obliquely incident on the cathode from the vacuum side; but
part of the effect may also be due to anisotropy of the cathode film.

Much larger polarization effects (~40–50%) occur for light obliquely
incident on tube cathodes.[89,90] These polarization effects are particularly
large when total internal reflection is used to improve tube response.[20,91]
At first sight, this seems surprising: the multiple-reflection techniques
allow essentially total absorption for both polarizations, so one might
expect equal responses, and no polarization sensitivity. The explanation[20]
lies in the distribution with depth of the absorbed photons.[92] Light
whose electric vector is normal to the plane of incidence penetrates more
deeply into the absorbing photocathode, so more of its photoelectrons are
produced close to the vacuum surface, where escape is more probable.
Thus, the response is higher for light of this polarization incident from
the glass or substrate side.

This fact was first discovered experimentally by Hora,[93] who, with his
co-workers, investigated the angular[93] and wavelength[93–95] dependences
of the polarization effect in Cs_3Sb,[93,95] S-20,[94] and other[93] cathodes.
The effect is smallest (10–20%) in the blue, and reaches a maximum
near the threshold wavelength. By measuring both the transmitted and
reflected components, they showed that the polarization effect is present

[87] B.A. Brice, M. Halwer, and R. Speiser, *J. Opt. Soc. Amer.* **40**, 768 (1950).

[88] E. P. Clancy, *J. Opt. Soc. Amer.* **42**, 357 (1952).

[89] S. A. Hoenig and A. Cutler III, *Appl. Opt.* **5**, 1091 (1966).

[90] S. V. Pepper, *Appl. Opt.* **8**, 1747 (1969).

[91] B. E. Rambo, Air Force Tech. Doc. Rep. ALTDR 64-19, April (1964).

[92] N. J. Harrick, *J. Opt. Soc. Amer.* **55**, 851 (1965).

[93] H. Hora, *Jenaer Jahrbuch*, part II, 514 (1960).

[94] G. Frischmuth-Hoffmann, P. Görlich, H. Hora, W. Heimann, and H. Marseille,
Z. Naturforsch. **15a**, 648 (1960). See also H. Hora and R. Kantlehner, *Solid State
Commun.* **4**, 557 (1966).

[95] H. Hora and R. Kantlehner, *Phys. Status Solidi* **17**, 59 (1966).

for the light actually absorbed (as is necessary, to explain the multiple-reflection experiments).

As would be expected, the opposite polarization effect is observed when the light is incident from the vacuum side.[96] A dependence on temperature has also been reported.[97] Unfortunately, the early workers[93-95] misinterpreted the effect as a vectorial interaction between the photon and the absorbing crystal, owing to an oversimplified treatment of the optics involved; when the optical physics is done correctly,[20,90,92] the data are explainable without recourse to such complications in the theory of photoemission.

Of course, apart from the different effects for different polarizations, there is a general reddening of the spectral response of transparent cathodes for obliquely incident light, because the light has been filtered by a longer path in the cathode material before reaching the vacuum side where escape is more probable. One can also expect some sensitivity to circular polarization, but no work has been done on this problem.

1.5. Photomultipliers and System Components

A photomultiplier is only a transducer; it must be followed by some equipment that measures its electrical output. In this section we shall consider some methods of doing this, and the advantages and disadvantages of each.

The output of a photomultiplier may be measured in several ways. A measurement of the average anode current is commonest; sometimes this is expressed as the average voltage ($V_{out} = I_A R_L$) developed by the anode current I_A across the load resistor R_L in Fig. 10. An equivalent method is to measure the charge delivered by the tube in a fixed interval of time, since the current is simply the rate of flow of charge. Formerly, this was done by integrating the charge Q on a capacitor C for the entire time (usually on the order of a minute), and measuring the voltage $V = Q/C$ developed; an alternative[1] is to measure the time required to accumulate a fixed charge. These "charge-integration" methods have the disadvantage of requiring extremely good insulation in the anode circuit, so that negligible charge is lost by leakage during a measurement (many seconds). Such low leakage is hard to achieve, particularly in a dusty or humid telescope dome. A modern variant requires charging a

[96] B. Petzel, *Phys. Status Solidi* **3**, K223 (1963).
[97] G. Frischmuth-Hoffmann and P. Görlich, *Phys. Status Solidi* **1**, K83 (1961).

much smaller capacitance, which is discharged whenever a preset voltage level is reached; the number of dischargings per second is thus a measure of the rate of charge accumulation, and the total number of such events is proportional to the total charge delivered in a fixed time. In this case the charge need be stored only a few milliseconds at most. A device that produces an output pulse for each discharge of the capacitor is called a "voltage-to-frequency converter," or VFC. The output pulses can be recorded in a counter to achieve an accurate, digital record of the total charge. At present, this is the most accurate means of measuring the integrated anode current.

All the above methods, which measure the rate of charge accumulation, will be called *direct-current* (dc) measurements.

An alternative method is to count the individual electron avalanches (pulses) arriving at the anode. At first sight this *pulse-counting* method appears more direct than the dc techniques. However, because the pulses have a wide range of sizes, the pulse-counter must decide which ones to count and which to ignore. Furthermore, at high light levels, pulse overlap produces nonlinear recording (see Section 1.4.2.2.2), which is not a problem in dc recording if the anode current is kept low enough. The two methods also differ in signal/noise ratio, as will be shown below in Section 1.5.2. (Other methods have been proposed, but they have no great advantage over these.[30]) In general, pulse counting has the advantage at very low light levels, and dc methods are better at high ones.

Traditionally, PMT noise has been analyzed in the frequency domain. However, photometric measurements usually cover a fixed time interval rather than a fixed frequency interval. Also, in the time domain we see individual pulses, whose heights depend on their source (photons, ions, cosmic rays, etc.), while most noise sources have flat (white) power spectra and consequently are indistinguishable in the frequency domain. Thus, it is more useful to study photomultiplier noise on the basis of the pulse-height distribution rather than the frequency distribution. The relations between the two approaches are discussed by Robben.[98]

1.5.1. Pulse-Height Distributions

1.5.1.1. Measurement.   Generally, two types of equipment are used to measure pulse heights. The more direct is the pulse-height analyzer (PHA), which may be regarded as an analog-to-digital converter coupled

---

[98] F. Robben, *Appl. Opt.* **10**, 776 (1971).

to a small computer. The digitized height of each pulse is used as an address in the computer's memory; the contents of this memory cell are then incremented by unity. After a large number of pulses have been treated in this fashion, the memory contains a statistical measurement of the number of pulses occurring in successive intervals of pulse height.

It is a good approximation to assume that pulses in a given small height interval occur at random; so, if the same distribution is remeasured many times, the numbers recorded in any given memory cell should follow Poisson statistics. Thus, the standard deviation to be expected for an observed count is just the square root of that count. As small pulses are much more numerous than large ones, a few seconds may suffice to obtain adequate statistics at small heights, but many hours may be needed to obtain precise information on the rare large ones.

Pulse-height analyzers have two disadvantages. First, they require a relatively long time (on the order of a microsecond) to measure the height of a pulse, compared to output pulse widths from photomultipliers (typically, a few nanoseconds). This means that some intervening device must be used to "stretch" the pulse in time; otherwise, the PMT pulse would go by before the PHA could determine its height, and an incorrect height measurement (and consequently an incorrect height distribution) would be obtained. Some published distributions[99] appear to suffer from this error.

Now, a simple passive RC network can be used[30] to stretch the pulses by the required factor ($\sim$100); but, as the total charge in a pulse is fixed, this reduces the apparent height (measured in current or voltage units) by the same factor. This makes all but the biggest pulses too small to measure. If, on the other hand, an amplifier is used to shape the pulses, there is the danger that it may introduce changes in the measured pulse-height distribution. For example, amplifier nonlinearity distorts the distribution, not only by putting pulses in the wrong height interval, but also by making the intervals unequal in width. The effect of nonlinear transformation of the independent variable of a distribution function is discussed in textbooks on statistics;[100] the effect on actual photomultiplier data can be quite striking, including the production of spurious maxima in the apparent distribution where none actually exists in the real pulse-height distribution.[31]

[99] R. G. Tull, *Appl. Opt.* **7**, 2023 (1968).

[100] See, for example, R. J. Trumpler and H. F. Weaver's, "Statistical Astronomy," sect. 1.13. Dover, New York, 1962.

The second problem with PHA's is the long time (several microseconds) they require to increment the selected memory cell. The analyzer is "dead" (i.e., unable to respond to another pulse) during this time. At modest count rates ($\sim$10^4/sec), a significant fraction of the elapsed clock time is dead time. This not only causes the apparent counting rate to be a nonlinear function of the true pulse rate (cf., Section 1.4.2.2.2); it also can grossly distort distributions from which a "background" has been subtracted. For example, suppose we want to find the distribution of pulses due only to light striking the cathode. We cannot observe this directly, for the dark pulses are always mixed in and measured along with the light pulses. Thus, if the number of dark pulses between heights h and $(h + dh)$ per unit time is $d(h)\, dh$, and the number of pulses in the same interval produced by a light source is $s(h)\, dh$, the distribution observed with the light on for a time t is

$$p(h) \cdot t_l = [s(h) + d(h)] \cdot t_l, \qquad (1.5.1)$$

where $t_l < t$ is the "live time." Now if we also observe the dark pulses for a time t, we obtain the distribution $d(h) \cdot t_d$, where $t_d < t$ is the live time for the dark measurement. Because $t_l < t_d$, the difference

$$p(h) \cdot t_l - d(h) \cdot t_d = s(h) \cdot t_l - d(h) \cdot (t_d - t_l) \qquad (1.5.2)$$

is clearly *not* proportional to $s(h)$, because of the $d(h)$ term. Because $t_d > t_l$, and $d(h)$ generally rises very rapidly at small h, the difference given by Eq. (1.5.2) can be driven to (or below!) zero at small h, creating a spurious maximum at moderate heights even if $s(h)$ is actually a monotonically decreasing function.

Clearly, it is necessary that all measurements be made for the same live time, not the same clock time.[101] Fortunately, most PHA's built today have live-timers. However, most simple pulse-counting systems used in photometry do not; so the above difficulty can still occur in measurements made with the discriminator-type systems described below.[101a]

The pulse-height distributions measured with an analyzer are sometimes called "differential pulse-height spectra." This is a misnomer, as they are (strictly speaking) frequency functions rather than spectra, and are "differential" only by contrast to the more properly named "integral" distributions, which include a broad interval of pulse height.

[101] C. W. McCutchen, *Phil. Mag.* (*8th ser.*) **2**, 113 (1957).

[101a] C. Smit and C. Th. J. Alkemade, *Appl. Sci. Res.* (*B*) **10**, 309 (1963).

The integral distributions are usually measured by counting all pulses greater than a threshold height h_t in a single-channel counter. The count rate is thus

$$P(h_t) = \int_{h_t}^{\infty} p(h)\, dh. \qquad (1.5.3)$$

In statistical terms, $P(h)$ is a (complementary) distribution function, while $p(h)$ is a frequency function. [For example, if $p(h)$ were the Gaussian error frequency function, $P(h)$ would be the complementary error function, usually denoted by erfc.]

The device that rejects, or discriminates against, the pulses with $h < h_t$, is called a *discriminator*. Often two discriminators are used in anticoincidence, so that only pulses with $h_1 < h < h_2$ are counted; the system is then described as a "window" counter, with lower and upper thresholds h_1 and h_2. Most discriminators are designed for speed rather than for a sharp cutoff, so the thresholds are actually somewhat fuzzy.

If an integral distribution $P(h)$ is measured for many closely spaced values of h, the differential frequency function $p(h)$ can be recovered by numerical differentiation. However, it may not be the same as the $p(h)$ measured directly with an analyzer. Aside from the obvious problem of statistical (counting) errors, there are several possible explanations for this.

(a) There may be nonlinearity in one system or the other (see above).

(b) The faster time resolution of the discriminator system may resolve correlated pulse pairs that are counted as one larger pulse by the PHA. This is especially true of afterpulses due to dynode glow feedback to the PMT cathode; these usually come less than a hundred nanoseconds after the parent pulse.[44]

(c) At moderate light levels, pulse overlap[70] may be severe enough to distort the PHA measurements, but not those from the faster discriminator.

(d) Impedance mismatches (usually at cable connectors) can produce an exponentially decaying train of echo pulses after a large pulse. Such a train has a frequency function of the form[30,31] $p(h) = a/h$; it may be recorded by a fast discriminator system, but the echo pulses are not usually resolved from the parent pulse by the slower PHA.

Finally, one must remember that the shape of the distribution depends on electrode voltages,[35,39,83] on the area of the cathode illuminated,[35,39] and the wavelength of the illuminating light,[35,83] as well as on the temperature,[39,102] overall voltage, and other operating conditions. The wavelength dependence is due partly to an interaction between the wavelength dependence of the cathode sensitivity map and the position dependence of the pulse-height distribution; partly to the variation of mean electron energy with wavelength; and partly to the appearance of multiple-photoelectron emission (due to pair production) in the ultraviolet.[35] The shape of the contribution from cosmic rays depends on tube orientation.[30,41]

Because of these complications, there are several tests for internal consistency that should be made during measurements of pulse heights. (a) The gain of any amplifier used should be changed, if possible, by a known factor k. Then if we use subscripts 1 and 2 to denote measures made before and after the gain change, respectively, we should have $P_1(h) = P_2(kh)$ and $p_1(h) = p_2(kh)/k$, for all $h^\dagger$ (this check is easy to make if k is a small integer). (b) The apparent light-pulse distribution $s(h)$ obtained by subtracting $d(h)$ from $p(h)$ should have a fixed shape, independent of light intensity. A convenient method of testing this is to find $s(h)$ for two light intensities differing by at least a factor of two, and plot $\log s(h)$ [or $\log S(h)$] against h; the two plots should coincide when displaced vertically. If only integral counts are obtained, the ratio $[S_1(h)/S_2(h)]$ should be independent of h (where now 1 and 2 refer to the two different light levels). The light source must be varied so as to keep its spectral content fixed, because of the dependence of $s(h)$ on color; varying the current through a lamp filament is thus *not* a suitable method unless a narrow spectral passband is used. (c) One should always make sure that very few small pulses are due to amplifier noise, pickup, line transients, etc. If many pulses are recorded with the PMT voltage off (or at a very low setting), the trouble must be eliminated, or higher dynode gain and higher discrimination settings must be used to get above the instrumental noise. This extraneous contribution to $d(h)$ rises precipitously as $h \to 0$,[30,104] so that a rather small increase in threshold reduces it from a horrendous level to an innocuous one.

[102] K. Haye, *J. Phys. Radium* **24**, 86 (1963).
[103] J. Rolfe and S. E. Moore, *Appl. Opt.* **9**, 63 (1970).
[104] G. C. Baldwin and S. I. Friedman, *Rev. Sci. Instrum.* **36**, 16 (1965).

---

† To a fair approximation,[30,31] a change in dynode gain produces a similar effect. Thus, an increase in tube voltage[98] or a drop in temperature[49] causes $P(h)$ to become stretched out toward higher h; but $p(h)$ becomes squashed down as well as stretched out, because the pulses are spread over more channels in the PHA.[30,103]

1.5.1.2. Interpretation. The pulse heights produced by various mechanisms were mentioned in Section 1.4.2, and are discussed at length in Ref. 30 and elsewhere; so only a brief review will be given here. We begin at the largest sizes and work down. Pulse heights are referred to mean light-pulse height

$$\bar{h}_s = \int_0^\infty h\, s(h)\, dh \Big/ \int_0^\infty s(h)\, dh. \qquad (1.5.4)$$

Most pulses with normalized heights greater than about 10 are Cerenkov pulses due to cosmic rays and radioactive decays. These occur only in end-window tubes with high-efficiency cathodes; they should not occur with either S-1 or opaque-backed cathodes. The CR component ($\sim$1–2/cm^2/minute) varies with orientation: very large ($>$100$\bar{h}_s$) pulses can occur in a horizontal tube; more, but smaller ones, occur in a tube facing up; and they are smallest in size and number in a face-down tube.[41] Occasionally, large power-line transients can cause very large pulses (e.g., when large electric motors are started up).

Pulses with heights greater than 2 or 3 and less than 10 or 20 are due to gas ions striking the cathode; a typical height[10] is about 4. They occur $\sim$1 µsec after light pulses, a few percent of the time; they also occur in the dark. They appear to make up most of the dark current in many tubes (except those with S-1 cathodes); they account for both the temperature-dependent and the time-dependent components of dark current.[30] Both gas ions and dynode glow should produce a contribution to $d(h)$ that is proportional to h^{-2} at smaller pulse heights, as is observed in many tubes.[30,39] Typically, some 10^2 ion pulses appear, at room temperature, per square centimeter of cathode area per second.

The expected distribution $s(h)$ due to single photoelectrons ($h \approx 1$) was discussed in Section 1.4.1.3.3. The actual distribution is more smeared out, partly because of dynode nonuniformity,[29] and partly because of small afterpulses produced in the dynode system,[30] which contribute a $1/h^2$ component at $h \ll 1$.

The pulse-height distribution due to thermionic or field emission from the cathode is similar to that caused by photoelectrons. It is not identical, however, owing to a different spatial distribution (most of the thermionic emission probably comes from local hot spots that do not contribute strongly to photoemission) and a lower initial electron energy ($\sim kT$, or $\sim$0.03 eV at 300 K), cf., the dependence of $s(h)$ on wavelength. Spontaneous emission from dynodes[30] contributes a $1/h$ component to $d(h)$; there is some indication that exposing a tube to a uniform gamma-ray flux produces such a component by direct stimulation of dynodes. Echo pulses[31] should also have a $1/h$ distribution.

At small ($h \ll 1$) pulse heights, the PMT should contribute only the $1/h$ and $1/h^2$ components mentioned above. However, the smallest pulses actually observed are usually due to amplifier noise. The simplest model for such noise is white, Gaussian noise, such as Johnson noise in the anode load resistor or in the input resistance of the preamplifier. (Johnson noise is just the Rayleigh–Jeans tail of blackbody radiation, observed at radio and audio frequencies. A resistor is an absorbing—hence "black"— body at these frequencies.) In fact, this noise is what makes photomultipliers necessary in the first place: without the very large gain provided by the dynodes, the individual photoelectrons would be buried in the thermal noise of the circuitry.

Obviously, circuit-noise pulses are important if they are comparable to other noise pulses in height and frequency ($\sim 10^3$/sec at room temperature, ~ 1/sec in cooled tubes). With a 1-MHz bandwidth, we are concerned about pulses that occur with probability 10^{-3} at most; at 100 MHZ, only 10^{-8} of the random noise pulses can be important. Thus if we had pure Gaussian noise, we would be concerned about events more than 3 standard deviations from the mean; but this region is notoriously non-Gaussian in real systems. Thus the noise of concern is likely to be due to rare events unrelated to thermal noise.

1.5.1.3. Time Distribution. Here we are concerned with the temporal randomness or correlation of pulses. We must treat light and dark pulses separately; and we should consider the two-dimensional correlation in a plot of pulse height against time of occurrences.

If photons enter the cathode "at random" (i.e., if the number of photons in a given time interval obeys Poisson statistics), and if photoelectrons are produced "at random" (i.e., each photon has the same, fixed probability of producing an emitted photoelectron), it can be shown[105] that the time distribution of photoelectrons also obeys Poisson statistics. Actually, because photons obey Bose–Einstein statistics, they tend to clump together instead of arriving independently and at random. The clumping occurs on a time-scale comparable to the coherence time of the light; for "white" light this is very short, but for quasi-monochromatic (e.g., laser) light it can exceed the resolution time of pulse-counting systems, so an excess of close pulse pairs can be seen. The clumping can also be detected as coincident (i.e., temporally unresolved)

[105] D. L. Fried, *Appl. Opt.* **4**, 79 (1965).

pairs of photoelectrons produced in two detectors illuminated by the same coherent wave. In laboratory experiments, the coincidences are usually detected by nuclear pulse-counting methods, and the wave is divided into two coherent parts by a beam-splitter; in the Hanbury–Brown "intensity-correlation" interferometer,[106] coincidences are detected by radio-frequency correlation techniques employing analog multipliers, and wave-front division is used to obtain two coherent beams. In both cases, the same principles apply. As this field has been reviewed recently,[107] no further mention will be made here. In any case, the deviations from Poisson statistics are quite small in ordinary photometric work.[48,49,98]

Although the light pulses are randomly distributed in time, to a good approximation, the dark pulses sometimes are not, especially in cooled tubes.[46a,48] Because each large radioactive or cosmic-ray event is followed by a host of small pulses,[43-46a] and because such events dominate the dark noise of cooled alkali-antimonide tubes,[41,47] the correlated clumps of dark pulses (especially in tubes with broad spectral response[46,48]) are probably due to such events. Presumably, the small daughter events are mostly single-electron photoresponses to induced window phosphorescence.[43,44] Unfortunately, no detailed study of the temporal and height dependence of these correlated pulses has yet been made.

The occurrence of clumps of pulses with a characteristic time scale τ should produce excess (nonwhite) noise at frequencies lower than $\sim 1/\tau$. PMT frequency-spectra have been studied by several authors.[98,108,109,109a] No excess noise has been reported at frequencies greater than 0.1 Hz ($\tau \approx 10$ sec), although one should expect some in the dark noise of cooled tubes with "nonstatistical" (i.e., correlated) dark noise[48] at frequencies up to perhaps 10^2–10^3 Hz. Whether the excess fluctuations observed for $\tau \geq 20$ sec should be regarded as "flicker noise" or merely a slight gain instability[98] is perhaps a matter of terminology. The "excess noise" level here corresponds to gain variations on the order of 10^{-3} with a time scale of a minute or so; small temperature fluctuations ($\sim 0.1\,^\circ$C) could play a part, as well as cesium migration and other drift or fatigue effects. Similar $1/f$ noise has been observed by Smit et al.[109a] in the dark count rate of an uncooled 1P28.

[106] R. Hanbury-Brown, J. Davis, L. R. Allen, and J. M. Rome, M.N.R.A.S. 137, 375, 393 (1967) and references therein.
[107] C. L. Mehta, Progr. Opt. 8, 375 (1970).
[108] A. H. Mikesell, Publ. U. S. Naval Obs., 2nd Ser. 17, 143 (1955).
[109] R. C. Schwantes, H. J. Hannam, and A. van der Ziel, J. Appl. Phys. 27, 573 (1956).
[109a] C. Smit, C. Th. J. Alkemade, and W. F. Muntjewerff, Physica 29, 41 (1963).

As is well known,[110] low-frequency noise can be avoided by "chopping." This usually means a rapid symmetrical alternation between the unknown signal source being measured, and a reference (usually, dark) signal. Many solid state detectors and amplifiers have considerable $1/f$ noise in the low audio-frequency range, so that chopping frequencies near 1 kHz are common. The effect of chopping is to frequency-shift the measurements from a region near 0 Hz (dc) to a region around the chopping frequency, where the $1/f$ noise is negligible. The chopped signal is then demodulated by synchronous rectification, which restores the dc baseband.

Chopping is advantageous in situations that are dominated by additive low-frequency noise. This is true of most photoconductive detectors, for example, which are nearly always operated with symmetrical, square-wave chopping between signal and dark states. In principle, multiplicative noise (e.g., gain variations) can also be eliminated by chopping, if the demodulation process takes ratios instead of differences, and if the reference is a standard source instead of darkness. However, the difficulty of establishing a constant reference source, and the complexity of taking ratios, have usually restricted this technique to a few specialized instruments, such as densitometers.

As $1/f$ noise is negligible in photomultipliers down to 0.1 Hz or so, the rapid chopping that is so useful with other detectors is not advantageous.[98] Instead, occasional dark or background readings, usually made by hand, may suffice to establish reference levels. This can be regarded as a very low-frequency chopping, with a very asymmetrical duty cycle.

Finally, in many applications a photomultiplier is not limited by either additive (dark) noise or multiplicative noise, but by photon noise. In this case, symmetrical chopping is disadvantageous, because it degrades the photoelectron statistics by discarding half of the events. Of course, one still has to establish zero (dark) and gain (reference) levels; but only a small fraction of the total time may be needed for this purpose. We shall take a closer look at photon-noise-limited detection in the next section.

1.5.2. Detection: the Signal/Noise Ratio

When the light falling on a detector is strong enough that statistical fluctuations in the number of detected photons are larger than the dark noise, the detector is described as *photon-noise limited*. Also, it is often

---

[110] R. H. Dicke, *Rev. Sci. Instrum.* **17**, 268 (1947).

possible to distinguish a photon-noise component, even when dark noise is not negligible. Thus we need a concise way of describing the photon noise.

For comparison, let us consider an idealized, perfect, photon detector: every photon received produces an output signal, and the output is zero in the dark. Clearly, the signal/noise ratio depends only on the statistical fluctuations in the incident beam of photons. If the beam is weak enough, photon correlations can be neglected and Poisson statistics are a good approximation. Thus if g is the average number of photons observed per unit time, the probability of observing k photons in a particular unit-time interval is given by Eq. (1.4.10), and the standard deviation (i.e., noise) of a series of such measurements is $g^{1/2}$. The ratio of signal g to noise $g^{1/2}$ is thus $g^{1/2}$. For example, to obtain a standard error of 1%, we must count photons until 10^4 of them have been detected, on the average.

Now let us make the slightly more realistic assumption that photons are detected with some reduced probability, $q < 1$; we can call q the *quantum efficiency* of this idealized detector. The average number of detected photons per unit time is $N = g \cdot q$; its standard deviation is (again assuming Poisson statistics) $N^{1/2}$, so the signal/noise ratio for the less efficient detector is $N^{1/2} = g^{1/2}q^{1/2}$. Thus the signal/noise ratio is reduced by the square root of the quantum efficiency. For example, a detector with an efficiency of $\frac{1}{4}$ gives half the signal/noise ratio of a perfect detector.

Now suppose we have a real photon-noise-limited detector, with a signal/noise ratio half that of the perfect detector. It is natural to describe this real detector as having an effective quantum efficiency of $\frac{1}{4}$. As we deal only with the signal/noise ratio, it does not matter whether the detector is operated as a photon counter, or whether it has an analog output. Thus the concept of an equivalent (or, "detective") quantum efficiency can be applied to dc as well as pulse-counting measurements with photomultipliers.

Obviously, the detective quantum efficiency (DQE) of a photomultiplier cannot be higher than the quantum efficiency of the tube's cathode, at a given wavelength. In fact, it is always less; for some electrons (typically ~20%) are lost in the dynodes, and not all photoelectrons are equally weighted[111] in the detection process. The weighting depends on the height of each anode pulse, and on the method of detection. As we shall

[111] W. A. Baum, *in* "Astronomical Techniques" (W. A. Hiltner, ed.), Chapter 1. Univ. of Chicago Press, Chicago, Illinois, 1962.

see, the DQE is typically about half the cathode quantum efficiency in photon-noise-limited PMT's.

1.5.2.1. DC vs. Pulse Counting[†]

1.5.2.1.1. GENERAL RELATIONS.   We now investigate the effects of the weighting functions corresponding to different detection schemes. In pulse counting, all pulses between two heights (or all above a threshold level) are counted equally (weight 1) and all others are ignored (weight 0). In dc or charge integration, each pulse is weighted by its height (charge). Each weighting function gives different results for signal and dark pulses, because of the different distribution with height. Thus, it is convenient to consider the two extreme cases of strong signal (where dark pulses are negligible) and weak signal (where the signal pulses are almost negligible), for each weighting function. Both of these limits are un-realistic, since in practice the dark pulses always dominate the small pulse end of the distribution, and signal pulses must make an appreciable contribution or the time required to make an observation is unreasonably long. Nevertheless, both cases are helpful in understanding practical situations.

In the following analysis, lower-case letters are used to denote pulse-height distributions, and upper-case symbols denote counting rates, or measurements integrated over a range of pulse heights (e.g., anode currents). The letters S, D, and P indicate signal (light alone), dark, and signal plus dark (i.e., the quantity actually measured with light on the tube), respectively. It is important to realize that the quantities analyzed are all *rates*, such as counts per unit time, or charge per unit time (current). If a quantity Q is measured for a time t, the total number of pulses observed will be proportional to Qt, and its standard deviation will be proportional to $(Qt)^{1/2}$; thus the statistical estimate (mean $\pm$ standard deviation) of the rate Q will be the observed total $[Qt \pm (\alpha Qt)^{1/2}]$ divided by t, or $[Q \pm (\alpha Q/t)^{1/2}]$. (The factor α absorbs the proportionality factors.)

If it is necessary to distinguish between rates estimated from a unit-time observation and rates estimated from an observation of duration t, the time interval (1 or t) will be used as a subscript. Obviously, the mean values are the same in either case, so this notation is applied only to estimated standard errors. In the above example, the expected error in Q estimated from an observation of unit time is $\sigma_{Q,1} = (\alpha Q)^{1/2}$, and the

---

[†] Most of the material in this section is taken from Ref. 30.

error in an estimate of the rate Q obtained from an observation of duration t is $\sigma_{Q,t} = (\alpha Q/t)^{1/2}$. Notice that the error in the estimated rate decreases with the square root of the observation time; to halve the error, we must observe four times as long.

In general we observe a pulse distribution

$$p(h) = s(h) + d(h), \qquad (1.5.5)$$

where s and d are, respectively, the signal and dark pulse-height distributions, and $s(h)$ is proportional to the amount of light falling on the tube per unit time. If our weighting function is $w(h)$, the quantity measured with the light on for a time t_{L} is

$$P \times t_{\mathrm{L}} = t_{\mathrm{L}} \times \int_0^\infty w(h) \times p(h)\, dh = (S + D) \times t_{\mathrm{L}}, \qquad (1.5.6)$$

where

$$S \equiv \int_0^\infty w(h)\, s(h)\, dh, \qquad (1.5.7)$$

and

$$D \equiv \int_0^\infty w(h)\, d(h)\, dh \qquad (1.5.8)$$

is the rate measured with the light off. (We assume that the parent distributions are independent of time, so that P, S, and D refer to unit times.)

Now consider the statistical fluctuation of a measurement of duration t. If there are on the average $[p(h) \times dh \times t]$ pulses with heights between h and $h + dh$, the variance in this quantity is $\sigma_p^2 \times t = [p(h) \times dh \times t]$. If there is no correlation between pulses of different heights, the variance in the measured quantity is

$$t\sigma_{P,1}^2 = t \int \{\partial P/\partial[p(h)]\}^2 \times \sigma_p^2 = t \int_0^\infty \{w(h)\}^2 \times p(h)\, dh, \qquad (1.5.9)$$

where the subscript 1 refers to unit time.

If the total observation time is t, of which the light is on for a fraction f and off for a fraction $1 - f$, the total measurement with the light on is $[P \times ft \pm (ft\sigma_P^2)^{1/2}]$; with the light off, it is $\{D \times (1 - f)t \pm [(1 - f) \times t \times \sigma_D^2]^{1/2}\}$. Hence, the variances of P and D are, respectively,

$$\sigma_{P,t}^2 = \sigma_{P,1}^2/(ft) \qquad (1.5.10)$$

and

$$\sigma_{D,t}^2 = \sigma_{D,1}^2/[(1-f)\times t]. \tag{1.5.11}$$

From Eq. (1.5.6), we see that the estimated light flux on the tube is proportional to

$$S = P - D, \tag{1.5.12}$$

so that

$$\sigma_{S,t}^2 = \sigma_{P,t}^2 + \sigma_{D,t}^2 = (\sigma_{P,1}^2/ft) + [\sigma_{D,t}^2/(1-f)t], \tag{1.5.13}$$

where

$$\sigma_{P,1}^2 = \int_0^\infty [w(h)]^2\, p(h)\, dh \tag{1.5.14}$$

and

$$\sigma_{D,1}^2 = \int_0^\infty [w(h)]^2\, d(h)\, dh. \tag{1.5.15}$$

The value of f which minimizes $\sigma_{S,t}^2$ is found by setting the derivative of Eq. (1.5.13) with respect to f equal to zero; this gives

$$(1-f)^2\sigma_{P,1}^2 = f^2\sigma_{D,1}^2 \tag{1.5.16}$$

or

$$[f/(1-f)] = \sigma_{P,1}/\sigma_{D,1}. \tag{1.5.17}$$

Hence,

$$f = \sigma_{P,1}/(\sigma_{P,1} + \sigma_{D,1}) = 1/[1 + (\sigma_{D,1}/\sigma_{P,1})]. \tag{1.5.18}$$

In the weak-signal limit, $p(h) \to d(h)$ so $(\sigma_{D,1}/\sigma_{P,1}) \to 1$ and $f \to \frac{1}{2}$; in the strong-signal limit, $p(h) \gg d(h)$ so $(\sigma_{D,1}/\sigma_{P,1}) \to 0$ and $f \to 1$. These results for the limiting cases are well known; McCutchen[101] even stated in 1957 that "the best use of a given total counting time requires that it should be divided between experiment and background runs as the square root of the recorded count rates," which is just Eq. (1.5.17) if we recall that $\sigma_{P,1} = P^{1/2}$ in a counting experiment (and similarly for D). However, because of the dependence of $\sigma_{D,1}$ and $\sigma_{P,1}$ on $w(h)$, the optimum value of f for other cases will depend on the weighting function used.

If the measurement is made against a background light level, as in astronomical photometry of faint stars against the night sky, the background distribution,

$$b(h) = s_B(h) + d(h), \tag{1.5.19}$$

must be substituted for $d(h)$ in Eqs. (1.5.5)-(1.5.18). Here, $s_B(h)$ is proportional to the background light. If $s_B(h)$ is larger than the signal $s(h)$, we again have the low signal value $f = \frac{1}{2}$, even if $s(h) \gg d(h)$, as may happen with a cooled photomultiplier. However, in this case the distributions $b(h)$ and $p(h)$ are similar apart from a scale factor, so that $(\sigma_{B,1}/\sigma_{P,1})$ is nearly independent of $w(h)$; this similarity introduces some features of the strong signal case.

We now examine the effects of different weighting functions in detail.

1.5.2.1.2. PULSE COUNTING. As has been pointed out above, the small-pulse end of $d(h)$ generally is at least as steep as $1/h$. Therefore, since the pulse-counting weighting function

$$w_{pc}(h) = \begin{cases} 0, & h < l, \\ 1, & l \le h \le u, \\ 0, & h > u, \end{cases} \tag{1.5.20}$$

converts the integrals of Eqs. (1.5.6)-(1.5.9) to the form

$$n_p(l, u) = \int_l^u p(h)\, dh, \tag{1.5.21}$$

we must have $l > 0$ to prevent divergence. Physically, the finite number of dynodes would in principle provide such a cutoff, but this is usually far below amplifier noise and load resistor noise. As a practical matter, a finite lower threshold is always required in pulse counting; the strong signal limit is physically unattainable for this weighting function.

What lower threshold l should be used to achieve maximum signal-to-noise ratio? The signal-to-noise ratio ϱ is

$$\varrho_t = S/\sigma_{S,t} = n_s(l, u) \times [\sigma_{S,t}^2 + \sigma_{D,t}^2]^{-1/2}$$
$$= t^{1/2} \times n_s(l, u)$$
$$\times \{(1/f)\, n_s(l, u) + \{(1/f) + [1/(1-f)]\} \times n_d(l, u)\}^{-1/2}, \tag{1.5.22}$$

where n_s and n_d are defined analogously to n_p [see also Eq. (1.5.21)]. Here we have decomposed the sum

$$n_p(l, u) = n_s(l, u) + n_d(l, u) \tag{1.5.23}$$

and collected the n_d terms. We already know from Eq. (1.5.18) that

$$f(l, u) = \frac{(n_s + n_d)^{1/2}}{(n_s + n_d)^{1/2} + (n_d)^{1/2}}. \tag{1.5.24}$$

We could substitute Eq. (1.5.24) into Eq. (1.5.22) and eventually solve for the optimum discriminator settings l and u in the general case. However, in practice it is usually desirable to set $f = \frac{1}{2}$ or $f \approx 1$; the small loss in ϱ is compensated by the increase of operational efficiency due to convenience.[101] In general, we select l and u to maximize ϱ for the weakest signal we wish to measure. We must keep l and u fixed for all signal levels, or [due to the dependence of $s(h)$ on wavelength] the effective spectral response will vary. Optimizing for weak signals means we have less than optimum signal-to-noise for strong ones, but this is not a problem since we still have much better signal-to-noise on strong signals than on weak ones.

Having selected a value for f, which should be $\frac{1}{2}$ for genuinely weak signals and should be closer to 1 for stronger ones, we must set

$$\partial \varrho(u, l)/\partial u = \partial \varrho(u, l)/\partial l = 0 \qquad (1.5.25)$$

to find optimum discriminator settings u and l. This gives

$$d(l)/s(l) = d(u)/s(u) = (1 - f) + 2(n_d/n_s). \qquad (1.5.26)$$

For the weak signal limit, both $[d(h)/s(h)]$ and $(n_d/n_s) \to \infty$, so the f term is negligible. This gives

$$d(l)/s(l) = d(u)/s(u) = 2(n_d/n_s), \qquad (1.5.27)$$

independent of f. (However, we already know we should pick $f = \frac{1}{2}$ in this case.) This condition is not easy to determine, since n_d and n_s are both functions of u and l.

The choice is simplified if we use the fact that both $d(h)$ and $s(h)$ are, in general, rapidly decreasing functions of h. Therefore, we are not far wrong to set

$$n_d(l, u) \approx n_d(l, \infty) = \int_l^\infty d(h)\, dh \qquad (1.5.28a)$$

and

$$n_s(l, u) \approx n_s(l, \infty) = \int_l^\infty s(h)\, dh. \qquad (1.5.28b)$$

The functions on the right are simply the usual "bias curves" or "integral pulse-height distributions." If we define

$$R(h) = 2s(h)/d(h) \qquad (1.5.29)$$

and

$$Q(l, u) = n_s(l, u)/n_d(l, u), \qquad (1.5.30)$$

Eq. (1.5.27) corresponds to

$$R(l) = R(u) = Q(l, u). \qquad (1.5.31)$$

Notice that $R(h)$ is twice the ratio of the signal and dark distributions, and $Q(l, u)$ is the ratio of signal counts to dark counts (sometimes erroneously called the signal-to-noise ratio). Equation (1.5.28) means that $Q(l, u) \approx Q(l, \infty)$, so

$$R(l_1) = Q(l_1, \infty) \qquad (1.5.32)$$

defines a good first approximation to l. Then

$$R(u_1) = R(l_1) \qquad (1.5.33)$$

defines an approximate upper cutoff, so we can determine a second approximation l_2 from

$$R(l_2) = Q(l_2, u_1) \qquad (1.5.34)$$

and so on. The process converges very rapidly; an example is given in the Appendix of Ref. 30.

Notice that the result is independent of the signal strength used to compute R and Q, as both sides of Eqs. (1.5.29)–(1.5.34) are proportional to S.

The first approximation l_1 corresponds to Morton's[112] condition that the slope of the signal pulse integral distribution be half that of the dark pulse distribution, when the curves are plotted on semilog paper. For,

$$d[\log n_d(h, \infty)]/dh = -d(h)/n_d(h, \infty) \qquad (1.5.35)$$

and similarly for $[n_s(h, \infty)]$, so that Morton's condition that

$$d[\log n_s(h, \infty)]/dh = \tfrac{1}{2}d[\log n_d(h, \infty)]/dh \qquad (1.5.36)$$

gives Eq. (1.5.32) if $h = l_1$.

If a single level discriminator is used instead of a window counter, $u = \infty$ and $l = l_1$.

In the case of moderately large signals, we may suppose that Q is

---

[112] G. A. Morton, *Appl. Opt.* **7**, 1 (1968).

very large, so that we can neglect the (n_d/n_s) term in Eq. (1.5.26). The optimum discriminator settings are then given by

$$d(l)/s(l) = d(u)/s(u) \approx (1 - f), \tag{1.5.37}$$

which is the fraction of the time we have reserved for counting dark pulses. We know that $d(h)/s(h)$ becomes very large as $h \to 0$ or $h \to \infty$; therefore, the only question regarding the applicability of Eq. (1.5.37) is whether the signal is indeed large enough to make $(n_d/n_s) \ll (1 - f)$, which is already a small quantity. Another way of saying this is that if the dark count is not negligible, we should spend more time measuring it; i.e., increase $(1 - f)$. The stronger the signal, the smaller we can make Eq. (1.5.37); in the limit of no dark pulses, we would have $l \to 0$ and $u \to \infty$, counting all the pulses. This would clearly give the maximum signal-to-noise ratio possible: we cannot do better than to count every photoelectron equally.

Even if our weakest signal is not strong, we may be able to use Eq. (1.5.37) to determine approximate discriminator settings l_1 and d_1, and again use an iterative technique to find the best values

$$\frac{d(l_k)}{s(l_k)} = \frac{d(u_k)}{s(u_k)} = (1 - f) + 2\,\frac{n_d(l_{k-1}, u_{k-1})}{n_s(l_{k-1}, u_{k-1})}. \tag{1.5.38}$$

All of the above assumed pulses are independent. However, both the signal and dark distributions contain afterpulses and other induced pulses such as photoemission from dynode glow, etc. These induced events are certainly not statistically independent of the primary events. Any correlation between pulses of heights h_1 and h_2 should appear as a cross-product term in the integrand of Eq. (1.5.9). Another way of treating this problem is to regard an induced pulse as increasing the weight of the parent pulse.

Let us consider the large afterpulses,[40] which may typically be about four times the height of a primary photoelectron pulse and occur with probability ~ 0.05 about 0.3 μsec after the primary pulse. If the counting equipment used is relatively slow, the two pulses will not be resolved and will be treated as one large pulse. Fast counters, on the other hand, will count both pulses. In this case we can regard 0.95 of the n_i independent pulses as having weight 1, and $0.05n_i$ as having weight 2, since these are counted twice. The observed count is then

$$n_{obs} = 0.95n_i + 2 \times 0.05n_i$$
$$= 1.05n_i, \tag{1.5.39}$$

and the variance [see Eq. (1.5.14)] is

$$\sigma_{obs}^2 = 0.95 n_i + 2^2 \times 0.05 n_i$$
$$= 1.15 n_i \approx 1.1 n_{obs}. \tag{1.5.40}$$

Thus the observed variance is about 10% larger than would be expected from the total count. Such a small deviation from ideal counting statistics would be hard to measure.

The large afterpulses could be rejected by the upper level discriminator in a fast system. In a slow system, on the other hand, they would appear (unresolved from their parent pulses) as legitimate signal pulses and should be counted. This comparison shows that the pulse height distributions should be measured with the same equipment that is used for light detection, since different systems will measure different pulse height distributions from the same tube.

The situation is much worse in the case of cosmic-ray afterpulses, since typically about 10 afterpulses may be produced per cosmic ray.[42-46] In this case $n_{obs} \approx 10 n_i$ and $\sigma_{obs}^2 \approx 10^2 n_i = 10 n_{obs}$, if the cosmic rays dominate the dark pulses. This agrees with Rodman and Smith's data[48] on refrigerated S-20 tubes, in which σ^2 was five to ten times higher than the total number of pulses counted per sample time. Their typical dark count rates of ~ 20 pulses/sec are a few dozen times the expected cosmic ray rate for a 2-in. (5-cm) tube; the agreement is very good if about half of Rodman and Smith's dark counts were due to cosmic rays, and the other half to spontaneous events originating within the tube.

Finally, one should realize that the $1/h^2$ component of both $s(h)$ and $d(h)$ is mostly due to daughter pulses that are not statistically independent; their numbers should fluctuate by about the same factors as the smaller number of parent pulses that caused them. Such large fluctuations are observed; in fact, the difficulty of obtaining reproducible results at small heights has prevented most investigators from publishing any data on this region. If the lower discriminator threshold is set too low, these small pulses will add to the effective weights of the parent pulses.

Clearly, the effective weighting function can be much more complicated than expected, if correlated pulses are present. The rapid increase in correlated pulses with extended red or uv response means that tubes with wider spectral response than actually needed should be avoided.

Before leaving the subject of pulse counting, we offer a prescription for adjusting a pulse-counting system to optimize signal/noise ratio. A practical method must allow for correlated pulses and other nonideal weighting effects, and should use the actual counting equipment used for photometry (i.e., should not require a PHA). Fortunately, experimental signal/noise functions have rather broad maxima for most tubes, so the exact discriminator setting is not critical. We assume that "strong" and "weak" constant-light sources are available, and that the PMT housing is light-tight, so that an accurate dark measurement can be made.

After allowing adequate warm-up or cool-down time to stabilize the (dark) PMT and electronics, turn the tube voltage down to a very low value (less than $\frac{1}{3}$ normal operating voltage), or off, to look for electronic noise. Run the lower discriminator level down until noise pulses appear in the counter; if a window counter is used, the upper edge should be set as high as possible. Find the discriminator level at which a few noise pulses per minute occur; then raise the discriminator to at least twice this level. Or, if no amplifier noise is seen, set the discriminator to a threshold level near $\frac{1}{2}$ V (most discriminators do not function reliably much below this level).

Having set the discriminator to a low but reliable level, increase the high voltage in steps of about 10% (50- or 100-V steps are usually convenient). At each step, count both the dark pulses and the (dark + light) pulses with the strong light on, each for a time $t \geq 10$ sec. (The strong light should produce a count rate much higher than the dark rate, but not so high as to cause fatigue or nonlinearity. About 10^4 counts/sec is suitable. A 1-mm pinhole placed 3 km from an ordinary 60-W incandescent lamp transmits about the right amount of light, for most tubes, so an attenuation of 10^6 is needed if the lamp is placed 3 m ($\sim$10 ft) away from such a pinhole. Thus the required strong light is really quite dim by laboratory standards.) Do not increase the tube voltage beyond the recommended maximum value, or the point at which the count rates exceed 10^5/sec, whichever comes first.

If the light was bright enough, the light count rate P should exceed the dark count rate D by a large factor at each voltage, so that $S = P - D \approx P$. Then a plot of $(P/D^{1/2})$ as a function of tube voltage should be a fair approximation to $(S/D^{1/2})$, the expected signal/noise ratio if all pulses are independent. The maximum in this function should be a fair first approximation to the optimum high voltage. In general, this value should be near the manufacturer's "typical" operating voltage for the tube. Excessively high voltages tend to increase dark noise; excessively low ones lead to poor photoelectron collection efficiency.

Set the high voltage 50–100 V above the indicated optimum, and reduce the light until the (light *minus* dark) count rate S is about half the dark rate D (weak light level). The high voltage, or the discriminator level, can now be adjusted in smaller steps ($\sim$2% in voltage, or $\sim$20% in discriminator level) to find the point that actually maximizes (S/σ_s), as estimated statistically from repeated measurements. For example, suppose we make n ($\geq$20) paired measurements of P and D at each setting. For each pair we compute $S_i = P_i - D_i$; the ratio of the mean ($\bar{S} = \sum S_i/n$) of the S_i to their standard deviation $\{s_S = [\sum (S_i - \bar{S})^2/(n-1)]^{1/2}\}$ is a good estimate of (S/σ_s) for weak signals. If 5 or 6 settings are used, and 20 bright and dark readings of 10 seconds' duration are made at each setting, the whole optimization can be done in less than 1 hr. Such a small investment of time is well worth while to achieve optimum results at low light levels.

If a window-type counter is used, the upper threshold setting is not critical, as very few very large pulses occur in any case. Usually, a setting that excludes 1% of the strong-light count (or 10 times the lower-threshold value, whichever is larger) is satisfactory.

Finally, if the maximum in (S/σ_s) is rather broad and flat, it may be desirable to pick an operating point slightly off the peak, so as to improve stability of operation. Maximum stability against gain changes occurs when $d[\ln P(h)]/d[\ln h] = h/P(dP/dh)$ is least, for a given light level; thus, one wants to operate at the flattest part of a log–log plot of count rate, $P(h)$, against discriminator level h. In general, this depends on light level, and cannot be optimized for both strong and weak lights. However, as "dark" or "background" checks are usually made more frequently than standard-source (gain) checks, it is usually best to optimize for gain stability at the higher light levels.

1.5.2.1.3. DIRECT CURRENT METHODS. In dc or charge-integration photometry,

$$w_{\mathrm{dc}}(h) = h. \tag{1.5.41}$$

Thus, the measured quantity per unit time is

$$P = \int_0^\infty h \times p(h)\, dh = \mu_{1,p} \tag{1.5.42}$$

with variance

$$\sigma_{p,1}^2 = \int_0^\infty h^2 \times p(h)\, dh = \mu_{2,p}. \tag{1.5.43}$$

The signal-to-noise ratio is therefore

$$\varrho_t = S/\sigma_{S,t} = \mu_{1,s} \times [\sigma_{S,t}^2 + \sigma_{D,t}^2]^{-1/2}$$
$$= (t)^{1/2}\mu_{1,s} \{(1/f)\mu_{2,s} + \{(1/f) + [1/(1-f)]\}\mu_{2,d}\}^{-1/2}, \tag{1.5.44}$$

where $\mu_{1,s} = S$ is the first moment of the signal pulse distribution, and the μ_2's are the second moments defined as in Eq. (1.5.43). Note that $\mu_{i,s}$ is proportional to S for all i.

In the weak signal limit we can ignore $\mu_{2,s}$ in Eq. (1.5.44) and have

$$\varrho_{\text{weak}} \rightarrow S(t/4\mu_{2,d})^{1/2} \tag{1.5.45}$$

for $f = \frac{1}{2}$. Here, $(2\mu_{2,d})^{1/2}$ is the dark-current noise; the dark current itself is $\mu_{1,d}$.

Compared to pulse counting, dc photometry discriminates against the small dark pulses originating in the dynodes. Thus, for spontaneous dynode emission, $d(h) \propto 1/h$ for $h < h_0$ and the dark current

$$\int_0^{h_0} h/h \; dh$$

is finite. Even the induced dynode pulses proportional to $1/h^2$ give a finite dark current, since each dynode contributes equally to the anode current and there are only a finite number of dynodes.

The noise contribution from dynodes is even smaller. Since spontaneous emission from D1 is amplified g times less than cathode emission, the noise power contributed by D1 emission is g^2 (typically 10–20) times less than that due to cathode emission. Even for induced emission, each dynode contributes g times less noise than its predecessor, so the total is finite and comes mainly from the first stages. Thus, the dark noise is dominated by the very large ion, cosmic-ray, and radioactive-decay pulses.

In a cooled tube, the dark *noise* can be due mainly to cosmic rays (and nearly independent of temperature), even though the dark *current* is mainly due to ions (and is still temperature-sensitive).[41] In computing the CR noise, the small afterpulses should be added to the weight (i.e., height) of the initial Cerenkov pulse in proportion to their own heights; at most, this doubles the noise computed from the giant pulses alone. Thus, although most of the dark current usually comes from the dynodes, most of the dark-current noise comes from the cathode. The dc value of dark current is therefore not a good indicator of the tube quality for low-level work, even if dc leakage is negligible. This is particularly true for a refrigerated end-on tube, where cosmic rays are relatively more important.

In the strong signal limit, we can ignore the $\mu_{2,d}$ term in Eq. (1.5.44) and have

$$\varrho_{\text{strong}} \rightarrow S(ft/\mu_{2,s})^{1/2}. \tag{1.5.46}$$

We can safely set $f \approx 1$ here, so

$$\varrho \approx [\mu_{1,s}/(\mu_{2,s})^{1/2}](t)^{1/2}. \tag{1.5.47}$$

An ideal quantum detector of efficiency q achieves a signal-to-noise ratio

$$\varrho_{\text{ideal}} = Nqt/\sigma_{Nqt} = Nqt/(Nqt)^{1/2} = (Nqt)^{1/2} \tag{1.5.48}$$

in observing a stream of N photons per second. If we normalize $s(h)$ so that

$$s(h) = Nq_k s_0(h), \tag{1.5.49}$$

where $s_0(h)$ is the probability distribution of signal pulses and q_k is an effective (i.e., allowing for lost electrons) cathode quantum efficiency, we see that Eq. (1.5.47) becomes

$$\varrho \approx (Nt)^{1/2} \times (q_k \mu_{1,s_0}^2/\mu_{2,s_0})^{1/2}. \tag{1.5.50}$$

Thus, the detective quantum efficiency of dc photometry is just the cathode quantum efficiency times a degradation factor,

$$\varDelta_{\text{dc}} = \mu_{1,s_0}^2/\mu_{2,s_0}, \tag{1.5.51}$$

which is the ratio of the square of the mean pulse height to the mean square pulse height. Since the detective quantum efficiency for pulse counting, in the strong signal limit, approaches the cathode quantum efficiency q_k, Eq. (1.5.51) also gives the ratio of dc to pulse-counting detective quantum efficiencies for strong signals.

Prescott[29] has shown that $s_0(h)$ probably belongs to a family of functions bounded by the Poisson distribution on one hand and by the exponential distribution on the other. Hence, we can use these two limiting forms to place bounds on $\varDelta_{\text{dc}}$.

For the exponential distribution e^{-h} we have $\mu_1 = 1$ and $\mu_2 = 2$, so $\varDelta_{\text{dc}} = \frac{1}{2}$, a result first published by Lynds and Aikens,[113] although it was probably discovered earlier by Baum. For the Poisson distribution, we have to pick the mean value. We will surely have an overestimate for $\varDelta_{\text{dc}}$ if we set $\mu_1 = g_1$, the gain of the first dynode; for we then neglect the additional broadening of $s_0(h)$ by multiplication statistics at the following dynodes. Since the Poisson distribution has the property that

$$\sigma^2 = \mu_2 - \mu_1^2 = \mu_1, \tag{1.5.52}$$

[113] C. R. Lynds and R. I. Aikens, *Publ. Astr. Soc. Pacific* **77**, 347 (1951).

we have

$$\Delta_{\text{dc}} \leqq \mu_1{}^2/(\mu_1{}^2 + \mu_1) = g_1/(g_1 + 1). \tag{1.5.53}$$

Hence, even for strong signals, pulse counting is more efficient than dc photometry. However, the pulse-counting advantage is smaller for tubes with narrower pulse-height distribution [contrary to the myth that says tubes with narrow $s(h)$ and/or $d(h)$ are more suitable for pulse counting]. Tubes such as the 1P21/931-A or ITT tubes, which have a more nearly Poisson signal pulse distribution, will therefore be slightly better for dc photometry than the EMI venetian blind tubes. These results agree with Baum's[111] experimental data, which showed a DQE advantage for pulse counting of about 1.3 for a 1P21 in the strong signal limit, and factors approaching 2.0 for other types. The relatively small advantage of pulse counting at high light levels agrees with Nakamura and Schwarz's statement[114] that "simple dc measurements are about as good as pulse counting for light levels above the dark current [equivalent] level."

Afterpulsing is more important for dc work than for pulse counting. If we again assume a 5% rate of afterpulses four times the average signal pulse height, we have an anode current proportional to

$$n_{\text{obs}} \approx 0.95n_i + (4 + 1)\times 0.05n_i = 1.2n_i, \tag{1.5.54}$$

where $(4 + 1)$ is the total height of an afterpulsing event, and n_i is the number of photoelectron pulses $(= Nq_kt)$. Ignoring the spread in pulse heights within each group, we have

$$\sigma_{\text{obs}}^2 \approx 0.95n_i + (4 + 1)^2\times 0.05n_i = 2.2n_i \approx 1.83n_{\text{obs}}. \tag{1.5.55}$$

Thus, a typical rate of afterpulsing can roughly double the relative noise power, and halve the DQE. This may explain why 1P21's usually give best dc signal/noise ratios at relatively low overall voltages (700–800 V): these relatively gassy tubes must be run at low potentials to keep ion production and afterpulsing to a minimum.[115]

One argument frequently offered[112] in favor of pulse counting is that it is relatively insensitive to gain drifts. This argument assumes that the discriminator level can be set at a point on $s(h)$ where nearly all signal pulses are counted, and where $ds/d(\ln h)$ is small, so that a small fractional change in effective discriminator level l produces a very small fractional

[114] J. K. Nakamura and S. E. Schwarz, *Appl. Opt.* **7**, 1073 (1968).
[115] R. W. Engstrom, *J. Opt. Soc. Amer.* **37**, 420 (1947).

change in $n(l, u)$. These assumptions can only be approximately met for strong signals, however. In the weak signal limit, we are always fighting the $1/h^2$ component of the dark pulses, so that a fractional change in l produces a comparable change in $n_d(l, u) \approx 1/l$. Thus, pulse counting and dc detection have similar sensitivities to gain changes, if the weak-signal advantage of pulse counting is to be realized. One can, however, choose to sacrifice weak-signal performance in order to obtain less sensitivity to gain changes at high light levels. Even so, the "peak-to-valley" ratio at the single-electron maximum of $p(h)$ is usually not very high, owing to the induced (dynode) pulses; so the decrease in gain sensitivity is rarely as large as a factor of ten (i.e., 10% change in gain will still produce a 1% change in signal).

1.5.2.2. Other Methods. Various other techniques have been proposed, involving still other weighting functions. Among these may be mentioned the measurement of the shot-noise power instead of the current at the anode. In this case the signal depends on the second moment of $p(h)$, and the noise depends on the fourth moment. Consequently, the shot-noise method is still less sensitive to the small dynode pulses (which are always a minor problem in dc detection), and still more sensitive to the large CR and ion pulses. The analysis has been carried out,[30] but will not be repeated here as it shows no large advantages for this method. This scheme is more complicated to use than either dc or pulse counting, and suffers from both the limited dynamic range (nonlinearity) of pulse counting (due to the nonlinear weighting), and the dc disadvantages of zero drift and analog (rather than digital) output.

Symmetrical chopping followed by synchronous detection has also been tried; as pointed out earlier, the lack of $1/f$ noise in photomultipliers deprives this technique of its main advantage. Of course, chopping does eliminate the zero-drift problems of most dc amplifiers; but this is better done by chopping the (filtered) anode current (as is done in some solid state operational amplifiers), rather than the light, which throws away half the photons. Chopping the light fixes f at $\frac{1}{2}$, since the dark current is observed during the half cycles when the light is cut off. No separate dark reading is then necessary. However, the value $f = \frac{1}{2}$ is inefficient at higher light levels. At moderate light levels, Nakamura and Schwarz[114] found little difference between synchronous detection and pulse counting; one should bear in mind that pulse counting may also reject an appreciable fraction of the light pulses, if the discriminator rejects most of the dark pulses.

Finally, one can ask what weighting function would maximize the signal/noise ratio ϱ. In the case of a noiseless photomultiplier [$d(h) \equiv 0$], we maximize the signal/noise ratio ϱ by counting all pulses equally. Also, in a noiseless tube with a background light level $b(h) \propto s(h)$, we readily see that all pulses should be counted equally.

In the case of the window counter, the condition $R(u) = R(l)$ shows that it is not the pulse height or pulse rate that is important, but the probability that a pulse is a signal pulse; if all pulses are equally likely to be signal pulses, all should be counted equally.

This suggests that we should take

$$w(h) = s(h)/p(h). \tag{1.5.56}$$

In the weak signal limit, $s \to 0$ and $p \to d$, so we may adopt

$$w_0(h) = s_0(h)/d(h), \tag{1.5.57}$$

where $s_0(h)$ is defined by Eq. (1.5.49). In fact, we can show that this function does maximize[30] the weak-signal signal-to-noise ratio ϱ. We may regard this result as an example of the principle of matched filtering, in which a weighting scheme proportional to the probability of success gives optimum detection.

However, the optimum weighting scheme, which would require detailed pulse-height analysis followed by computer processing, has hardly any advantage over pulse counting. Clearly, it has maximum advantage when $s(h)$ and $d(h)$ have very different shapes; for if they had the same shape, we should simply count all pulses (equal weighting) to achieve maximum ϱ. So, we shall consider an idealized tube in which

$$s_0(h) = e^{-h}$$

and

$$d(h) = \delta \times h^{-2}$$

for $h \leq h_{\max}$; we suppose $h_{\max} \approx 4$ if ion pulses are the largest dark pulses, and $h_{\max} \approx 10\text{--}100$ if cosmic-ray and gamma-ray pulses are important.[†] We adopt these forms for $s(h)$ and $d(h)$ because they are typical of many real tubes, and because this choice makes s and d very different in shape and hence produces large changes for different weighting functions. For tubes (such as the ITT tubes) in which $s(h)$ and $d(h)$

[†] In some of the following we shall take $h_{\max} \approx \infty$, where only a small error is involved.

are similar, we have essentially the large signal or background situation, in which different weighting functions produce rather similar results, and pulse counting gives the best results.

For pulse counting, we find $l_1 = \frac{1}{2}$; the successive approximations are $u_1 = 5.19$; $l_2 = 0.520$; $u_2 = 5.07$; $l_3 = 0.520$. The corresponding values of ϱ^2 as a function of l and u are

$$\varrho_{pc}^2(\tfrac{1}{2}, \infty) = tN^2q_k{}^2/8e\delta = (tN^2q_k{}^2/\delta)\times 0.046 \qquad (1.5.58a)$$

and

$$\varrho_{pc}^2(0.52, 5.07) = (tN^2q_k{}^2/\delta)\times 0.050. \qquad (1.5.58b)$$

Thus, a window counter, used optimally, would give about 10% higher efficiency than a simple discriminator. In order to reject the optimum amount of dark noise, we can only count about 60% of the actual signal pulses; if a wide dynamic range is required, the detective quantum efficiency for strong signals cannot exceed $0.6q_k$. For weak signals, the noise equivalent input is about

$$N_{0,pc} = (Se\delta)^{1/2}/q_k \qquad (1.5.59)$$

for the simple discriminator, and about 5% lower for the optimum window counter.

With dc detection, Eq. (1.5.45) gives

$$\varrho_{dc} = Nq_k/(t/4\,\delta h_{max})^{1/2}, \qquad (1.5.60)$$

so

$$N_{0,dc} = 2(\delta h_{max})^{1/2}/q_k. \qquad (1.5.61)$$

Thus, the dc performance for weak signals depends on the largest dark pulses, and may be either better or worse than pulse counting, depending on whether h_{max} is less or greater than $2e \approx 5.4$. Nakamura and Schwarz compared dc and pulse counting in an EMI 9558 at $-45°C$, and found only about a factor of two difference in noise equivalent input. This gives $h_{max} \approx 20$, which seems reasonable since cosmic-ray noise dominates under these conditions. As mentioned before, $q = q_k/2$ for this case at large signals. Thus, pulse counting is only 20% more efficient than dc detection in the large signal limit, if the discriminator is optimized for weak signals. (We must use the same discriminator setting for both; because the pulse-height dependences of spectral and spatial response, data taken at different discriminator settings are not directly comparable.)

Now let us look at the results expected for the optimum weighting function $w_0(h) = h^2 e^{-h}/\delta$. We have

$$W_0 = \int_0^\infty S_0{}^2(h)/d(h)\, dh = \int_0^{h_{\max}} (e^{-2h}/\delta h^{-2})\, dh = 1/4\delta, \quad (1.5.62)$$

so

$$\varrho_{\mathrm{opt}}^2 = N^2 q_k{}^2 t/16\delta \qquad (1.5.63)$$

and[†]

$$N_{0,\mathrm{opt}} = 4(\delta)^{1/2}/q_k. \qquad (1.5.64)$$

Thus, in the case we have considered, the advantage of using optimum weighting is small; the noise equivalent input is lower by a factor of only $(e/2)^{1/2} = 1.17$ than in pulse counting.

The price paid for optimum weak-signal detection is some decrease in strong signal detective quantum efficiency. In the strong signal limit,

$$\sigma_{S,t}^2 = \sigma_{S,1}^2/ft = \int_0^\infty w_0{}^2(h) \times s(h)\, dh/ft$$

$$= \int_0^\infty (h^2 e^{-h})^2 N q_k e^{-h}\, dh/ft\delta^2$$

$$= N q_k/ft\delta^2 \int_0^\infty h^4 e^{-3h}\, dh = N q_k/ft\delta^2 \times 8/81, \quad (1.5.65)$$

and

$$S = \int_0^\infty w_0(h)\, s(h)\, dh$$

$$= \int_0^\infty S_0(h)/d(h) \times N q_k S_0(h)\, dh$$

$$= N q_k \times W_0 = N q_k/4\delta. \qquad (1.5.66)$$

So, for $f \approx 1$,

$$\varrho^2 \approx S^2/\sigma_{S,t}^2 = (N^2 q_k{}^2/16\delta^2)/(8N q_k/81 t\delta^2) = N q_k t \times 81/128. \quad (1.5.67)$$

The detective quantum efficiency for large signals is thus $(81/128)q_k = 0.633 q_k$. For the model we have chosen, this represents less degradation at high light levels than for any of the other detection methods (weighting functions). Thus, optimum weighting should produce better results at

---

[†] Equation (1.5.64) was derived assuming $h_{\max} = \infty$. The error made is approximately $0.5 h_{\max}^2 \exp(-2h_{\max})$, which is ~ 0.01 for $h_{\max} = 3$ and 0.003 for $h_{\max} = 4$.

all light levels. The improvement in going from pulse counting to optimum detection is relatively small, however.

It is desirable to monitor the dc anode current even in non-dc detection, as gain drifts on the order of a factor of two may occur at microampere anode currents, due to dynode fatigue. Such large signals usually produce a temporary rise in dark noise, also.

A major limitation of any nonlinear weighting system is limited dynamic range; when a significant fraction (say, 1%) of pulses overlap, a similar degree of nonlinearity results. If we try to extend the range by switching over to dc methods for strong signals, we must remember that the apparent spectral response will change when we change weighting functions. The same problem occurs if we try to avoid saturation and fatigue effects by changing the tube voltage; the apparent change in red/blue response ratio can be several percent.

Some detection schemes involve combinations of techniques, such as those involving chopping and synchronous detection, or pulse counting followed by a rate meter and analog recording.[116] Apart from the loss of photons if chopping is used, such methods inherently offer the same signal/noise ratio as the simpler techniques, depending on which moment of the pulse-height distribution is measured (zeroth for pulse counting; first for current measurements; etc.).

1.5.2.3. Analog Versus Digital Recording. When high precision ($\sim$1% or better) is required, or when a large number of measures are to be averaged together to improve the signal/noise ratio, there are substantial advantages in digitizing the signal and recording all data in digital form. Obviously, all data must eventually be reduced to digital form for quantitative analysis; but many people, especially spectroscopists, like to see a graphical (analog) display. There is no doubt that a strip-chart display is valuable as a gross indication of equipment stability or malfunction, but such a qualitative diagnosis does not require very high resolution or linearity. The range of signal/noise ratios that can usefully be displayed on a chart only extends from 5:1 up to perhaps 100–200 to one, however.[103] At lower S/N, the signal is not visible in the noise; above this range, the noise becomes invisible against the signal.

Even in the useful S/N range of analog recording, it is inefficient to try to recover numerical data from a chart record.[117] Even an experienced worker can extract only about half or a third of the original information

[116] D. L. Akins, S. E. Schwarz, and C. B. Moore, *Rev. Sci. Instrum.* **39**, 715 (1968).

[117] A. T. Young, *Observatory* **88**, 151 (1968).

from a chart record. That is, the error of reading the chart is at least as large as the standard deviation of the mean deflection obtained by averaging over a centimeter of chart, even if the person reading the chart has years of experience; a novice makes much larger errors.

The chart-reading error appears to be independent of the filter used to smooth the analog record, however; a given steady signal can be read with equal accuracy, whether recorded with a short or a long time constant. Thus, provided that the filtering does not remove information, the observer may choose whatever smoothing he likes.

In addition to the substantial errors introduced by trying to digitize an analog record manually, one must consider the economics involved. A digitizer and printer or punch cost less than the annual salary of a chartmeasurer, as well as being faster and more reliable.

Finally, a digital system has a better filter function than an analog system with RC filtering. The digital system, which integrates for a time τ, not only averages symmetrically and uniformly in the time domain (in contrast to the one-sided and unequal weighting of the RC filter); it also gives statistically independent samples, instead of the partially correlated analog data, which are asymmetrically distorted by the "memory" of earlier values. On the other hand, one must remember that the equivalent bandwidth Δf of the RC filter is $1/(4RC)$, while that of the integrator is $1/(2\tau)$; thus, the two systems include the same noise bandwidth only if $RC = \tau/2$, not τ as is often erroneously assumed. The noise bandwidths and other characteristics of these and other filters are well discussed by Robben.[98]

To sum up, it is best to use a digital system to record data for analysis, while maintaining an analog record as a crude real-time indicator of data quality. (We may add that the advantages of digitized data have even been recognized in the analysis of photographic spectra.[61,118])

[118] G. I. Thompson, *Publ. Roy. Obs. Edinburgh* **5**, no. 12, 245 (1967); **7**, no. 2, 19 (1970). See also W. K. Bonsack, *Astron. Astrophys.* **15**, 374 (1971).

2. OTHER COMPONENTS IN PHOTOMETRIC SYSTEMS*

2.1. Optical Systems

2.1.1. The Telescope and Atmosphere

2.1.1.1. The Structure of a Star Image. In measuring the light of a star, a small diaphragm is placed in the focal plane of the telescope to exclude, as far as possible, unwanted light from neighboring stars and from the sky. The smaller the diaphragm, the better this interference can be rejected. However, if too small an aperture is used, a significant fraction of the measured star's light is also excluded. Because this excluded light depends on variable atmospheric and instrumental factors, it cannot be accurately corrected for, and errors will result in the measured data. Thus there is an optimum diaphragm size, which depends on the image structure and the precision and accuracy required.

2.1.1.1.1. THE TELESCOPE ALONE: DIFFRACTION AND STRAY LIGHT. The diffraction pattern in the focal plane of a perfect telescope with a uniform circular aperture (pupil) is proportional to

$$I(r/\pi) = [2J_1(r)/r]^2, \tag{2.1.1a}$$

where r is the distance from the center of the image in (angular) units of λ/D rad; λ is the wavelength of observation; D is the diameter of the entrance pupil, and J_1 is the Bessel function of the first kind. However, most telescopes used for photometry have an annular aperture, obstructed by a secondary mirror. If the fraction of the diameter obscured is t, the ideal diffraction pattern has the form

$$
\begin{aligned}
I(r/\pi) &= 4[J_1(r)/r - t^2 J_1(tr)/(tr)]^2/(1 - t^2)^2 \\
&= 4[J_1(r) - tJ_1(tr)]^2/[r(1 - t^2)]^2.
\end{aligned} \tag{2.1.1b}
$$

---

* Part 2 is by Andrew T. Young.

Because $J_1(x)$ decreases as $x^{-1/2}$ for large x, the asymptotic behavior of Eq. (2.1.1) is a decrease as r^{-3}. At first sight this makes the outer parts of the diffraction pattern seem unimportant. However, the area of a circle with radius r increases as r^2; so the total amount of light outside a focal-plane diaphragm of radius r decreases only as $1/r$. In fact, a good approximation to the fractional excluded energy[1] is

$$X(r) \approx [5r(1 - t)]^{-1}. \tag{2.1.2}$$

As a common example, consider a 40-cm (16-in.) telescope with 40% central obscuration, used at visual wavelengths ($\lambda = 5000$ Å): the diaphragm size needed to include 99% of the light is 17 arcsec in diameter, neglecting any imperfections due to seeing or instrumental imperfections. In practice, a diaphragm of at least 20 arcsec diameter would be required. If we want to reduce the excluded light by a factor of 2, we must double the focal-plane aperture. If we want the diffraction error to be less than 0.002 mag, the aperture must be at least 1.4 arcmin across; to reach this small an error with a 15-arcsec diaphragm requires at least a 250-cm (100-in.) reflector with excellent seeing! Furthermore, actual measurements of excluded energy can be 10–20 times larger than Eq. (2.1.2) indicates. The importance of large telescopes for accurate photometry is not generally appreciated.

Kormendy[1a] has shown that diffraction from a power-law distribution of dust (or pinholes in the mirror coatings) can produce a slower fall in brightness than the inverse-cube law of Eq. (2.1.1). For radii larger than a minute of arc, he finds image brightnesses decreasing more slowly than r^{-2}. Obviously, such a decrease cannot continue indefinitely, for the encircled energy integral would diverge. For apertures a few tens of seconds of arc across, Kormendy's data are close to an inverse-cube law, but the actual values are 100 times brighter than the values expected for a clean circular aperture. Thus, for the apertures typically used in stellar photometry, the above discussion is qualitatively correct, but the effects may be much more severe than Eq. (2.1.2) indicates.

Fortunately, it can be shown[1] that small optical aberrations and focus errors do not greatly increase the light loss. The main effect of "seeing" is to extend the outer isophotes of the image by the diameter of the "seeing disk," so the required diaphragm size is increased by a few arcseconds at least. Unfortunately, there tends to be an inverse correlation between

---

[1] A. T. Young, *Appl. Opt.* **9**, 1874 (1970).
[1a] J. Kormendy, *Astron. J.* **78**, 255 (1973).

seeing and transparency at most sites,[†] so that 5–10 arcsec seeing (or even worse) is common on good photometric nights. Thus even for moderately large apertures, it is generally inadvisable to use diaphragms smaller than 20 arcsec. Finally, photometry of extended objects (planets, nebulae, etc.) also requires the use of larger diaphragms than for stars, if systematic errors between extended and point sources are to be avoided.

In faint-star photometry, sky noise (both photon noise and real brightness variations in airglow) is usually the dominant error source, so it is important not only to use a small diaphragm, but also to ensure that the telescope does not increase the background brightness. Of course, light scattered from dirty optics can be a problem, but can be avoided by using dust covers, by refraining from smoking in the dome, and by frequent cleaning. Cassegrain reflectors must also be adequately shielded so that direct sky light does not reach the focal plane (i.e., the sky must not be visible around the secondary mirror from any point in the useful field). Several computer algorithms are now available[2] to design baffles that produce minimum obscuration of the pupil—an important consideration, as Eq. (2.1.2) shows. The obscuration rises rapidly as the field is increased, so wide-field designs (e.g., Ritchey–Chretien) are not well suited to photometry.

Finally, in telescopes with openwork tubes, it sometimes happens that stray light from indicator lamps, etc., can shine on baffles or other parts of the telescope visible from the focal plane. This can be a special problem at red and infrared wavelengths, where a light that appears very dim to the eye may actually be quite bright.

2.1.1.1.2. ATMOSPHERIC EFFECTS: SEEING, SCINTILLATION, AND DISPERSION. The simplest of the atmospheric effects is dispersion (chromatic differential refraction), so we treat it first. Because the refractivity $(n-1)$ of air varies about 2% across the visible spectrum, the general astronomical refraction varies with wavelength. In good seeing, the star image is visibly spread out into a vertical spectrum, with the blue end up. If the image is visually centered in the diaphragm, the blue (and especially

[2] A. T. Young, *Appl. Opt.* **6**, 1063 (1967). See also R. Prescott, *ibid.* **7**, 479 (1968); A. Cornejo and D. Malacara, Bol. Ton. y Tac. No. 30, p. 246 (Oct. 1968).

[†] There is some physical basis for this: the best transparency occurs on the driest nights; but low atmospheric water vapor allows strong radiational cooling to develop large temperature differences near the telescope.

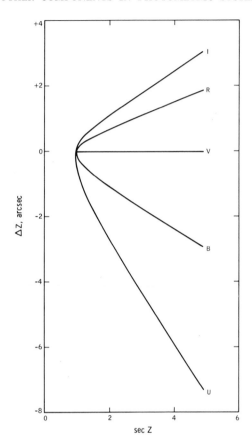

FIG. 1. Atmospheric dispersion (displacement, relative to visual image) for some common photometric bands, as functions of air mass. Z is the zenith angle.

the ultraviolet!) image is decentered. Even in the near infrared, appreciable displacements occur at moderate air masses (see Fig. 1).

The photometric error due to decentering is the difference between the light shifted out of the diaphragm on one side, and the light shifted in on the other. Both the difference in surface brightness and the area shifted are (to first order) proportional to the decentering Δr, so the total error varies like $(\Delta r)^2$. A rough estimate suggests that the decentering error is

$$\Delta I(\Delta r) \approx 10X(r) \cdot (\Delta r/r)^2, \qquad (2.1.3)$$

where $X(r)$ is the excluded light given by Eq. (2.1.2). The lost light is doubled if $\Delta I(\Delta r) = X(r)$, or $(\Delta r/r)^2 \approx 0.1$, which corresponds to

$(\Delta r/r) \approx 0.3$. For the previous example of a 40-cm (16-in.) telescope using a 20-arcsec diaphragm, a decentering error of about 3 arcsec excludes an additional 1%, or 0.01 mag. This error occurs at about 2 air masses for the U band of the UBV system (cf., Fig. 1). Because the dispersion is proportional to tan z, the photometric error varies as $\tan^2 z = \sec^2 z - 1$.

The effects of atmospheric turbulence are quite complex. A good understanding of both atmospheric turbulence and the resulting optical effects has been obtained by meteorologists and physicists in the last decade or so; only a few results and reviews will be cited here. The basic theory appears in Tatarski's monograph,[3] which combined the Kolmogorov turbulence spectrum (now abundantly confirmed by direct measurements) with a useful perturbation treatment of wave propagation through turbulence. The theory has been reviewed[4] and the results discussed[5] in two very clear technical articles, as well as a more popular account,[6] to which the reader is referred. An earlier phenomenological review[7] is also quite readable, and is still fairly accurate.

The seeing is due to random wave-front (phase) variations over the telescope's entrance pupil. With small apertures the major effect is a wave-front tilt, appearing as an image displacement. Nearby turbulence affects the whole field, but distant turbulence causes different image points to move with respect to each other. A large aperture averages the different image displacements of many turbulent elements, so image blur is seen instead of image motion. In monochromatic light, the blur is resolved into a host of rapidly moving "speckles" the size of the Airy disk, which collectively make up the seeing disk of a point source.[8] These speckles are averaged out in a long-exposure measurement; the mean point-spread function is roughly (but not exactly) Gaussian, and can be described in terms of an atmospheric modulation-transfer function.[9] The instantaneous image[10] cannot be described by an atmospheric

[3] V. I. Tatarski, "Wave Propagation in a Turbulent Medium." McGraw-Hill, New York, 1961 (reprinted in 1967 by Dover, New York).

[4] R. W. Lee and J. C. Harp, *Proc. IEEE* **57**, 375 (1969).

[5] R. W. Lee, *Radio Sci.* **4**, 1211 (1969).

[6] A. T. Young, *Sky Telescope* **42**, 139 (1971).

[7] H. Elsässer, *Naturwissenschaften* **47**, 6 (1960).

[8] A. Labeyrie, *Astron. Astrophys.* **6**, 85 (1970). See also D. Y. Gezari, A. Labeyrie, and R. V. Stachnik, *Astrophys. J.* **173**, L1 (1972).

[9] R. E. Hufnagel and N. R. Stanley, *J. Opt. Soc. Amer.* **54**, 52 (1964).

[10] D. L. Fried, *J. Opt. Soc. Amer.* **56**, 1372 (1966).

MTF, however, because it depends on the telescope pupil in a nonlinear way (cf., Labeyrie's "speckles"[8]). As the dividing line between "long" and "short" exposures is the time required for the atmospheric turbulence to be blown past the pupil by the wind ($\sim$10 m/sec), most photometric observations are "long" exposures. In this case, the "atmospheric MTF" is a valid concept, leading to the conclusion that the outer parts of the image are still dominated by diffraction.[1]

Instead of regarding the time-averaged image as the two-dimensional convolution of the seeing disk and the ideal diffraction pattern, one may estimate the light loss due to seeing from Eq. (2.1.3), if a suitable mean square light displacement Δr^2 is used. Classical arguments suggest that Δr^2 (and hence the light loss) should grow linearly with air mass, but modern seeing theory indicates a $(\sec z)^{1.2}$ power law; observational data suggest about an 0.9-power law.[11] To avoid systematic errors, the focal-plane aperture must be large enough to exclude negligible light not only in the zenith, but also at the largest air masses used, allowing for both dispersion and seeing.

While seeing is largely due to phase fluctuations over the pupil, scintillation is due to amplitude fluctuations. The theory of astronomical scintillation has been worked out in sufficient detail to allow accurate calculation of the temporal power spectra observed with any circular, annular, or rectangular aperture, using any optical bandwidth, for planets as well as stars.[12] The results for small apertures depend strongly on the vertical distribution of turbulence, the angular size of extended objects (planets or satellites), and the wavelength range used.[11,12]

However, for the larger telescopes normally used in photometry, the situation is much simpler; to a fair approximation, the spectrum is nearly flat up to a frequency

$$f_{\mathrm{c}} = V_{\perp}/\pi D, \tag{2.1.4}$$

where D is the diameter of the telescope and $V_{\perp}$ is the speed at which turbulence crosses the line of sight. At moderate zenith distance z, $V_{\perp}$ is primarily determined by winds at $h_{\mathrm{eff}} = 5$–20 km height (typically 30 m/sec); but a component

$$V_{\mathrm{rot}} = (2\pi/86400)h_{\mathrm{eff}} M \cos \delta$$
$$\approx 0.73 \times 10^{-4} h_{\mathrm{eff}}/(\sin \phi \tan \delta + \cos \phi \cos H), \tag{2.1.5}$$

[11] A. T. Young, *J. Opt. Soc. Amer.* **60**, 1495 (1970).
[12] A. T. Young, *Appl. Opt.* **8**, 869 (1969).

due to the earth's rotation, is appreciable near the horizon. [Here, H and δ are the hour angle and declination of the star, ϕ is the observer's latitude, and $M \approx \sec z$ is the air mass. As $h_{\text{eff}} \approx 10^4$ m, Eq. (2.1.5) is roughly $(M \cos \delta)$ m/sec, which is usually negligible.]

At frequencies below f_c, the rms error due to scintillation noise is[12,13]

$$\sigma = S_0 D^{-2/3} M^p \exp(-h/h_0)(\Delta f)^{1/2}, \qquad (2.1.6)$$

where D is the diameter of the telescope, Δf is the bandwidth $(= 1/(4\tau)$ for an integration of τ sec[13a]), h is the observer's height above sea level, and $h_0 = 8000$ m is the atmospheric scale height. The exponent $p = \frac{3}{2}$ at azimuths perpendicular to winds at heights around h_{eff}, and $p = 2$ when looking along the wind direction.[12] Typical values of S_0 are 0.05 if D is in inches, or 0.09 for D in centimeters.[13] As the refractivity of air is nearly independent of wavelength, so is S_0; this weak dependence can be seen in simultaneous measurements at two wavelengths,[11] but other chromatic effects (largely due to dispersion, which causes rays of different colors to sample different pieces of turbulent air) are also important if small apertures are used.[12] Table I[†] shows the standard error to be expected from scintillation noise for some typical circumstances, and the approximate B magnitude at which scintillation noise and photon noise are equal. Although the latter figures depend somewhat on variable factors such as S_0, optical transmission, and detective quantum efficiency, the relative values do not. Clearly, scintillation noise is important to at least tenth magnitude if either large apertures or large zenith distances are used.

At frequencies above f_c, the scintillation noise rises rapidly with central obstruction of the aperture,[13] although the low-frequency scintillation does not. Typical obscuration ratios (~ 0.4) double the high-frequency noise. This may be a problem in observing occultation curves and other high-speed phenomena.

One peculiarity of scintillation is that the intensity is log-normally distributed.[3,12] Thus, in strong scintillation, the intensity distribution is very asymmetrical, with large narrow spikes. Although the logarithm of the intensity (e.g., stellar magnitude) is normally distributed, the mean log is not the log of the mean, so systematic errors can occur in least

[13] A. T. Young, *Astron. J.* **72**, 747 (1967).
[13a] F. Robben, *Appl. Opt.* **10**, 776 (1971).

---

† This table is slightly revised from Tables II and IV of Ref. 13.

TABLE I. Standard Error (Magnitudes) of a 10-sec Integration, Due to Scintillation Noise; and B Magnitude at Which Scintillation Equals Photon Noise.[a]

Aperture	Air mass				
	$M = 1$	$M = 2$		$M = 3$	
Scintillation noise					
16 in. (40 cm)	0.0013	0.0038	0.0054	0.007	0.012
36 in. (90 cm)	0.0008	0.0022	0.0032	0.004	0.007
60 in. (150 cm)	0.0006	0.0016	0.0022	0.003	0.005
B magnitude					
16 in. (40 cm)	7.0	8.9	9.6	9.9	11.0
36 in. (90 cm)	7.6	9.5	10.2	10.4	11.6
60 in. (150 cm)	7.9	10.9	11.6	10.8	12.0

[a] For air mass $M > 1$, two values are given: the first corresponds to $p = \frac{3}{2}$ in Eq. (2.1.6) (crosswind azimuth), and the second is for $p = 2$. Values shown are for sea level; at $h = 2000$ m, multiply errors by 0.78, and subtract about 0.3 mag from crossover values.

squares analyses of data. If intensity units are used, the non-Gaussian distribution makes least squares invalid; but if magnitudes are used, the wrong mean values result. This can be particularly troublesome in absolute comparisons with a standard lamp, which require a small telescope aperture and hence large scintillation noise. Another problem occurs if nonlinear detection (such as pulse counting) is used, for the large spikes can go well into the nonlinear region even if the average intensity does not. Unfortunately, the spikes occur on a short time scale (milliseconds) and are usually smoothed out on an analog record, so this problem can be much worse than it appears.

2.1.1.2. Guiding, Flexure, and Field Lenses.   Although Eq. (2.1.3) shows that small guiding errors should not greatly change the amount of light passing through the focal-plane diaphragm, there is still the problem of keeping the PMT output constant. If the photocathode is placed directly behind the diaphragm, image motion moves the illuminated spot on the tube; the large spatial variations in PMT responsivity then cause large output variations, even if the light passing through the diaphragm remains constant. The usual solution is to image the telescope pupil on the cathode.

This can be done either by using the diaphragm as a pinhole camera[13b] (which requires placing a much larger photocathode far from the diaphragm), or using a field lens to produce a small image of the pupil. This lens is usually called a *Fabry lens*; a zero-order account of the theory is given by Michlovic.[14] At first, it would seem that a very small image could be produced if a short-focus lens were used, so that a very small cathode area (with small dark current) could be used. However, the field lens converts spatial changes into angular ones; the light from an off-center star strikes the tube face obliquely. Thus one has traded the spatial sensitivity variations for the (usually much smaller) angular ones (not only at the photocathode, but also the increased Fresnel-reflection losses at oblique incidence on lens and window surfaces). The shorter the focal length and the smaller the pupillary image, the larger these angular effects will be; eventually they become unacceptable. If an achromat is used, variable uv absorption across the lens (depending on the local thickness of the cement and the higher-index element) may cause problems.

Finally, one must put the lens far enough from the focal plane so that the inevitable dust specks do not cause significant shadowing. For example, a 50-μm dust speck removes 1% of the light from a 0.5-mm beam— say, 6 mm from the focus of an unobstructed $f/12$ system. If we want to be safer and keep 100-μm dust down to 0.1% loss, the beam must have expanded to over 3-mm diameter, and the lens must be put some 40 mm from the focal plane.

Even if a field lens is used, flexure (in either the telescope tube, the mirror supports, or the photometer or its mounting) will cause the image of the pupil to move relative to the tube cathode. In this case, the field lens converts an angular misalignment of the telescope and photometer optical axes into a spatial displacement of the image on the PMT. Of course, if the tube itself is not firmly mounted, it may move with respect to the image as the photometer moves with respect to gravity. These problems are minimized if the tube and image are aligned for maximum response. Notice that a standard source in the photometer does *not* provide any compensation for flexure problems.

The spatial/angular error conversion of a field lens can be avoided if an integrating sphere (or some similar multiply reflecting cavity) is used

---

[13b] P. Guthnick and R. Prager, *Veröff. Kgl. Sternwarte Berlin*, Band I, Heft 1 (1914); Band II, Heft 3 (1918).

[14] Joe Michlovic, *Appl. Opt.* **11**, 490 (1972).

to homogenize the cathode illumination. Simple diffusing screens (opal or ground glass) are not satisfactory: they produce considerable light loss and poor homogeneity, and may introduce position-dependent polarization effects. A convenient integrating cavity is described by Blackwell *et al.*[15] The image can also be scrambled by passing through a light pipe, such as a quartz rod cemented to the tube face; the pipe must be long enough for most rays to be internally reflected several times, and small enough in diameter to break the pupil into several differently reflected parts for each field point.

2.1.1.3. Polarization Effects. Unfortunately, the light of many astronomical objects is partially polarized, either by transmission through anisotropic media (interstellar reddening), reflection (some nebulae; planets), or even intrinsically (emission from within magnetic fields). Because of the polarization sensitivity of PMT's, especially if reflective enhancement techniques are used, the apparent brightness of an object may depend on the interaction between its polarization and that of the instrument. Finally, most telescopes alter the state of polarization of light, and may even do so in a position-dependent way. Thus, these effects may become quite complicated.

The basic optical phenomena involved are described in optics textbooks, so only a few reminders are given here. Oblique reflection from metallic surfaces not only produces a small linear polarization (typically $\sim 1\%$) in incident unpolarized light, but partially converts incident linear (or circular) polarization into circular (or linear) polarization by introducing a phase lag between two principal components. The phase lag also occurs on total reflection in prisms. Oblique transmission through lenses and windows produces polarization by Fresnel reflection losses; and even normal-incidence transmission can produce or alter polarization in many anisotropic media (many crystals, or isotropic materials under strain).

Probably the best policy is to measure the state of polarization of the incident light, and the polarizing properties of the instrument. However, this is both difficult and tedious, so we try to make the measurement insensitive to polarization. This means making both the instrumental transmission and the detector response independent of polarization.

A minimum requirement for constant transmission is a symmetrical, on-axis optical system. However, even reflection at normal incidence on crystalline metals may cause some polarization, especially if the substrate

[15] D. E. Blackwell, A. D. Petford, and S. H. McCrea, *Observatory* **85**, 21 (1965).

causes a preferred orientation of the crystals. Such orientation can be induced by anisotropy in the surface layers of polished glass, owing to the mechanical strain of polishing. In any case, 45° reflections in a coudé system can produce large position-dependent changes.[16]

To overcome detector polarization, a so-called "depolarizer" is used in front of the detector (and any strongly polarizing optical elements). This device does not really remove polarization, but merely modulates it rapidly across the instrumental passband. To overcome the mono-chromatic polarization, one must modulate the polarization rapidly across the instrumental pupil.[17]

If a Fabry lens is used, an off-axis star can suffer polarization by oblique incidence on the lens and subsequent window surfaces. If an image scrambler is used, it should ensure many symmetrical reflections; for example, the white coating used by Blackwell *et al.*[15] produces 5–10% polarization of natural light at oblique incidence, and only partial de-polarization of polarized light on oblique reflection.[18] However, multiple reflections may cause an unacceptable light loss.

Polarization effects are often neglected entirely, on the ground that the polarizations of both the instrument and most objects are "small," so their product is a second-order effect and hence negligible. This view is valid for bright stars if 1% accuracy is adequate. It is not valid for a variety of interesting objects with relatively large polarization (several percent or more), or for really accurate photometry of normal objects.

2.1.2. Filters and Spectrographs

2.1.2.1. Thermal Effects. Two types of thermal effects occur: wave-length shifts and transmission changes. Of course the thermal wavelength shifts in spectrographs have been notorious in radial-velocity work for a long time; they are due both to changes in the dispersive element (prism or grating) and to mechanical displacements (e.g., of entrance and exit slits) caused by expansion of structural members. (Flexure can cause similar problems.) Except at high spectral resolutions, these effects are not usually serious in spectrophotometry.

However, filters generally exhibit much larger temperature shifts, particularly at the short-wavelength side of their passbands. A rather

[16] M. Marin, *Rev. Opt.* **44**, 115 (1965).

[17] L. A. Rahn, P. A. Temple, and C. E. Hathaway, *Appl. Spectrosc.* **25**, 675 (1971). See also R. B. Kay and R. J. Holland, *Appl. Opt.* **10**, 1587 (1971).

[18] D. C. Carmer and M. E. Bair, *Appl. Opt.* **8**, 1597 (1969).

complete review[19] of available data shows remarkable similarities among all types of absorbing filters—glass, dyed gelatin, liquid solutions, etc. Typical shifts are 0.5 Å/°C at 3000 Å, increasing to 5 Å/°C or more near 1-μm wavelength. An increase in temperature extends the absorbing region to longer wavelengths. This behavior is just like the extension of red-tail absorption and response in photocathodes with increasing temperature: thermal excitation within the material allows less-energetic (redder) photons to be absorbed.

In fact, the temperature-coefficient data are readily explained by a simple model. Suppose the filter material has a band gap energy E_G; at $T = 0$ K, we expect a sharp absorption edge at $\lambda_G = hc/E_G$. At a temperature T, we expect the absorption coefficient at $\lambda > \lambda_G$ to fall off exponentially with decreasing photon energy, decreasing a factor of $e = 2.718\ldots$ for each decrease of kT in hc/λ. Thus at a photon energy

$$E = E_G - \alpha kT, \tag{2.1.7a}$$

we expect an absorption coefficient $K \approx e^{-\alpha}$. Now, most filters are used in thicknesses t of a few millimeters, so the absorption coefficient at the "cutoff"—or $(1/e)$—transmission wavelength λ_c is about $\varkappa \approx \lambda_c/t = 2.5 \times 10^{-4}$ for $t = 2$ mm and $\lambda_c = 5000$ Å. Hence $\alpha \approx -\ln \varkappa \approx 8$, and is relatively insensitive to the exact value of $\varkappa$ because only its logarithm is involved.

We can now write

$$E_c = hc/\lambda_c = E_G - \alpha kT; \tag{2.1.7b}$$

differentiation gives

$$-hc/\lambda_c{}^2 \, d\lambda = -\alpha k \, dT, \tag{2.1.8}$$

or

$$d\lambda/dT = \lambda_c{}^2(\alpha k/hc) \approx 5 \times 10^{-9}\lambda_c{}^2 \tag{2.1.9}$$

for λ in angstroms, T in Kelvin or Celsius degrees, and $\alpha = 8$. Figure 2 shows Eq. (2.1.9) plotted with data from Ref. 19 and the Schott factory; the agreement is quite satisfactory. Equation (2.1.9) also shows why the temperature coefficient is usually independent of T.

Because we always use absorbing filters in the far "tail" region, the factor α is large, and temperature stabilization of filters is even more important than photocathode temperature regulation.

[19] A. T. Young, *Mon. Notices Roy. Astron. Soc.* **135**, 175 (1967). See also R. A. Botsula. *Astron. Tsirk.* No. 593, p. 5 (1970); F. H. Gieseking, *Messtechnik* **10/69**, p. 240 (1969),

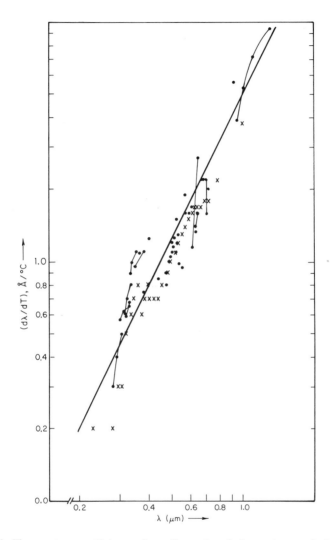

FIG. 2. Temperature coefficients of cutoff wavelength for various optical and filter glasses, from A. T. Young, *Mon. Notices Roy. Astron. Soc.* **135**, 175 (1967) [denoted by ●] and from data supplied by the Schott glass factory [denoted by ×]. [Individual glasses are identified in Young (above reference).] The diagonal line is Eq. (2.1.9), calculated for a 2 mm thickness. Generally, thicker samples lie above this line, and thinner ones below it.

We remark that colored solutions in which thermochemical changes occur can show much larger effects ($\sim$10 Å/°C).[20] On the other hand, interference filters show temperature coefficients about five times smaller.[19] (This is understandable since their thermal effects are primarily due to expansion of the dielectric layers, and should therefore be closer to kT without the large multiplier α.) Unfortunately, interference filters are strongly affected by changes in humidity.[20a] Passband shifts on the order of 50 Å between dry and saturated air at 20°C can occur. The effect is reduced by age, and by storage under dry conditions.

In general, the transmission of absorbing filters decreases with increasing temperature as the absorption bands widen out. The transmission of interference filters may decrease in the passband and increase in the wings if the temperature changes produce a mismatch in the resonant structure at the (temperature-shifted) peak wavelength, but this effect is usually small. On the other hand, temperature changes in spectrographs can cause large (up to 4%/°C) changes in throughput, mainly by altering the widths of entrance and exit slits.[21]

Finally, one should expect, on physical grounds, that the reflectance of metal mirrors should be temperature-dependent. This should be especially true in instruments such as coudé spectrographs where a large number of mirrors cause severe reflection losses, and in spectral regions where the mirror material has an absorption band (e.g., silver near 3150 Å; aluminum near 8000 Å; gold and copper throughout the visible region). We know of no measurements; but this effect should be looked for[†].

2.1.2.2. **Unwanted Wavelengths and Stray Light.**   No filter or spectrometer is perfect; there is always some energy transmitted outside the intended passband. In accurate spectrophotometry, such "leaks" can be quite severe. If a narrow passband is used, even very low transmission elsewhere can contribute a large unwanted signal. For example, suppose an interference filter with a 20-Å passband is used; then a transmission of

[20] A. L. Olsen, L. W. Nichols, and A. K. Rogers, *Appl. Opt.* **11**, 954 (1972).
[20a] J. Schild, A. Stendel, and H. Walther, *J. Phys.*, **28**, C2-276 (1967).
[21] A. Ozolins, W. C. Lineberger, and F. E. Niles, *Rev. Sci. Instrum.* **39**, 1039 (1968).

[†] Theoretical calculations by K. Ujihara [*IEEE J. Quantum Electron.* **QE-8**, 567 (1972) and *J. Appl. Phys.* **43**, 2376 (1972)] show that the absorption of metal mirrors is roughly proportional to the absolute temperature. Thus the output of a system with multiple reflections will change by 1%/°C if it transmits 23%.

only 10^{-4} (0.01%) over a 2000-Å region can contribute a 1% spurious signal from an equal-energy source. If the detector has a much higher response in this "leak" region, or if the source spectrum is much brighter there, the "leak" signal can easily become tens of percent. Thus a filter or spectrometer that claims "less than 0.1% transmission" outside the nominal passband may be utterly worthless, depending on the source spectrum and detector response. This also emphasizes the importance of using a detector that does not respond outside the spectral range required.

A particular problem with absorbing filters is unwanted passbands. In general, all transparent materials absorb somewhere in the infrared because of molecular vibrational transitions, and in the ultraviolet (for a photon energy greater than E_G) because of electronic transitions. Most colored glasses absorb at intermediate wavelengths because of impurity absorbers, which create electronic states in the band gap. Thus "blue" or "ultraviolet" filters have a large absorption band in the green and red regions, but are even more transparent ($\sim 100\%$) in the near infrared than in the short-wavelength passband (~ 60–70%). Even in the visible absorption band, these filters have appreciable transmission (10^{-3}–10^{-4}). When used with an ordinary Cs_3Sb photocathode, these filters give a far-red "leak" that can be quite offensive, particularly if red stars are observed with a long-tailed photocathode.[21a] The newer red-sensitive cathodes make the "red-leak" problem more severe; various absorbing or interference-type filters can be used to block the "leak," but they modify the response in the blue as well. Although the leak can be represented by various interpolation formulas if it is small, it is generally best to use a blue-absorber to measure the leak accurately (making corrections for reflection and absorption losses in the leak-isolating absorber as well[21a]).

Although filters can be very opaque to short wavelengths (below the uv absorption edge), a photometric instrument can show appreciable response to such light. This "blue leak" is due to fluorescence of the filters; some standard sharp-cutoff filters are even used as standards of fluorescence.[22] The fluorescence usually occurs in a broad band some 2000 Å to the red of the filter's absorption edge. It can be excited by almost any wavelength that is strongly absorbed by the filter (uv is not required); for example, green light causes red filters to fluoresce.[23]

[21a] C.-Y. Shao and A. T. Young, *Astron. J.* **70**, 726 (1965).

[22] C. S. French, *Appl. Opt.* **4**, 514 (1965).

[23] Loren N. Pfeiffer and J. F. Porter, Jr., *Appl. Opt.* **3**, 317 (1964).

Fluorescence is strongly affected by impurities, and may vary strongly from sample to sample of the same glass, or even across the face of a single piece.[22] Lowering the temperature 40°C can double the intensity of fluorescence.[24] Some of the factors influencing glass and quartz fluorescence are discussed by Pringsheim.[25]

Fluorescence can be important if the filter is close to the photocathode, so that a large solid angle of the (nearly isotropically) emitted radiation is intercepted by the tube. If two or more filters are used in series, their order makes a large difference; Pfeiffer and Porter[23] reduced their fluorescent "leak" by a factor of 2000 simply by reversing the order of two filters.

Finally, one must be careful that stray light cannot work its way around the edges of a loosely mounted filter. Such light leaks are easily blocked by using opaque O-rings, or cements.

Spectrometers have similar leakage problems, not only due to light scattered from dust on optical elements, but also from overlapping orders in grating instruments. Order separation may be done by crossed dispersion (usually using a prism), or by absorbing filters. Another problem in some grating instruments is scattered or redispersed light from the zero-order image, or doubly dispersed light from other orders[26] (leading to the observation of pseudo-half-order spectra.[27] It is unusual for a single monochromator (even if double-passed) to achieve 0.1% scattered light; many astronomical spectrographs[28] have 5–15% contamination. The stray light problems are probably one cause of poor agreement among various sets of scanner observations; a double monochromator should be used for such work.

2.1.2.3. Polarization and Knife-Edge Effects.

As is well known,[29] diffraction gratings can produce very large (up to 90%) polarization, especially near the grating anomalies.[30-32] The ratio of spectrometer throughputs for the two polarizations can vary by a factor $\sim$100 in a moderate

[24] G. Kuwabara, *J. Phys. Soc. Japan* **9**, 992 (1954).

[25] P. Pringsheim, "Fluorescence and Phosphorescence," pp. 506–507. Wiley (Interscience), New York, 1949.

[26] J. K. Pribram and C. M. Penchina, *Appl. Opt.* **7**, 2005 (1968). See also J. J. Mitteldorf and D. O. Landon, *Appl. Opt.* **7**, 1431 (1968).

[27] N. R. Butler, *Appl. Opt.* **9**, 1475 (1970).

[28] R. F. Griffin, *Mon. Notices Roy. Astron. Soc.* **143**, 319, 349, 361 (1969).

[29] D. C. Hammer, E. T. Arakawa, and R. D. Birkhoff, *Appl. Opt.* **11**, 79 (1964).

[30] F. C. Evering, Jr. and Marie-Jose Aujouannet, *Appl. Opt.* **8**, 710 (1969).

[31] J. B. Breckinridge, *Appl. Opt.* **10**, 286 (1971).

[32] C. H. Palmer and H. W. Le Brun, *Appl. Opt.* **11**, 907 (1972).

wavelength interval.[31] The spectrograph slits can also introduce some polarization if they are only a few wavelengths wide; the effect for a slit of width w is on the order of λ/w, and is usually much smaller than the grating effects. Prism spectrometers produce smaller polarizations, mostly by Fresnel reflection losses. If a large instrumental polarization is acceptable, the overall sensitivity can be improved by arranging the planes of polarization favored by both the optics and the detector to coincide.

Many filters also produce, or alter the state of, polarization. Colored glass filters may have appreciable strain birefringence, and even "neutral" filters can produce polarization.[33] Because of polarization and reflection effects, the effective transmission of two filters in series is *not*, in general, equal to the product of their apparent transmission measured separately. Reflections between filters, or between them and other optical elements, can be eliminated by tilting the filters a few degrees so that multiply reflected light is deflected from the detector; this introduces polarization, however.

Interference filters are a particular problem because of their high reflectance, but they cannot be tilted without shifting (and, in non-parallel light, broadening) the passband. If they are placed near a shiny focal-plane diaphragm, large (several percent) reflections can occur between the filter and the diaphragm. We have even seen the apparent brightness of a star decrease when a larger aperture was used, because some of the light reflected from the back of a small diaphragm was sent back out of the telescope through a large one. As this doubly reflected light is strongest near the half-intensity wavelengths of the filter, the instrumental bandwidth is broader when a small diaphragm is used. Appreciable reflections can also occur between filters and windows, field lens, and phototube surfaces. Thus broad-band antireflection coatings are useful, not only to reduce losses, but also to reduce some of these multiple reflections.

In addition to the polarization mentioned above, narrow slits in high-resolution scanners can produce increased scintillation noise by acting as Foucault knife-edges, so that the atmospheric phase variations across the pupil appear as very strong intensity variations (schlieren). When a large part of the star image is excluded by the slit, only a few small bright patches appear in the pupil, and the scintillation noise is greatly increased. Furthermore, if the pupil is imaged on the photocathode, additional

[33] J. L. Kohl, L. J. Curtis, D. A. Chojnacki, and R. M. Schechtman, *Appl. Opt.* **10**, 34 (1971).

modulation occurs as the bright spots move over the nonuniform tube face; and this modulation cannot be removed—in fact, it is increased—by taking the ratio of the scanning photomultiplier signal to that of a fixed "reference" detector nearby, because the two detectors have different nonuniformities. However, if an image scrambler is used, a reference detector can eliminate most of the scintillation noise (provided it is within a few hundred angstroms of the scanning detector).

Finally, both interference and colored glass filters may show serious nonuniformity. In the former, the main problem is pinholes and thickness variations in the coatings; in the latter, bubbles, "seeds," and striae cause similar problems. The only real solution, as for detector nonuniformity, is to image the pupil on the filter. Failing this, only optical-quality glass should be used (not the Corning glasses, originally intended for railway signal lamps).

2.2. Calibration Problems and Standard Sources

2.2.1. Light Sources

Because every photometer suffers from variations in response, a constant reference light source must be measured from time to time. An astronomical photometer has not only to contend with time and temperature variations, but also positional effects due to magnetic and gravitational fields. Thus it is wise to measure a standard source along with each star.

2.2.1.1. Standard Lamps (Absolute Calibration).
In addition to the reference source that is required to eliminate the above effects, we sometimes need a source for absolute energy calibration. This can be either a black body of known temperature and area, or a calibrated standard lamp. The construction and calibration of such sources lie outside the scope of this article; but for many purposes, commercially available tungsten-filament lamps are suitable and convenient.[34] However, certain precautions must be observed in their use.

Obviously, the conditions under which the standard lamp was calibrated at NBS must be duplicated as exactly as possible. This means not only the operating voltage and current, but also the orientation of the lamp with respect to gravity (filament sag can seriously affect its resistance, temperature, and light output) and the line of sight (output

---

[34] Ralph Stair, W. E. Schneider, and J. K. Jackson, *Appl. Opt.* **2**, 1151 (1963).

is not isotropic). Furthermore, if halogen-filled lamps are used, measurements should cover the same spectral intervals (width as well as position) as were used for lamp calibration, because some spectral regions contain strong molecular absorption bands (readily seen as a purplish tint within the bulb when an iodine-filled lamp is turned off but still hot, for example). Finally, one expects some variation of light output with ambient temperature, owing to conductive and convective cooling of the filament, etc.

Tungsten-filament lamps last longer and have more constant output when operated on ac instead of dc power, owing to electromigration; filament irregularities are smoothed out by ac operation, but grow larger with dc.[35] Square-wave supplies are commercially available, which provide the constant power input and accuracy of regulation of a dc source, combined with the lamp stability of ac operation.

Several serious problems occur in comparing standard lamps with stars. First of all, standard lamps are so bright that they must be placed a considerable distance from the telescope to maintain a reasonable dynamic range in the comparison. Secondly, in order for the lamp to look like a star to the optics, it must also be placed far from the telescope. Not only must it be placed many focal lengths away (to bring the image near the focal plane of the telescope); it must also be far enough away to subtend a small angular size, or the diaphragm problem mentioned in Section 2.1.1.1.1 will become severe.

These problems typically require a lamp-to-telescope distance of hundreds of meters. For example, consider a 10-cm, $f/10$ telescope (1-m focal length). For a distant object, the focal length f is related to the object distance D and the deviation from normal (infinity) focus Δf by

$$\Delta f/f \approx f/D. \qquad (2.2.1)$$

Thus a lamp 100 m ($100f$) away is imaged 1 cm ($f/100$) from the normal focal plane. If a conventional photometer is moved to refocus on the lamp, the pupil's image on the detector will be 2% smaller than normal. With typical (large) cathode nonuniformity, this will produce a systematic (and wavelength-dependent) error on the order of 1%. (An additional error occurs with Cassegrain optics, because the secondary casts a larger shadow on the primary if the source is at a finite distance, owing to perspective effects.)

[35] D. O'Boyle, *J. Appl. Phys.* **36**, 2849 (1965).

The apparent size of a 1-cm lamp filament at 100 m is 10^{-4} rad or 20 arcsec. This may cause an appreciable excluded-energy problem, unless a large focal-plane diaphragm is used. Thus both the angular size problem and the focus problem require that the lamp be at least 100 m from the telescope (preferably, a good deal more, if we want to keep systematic errors well below 1%).

On the other hand, a lamp 100 m away will suffer appreciable horizontal extinction in the intervening air. If the lamp is placed farther away to reduce the foregoing problems, the horizontal extinction rises proportionally. The usual method of estimating horizontal extinction has been to measure the vertical extinction astronomically, and scale the result by the amount of air in each path. Since the scale height of the atmosphere is 8 km, and 100 m is 1/80 of this, the adopted horizontal extinction would be 1/80 or 0.0125 of the zenith extinction; at 400 m, it would be 5% of the vertical extinction, or about a 1% correction near 5000 Å.

Unfortunately, this method, which assumes the atmospheric opacity is uniformly mixed, may be seriously in error. The scale height of "dust" (aerosol) extinction is typically only 1 km or so.[36,37] Thus the conventional method underestimates this contribution to the horizontal extinction by nearly an order of magnitude. At the longer visible wavelength, the aerosol component can be a third or more of the total vertical extinction,[38] so the horizontal extinction may be underestimated by a factor of 8/3 (say, 0.02 mag) or more. A similar error occurs in the near infrared, where water vapor (another small-scale-height component) is important. On the other hand, below 3400 Å where ozone is important, the opposite error occurs, because it is mainly at great altitudes and has little effect near the ground (except in smog near cities). For example, at 3200 Å ozone typically[38] contributes a quarter of a magnitude, so the uniform scaling would overestimate horizontal extinction by some 0.03 mag at 100 m, or over 0.1 mag at 400 m.

If we try to diminish the horizontal extinction by scaling down all linear dimensions, including the telescope aperture, we have two problems. First, the dynamic range goes up, because the closer lamp looks still brighter (at 100 m, a 200-W lamp[34] looks almost as bright as the full moon). Second, the smaller telescope suffers from much worse scintilla-

[36] L. Elterman, R. Wexler, and D. T. Chang, *Appl. Opt.* **8**, 893 (1969).

[37] A. E. S. Green, A. Deepak, and B. J. Lipofsky, *Appl. Opt.* **10**, 1263 (1971).

[38] C. W. Allen, "Astrophysical Quantities," 2nd ed., p. 122. Athlone Press, Univ. of London, London, 1963.

tion noise.[12,38a] Thus we are caught in at least a three-way bind: there is no way to alleviate all the problems at once.

Finally, a major problem in comparing the relatively cool terrestrial sources with the hotter, bluer stars, is due to unwanted wavelengths (see Section 2.1.2.2). Any spectral impurity generally tends to make the lamp look bluer, and the stars, redder; stray light tends to decrease the apparent size of the Balmer discontinuity as well. This problem can be avoided by careful filter blocking, or by using double monochromators, however.

The problems of absolute calibration generally cannot be solved by using NBS-calibrated photomultipliers, both because of difficulties in reproducing the calibration conditions, and because of the poor stability of PMT's (compared to that of standard lamps). Nor can the problem be solved by using heat detectors (e.g., thermopiles); for they frequently show spectrally selective response (owing to imperfect blackness), and have severe spatial nonuniformities (response variations $\sim 5:1$ or more),[39] as well as very low sensitivity.

2.2.1.2. Radioactive Sources (Relative Calibration).

For many purposes, one needs only a light source of uncalibrated brightness, which remains constant (within the required accuracy) for a few hours or a few days. Tungsten-filament lamps are not suitable for use in astronomical photometers because they vary with the direction of gravity (filament sag). However, a variety of radioisotope-excited sources have been used, from a spot of radium (watch-dial) paint to sealed phosphor and Čerenkov sources.

The commonest sources use a radioactively excited phosphor. Usually, a fairly weak beta emitter is used, such as H^3 or C^{14}. The fast electrons enter the phosphor crystals and excite numerous electron–hole pairs; some of these decay radiatively, producing light in one or more bands, usually a few hundred angstroms wide. As radiative decays compete with phonons for the excitation energy, it is not surprising that temperature effects, comparable to those in photomultipliers and filters, also occur in phosphors.[40–43] In all cases, the spectral distributions

[38a] A. H. Mikesell, Publ. U. S. Naval Obs., 2nd Ser. **17**, 143 (1955).

[39] R. Stair, W. E. Schneider, W. R. Waters, and J. K. Jackson, *Appl. Opt.* **4**, 703 (1965).

[40] V. M. Morozov, A. D. Bolyunova, and M. A. Ermolaev, *Izv. Akad. Nauk USSR, Geophys. Ser.* 840–844 (1962).

[41] H. V. Blacker and M. Gadsden, *Planet. Space Sci.* **14**, 921 (1966).

[42] P. S. Kutuzov and K. L. Mench, *Sov. Astron.* **10**, 703 (1967).

[43] A. T. Young and W. M. Irvine, *Astron. J.* **72**, 945 (1967).

change in both shape and magnitude, with coefficients from a few tenths of a percent to over 1%/deg. Because the light sources' temperature coefficients are of the same order as those of photomultipliers and filters, they cannot be used to standardize the response of the photometer unless they are themselves temperature-stabilized. Furthermore, phosphors often display hysteresis of several percent when cycled in temperature, and usually show long-term variations in response that can exceed a factor of two in a year (depending on wavelength, again).

In order to avoid these deficiencies of phosphor-type standard sources, van Albada and Borgman[44] began the use of Čerenkov radiation, produced in quartz by the relativistic ($\sim$2 MeV) β-decay of yttrium-90, a short-lived daughter of strontium-90. Subsequently, such sources have been used at McDonald Observatory and at Geneva.[45] These sources have the advantage of a nearly white (i.e., constant power per unit frequency interval) spectrum, so they can be placed ahead of the filters to detect changes in instrumental color response; and a much smaller temperature coefficient than phosphors, due to the small changes in refractive index of quartz with temperature. However, they also present problems. First, the Čerenkov mechanism is so inefficient that relatively large amounts of the hazardous Sr-90 must be used (5–1) mC/source). [Even our 10-mC sources appeared only as bright as a 9th-mag star in the 80-cm (36-in.) telescope.] Because the 2-MeV Bremsstrahlung from the source in both intense and penetrating, shielding is required to protect both the astronomer and the photomultiplier. At McDonald, we found that about 2 cm of lead were required to keep the gamma-ray-induced dark noise down to an acceptable level. This meant that, in order to be sure of calibrating the tube's light response instead of its gamma-ray response, the light had to be introduced by a mirror. We also found that some of the light was coming from phosphorescence of the epoxy cement used to seal the quartz window in place; such phosphorescence reintroduces all the problems we sought to avoid by using a Čerenkov source in the first place. Finally, the quartz must be extremely pure to avoid radiation-induced coloration.

In spite of these problems, a reliable standard source is required, as photomultipliers may show gain changes of 1%/hr, even when operated under constant conditions for a long time.[45a] One possible solution to

[44] T. S. van Albada and J. Borgman, *Astrophys. J.* **132**, 511 (1960).

[45] E. Peytremann, *Publ. Obs. Genève, Ser. A* No. 69 (1964).

[45a] D. W. Latham, SAO Spec. Rep. No. 321, p. 2-27 (1970).

the standardization problem is a simple phosphor light source in the photometer (to remove short-term temporal and positional effects), supplemented by a regulated standard or reference lamp, fixed with respect to the ground (to remove longer-term variations). It seems unwise to rely on the constancy of the instrument alone, even during a single night, because temperature effects and aging not only affect the components in the photometer head, but also can cause gain and zero-point drifts in amplifiers, discriminators, power supplies, etc. Of course, if careful measurements show the rate of drift to be very slow, standardizations may be spaced further apart in time; a stable instrument thus allows more efficient use of telescope time.

Finally, a source in the photometer, even if unregulated, is useful for checking the operation of the instrument, and for measuring positional (magnetic and gravitational) effects due to motion of the telescope or dome. However, the source gives only qualitative information, even in these cases, unless it illuminates (as nearly as possible) the *same* area of the photocathode as the starlight.

2.2.1.3. Stars. The work of McCullough[46] and the Lowell solar-brightness program[47] show that ordinary main-sequence stars are excellent (though poorly calibrated) standard lamps. For some programs, direct intercomparisons with stars provide all the calibration necessary. Usually, however, the comparisons involve the atmospheric extinction—a complex problem discussed in Section 3.1.

In general, because of scintillation and atmospheric variations, it is *not* wise to use stars to test the photometer for proper operation, nor to determine its characteristics; better data can be obtained in the laboratory without wasting telescope time. However, stars can (and should) be used to measure interactions between telescope and photometer, such as the uniformity of response within a focal-plane diaphragm, or the variation of excluded energy with diaphragm size.

2.2.2. Electronic Systems

Twenty years ago, a photometrist was someone who could build a good dc amplifier. Today, thanks to substantial markets in areas like Raman scattering and nuclear physics, the astronomer can buy nearly

[46] James R. McCullough, *Astron. J.* **69**, 251 (1964).
[47] H. L. Johnson and B. Iriarte, *Lowell Obs. Bull.* **4**, 99 (1959); K. Serkowski, *ibid.* **5**, 157 (1961); and M. Jerzykiewicz and K. Serkowski, *ibid.* **6**, 295 (1966).

everything he needs, for either dc or pulse-counting work, off the shelf. However, he still may demand better linearity of his equipment, or use it in harsher environments, than normal users; so he still needs to test the accuracy of his equipment under working conditions. A skeptical attitude toward manufacturers' specifications is often beneficial.

2.2.2.1. Calibration. All photometers should be checked for linearity, especially at the higher light levels. Several methods of measuring photometric nonlinearity[48,49] have already been cited.

In addition, it is desirable to verify independently the linearity of analog components, such as amplifiers. Such measurements, which are purely electrical, can be carried out with very high accuracy. The usual method is to replace the PMT and its anode load resistor with a high-precision voltage divider and stable voltage source. If the input resistance of the amplifier is normally used as the load resistor, the voltage divider must have much less resistance, e.g., to test a 1-MΩ amplifier to 0.1%, the total divider resistance should not exceed a few thousand ohms. It is best to take several readings between zero and full scale, and fit a straight line to the data by least squares; then the residuals provide an estimate of the precision achieved, as well as indicating any local deviations from linearity. The lazy practice of checking only half- and full-scale points is not so reliable. Also, if possible, it is desirable to check the linearity somewhat beyond the normal "full-scale" region, as the instantaneous signal may greatly exceed the time-averaged value, owing to scintillation or other noise. Most feedback-linearized systems are quite linear, up to the point where some component saturates; we have seen saturation occur dangerously near the top of the normal working range (or even fall within it, under unfavorable temperature or line-voltage conditions).

If the system is linear, there is a well-defined proportionality constant, the *gain*, between output and input. Direct-current systems usually have several different gain settings, which may be achieved by altering the PMT load resistor, the amplifier feedback ratios, etc. The ratios of these gains must be measured accurately, usually to 0.1% (0.001 mag) or better. Sometimes (e.g., feedback ratios) this can be done by an electrical measurement similar to a linearity calibration; sometimes (e.g., load resistors) it requires looking at a fixed light source and measuring

[48] H. E. Bennett, *Appl. Opt.* **5**, 1265 (1966). See also K. D. Mielenz and K. L. Eckerle, *ibid.* **11**, 594 (1972).

[49] R. C. Hawes, *Appl. Opt.* **10**, 1246 (1971).

the ratio of system outputs for different gain settings. A reliable standard source is invaluable for this (stars should not be used); and it is quite helpful to have one more digit available than is normally needed. For example, if we normally work to 0.001 mag, 3 significant figures suffice; but to read 0.1 of full scale to 3 significant figures may require 4-place full-scale digitization. (The advantage of digital over analog readout is again apparent.)

Calibration should be repeated several times per year, as components such as load resistors show long-term changes at the 0.1% level (as well as appreciable temperature coefficients).

2.2.2.2. Environmental Problems. Most electronic equipment is intended for use in an air-conditioned laboratory with reasonably well-regulated power lines. Such equipment is usually much less happy in a remote observatory, where extremes of temperature and humidity are common, and where line-voltage surges, dropouts, and noise due to large motors may occur.

These conditions may not only degrade the stability of a photometric instrument, but may also cause complete failure of electronic units. High temperature and humidity are destructive to nearly all electronic components, causing rapid deterioration and premature failures. Semi-conducting components such as transistors and integrated circuits are prone to thermal runaways (due to increased leakage or "dark" currents) at high ambient temperatures, leading to complete destruction; adequate cooling and thermal overload protection are essential to reliable operation. On the other hand, these same components may fail to operate reliably at low temperatures, because decreased conductivity can lower analog or logic signals below acceptable levels. In general, silicon devices are less temperature-sensitive than germanium, so the former should be preferred for operation in uncontrolled environments.

Solid-state devices are also readily destroyed by brief (even micro-second) voltage overloads that would not affect vacuum tubes. These may be difficult to track down, because a high, narrow spike may be invisible on an oscilloscope trace. Even nondestructive noise spikes can produce spurious counts in pulse-counting systems. Some protection can be provided by using appropriate filters and limiters. High-frequency noise is best suppressed at the source; fast switching devices such as silicon-controlled rectifiers and brushes on dc motors, are common troublemakers.

Every component is temperature-sensitive, but good compensation can sometimes be obtained over a moderate temperature range (e.g., by use of matched resistors in voltage dividers). A particular problem occurs with the voltage-reference tubes used in most high-voltage supplies and some dc amplifiers; they may become unstable in some temperature range. However, they have enormously smaller temperature coefficients than Zener diodes (which are also quite noisy, especially at low currents). Temperature-sensitive units can usually be identified in a few minutes by blowing cold air on them, or cooling the case with a few small pieces of dry ice. Critical units or components may have to be thermostatically isolated from ambient temperature changes. This can often be done with small heaters, as telescope domes are usually too cold rather than too warm for the electronics. However, excessive heating can spoil the seeing within the dome.

Sometimes very peculiar interactions occur. For example, an amplifier that operates reliably in a rack may drift irregularly if mounted on the telescope, owing to internal convection currents that produce a stable steady-state temperature distribution in a fixed orientation and protected location, but that cause continual changes when exposed to drafts and changes of orientation. Also, we have seen an operational amplifier show large zero-point drifts when any magnetic object, even a screwdriver, was brought near it. (This was a chopper-stabilized amplifier whose carrier frequency changed when magnetic material was brought near enough to change the inductance of a ferrite core in the oscillator; internal phase shifts then altered the zero offset.) Such an amplifier could not be used on a moving telescope, or near moving parts in a photometer, without magnetic shielding. Two such amplifiers showed mutual interactions that depended on relative orientation. Such experiences lead one to anticipate "black magic" effects in electronics as well as in photomultipliers!

Finally, a major hazard at most dry-climate observatories is dust. This not only spoils the optics and gets into moving parts in the telescope and photometer; it can cause leakage or arcing, and impede heat transfer, in electronic components. It leads to slide-wire wear (and consequently nonlinear and/or noisy operation) in the self-balancing potentiometers commonly used in strip-chart recorders. Dust causes malfunctions such as intermittent dropouts in magnetic, printed, or punched tape recording systems; these errors are not usually detected until one tries to analyze the data, when it is too late. Only regular cleaning and lubrication, and replacement of worn parts, can keep these (and other) electromechanical devices working reliably.

2.3. Principles of Photometer Design

Perhaps the best words of advice we can offer an astronomer who wants to design a photometer are, "Do not." It is a mistake to think that a Ph.D. qualifies one to design instruments—that is a job an engineer can do much better. Astronomers and physicists tend to think in terms of idealized concepts, without considering such messy realities as friction, errors of fabrication and assembly, and flexure of "rigid" parts under gravity and screw tension.

In addition to designing instruments that look good on paper, but work out poorly in practice, scientists tend to design overly complex instruments. It is a remarkable fact that the actual cost of a good-sized instrument is closely proportional to the total number of pieces. A typical factor at present is \$ 50/piece, counting every nut, bolt, and washer as a separate piece; and it is surprisingly easy for a design to reach several hundred pieces. A good engineer can usually produce a design with at least 30% fewer pieces than can an astronomer; so the properly designed instrument will be substantially cheaper to build, as well as better working.

Finally, the experienced designer will produce an instrument that is easy to adjust and use; unnecessary degrees of freedom will be eliminated, so that fewer adjustments are necessary and fewer misadjustments are possible; also the system can be assembled and operated by a normal human being, without requiring double-jointed elbows, or producing chronically bruised knuckles, pinched fingers, and gashed heads. The natural results will be better use of observing time, and fewer mistakes made at the telescope.

It should also be pointed out, however, that the complex of optical, mechanical, thermal, and electromagnetic problems encountered in pho-tometer design may require attention from several engineers with ex-pertise in these different areas, and these require coordination. For example, holes drilled for electrical feed-throughs are a common source of stray light. In order to avoid such problems, the observer will do well to serve as the final coordinator of all design efforts.

The astronomer who is reluctant to pay for engineering advice should realize that it can pay for itself many times over, during a few years of operation, by savings of telescope time, travel expenses to and from the observatory, salaries of assistants and computer costs in data reduction and analysis, and so forth—not to mention the effect on his professional reputation of publishing good data rather than garbage. Those who insist on using students as cheap labor should at least use engineering students instead of astronomy or physics students for design and construction work.

3. OBSERVATIONAL TECHNIQUE AND DATA REDUCTION*

3.1. Atmospheric Extinction

3.1.1. Introduction

"Die enorme Bedeutung der Extinktion für die astrophysikalische Forschung ist offensichtlich... ."

"Die Lehre von der Extinktion ist deshalb eine der Grundlehren der Astrophysik."

With these strong words, Schoenberg[1] introduced his discussion of extinction in the *Handbuch der Astrophysik*. Perhaps because his treatment is so impressive, little has been done since. However, the intervening forty years have brought photoelectric techniques of high precision into general use. These techniques not only allow a more accurate experimental study of extinction; they also demand a reexamination of the theoretical basis of extinction corrections, if the full accuracy of the best observations (a few thousandths of a magnitude) is to survive the process of data reduction.

Unfortunately, many observers believe that it is difficult to determine the extinction accurately, and that an accurate measurement is not necessary anyway, because mean values are adequate for photometric nights, especially in differential work. These opinions are primarily based on Stebbins and Whitford's[2] assertion that "it is impractical to determine the extinction thoroughly and accomplish anything else," and Hiltner's[3] statement that "the extinction is less variable than the accuracy with which it can be determined during a night when the extinction observations are interspersed with a significant number of observations on pro-

---

[1] E. Schoenberg, *in* "Handbuch der Astrophysik," Band II, p. 1. Springer, Berlin, 1929.

[2] Joel Stebbins and A. E. Whitford, *Astrophys. J.* **102**, 318 (1945).

[3] W. A. Hiltner, *Astrophys. J. Suppl.* **2**, 389 (1956).

---

* Part 3 is by Andrew T. Young

gram stars." Also, there is a widespread belief that stars at very large air masses must be observed in order to determine the extinction accurately.

The purpose of this part is to show how accurate extinction measurements can be made easily and without going to great air masses, and that such measurements are in fact desirable. A thorough discussion of the pitfalls of extinction measurement and correction is given, both from the observational side and from the point of view of making accurate reductions.

In order to optimize the extinction determination, it is necessary to understand the random and systematic sources of error in the observational data, and in particular their dependence on air mass. Many of these have been described in detail in previous sections, and will only be mentioned briefly here.

3.1.2. Basic Error Analyses

Initially, let us examine an idealized situation without systematic errors. We assume that the random error of a photometric measurement is a function of air mass alone, since it is common experience that the error is an increasing function of air mass.

For convenience, we shall at first assume the standard error of measurement ε is proportional to some power p of the air mass M. (More general laws will be considered later.) Thus if

$$\varepsilon = \varepsilon_0 M^p, \tag{3.1.1}$$

the statistical weight of a measurement is

$$w \propto \varepsilon^{-2} \propto M^{-2p}. \tag{3.1.2}$$

Although, in practice, this weighting is not usually taken into account, giving all data equal weight is equivalent to assuming the errors are independent of air mass, which is contrary to experience.

3.1.2.1. Two Measurements.   To illustrate our general approach, suppose we try to find the extinction from just two measurements of a star, one in the zenith and the other at air mass M. If the observed magnitudes are m_1 and m_M, the extinction coefficient A is

$$A = (m_M - m_1)/(M - 1). \tag{3.1.3}$$

The error to be expected in A can be found from the law of propagation of errors, assuming the two measurements to be independent:

$$\sigma_A{}^2 = [\sigma_1 \, \partial A/\partial m_1]^2 + [\sigma_M \, \partial A/\partial m_M]^2. \tag{3.1.4}$$

Equation (3.1.1) gives σ_1 and σ_M, so we find

$$\sigma_A{}^2 = [\varepsilon_0{}^2/(M-1)^2](M^{2p} + 1). \tag{3.1.5}$$

Figure 1 shows the graph of Eq. (3.1.5) for a few values of p. For $p > 1$, there is a value of M for which $\sigma_A{}^2$ is a minimum, i.e., from which the extinction is most precisely determined. The condition for this mini-

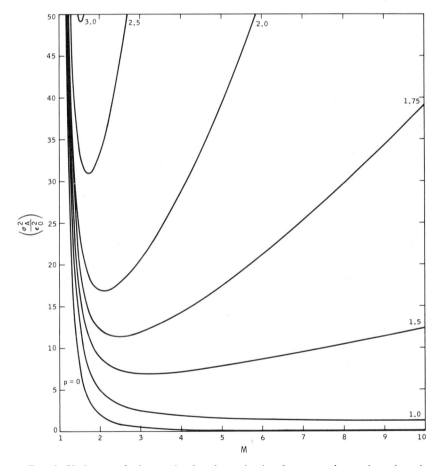

FIG. 1. Variance $\sigma_A{}^2$ of an extinction determination from two observations, in units of the variance of the zenith measurement $\varepsilon_0{}^2$. The error of one observation at M air masses is assumed equal to $\varepsilon_0 M^p$ [see Eq. (3.1.5) in the text].

mum is

$$\partial(\sigma_A{}^2)/\partial M = 0 \tag{3.1.6}$$

or

$$0 = \varepsilon_0{}^2\left[\frac{2pM^{2p-1}}{(M-1)^2} - \frac{2(M^{2p}+1)}{(M-1)^3}\right], \tag{3.1.7}$$

which reduces to

$$(p-1)M^{2p} - pM^{2p-1} - 1 = 0. \tag{3.1.8}$$

Equation (3.1.8) has no root for $p \leq 1$, so in this case M should be as large as\ possible. However, we shall show below that $p \approx 2$ in most cases; for $p = 2$, Eq. (3.1.8) has a root near 2.1. Thus, in this case, the extinction is best determined if the low-altitude observation is made near two air masses. At larger values, the rapid increase in σ_M more than offsets the advantage of increasing $(M-1)$; in the example, $\sigma_A{}^2$ is twice the minimum value if $M = 4.5$ air masses.

The above analysis indicates both the method to be used, and an important conclusion: that best results may be obtained if very large air masses are avoided.

3.1.2.2. n Measurements. From a statistical point of view, we must avoid extremely large air masses because such observations have very low weight. Thus we can improve the extinction determination if we devote more time to the observations at large M, thereby increasing the weight of the low-altitude data.

The analysis is quite similar to that used in pulse-counting to determine the optimum distribution of observing time between background and signal (see Part 1, Section 1.5.2).[3a,b]

We thus suppose that a fraction f of the total observing time is spent at the larger air mass M_2, and a fraction $(1-f)$ is spent at the smaller air mass M_1 (which need not, in general, be unity). If there are a total of n observations of equal duration, and ε_0 is again the standard error of a single observation in the zenith, then the *mean* magnitude observed near the zenith is m_1 with standard error σ_1, where

$$\sigma_1 = \varepsilon_0 M_1{}^p/[(1-f)n]^{1/2} \tag{3.1.9}$$

and similarly

$$\sigma_2 = \varepsilon_0 M_2{}^p/(fn)^{1/2}. \tag{3.1.10}$$

[3a] A. T. Young, *Appl. Opt.* **8**, 2431 (1969).
[3b] C. W. McCutchen, *Phil. Mag. (8th Ser.)* **2**, 113 (1957).

Since

$$A = (m_2 - m_1)/(M_2 - M_1), \tag{3.1.11}$$

we have

$$\sigma_A{}^2 = \frac{1}{n} \left(\frac{\varepsilon_0}{M_2 - M_1} \right)^2 \left[\frac{M_1^{2p}}{(1-f)} + \frac{M_2^{2p}}{f} \right]. \tag{3.1.12}$$

Now, in order to minimize σ_A, we have, in addition to the analog of Eq. (3.1.6), the condition

$$\partial(\sigma_A{}^2)/\partial f = 0. \tag{3.1.13}$$

The former yields

$$(p-1)(1-f)M_2^{2p} - p(1-f)M_1 M_2^{2p-1} - f M_1^{2p} = 0 \tag{3.1.14}$$

and the latter gives

$$f^2 M_1^{2p} - (1-f)^2 M_2^{2p} = 0. \tag{3.1.15}$$

In general, σ_A will be smallest when M_1 is as small as possible, i.e., $M_1 = 1$. This ideal can only be approached in practice; however, it can be approached rather closely, as $M_1 < 1.1$ within $24°$ of the zenith. We therefore adopt $M_1 \equiv 1$ for the next part of the discussion. Even with this simplification, the conditions for minimum σ_A are rather complex

$$(p-1)(1-f)M_2^{2p} - p(1-f)M_2^{2p-1} - f = 0 \tag{3.1.16}$$

and

$$f^2 - (1-f)^2 M_2^{2p} = 0. \tag{3.1.17}$$

If we solve Eq. (3.1.17) for M_2, we find

$$M_2 = [f/(1-f)]^{1/p}. \tag{3.1.18}$$

We can then replace M_2 in Eq. (3.1.16) with Eq. (3.1.18), collect all terms involving p on one side, and raise the result to the pth power to find

$$[(pf-1)/pf]^p = (1-f)/f = M_2^{-p} \tag{3.1.19}$$

or

$$(pf-1)^p - p^p f^{p-1}(1-f) = 0. \tag{3.1.20}$$

We can solve Eqs. (3.1.18)–(3.1.20) exactly only for simple rational values of p. For example, for $p = \frac{3}{2}$ we readily find $f = \frac{8}{9}$, $M_2 = 4$; for $p = 2$, we obtain $f = (2 + 2^{1/2})/4 = 0.854$, $M_2 = 1 + 2^{1/2} = 2.414$.

In general, f can be found by solving Eq. (3.1.20) for $(1 - f)$ and iterating

$$f_{k+1} = 1 - f_k[(pf_k - 1)/pf_k]^p. \tag{3.1.21}$$

Then M_2 follows from Eq. (3.1.18). The resulting values of M and f are shown as functions of p in Fig. 2, along with the corresponding values of $(\sigma_A n^{1/2}/\varepsilon_0)$.

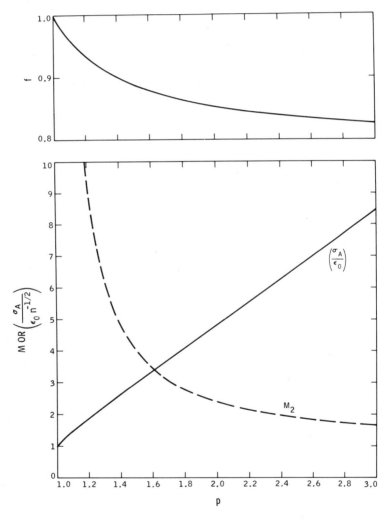

FIG. 2. Values of M_2 and f for optimum extinction determination, as functions of p. The standard deviation σ_A of the extinction coefficient is also shown for these optimum conditions, in units of $(\varepsilon_0/n^{1/2})$.

In order to show how rapidly the quality of the extinction measurement deteriorates as f and M_2 deviate from their optimum values, we have computed contours of constant σ_A^2 (in units of ε_0^2/n) in the (f, M_2) plane. To show the results in terms of observational variables, we have plotted the contours in the (f, z) plane, assuming $M = \sec z$. These plots (see Figs. 3 and 4) show that the observations should be planned carefully in order to determine A precisely. In general, values of f near 0.8–0.9 and values of M near 2–4 (or values of z from 60 to 75°) give the best results. In particular, stars within about 10° of the horizon should be avoided.

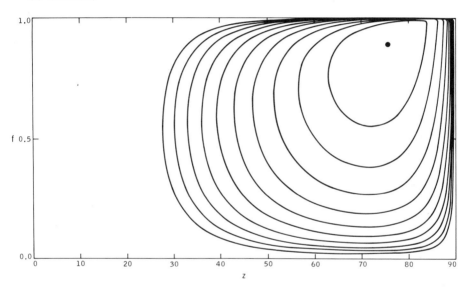

FIG. 3. Contours of equal extinction-coefficient variance σ_A^2 in units of the minimum variance, for $p = 1.5$. The contours are separated by a factor of $2^{1/2}$. Thus the weight of an extinction determination of the innermost contour is 0.707 of the weight of an extinction determination under optimum conditions (indicated by the large dot); the weight on the second contour is half the optimum; and, in general, the weight of an extinction determination along the jth contour out from the center is $1/2^{(j/2)}$ of optimum. The cut through the contoured surface along the line $f = 0.5$ corresponds to the curve for $p = 1.5$ in Fig. 1.

The following properties of Figs. 3 and 4 are of practical interest: (1) The contours are more closely spaced on the large z side of the optimum than the small z side. Thus it is safer to err in the direction of lower air masses; a 10° error in altitude is less harmful in this direction. (2) At smaller than optimum air masses, the best value of f approaches 0.5.

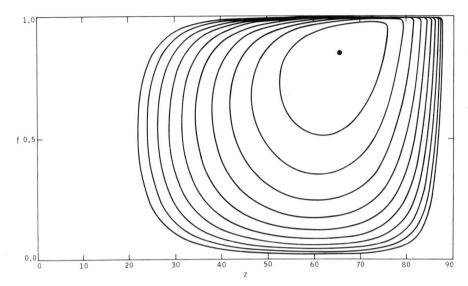

FIG. 4. Contours as in Fig. 3, but for $p = 2.0$.

Thus, (3) the common practice of observing each extinction star once on the meridian, and once each at 60 to 70° zenith distance in the east and in the west (giving $f = \frac{2}{3}$) gives a result not far from optimum. (4) The innermost contour comes very near to the line $f = 1$ in the neighborhood of $z = 60$–70°. Thus it is practically impossible to have too many observations in the 2–3 air-mass range, if a few extinction stars are observed near the zenith.

Of course, the above analysis is rather idealized. In order to investigate the *actual* dependence of ε on M, and hence to determine realistic values of f and z to be used in planning observing programs, we consider the actual behavior and magnitude of both random and systematic errors in the next sections. We can then reexamine the optimization problem in more realistic terms.

3.1.3. Random Errors in Photometry

We shall consider sources of random error in order of increasing dependence on zenith distance, beginning with errors independent of air mass.

3.1.3.1. Fixed Errors.

The fixed random errors are primarily instrumental, and can in principle be eliminated; but in actual practice, they are usually significant. The most obvious instrumental errors are random variations in power supplies and amplifiers (either intrinsic, or due to line-voltage or line-frequency variations); these can be reduced to negligible levels by using well-regulated and stable units and (if line stability is a problem) line regulators—preferably of the electronic servo-regulator type rather than the simple transformer type, which are very sensitive to frequency and waveform changes.

Random changes in photomultiplier dynode gain may be caused by cesium migration within the tube. These changes are usually rather slow, and may be measured by observing a standard source. However, such sources are often so faint that photon noise from the source becomes an important cause of random error. Random dynode voltage variations may also be caused by noisy components, such as Zener diodes or carbon resistors.

Components that work well at room temperature may become noisy at low temperatures. Leakage is often a problem if water vapor is allowed to condense on sockets at low temperatures. Silica gel is a much poorer drying agent than is generally supposed; if it is used, it should be kept cold, as warm silica gel may easily have a higher water vapor pressure than a cold socket, and thus act as a source of water rather than a sink. Silica gel has also been found to emit light spontaneously, so it should be optically isolated from the photomultiplier.

Random photomultiplier temperature variations affect dynode gain and cathode spectral response. These variations are usually large and significant in dry-ice-cooled boxes,[3c] but can be much reduced by either (a) using a heat-transfer liquid such as ethyl acetate or Freon-11 on the dry ice; (b) using water ice instead of dry ice; (c) using a servo-controlled refrigeration system; or (d) using no cooling at all. Method (a) should be used only with very noisy photomultipliers which require extreme cooling, such as 1P21s and tubes with S-1 cathodes. Method (b) is satisfactory with quiet end-on tubes such as the EMI 6256, in which the noise reaches its minimum value near $0°C$;[3d] this method has been used at the Cape Observatory with good results. It is preferable to (a) because water has much higher heat capacity and latent heat of fusion than dry ice; also, there is no need to vent escaping gas, so the box can be completely closed and spillage is eliminated. Unfortunately, many cold boxes are not watertight. If these are replaced by servo-regulated systems (c), the dead-band of the system is the source of random temperature variations. If

[3c] A. T. Young, *Appl. Opt.* **2**, 51 (1963).
[3d] A. T. Young, *Rev. Sci. Instrum.* **36**, 394 (1967).

cooling is abandoned altogether, the ambient temperature should be monitored, so that corrections for temperature effects can be applied. This method (d) has been used very successfully by Stock at Cerro Tololo, and is best used where the ambient temperature is nearly constant through the night.

Generally, dark noise is negligible for the bright stars used to measure extinction. However, in some photometers a spurious "dark" noise may be caused by light leaks, especially from indicator or reticle-illumination lamps within the photometer head.

A final instrumental effect is the constant reading error. This is on the order of 0.001 of full scale on strip charts,[3e] but may become a larger fraction of a magnitude if small deflections are used. In digital systems, this error is $1/(12)^{1/2}$ of the least count (root mean square), and can be kept negligible by the use of four or more significant digits.

All the above errors can be measured by repeated observations of a constant light source during several hours. If a standard source is not part of the photometer, a well-regulated incandescent lamp can be used.

One constant source of error that is often overlooked is the uncertainty in standard-star values, if these are used to determine the extinction.[4] For example, the internal errors of the 108 standards which define the UBV system[5] are about 0.04 mag (standard error) per observation in V; even for stars with 5 or more observations, the estimated standard error is 0.018 mag. Considering that transformations between instrumental and standard systems are never perfect, owing to differences in passbands (and possibly to changes in the stars themselves since the standards were set up), we may adopt a constant rms error of 0.02 mag/star if Hardie's method is used. We shall return to this later.

3.1.3.2. Errors Nearly Proportional to Air Mass. The most obvious error source proportional to M is a random change in the extinction coefficient itself. That is, if A changes by ΔA, the magnitude of a star at M air masses changes by $M \Delta A$. However, it is not clear that there is any reason to expect random changes in A itself. Rather, one should expect the variable (aerosol) component of atmospheric opacity to be carried along by the atmospheric turbulence as a passive additive. In this case, the autocorrelation function of this component should decrease

[3e] A. T. Young, *Observatory* **88**, 151 (1968).

[4] R. H. Hardie, *in* "Astronomical Techniques" (W. A. Hiltner, ed.), Chapter 8, p. 178. Univ. of Chicago Press, Chicago, Illinois, 1962.

[5] H. L. Johnson and D. L. Harris, *Astrophys. J.* **120**, 196 (1954).

with the $\frac{2}{3}$ power of the distance;[5a] because of this correlation, deviations from the mean extinction should grow about like $M^{5/6}$. A full analysis of this problem depends on a knowledge of the outer scale of the turbulence and its variation with height, and will not be attempted here.

Near the zenith, sky brightness varies almost proportionally to M. This may be appreciable on moonlit nights, especially if large focal-plane apertures are used, or in daytime work. However, if the sky is good and sky measures are made for each star, only the photon noise from the sky should contribute a random error; this varies about as $M^{1/2}$ near the zenith. At large zenith distances, the atmosphere becomes optically thick and sky noise is nearly constant.

3.1.3.3. Errors Proportional to Higher Powers of M.   As shown in Section 2.1.1.1.2, the low-frequency component of scintillation noise varies as the $\frac{3}{2}$ power of M near the meridian (at right angles to the upper-air winds), and as the square of M near the prime vertical (looking parallel to the wind direction). In general, the scintillation noise ε_S is given by

$$\varepsilon_S{}^2 = S^2 \sec^4 z [1 + \tan^2 z \sin^2(\theta - \theta_0)]^{-1/2}$$
$$= S^2 M^4 [1 + (M^2 - 1) \sin^2(\theta - \theta_0)]^{-1/2}, \qquad (3.1.22)$$

where θ and θ_0 are the star and wind azimuths, and we again use $M \approx \sec z$. (The dependence of S on telescope aperture and other parameters was discussed previously.)

For large M, Eq. (3.1.22) is proportional to M^3, except in a small range of θ near θ_0. Thus, on the average, ε_S is proportional to $M^{3/2}$. If we average Eq. (3.1.22) over θ, we find

$$\overline{\varepsilon_S{}^2} = (1/2\pi) \int_0^{2\pi} \varepsilon_S{}^2(\theta) \, d\theta$$

$$= (2/\pi) S^2 M^4 \int_0^{\pi/2} [1 + (M^2 - 1) \sin^2 \theta]^{-1/2} \, d\theta$$

$$= (2/\pi) S^2 M^3 K[(M^2 - 1)^{1/2}/M], \qquad (3.1.23)$$

where $K(x)$ is the complete elliptic integral of the first kind. Figure 5 shows the behavior of $\overline{\varepsilon_S{}^2}/S^2$ as a function of M, together with M^3 and M^4 for comparison. For moderate air masses, the scintillation standard

[5a] V. I. Tatarski, "Wave Propagation in a Turbulent Medium." McGraw-Hill, New York, 1961 (reprinted in 1967 by Dover, New York).

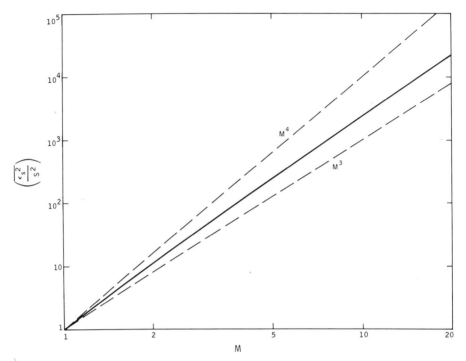

FIG. 5. Azimuth-averaged low-frequency scintillation noise as a function of air mass (log–log plot). Cubic and fourth-power laws are shown for comparison.

error is proportional to the 1.75 power of the air mass, but becomes asymptotically proportional to the $\frac{3}{2}$ power for very large air masses.

3.1.3.4. Errors Exponential in M. Since the extinction makes a star grow fainter exponentially with M, the noise due to photon statistics grows exponentially with M. For, the observed brightness of a star of apparent magnitude m is

$$I = I_0 \cdot 10^{-0.4m}, \tag{3.1.24a}$$

so we have

$$\sigma_I = CI^{1/2} = C' \cdot 10^{-0.2m}, \tag{3.1.24b}$$

for the photon noise in intensity units. The photon-noise error, in magnitudes, is

$$\begin{aligned} \varepsilon_m &= -2.5 \log_{10}[(I + \sigma_I)/I] \\ &= 1.08574 \ln(1 + \sigma_I/I) \\ &\approx 1.08574\sigma_I/I, \end{aligned} \tag{3.1.25}$$

if $\sigma_I \ll I$. Thus

$$\varepsilon_m \approx (\text{const})10^{+0.2m} = (\text{const}) \exp(0.46052m). \qquad (3.1.26)$$

However, the apparent magnitude m is a function of air mass M

$$m = m_0 + A \cdot M, \qquad (3.1.27)$$

where A is the extinction coefficient and m_0 is the extra-atmospheric magnitude. Thus,

$$\varepsilon_m = (\text{const})10^{+0.2A \cdot M} = (\text{const}) \exp(0.46AM), \qquad (3.1.28)$$

where we absorb the m_0 term in the exponent into the constant.

In UBV photometry, photon noise is generally smaller than scintillation noise for naked-eye and most HD stars at moderate air masses (cf., Table I of Part 2). However, for very large air masses, or for narrow-band systems—especially those using the low-efficiency S-1 cathode—photon noise may be important. This is generally true in spectrophotometry, where passbands as narrow as 20–50 Å are often used.

Because of the importance of this special case, we have carried through the error analysis for pure photon noise (see Fig. 6). If we have

$$\varepsilon(m_0, M) = \varepsilon_0 \exp[0.46(m_0 + AM)] = C \exp(0.46AM) \qquad (3.1.29)$$

in place of Eq. (3.1.1), then

$$\sigma_A{}^2 = \left(\frac{C}{M_2 - M_1}\right)^2 \left[\frac{\exp(0.46AM_1)}{f^{1/2}} + \frac{\exp(0.46AM_2)}{(1-f)^{1/2}}\right]. \qquad (3.1.30)$$

The minimization with respect to M_2 gives

$$[0.46A(M_2 - M_1) - 2] \exp[0.46A(M_2 - M_1)] = 2\left(\frac{f}{1-f}\right)^{1/2}, \qquad (3.1.31)$$

and that with respect to f gives

$$\exp[0.46A(M_2 - M_1)] = [f/(1-f)]^{3/2}. \qquad (3.1.32)$$

The exponential factor can be eliminated by taking the ratio of Eq. (3.1.31) and (3.1.32), which leads to

$$0.46A(M_2 - M_1) = 2/f, \qquad (3.1.33)$$

or

$$M_2 = M_1 + (2/0.46Af). \qquad (3.1.34)$$

Combining Eqs. (3.1.32) and (3.1.33) we have

$$[f/(1-f)]^{3/2} = \exp(2/f), \qquad (3.1.35)$$

whose solution is $f = 0.8322967191\ldots$, independent of A. (Note that this value is similar to those found for moderate power laws.) The optimum value of M_2 then follows from Eq. (3.1.34). In general, this value is rather large; for $A \leq 1.0$, $M_2 > 6$ for $M_1 = 1$.

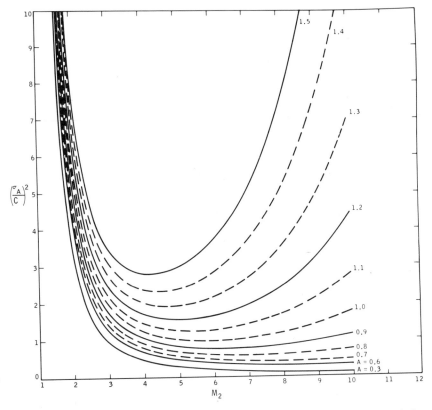

FIG. 6. Curves analogous to Fig. 1 for $f = \frac{1}{2}$ and noise due only to photon statistics. The parameter is the extinction coefficient A in magnitudes per air mass.

In practice, A is usually so small that scintillation noise dominates at moderate air masses. Of course, if photon noise is important, only the very brightest stars in the sky should be used to determine the extinction, in order to keep the observational errors low.

3.1.4. Systematic Errors in Photometry

We may generally distinguish between position-dependent and time-dependent systematic effects, although this distinction is not always clear-cut. For example, temperature effects usually show up as time-dependent errors, as the ambient temperature decreases during the night. However, instrumental temperature effects may in some cases show a positional dependence. For example, consider the case of a very poorly constructed "cold box" made of balsa wood, with only the photocell housing made of metal. When the telescope was turned so the dry ice lay against the metal cell housing, the cell rapidly cooled some tens of degrees; when the telescope pointed in other directions, the cell became much warmer. The result was large (several tenths of a magnitude) zero-point shifts, partially correlated with position.

Of course, it is absolutely necessary to have a linear photometer; otherwise, nonlinear or even "fatigue" effects, which depend on the apparent brightness of a star, will cause systematic errors correlated with air mass—especially on the brightest stars observed, which are the ones used for measuring the extinction. One should be careful about "correcting" for instrumental nonlinearities, as these corrections are likely to show significant variations with temperature, line voltage, scintillation, and other variables.

We now consider purely position-dependent and time-dependent systematic errors, in turn. (Systematic errors inherent in observational or reduction techniques will be deferred to the following section, as they are generally related to each other.)

3.1.4.1. Position-Dependent Errors. Positional systematic effects are of two types: instrumental and atmospheric. The latter are usually a function of air mass, and are therefore more difficult to detect from ordinary photometric observations.

3.1.4.1.1. MAGNETIC EFFECTS. The best-known instrumental effect is that of external magnetic fields on the photomultiplier. From data published by the manufacturers, one can estimate the magnitude of the effect for different types (see Table I). No data are available for the close-spaced EMI tubes such as the 6256 or 9502; these should be somewhat less sensitive to magnetic fields than the wide-spaced 6097. Also, no data are available on the ITT tubes (FW-118, -130, etc.), but experience indicates they are similar in magnetic sensitivity to the EMI box-and-grid tubes (9524, etc.). In some cases, the effect differs by an

TABLE I. Magnetic Effects in Photomultipliers

Type	Dynode system	Response change to 1-G field (%)	Field required for 1% response change (G)
RCA 1P21	Focused, squirrel-cage	5	0.2
RCA 8575	Linear focused	40	0.15
EMI 6097	Unfocused, venetian-blind	5	0.2
EMI 9524	Unfocused, box-and-grid	10	0.4

order of magnitude from one field orientation to another; the table is based on the direction of maximum effect.

The earth's field is on the order of $\frac{1}{3}$ to $\frac{1}{2}$ G, depending on the location of the observer. However, owing to magnetization of the telescope and dome, considerably larger fields may occur at the photometer. In general, fields must be kept below 0.1 G. The adequacy of magnetic shielding can be checked by looking at a standard source while moving either telescope or dome.

3.1.4.1.2. GRAVITATIONAL EFFECTS. Gravitational effects may result from flexure in the telescope or photometer, or even within the photomultiplier. For example, Moreno[6] has reported a position-dependent error within the photometer amounting to 0.02 mag.

Flexure effects within a photomultiplier may be due to sagging of grid wires or motion of the entire multiplier structure within the envelope; a tube that "clunks" when shaken gently from side to side should be rejected. Such flexures can be detected by moving the telescope while looking at a standard source, and cannot be distinguished from magnetic effects without further tests. However, some standard sources may also show gravitational effects: loose particles of phosphor may migrate within sealed, radioactively excited sources.

3.1.4.1.3. DIFFRACTION PLUS SEEING AND DISPERSION. Another important class of errors results from the exclusion of some of the measured star's light by the focal-plane diaphragm. These effects are discussed in Section 2.1.1.

[6] H. Moreno, *Astron. Astrophys.* **12**, 442 (1971).

3.1.4.1.4. REFRACTION EFFECTS.   Dispersion is not the only source of systematic error due to atmospheric refraction. In principle, one should also consider the foreshortening of the telescope aperture at large zenith distances, somewhat like the foreshortening of a penny seen obliquely at the bottom of a glass of water. This causes the vertical extent of the extra-atmospheric bundle of rays intercepted by the telescope to be reduced by a factor equal to the cosine of the atmospheric refraction. However, even at the horizon, the refraction is only about half a degree, so this factor always lies between unity and cos 35 arcmin = 0.99995. The maximum systematic error is thus 0.00005 mag, so this effect can safely be neglected.

A more serious error of the same general kind arises because all objects have a finite angular extent. In this case, the differential refraction between upper and lower limbs reduces the solid angle subtended by the object. Since no optical system (such as the atmosphere) can increase the surface brightness of an object, the total stellar magnitude becomes fainter.[†] In the flat-earth approximation, the refraction is $r(z) = (n - 1) \tan z$, and the compression of a small disk of diameter D is

$$\Delta D = D(n - 1)\, d(\tan z)/dz$$
$$= D(n - 1) \sec^2 z = D \cdot M^2(n - 1), \qquad (3.1.36)$$

where n is the atmospheric refractive index. Since $(n - 1) \approx 3 \times 10^{-4}$, the *fractional* change, which is nearly equal to the error in magnitudes, is

$$\Delta D/D \approx (3 \times 10^{-4})M^2. \qquad (3.1.37)$$

This amounts to a hundredth of a magnitude at about 6 air masses, and rapidly increases beyond that. For visual light, where the extinction is about 0.2 mag per air mass, this effect is equivalent to an air-mass error of 0.05 at $z = 80°$. Thus, the contribution of the refraction effect to the apparent extinction is quite comparable to many of the high-order correction terms ordinarily included in air-mass calculations. From an optical point of view, it may be regarded as a weak barrel-distortion of the entire sky; in effect, the atmosphere acts like a weak fish-eye lens, with its axis vertical.

[†] This effect is an example of the differential refraction that dims a star during an occultation by a planet; looking toward the earth's horizon is looking at the limb of our planet.

3.1.4.2. Time-dependent Errors

3.1.4.2.1. INSTRUMENTAL DRIFT. Systematic errors of this type are often associated with instrumental temperature drifts. Temperature effects in photomultipliers and in filters have already been discussed. As was emphasized earlier, most photomultipliers have long thermal time constants; typically, 3 or 4 hr should elapse between the beginning of cooling and the beginning of observations. The only real solution is to use a reliable standard source (preferably, thermostatically controlled—see Section 2.2).

3.1.4.2.2. VARIABLE EXTINCTION. A common time-dependent problem is the slow change of the extinction coefficient itself. This usually appears as a gradual decrease in the extinction during the night, owing to a slow fallout of the aerosol component. The reverse effect appears during the day, as convection caused by solar heating mixes aerosols produced near ground level into the lowest few kilometers of the atmosphere. This is visible as the growth in size and brightness of the solar aureole from morning until late afternoon.[7] Sometimes the extinction change can be represented as a linear function of time, at least over a few hours. In general, the only safe policy is to observe enough extinction stars to maintain a nearly continuous check on the course of the extinction coefficient.

A convenient method has been described by Young and Irvine.[7a] In this program, each measurement is weighted by $1/\sec z$, so that the residuals have the dimensions of an extinction coefficient (i.e., magnitudes per air mass). Thus, a simple plot of residual vs. time gives a picture of the deviations, if any, from the mean extinction. Because data from several nights are reduced together, any star observed more than once serves as an extinction star, so that a good record of any variation is obtained. Nights with variable extinction automatically receive lower weight than those with constant extinction. Finally, one can use the time-dependent residual plots to interpolate an improved extinction coefficient, and correct the results. A somewhat similar procedure is used by Nikonov and Nikonova,[8] and by Rufener.[9]

A problem closely related to temporal changes in extinction is an azimuth dependence. One must first of all be careful to distinguish between

[7] A. T. Young and L. G. Young, *Sky Telescope* **43**, 140 (1972).

[7a] A. T. Young and W. M. Irvine, *Astron. J.* **72**, 945 (1967).

[8] V. B. Nikonov and E. K. Nikonova, *Izv. Krim. Astrofiz. Obs.* **9**, 41 (1952).

[9] F. Rufener, *Publ. Obs. Genève, Ser. A* No. 66 (1964).

a genuine positional variation in extinction, and a positional variation in instrumental response, such as might be due to magnetic or gravitational effects. In particular, apparent gradients or asymmetries are less likely to occur in uniform terrain than near cities or bodies of water, and are likely to be due to instrumental rather than atmospheric effects if there is no obvious geographic cause near at hand. As the variable extinction is due to aerosols, which usually[9a,b] have scale heights less than 3 km, more typically close to 1.3 km, even in the daytime, any real asymmetry must be quite local—say, within 10 km of the telescope. Unless maintained by local sources and sinks, any irregularities will simply be carried away by the wind in a short time. For example, assume a patch of aerosol at 2-km height and a 3-m/sec wind, which is typical at this height. This patch moves from the zenith to two air masses in 19 min, to three air masses in 31 min, and to four air masses in 43 min. Thus, the time required to pass across the whole useful area of sky is typically an hour or less. One must therefore regard with suspicion reports of general extinction gradients lasting several hours at remote locations; an instrumental problem seems much more likely.

Even if a time-varying extinction is due to advection of air with different aerosol content, not much asymmetry across the sky is to be expected. For example, assume the rather large extinction change of 0.02 mag/air mass per hour. With the same wind model (3 m/sec $\approx$ 10 km/hr), the horizontal separation between points at sec $z = 2.0$ at 2 km height on opposite sides of the sky is only 7 km, which is traversed in about 40 min. Thus, the difference in extinction coefficient between these two points would be less than 0.015 mag/air mass, even if the aerosol were all concentrated at 2-km height. As the aerosol is usually much more concentrated toward the ground, the east–west extinction gradient would generally be much less than 0.01 mag/air mass per (effective) scale height, even in this rather extreme case. Thus, it is sufficiently good to assume an extinction that is uniform over the sky, even when appreciable variations occur with time.

At first, it seems difficult to distinguish among a changing extinction coefficient, an east–west asymmetry, and an instrumental drift, as all three are strongly correlated with time. However, it appears possible, at least in principle, to diagnose the problem from the shape of the Bouguer plot. Consider first a uniform east–west gradient in the ex-

[9a] L. Elterman, R. Wexler, and D. T. Chang, *Appl. Opt.* **8**, 893 (1969).
[9b] A. E. S. Green, A. Deepak, and B. J. Lipofsky, *Appl. Opt.* **10**, 1263 (1971).

tinction. Assume the extinction coefficient varies linearly with horizontal distance; then, since the horizontal displacement of the line of sight at each height is proportional to tan z (in the direction of the gradient), the apparent magnitude of a star is

$$m = m_0 + (A_0 + A_1 \tan z) \sec z \qquad (3.1.38)$$

in the flat-earth approximation. Here m_0 is the outside-the-atmosphere magnitude. (This formula also assumes the star is confined to the prime vertical, i.e., it is exact only for observatories at the terrestrial equator. However, it illustrates the phenomenon fairly well, even for moderate

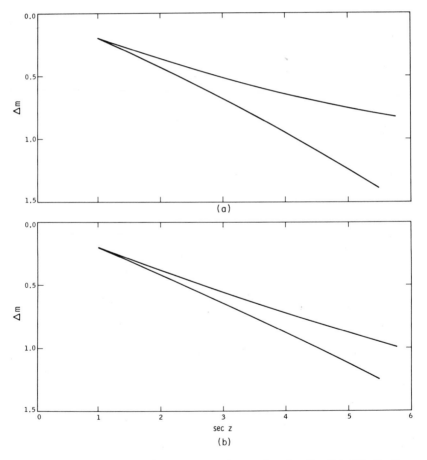

FIG. 7. Bouguer plots for a horizontal extinction gradient [see Eq. (3.1.38)]. Gradients of 0.01 and 0.005 mag/air mass/scale height are shown. (a) $A_0 = 0.2$, $A_1 = 0.01$; (b) $A_0 = 0.2$, $A_1 = 0.005$.

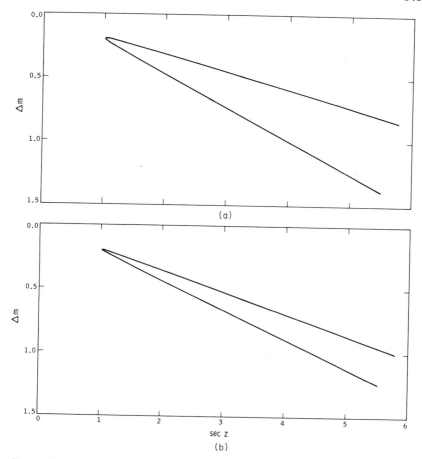

FIG. 8. Bouguer plots for a time-varying extinction coefficient [see Eq. (3.1.39)]. Variations of 0.01 and 0.005 mag/air mass/hr are shown. (a) $A_0 = 0.2$, $A_1 = 0.01$; (b) $A_0 = 0.2$, $A_1 = 0.005$.

latitudes.) The results are shown in Fig. 7. Alternatively, suppose the extinction coefficient varies linearly with time, but is the same all over the sky. Then

$$\Delta m = m - m_0 = (A_0 + A_1 t) \sec z. \qquad (3.1.39)$$

With the same assumptions as before, the hour angle t is just the zenith distance z (see Fig. 8). Finally, an instrumental drift linear in time gives (Fig. 9)

$$\Delta m = A_0 \sec z + A_1 t. \qquad (3.1.40)$$

The three alternatives may be distinguished by the shapes of their Bou-

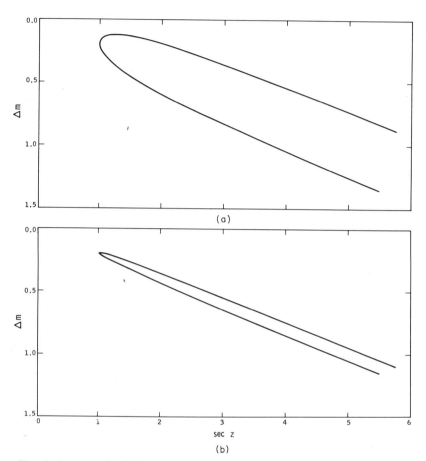

FIG. 9. Bouguer plots for instrumental drifts of 0.05 and 0.01 mag/hr [see Eq. (3.1.40)]. (a) $A_0 = 0.2$, $A_1 = 0.05$; (b) $A_0 = 0.2$, $A_1 = 0.01$.

guer curves, especially near the meridian. The east–west gradient gives two branches convex toward each other, meeting at a cusp with a common tangent. The time-dependent extinction coefficient gives two nearly straight lines, meeting at an angle. The instrumental drift gives two nearly parallel curves, concave toward each other, and joined by a rounded bend.

The time-dependent curves look more familiar to observers than the curve with asymmetry in the extinction. In particular, Fig. 8 (time-varying extinction coefficient) closely fits Code's[10] description: "At Cape

[10] A. D. Code, in Proc. NSF Astronom. Photoelec. Conf. (J. B. Irwin, ed.), p. 79, Aug. 13-Sept. 1, 1953.

Town during the winter there was almost a consistent extinction curve, or rather, curves. On the east side of the meridian one nice straight extinction curve and on the west side another nice straight but steeper extinction curve. You could almost count on this difference which was of the order of a tenth of a magnitude." (The increase of extinction at night near the city in winter is probably due to the low-grade coal that is burned for space heating in South Africa.)

Obviously, it is easier to distinguish among the three effects just discussed if several extinction stars, well distributed in hour angle, are used. For example, a drifting zero point produces the same magnitude residuals at small and large air masses, but a changing extinction coefficient affects low-altitude data more than high ones. (However, the differences may be obscured by other errors at large z.) Also, the simple linear changes shown in Figs. 7–9 are eliminated if the observations are symmetrically disposed with respect to the meridian, i.e., the least-squares value of the extinction coefficient is just A_0 in each case. In actual fact, things do not change linearly, and observations are not made symmetrically, so some systematic error will result. Furthermore, other related effects may contribute to systematic errors; for example, any north–south asymmetry in extinction would, in general, lead to a wrong value of A.

As is well known, the least-squares analysis of data leads to wrong answers (i.e., systematic errors in the results) not only if there are systematic errors in the observations, but also if the mathematical model used does not conform to the real situation. This type of error is so important that it deserves a separate section.

3.1.5. Errors in Reduction Methods

As there are many possible errors that can be introduced by inappropriate methods of analysis, we shall try to cover the most general ones first, and then treat a few specific examples of more specialized errors.

3.1.5.1. Errors in the Extinction Model. These errors may be classified as either monochromatic or wide band.

3.1.5.1.1. VIOLATION OF BOUGUER'S LAW. The refraction effects discussed in Section 3.1.4.1.4 constitute a diminution in measured brightness that is not included in Bouguer's law. We hesitate to call them "extinction," as they are not due to absorption or scattering. Because they do not represent the removal of energy from a beam of light, but rather a change in the area and solid angle occupied by the beam, we have treated them as an observational error. They do not vary with the zenith extinction. It seems best to avoid them by keeping to small air masses.

3.1.5.1.2. ERRORS IN CALCULATING THE AIR MASS. For a long time, Bemporad's work[11] on the air mass, which includes curvature of the atmosphere and curvature of the ray path due to refraction, has been accepted as definitive. This is true only insofar as the extinction is proportional to the actual mass of air traversed by the ray, as is true for Rayleigh scattering. However, large contributions to the extinction are made by aerosols, water vapor, and ozone, which are by no means uniformly mixed. For example, in red light over half the extinction is due to aerosols, which typically have a scale height on the order of a kilometer.[9a,b] Thus, Bemporad's tables (and interpolation formulas based on them, such as Hardie's[4]) must grossly overestimate the curvature corrections for red light: the scattering material is a thinner (flatter) layer than assumed, so sec z is a better approximation here. On the other hand, ozone contributes significantly between 5000 and 7000 Å, and below $\lambda 3400$, but is mostly concentrated near 30-km height. Thus Bemporad's work tends to *under*estimate the corrections to sec z in this case. Furthermore, volcanic eruptions have occasionally deposited significant amounts of aerosols in the stratosphere; these require an "air-mass" modification like ozone. Lest it be supposed that these effects are minor, we must point out that the vertical distribution of ozone[12] can be inferred, on a practical basis, from a method that depends on deviations of the ozone extinction from a sec z law.[†]

In principle it would be possible to generate modified air-mass (perhaps one should say "extinction-mass") tables for each wavelength, given the vertical distributions and extinction coefficients for all the important components. Unfortunately, the reason extinction measurements are

---

[11] A. Bemporad, *Zur Theorie der Extinktion des Lichtes in der Erdatmosphäre*, Heidelberg Mitt. No. 4 (1904).

[12] A. E. S. Green, "The Middle Ultraviolet: Its Science and Technology," p. 94. Wiley, New York, 1966.

[†] In fact, even Bemporad's investigation of the extinction was largely motivated by hopes of using it to study the constitution of the atmosphere: "Es liegt nun nahe, dass man in dieser Beziehung noch mehr von der Extinktion erwarten kann, welche im Zusammenhang mit dem atmosphärischen Zustand unvergleichlich grössere Veränderungen als die Refraktion erleidet." (Ref. 11, p. 1.)

necessary in the first place is that major components like ozone and dust are variable, so that the corrections to Bemporad's theory are also variable.

However, Abbot et al.[12a] have attempted to correct Bemporad's air masses by assuming that the non-Rayleigh component of extinction is in a very thin layer near the ground, so that it can be represented by the simple sec z approximation.

In order to indicate the approximate size of such corrections, it suffices to use just the leading terms in the theory, and see how much these are affected. For instance, the air mass for an exponential spherical atmosphere of radius R and scale height h is very nearly[13]

$$M(z) = \left(\frac{\pi R}{2h}\right)^{1/2} \exp\left(\frac{R \cos^2 z}{2h}\right) \operatorname{erfc}\left[\left(\frac{R \cos^2 z}{2h}\right)^{1/2}\right], \quad (3.1.41)$$

neglecting refraction. Since the ray curvature due to refraction is only about $\frac{1}{6}$ of the curvature of the earth,[14] we can allow approximately for refraction simply by increasing R by about $\frac{1}{6}$. To show the accuracy of this approximation, Fig. 10 compares $[M(z) - \sec z]$ computed from Eq. (3.1.41) (with $R = 7400$ km, and $h = 8$ km) with the corresponding values from Bemporad (which were subsequently reproduced by Schoenberg[1] and Allen[14a]). For comparison, values calculated from Hardie's[4] interpolation formula are also given. The exponential model can be made to fit Bemporad's results very accurately by fudging the radius of curvature slightly; $R = 8500$ km gives errors of only a few thousandths of an air mass up to sec $z = 10$, although slightly better results are obtained at small air masses with even larger values of R. Because of the excellent fit over a wide range, the ($h = 8$, $R = 8500$) model will be used for comparisons with ozone and aerosol models having a different vertical distribution of extinction.

For the aerosol model, an aerosol with scale height of 1 km is assumed to contribute $\frac{2}{3}$ of the total extinction. The extinction-mass *difference* between this model and the standard model is shown in Fig. 11. In this

[12a] C. G. Abbot, F. E. Fowle, and L. B. Aldrich, *Ann. Astrophys. Obs. Smithsonian Inst.* **4**, 335 (1922).

[13] A. T. Young, *Icarus* **11**, 1 (1969). See also Ref. 12, p. 147.

[14] S. Newcomb, "A Compendium of Spherical Astronomy," p. 199. Macmillan, New York, 1909.

[14a] C. W. Allen, "Astrophysical Quantities," 2nd ed., p. 122. Univ. of London, London, 1963.

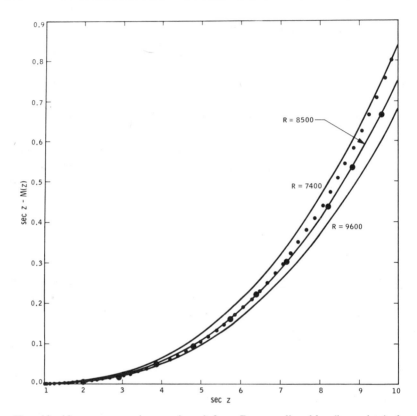

FIG. 10. Air-mass corrections to (sec z) from Bemporad's tables (large dots), from Hardie's interpolation formula (small dots), and values calculated for exponential atmospheres with 8-km scale height and radii of 7400, 8500, and 9600 km (solid curves).

case, the standard (Bemporad) model is in error by 0.01 air mass at sec $z = 2.65$, and by 0.05 at sec $z = 4.6$; a 1% systematic error is made in the extinction if an observation in the zenith is combined with one at sec $z = 3.74$, owing solely to the deviation of the effective air mass from that for pure air.

For the ozone model, it suffices to assume a thin layer at height h^* ≈ 30 km. In this case, the air path through the ozone layer is simply sec ζ, where ζ is the *local* zenith angle where the line of sight intersects the layer (see Fig. 1 of Ref. 13). Again absorbing refraction effects into the enlarged value for R, we have

$$\sec \zeta = (R + h^*)[(R + h^*)^2 - (R \sin z)^2]^{-1/2}. \qquad (3.1.42)$$

In this case, let us assume the ozone contributes one quarter of the total extinction, as it does near $\lambda 3200$, and again in the Chappuis bands near $\lambda 5800$. The deviations of this model from the standard model are also shown in Fig. 11. The Bemporad values are in error by about the same amount as before, but in the opposite sense.

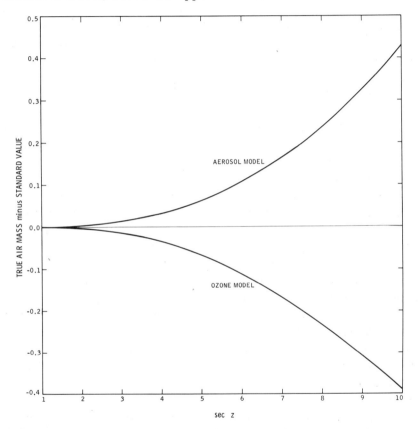

FIG. 11. Differences in effective air mass between the aerosol and ozone models described in the text, and the exponential model ($R = 8500$, $h = 8$) that closely matches Bemporad's theory.

Because the Bemporad values are in error in opposite directions, depending on whether ozone or low-level aerosols are more important, they are a good compromise. However, comparison of Fig. 10 with Fig. 11 shows that the air masses for different wavelengths can differ from each other by as much as the Bemporad air masses themselves differ from sec z. The implied high precision of the tables, and of high-order interpolation formulae such as Hardie's is therefore largely illusory.[†] The problem is not alleviated by going from sea level to high-altitude observatories, because the ozone (being almost entirely contained above the stratosphere) becomes relatively *more* important as the Rayleigh and dust contributions diminish.

It does not seem practical to devise wavelength-dependent air-mass tables, as these would depend on both the (unknown) amounts and vertical distributions of the variable components. The only way to avoid systematic errors is to stay in the region where the terms depending on the vertical distributions remain unimportant. This means keeping to air masses less than about 2.3 if the wavelength variations in $(M - 1)$ are to be kept below 1% (see Fig. 11). By sec $z = 4$, the wavelength-dependent differences in $(M - 1)$ exceed 5%, which corresponds to a systematic error of at least 0.01 mag in reducing V to outside the atmosphere, and even larger errors at shorter wavelengths. This may partly account for the large systematic differences between different observers' absolute-energy calibrations of stars in the Balmer continuum, which practically coincides with the region where ozone absorption is important.

A further cause of systematic error is misuse of Bemporad's air-mass tables. It must be remembered that the argument in these tables is *apparent*, not true, zenith distance. Thus, if the true zenith distance is calculated from the time and the coordinates of star and observer, the refraction should be added before entering the tables. Figure 12a shows the air-mass error committed by using the true zenith distance, instead of the refracted zenith distance, as the argument in Bemporad's tables. The error reaches 0.01 at about three and a half air masses.[‡]

However, in computer work it is often convenient to use the true zenith distance. In this case the correction (sec z — air mass) has the form

---

[†] It is clear that, in spite of subsequent opinions to the contrary, Bemporad regarded his work as anything but definitive. In the introduction to his paper, he says: "Ich brauche kaum zu betonen, dass die hier vorgeschlagene Theorie nur als eine erste Annäherung der Auflösung eines sehr verwickelten Problems anzusehen ist." He was also aware of the aerosol problem. After mentioning the then-current idea that poor transparency of the lower atmosphere was an abnormal condition, he says: "Obwohl wir nicht ganz dieser Meinung sind und lieber eine geringere Durchlässigkeit der unteren Schichten fast als normal ansehen möchten, so werden wir doch in der allgemeinen Entwicklung der Theorie die Hypothese der Konstanz des spezifischen Absorptionsvermögens beibehalten, aber nur in dem Sinne, dass dies eine erste Annäherung für die Auflösung des Problems bildet." (Ref. 11, pp. 1 and 4, respectively.)

[‡] Bemporad's corrections for temperature and pressure are often overlooked also, though they can readily exceed 1% of the air mass.

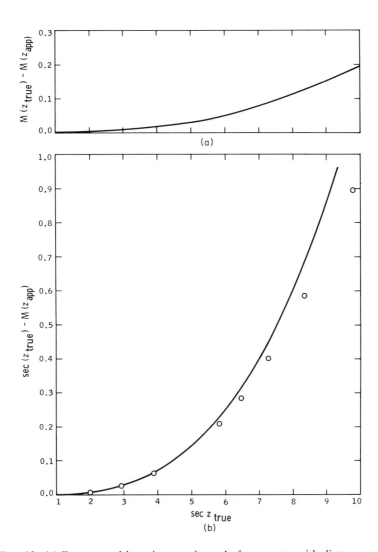

FIG. 12. (a) Error caused by using true instead of apparent zenith distance as argument in Bemporad's tables, as a function of sec z_{true}; (b) Points: difference between (sec z_{true}) and actual air mass from Bemporad's tables, as a function of sec z_{true}. (The apparent zenith distance, including refraction, is used as argument to the tables.) The continuous curve is Eq. (3.1.43) in the text. Note that both the abscissa and the ordinate differ from those of Fig. 10.

shown in Fig. 12b, where Bemporad's values are plotted against the secant of the *true* zenith distance. Up to sec $z_{true} = 4$, these values are well represented by the very simple formula

$$M(z_{true}) = \sec z[1 - 0.0012(\sec^2 z - 1)], \qquad (3.1.43)$$

which has been used by Young and Irvine[7a] for this purpose. [The same form has also been used by Rufener[9] for $M(z_{app})$.] From the foregoing discussion, it should be clear that air masses greater than 4 should be avoided because of large random and systematic errors; therefore this formula is good enough for all practical work.[†]

3.1.5.1.3. PROBLEMS CONNECTED WITH COLORS. Two classes of errors are included here: (a) those associated with the measurement and reduction of colors instead of magnitudes, and (b) those due to bandwidth effects, which appear as color-dependent terms in the extinction.

It is often claimed that there are advantages to forming colors from the raw observations, and treating these as the observed quantities in the extinction corrections, instead of reducing magnitudes to outside the atmosphere and then forming colors from their differences. Some of the supposed advantages are fictitious; but some real advantages have escaped notice.

In the first place, it is essential that the observed quantities really be colors. This requires all the different wavelengths to be measured through the same air mass. This is most easily done if simultaneous observations are made with a multichannel instrument, which has the further advantage that most of the scintillation noise, which is strongly correlated between bands when small zenith angles and large apertures are used, cancels out. It should thus be possible to measure colors to a thousandth of a magnitude, for bright stars. However, the color zero-points then

---

[14b] J. D. Forbes, *Phil. Trans.* **132**, 225 (1842).

[†] Abbot *et al.*[12a] (p. 344) arrived at the same conclusion: "On account of the uncertainty which attends the theory of the determination of air masses, when zenith distances exceeding 75° are in question, we conceive that it will be better to confine our observations... to the range of air masses less than 4..." Similarly, Forbes (Ref. 14b, p. 235) remarks that "...we would do well to avoid much use of observations near the horizon... any law of extinction will, therefore, be better determined from multiplied observations at elevations above 15°, than by those nearer the horizon..." Both of these authors stress the importance of adhering to the region where the simple sec z approximation is adequate.

depend on gain drifts between different photomultipliers, and must be controlled by using a broad-band standard source with good spectral stability (not a phosphor source). If a single photomultiplier is used with filters, one must observe in forward and then in backward sequence through the filters, and then interpolate all colors to a common time. This is inconvenient if more than 3 or 4 filters are used; furthermore, some efficiency is lost if gain changes are required in going from one filter to another, because all changes must be made twice. If many filters are used, linear interpolation between two deflections several minutes apart may not be adequate at the large air masses.

In *UBV* photometry, it is customary to measure in the order (*B*, *V*, *U*, red leak) to keep most gain changes unidirectional. The *B* and *U* deflections typically differ by about 1 min in time. Figure 13 shows the air-mass change in 1 min for several common situations, as a function of air mass. At moderate air masses, the change amounts to several hundredths of an air mass. This is already an important effect, but we know

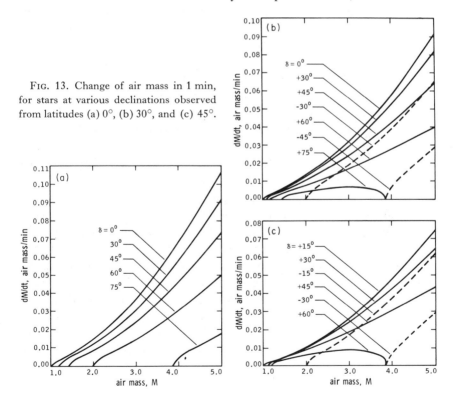

FIG. 13. Change of air mass in 1 min, for stars at various declinations observed from latitudes (a) 0°, (b) 30°, and (c) 45°.

of one instance in which very faint objects were observed, so that long integrations were necessary, together with equally long sky measurements. In this case, 10 min or more were spent on each filter, and the systematic errors that resulted from reducing "colors" instead of magnitudes sometimes exceeded a tenth of a magnitude! The same problem arises in spectrum scanning, where 10–20 min may be required for a complete scan. In such cases, it is essential to use the correct air mass for each color, as was done in the program described by Young and Irvine.[7a] Of course, beyond about sec $z = 3$, even simultaneous observations in different colors can have significantly different air masses (see Fig. 11).

One supposed advantage of using colors instead of magnitudes is that the instrumental color zero points are more stable than the magnitude zero. However, in fact, the colors are typically[15] more stable by a factor of only 2 or 3, and the variations should in any case be measured and removed with a reliable standard source. Thus this argument works only for uncontrolled, unstable instruments. If the instrument is stable, individual magnitudes should be more precise than their differences (colors) by a factor of $2^{1/2}$, except in the important case of simultaneous measurements where cancellation of scintillation noise is possible.

Another argument is that the variations in extinction are strongly correlated in neighboring colors, so that the extinction coefficients for colors vary less than for magnitudes. Granted; but so what? If, say, an ultraviolet magnitude is needed, the sum of the small errors in V, $B - V$, and $U - B$ is the same as the original error in the U magnitude; alternatively, if colors are wanted, the (correlated) magnitude errors are just as diminished by differencing after extinction correction as before, provided that the correction is accurate. Thus the results should be the same whether the data are reduced as magnitudes or as colors.

The choice whether to reduce magnitudes or colors therefore depends primarily on (1) the quality and type of the instrumentation, and (2) the way in which observations in different colors are disposed in time— sequentially, symmetrically, or simultaneously. Good results can be obtained only if the reduction program is appropriate to the data at hand.

The second class of color problems has to do with bandwidth effects. Within each filter passband, some wavelengths suffer more extinction than others. Thus some stars (generally, bluer ones) suffer more extinction than others. Furthermore, at greater air masses the more susceptible rays are more completely removed, so that the incremental ex-

---

[15] A. T. Young, *Mon. Notices Roy. Astron. Soc.* **135**, 175 (1967).

tinction per air mass becomes less and less (the Forbes[14b] effect); in modern terms, this is a curve-of-growth phenomenon. These problems were treated theoretically by King[16] twenty years ago, but the theory is generally ignored. Instead, it is customary to represent the extinction by a formula like

$$m = m_0 + (A_0 + A_1 C) \cdot M, \tag{3.1.44}$$

where m may be either a magnitude or a color, C is a color, and M is the air mass.

To see how this is related to King's theory,[16] we rewrite King's Eq. (21) for the extinction as

$$\Delta m = 0.543 w^2 \left(\lambda^2 \frac{I''}{I} \right)_0 - \left\{ 1 + \frac{n(n+1)}{2} w^2 - nNw^2 \right\} A_0 M$$
$$+ \frac{n^2}{2.17} w^2 (A_0 M)^2, \tag{3.1.45}$$

where primes denote wavelength derivatives; the zero subscript refers to evaluations at the effective wavelength λ_0;

$$w = \mu_2 / \lambda_0 \tag{3.1.46}$$

is the normalized (rms fractional) bandwidth defined by King's Eq. (8);

$$n = -(d \ln A / d \ln \lambda)_0 \tag{3.1.47}$$

is (minus) the logarithmic gradient of the monochromatic extinction $A(\lambda)$ at λ_0;

$$N = (d \ln I / d \ln \lambda)_0 \tag{3.1.48}$$

is the logarithmic gradient of the star's spectrum $I(\lambda)$ at λ_0; and $M \approx \sec z$ is the air mass. Thus, n and N represent the atmospheric reddening and the color of the star, respectively. As King points out, the first term is the color equation between monochromatic and wideband magnitudes at λ_0, and can be dropped. We then have

$$\Delta m = M \{ A_0 [1 + w^2 n(n+1)/2] - w^2 n A_0 [N + (n/2.17) A_0 M] \}. \tag{3.1.49}$$

Now, w^2 is 0.007 for B and 0.004 for V; and n is *at most* 4, for pure Rayleigh scattering, so that $n(n+1)/2 \leq 10$. Thus, the second term in

---

[16] Ivan King, *Astron. J.* **57**, 253 (1952).

the first square brackets is, at most, 0.07, and usually much smaller. [Notice that its effect is to produce a shift in the effective wavelength at which the extinction is measured; it plays a role, with respect to the extinction, similar to that played by King's dropped term in $(\lambda^2 I''/I)$, with respect to magnitudes.] We need not drop this term, if we write

$$A_0{}^* = A_0[1 + w^2 n(n + 1)/2].\qquad(3.1.50)$$

We now have the problem of expressing n and N in terms of measurable quantities. To do this, suppose that we have measurements in a *second* band centered at $\lambda_1 < \lambda_0$, but nearby λ_0. (We suppose $\lambda_1 < \lambda_0$ for definiteness in the choice of signs; the opposite inequality could also be used for the derivation.)

We now approximate the required logarithmic derivatives by finite differences, a procedure that should not be far wrong if λ_1 is not far from λ_0:

$$N = \frac{d(\ln I)}{d(\ln \lambda)} = \frac{d(\log I)}{d(\log \lambda)} \approx \frac{\log(I_0/I_1)}{\log(\lambda_0/\lambda_1)} = \frac{0.4(m_1 - m_0)}{\log(\lambda_0/\lambda_1)},\qquad(3.1.51a)$$

or

$$N = 0.4\, \Delta X/\log(\lambda_0/\lambda_1)\qquad(3.1.51b)$$

in the notation of Young and Irvine;[7a] ΔX is the star's extra-atmospheric color index.

Similarly,

$$n = \frac{-d(\ln A)}{d(\ln \lambda)} = \frac{-\lambda_0}{A_0}\left(\frac{dA}{d\lambda}\right)_0 \approx \frac{-\lambda_0}{A_0}\left(\frac{A_0 - A_1}{\lambda_0 - \lambda_1}\right)$$

$$= \frac{\lambda_0}{(\lambda_0 - \lambda_1)}\cdot\frac{(A_1 - A_0)}{A_0} = \frac{\lambda_0}{(\lambda_0 - \lambda_1)}\cdot\left(\frac{\Delta A}{A_0}\right).\qquad(3.1.52)$$

[Notice that in both ΔX and ΔA the shorter-wavelength item comes first, so that red stars have positive (ΔX) colors and N, and a reddening atmosphere has positive ΔA and n.]

Now we use these approximations in Eq. (3.1.49):

$$\Delta m = M\left\{A_0{}^* - \frac{w^2 A_0 \lambda_0\, \Delta A}{(\lambda_0 - \lambda_1)A_0}\left[\frac{0.4\, \Delta X}{\log(\lambda_0/\lambda_1)} + \frac{\lambda_0\, \Delta A}{2.17(\lambda_0 - \lambda_1)A_0}\cdot A_0 M\right]\right\}$$

$$= M\left\{A_0{}^* - w^2\frac{\lambda_0}{(\lambda_0 - \lambda_1)}\left[\frac{0.4}{\log(\lambda_0/\lambda_1)}\right]\right.$$

$$\left.\times \Delta A\left[\Delta X + \frac{\lambda_0 \log(\lambda_0/\lambda_1)}{0.868(\lambda_0 - \lambda_1)}(\Delta A)M\right]\right\}.\qquad(3.1.53)$$

Let us call

$$\frac{0.4w^2\lambda_0}{(\lambda_0 - \lambda_1)\log(\lambda_0/\lambda_1)} = W. \qquad (3.1.54)$$

We can simplify these last two equations by converting $\log_{10}(\lambda_0/\lambda_1)$ to $\ln(\lambda_0/\lambda_1)$ and expanding, using the assumption that $(\lambda_0 - \lambda_1) \ll \lambda_0$:

$$\log\left(\frac{\lambda_0}{\lambda_1}\right) = (\log_{10} e) \ln\left(\frac{\lambda_0}{\lambda_1}\right) = 0.434 \ln\left[1 + \frac{(\lambda_0 - \lambda_1)}{\lambda_1}\right]$$

$$\approx 0.434\left[\frac{\lambda_0 - \lambda_1}{\lambda_1}\right] \approx 0.434\left[\frac{\lambda_0 - \lambda_1}{\lambda_0}\right]. \qquad (3.1.55)$$

Then, to this degree of approximation,

$$W \approx \frac{0.4w^2}{0.434}\left(\frac{\lambda_0}{\lambda_0 - \lambda_1}\right)^2 = \frac{w^2}{1.086}\left(\frac{\lambda_0}{\lambda_0 - \lambda_1}\right)^2, \qquad (3.1.56)$$

and Eq. (3.1.53) becomes

$$\Delta m = M\{A_0{}^* - W \cdot \Delta A[\Delta X + M\,\Delta A/2]\}. \qquad (3.1.57)$$

As a practical matter, we must employ the approximation $\Delta A \approx \Delta A^* = A_1{}^* - A_0{}^*$; if we drop the subscripts and superscripts, this is Eq. (4) of Young and Irvine.[7a] Comparison with the commonly used Eq. (3.1.44) shows that (setting $C = \Delta X$) the latter ignores both the Forbes effect [represented by the last term in Eq. (3.1.57)] and variations in atmospheric reddening (i.e., $W \cdot \Delta A$ is treated as a constant A_1).

There have been arguments as to whether the color C in Eq. (3.1.44) should be the observed or the extra-atmospheric color of the star; the foregoing linear theory shows that the *mean* of these should be used, to represent the Forbes effect properly. Figure 14 shows the results of a numerical experiment to test this idea. (The figure is similar to Fig. 1 of Hardie.[17]) The energy distributions of a number of stars were multiplied by typical instrumental response functions for the B and V bands, both without any atmospheric extinction, and with various amounts of extinction corresponding to 1.0, 1.5, . . . , 5.0 air masses for a standard atmospheric extinction model.[14a] The integrated responses were converted to magnitudes, and the total extinction computed as the magnitude difference between the value for M air masses and the extra-atmospheric value. The effective extinction coefficient (the ordinate in Fig. 14) was then computed as the ratio of the total extinction to the total air mass M. Thus, for each star there are nine extinction values, which are plotted

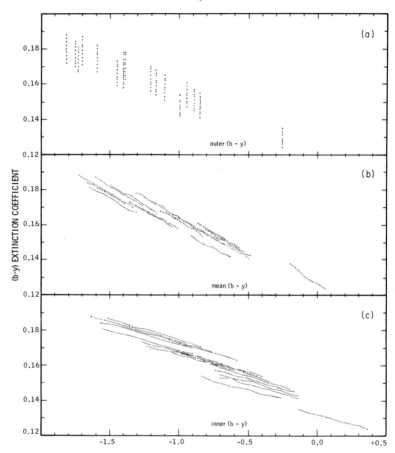

FIG. 14. Effective extinction coefficient (mag/air mass) found from numerical integrations, as a function of (a) extra-atmospheric color of star; (c) observed color; and (b) mean color (see text for explanation).

as a function of the instrumental color outside the atmosphere (a); the apparent color inside the atmosphere, at M air masses (c); and the mean of these two colors (b). In the latter two cases, successive "observations" of each star at different air masses are joined by a straight line. The mean color gives the best results, both in terms of minimum dispersion about the mean relation and also in the sense that it gives each star an atmospheric "reddening line" most nearly parallel to the general relation for all stars together, which means that the extrapolated extra-atmospheric value for a star is independent of the air mass at which it was observed. Notice, however, that there is some intrinsic spread between different stars; i.e., two stars can have the same color but different extinction coefficients. This is due to two factors: (1) higher-order terms, omitted from the theory; and (2) an imperfect correlation between $(b - y)$ color and the spectral gradients at the effective wavelengths of the bands, owing to blanketing, rotation, microturbulence, and other effects. (Similar results were found for other colors and magnitudes.)

The Forbes effect is appreciable, especially in the ultraviolet magnitude. Figure 15 shows extinction curves from the numerical integrations, together with Hardie's observations.[17] Most of the curvature of the extinction lines occurs at *small* air masses, which explains why Hardie's data do not show the effect between 2 and 6 air masses. This is due to the increasing monochromaticity at large air masses, after the shorter

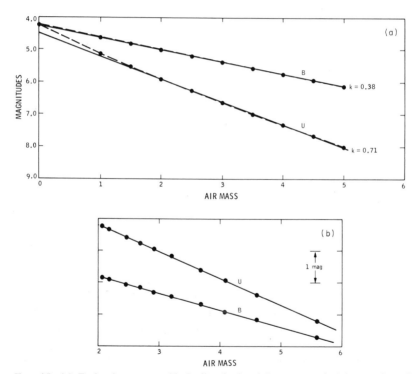

FIG. 15. (a) Extinction curves (dashed) calculated from numerical integrations for a B3 star. The solid straight lines are drawn to fit the points between 2 and 5 air masses; (b) Hardie's observed extinction curves at low altitudes for π^5 Ori.

[17] R. H. Hardie, *in* "Spectral Classification and Multicolour Photometry," p. 243. IAU Symp. No. 24 held in Saltsjöbaden, Sweden, 17–21 August 1964. Academic Press, New York, 1966.

wavelengths have been almost completely removed. However, as can be seen from the figure, the straight line fit to the 2–5 air-mass region gives systematic errors in the extrapolated extra-atmospheric magnitudes of 0.03 mag in B, and about 0.25 mag in U. A straight line passed through the calculated points at 2.0 and 5.0 air masses misses the point at $M = 3.5$ by only 0.008 mag in B and 0.034 mag in U, which corresponds to air-mass errors of 0.02 and 0.05 in B and U, respectively. The rms residual from a linear least-squares fit would be less than a hundredth of a magnitude for both colors in this region. Thus, although a straight line is an excellent representation of the large-air-mass data, it involves large errors in extrapolating to zero air mass.

With respect to the parameters W and w^2, we might add that for B and V, $\lambda_0/\Delta\lambda$ is about 5 or 6; then $W \approx 30w^2$. King gives $w^2 = 0.007$ for B and 0.004 for V, based on rather poor values of the response functions; the problem is further complicated by the Chappuis bands of ozone, which greatly reduce the effective value of n across the V band. Anyway, one might expect $W \approx 0.21$ or so for B, and something less than 0.12 for V. Our Agassiz Station reductions give $W_B = 0.26$ fairly consistently, and $W_V \approx 0.03$; the numerically integrated values were reduced, using a program based on this theory,[7a] and gave $W_B = 0.21$, $W_V = 0.026$. (One effect of the ozone is to make the extinction larger in V, which reduces the apparent $(B - V)$ extinction gradient below its true value across the B band, so it is not surprising that W_B comes out a bit larger than expected.)

In common practice, the parameter A_1 in Eq. (3.1.44) is generally assumed to be either a free parameter that is solved for separately on each night, or a fixed constant whose value is found once for all time. However, the theory [Eq. (3.1.57)] shows that it should depend on the reddening power of the atmosphere ΔA. If the parameter A_1 is free, it can soak up and conceal various systematic errors. On the other hand, if it is fixed, it introduces systematic errors in the results whenever the atmospheric extinction law changes (due to seasonal or random effects such as volcanic eruptions[18]). If the variable part of the extinction were neutral, ΔA would be constant; but it is not, as the variable color of the setting sun vividly demonstrates. Only by making the mathematical representation of extinction conform to the physical situation can we hope to obtain accurate results.

A computer program that does represent the color terms according to

[18] G. de Vaucouleurs, *Publ. Astron. Soc. Pacific* **77**, 5 (1965).

this theory has been used and described by Young and Irvine.[7a] The
added complexity of using a correct representation of the physics of
extinction is so small, compared to the capabilities of modern computers,
that simplicity is no longer an acceptable excuse for inaccurate results.
One must bear in mind that an incorrectly formulated least-squares
program will absorb systematic errors into whatever adjustable parameters
it has available. Thus, particularly if there are only a few observations
for each disposable parameter, systematically wrong values will be found
that may fit the data quite well. Small residuals (high precision) do not
mean small errors (high accuracy).

3.1.5.2. Errors Due to Misuse of Least Squares.   It should be remembered that the expectation values of the parameters in a least-squares
solution are their true values only if certain assumptions are satisfied.
In particular, the errors in the observations are supposed to be normally
distributed, with mean zero. If the resulting estimate is to be efficient,
the equations of condition must be weighted, so that the residuals are
all drawn from the same parent population (i.e., they must all have the
same probable error). These conditions are often violated in practice.

First of all, the observed quantities are intensities, not magnitudes.
In the presence of any experimental error, the mean intensity does not
correspond to the mean magnitude, because the logarithm of the mean
of a random variable is not equal to the mean of its logarithm. It is clear
from the law of conservation of energy that scintillation can redistribute
the energy in the shadow pattern, but cannot alter the mean intensity.
In fact, it is well known that scintillation causes the logarithm of the
intensity (i.e., magnitudes) to be normally distributed, and that in this
case the difference in the means is

$$\log\langle I\rangle - \langle \log I\rangle = \sigma^2/2, \qquad (3.1.58)$$

where the angular brackets denote averaging, and σ^2 is the variance of
$\log I$. Natural logarithms are close enough to magnitudes that we can
also consider σ^2 to be the variance of the magnitude estimates. For
example, if $\sigma_m = 0.1$ mag, Eq. (3.1.58) tells us that the mean magnitude
differs from the true magnitude of the star by 0.005 mag, which begins
to be important. This sets a practical lower limit to the duration of photometric observations. For example, if we wish to use a 40-cm (16-in.)
telescope at 3 air masses, Table I of Part 2 shows that about a 10-sec
integration is required to achieve rms errors of 0.01 mag on bright stars,

which means significant systematic errors will appear if integrations shorter than 0.1 sec are used.

At the other extreme, the errors on faint stars are dominated by photon noise and sky noise, which may easily exceed 0.1 mag. Furthermore, these errors are not distributed normally (or even symmetrically) in magnitude, which leads to further systematic errors. It appears that the photometrist's only hope in such cases is to observe faint stars only near the meridian, so that a long enough integration can be achieved to reduce the random (and consequent systematic) errors to an acceptable level, while keeping the air mass practically constant. Otherwise, the whole quasi-linear system of equations using magnitudes must be replaced by a system of (exponential!) equations in terms of intensities. Without such precautions, the mean magnitudes will depart more and more from a Pogson scale at the faint end. For example, adding 30% to the intensity decreases its magnitude by 0.29, but subtracting 30% increases the magnitude by 0.39; the mean intensity is correct, but the mean magnitude is 0.05 too faint. Again, the systematic error rises with the square of the noise level.

Finally, there is the question of weighting the observations. Because of the rapid increase in errors with $\sec z$, low-altitude observations have very low weight. For example, if the errors increase like $(\sec z)^2$, the weight of an observation is proportional to $(\sec z)^{-4}$. Thus, at two air masses, we have only 0.06 of the weight of a zenith star; at three, the weight is down to 0.012. These observations provide information about the extinction, but are worthless for determining the actual brightness of a star. Even at $z = 45°$, the weight is down by a factor of four. Clearly, it is essential that a realistic weighting system be used, and that all program stars be observed as near the meridian as possible.

3.1.5.3. Errors Due to "Standard" Stars.

Hardie[4,19] has advocated the use of standard stars, whose extra-atmospheric values on the instrumental system are known, for quick measurements of the extinction. This works beautifully *if* these values are accurately known. Both the extinction and the instrumental magnitudes are determined simultaneously from many nights' data in the method of Young and Irvine,[7a] and this is also satisfactory. A similar method has been used by Rufener,[9] by Weaver,[20] and others.

[19] R. H. Hardie, *Astrophys. J.* **130**, 663 (1959).
[20] H. F. Weaver, *Astrophys. J.* **116**, 612 (1952).

However, some observers have relied on the *published* magnitudes of *UBV* standard stars, rather than determining good instrumental values themselves. They then solve for the transformation coefficients between instrumental and standard systems, as well as for the extinction coefficients. Unfortunately, the *UBV* "standards" contain mean errors per star on the order of 0.02 mag; furthermore, the usual linear transformations are imperfect, especially in the ultraviolet, and may contribute similar errors of a systematic nature. If a large number of standard stars were observed, their errors should average out. (For example, Johnson and Harris[5] say that "it is best...to use at least 20 stars of all types....") However, the standard stars are usually regarded as a "shortcut" to extinction; we have seen as few as four or five stars used altogether. In such cases, the *accidental* errors in the published values for the particular stars used become appreciable *systematic* errors in the extinction and the transformation coefficients, and consequently in the final results. Recently, Moreno[6] has shown that the errors introduced by incorrect extinction coefficients are not transformed away by adjusting the transformation coefficients. Here again is a situation in which the least-squares process absorbs systematic effects, producing small internal but large external errors.

3.1.6. Actual Error Laws

From the foregoing discussion, it is clear that unavoidable systematic errors that depend on variable atmospheric conditions (seeing; vertical distribution of absorbers) become important in the neighborhood of 3 or 4 air masses, so that larger air masses should be avoided in work of the best quality. Within this range, it is still important to know how the random errors actually vary with air mass; not only to plan observations for maximum efficiency, but also to assign realistic weights in reducing the data.

Very few observers have bothered to investigate their errors as a function of air mass. Rufener[9] has given a very thorough discussion of errors in the Geneva photometry. His errors due to scintillation (from his Fig. 9) are replotted in log–log coordinates in Fig. 16. The lines, which have been drawn by eye to fit these data, have very nearly slope 2; that is, the scintillation errors grow like $(\sec z)^2$, as expected. In addition, he finds a root mean square fixed error of about 0.006 mag, due to instrumental uncertainties (mainly due to nonlinearity, and gain changes). These appear to be the principal sources of error in determining the

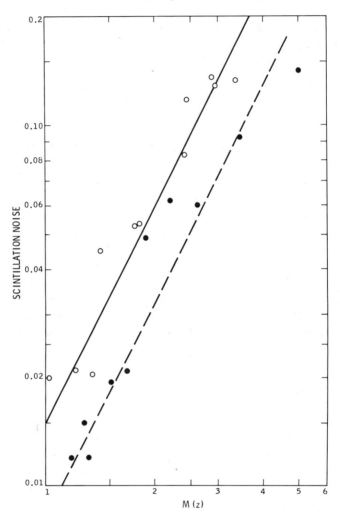

FIG. 16. Scintillation noise at Geneva, in units of $\Delta d/d$, on nights of strong (open circles and upper curve) and weak (filled circles and lower curve) scintillation [taken from Fig. 9 of F. Rufener, *Publ. Obs. Geneve, Ser. A.* No. 66 (1964)].

extinction at Geneva, apart from time variations of the extinction coefficient, which are carefully monitored. Evidently the scintillation noise dominates.

Another example of careful work is given by Stock,[21] who also found the errors to increase with the square of the airmass, and derived from

[21] J. Stock, *Vistas Astron.* **11**, 127 (1968).

this the condition $M_2 = 2.1$ [as pointed out in the discussion following Eq. (3.1.8)].

We had also come to this conclusion at about the same time;[22] a slide of Fig. 1 was shown at the December, 1966, AAS meeting. Immediately afterward, Hardie told us that he had found a contrary result experimentally: the extinction was most precisely determined for very large values of M_2, rather than showing an optimum near 2 or 3 air masses. We now believe this is due to Hardie's method of determining the extinction by using the published values of UBV for standard stars. For, in this case, the (constant) errors in the standards themselves are dominant at small-to-moderate air masses, so that $p \approx 0$ (cf., Fig. 1). The effective error law would then consist of this large constant term, in addition to the usual (sec z)^2 term for scintillation. The weighting scheme used in determining the *extinction* should reflect the effect of this constant error term; however, observations of bright *program* stars should still receive the (sec z)$^{-4}$ weights appropriate to scintillation noise alone.

As a final example, Fig. 17 shows estimates of the error law from data taken at Agassiz Station and reduced with the program described previously.[7a] The original data were taken in pairs of 20-sec deflections; but because the efficiency of reading the chart paper is only[3e] about 25%, the scintillation noise should be that for a 10-sec deflection, which is 0.001 mag at the zenith for a 60-cm (24-in.) telescope (see Part 2, Table I). The root mean square residual, expressed in magnitudes, was computed for data grouped into four intervals of air mass: 1.0–1.2, 1.2–1.5, 1.5–2.0, and 2.0–3.0; only stars observed five or more times were used, so that errors in their instrumental magnitudes should not strongly affect the results. Also, only nights with no marked systematic run of residuals were used. The December, 1960, data show rather large errors, roughly proportional to air mass. This suggests that small variations in the extinction coefficient with time were the main source of error, a conclusion supported by considerable systematic runs in the residuals on about half the nights of that month. On the other hand, the March 1961 data are much better, although they are still far above the level of pure scintillation noise. A constant error of about 0.007 mag seems to dominate these observations at small air masses.

As a practical matter, the above studies suggest that, under good conditions, the error law can be approximately represented by a combination of constant and scintillation errors. Of course, the scintillation

[22] A. T. Young, *Astron. J.* **72**, 328 (1967).

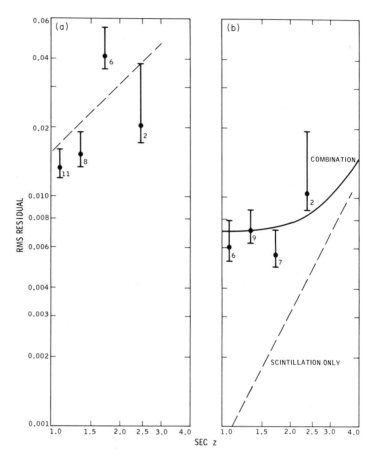

FIG. 17. rms errors in V magnitude vs. air mass, for two periods at Agassiz Station. The data were taken with the 60-cm (24-in.) Clark reflector. (a) December, 1960. The errors are large, and increase roughly proportionally to the air mass, as indicated by the dashed line. (b) March 1961. The errors are smaller, and seem to be composed of a constant component ($\sim$0.007) plus scintillation (shown by the dashed line); the heavy curve is the combination of these terms. (The "error bars" represent 50% confidence intervals.) The number of observations included in each point is indicated beside the "error bar."

component may be negligible for large apertures (say, over 150 cm); but most photometry is done with smaller instruments. However, regardless of the shape of the random error law, its dependence on air mass can be found from the observations. Then, given the actual error law, how can we find the optimum air mass for extinction measurement?

Let the random error law be some function $\sigma(M)$, which may be given numerically or graphically. In the simplest case, Eq. (3.1.4) tells us that variance in the extinction coefficient derived from two observations at M_1 and M_2 is just

$$\sigma_A{}^2 = [\sigma^2(M_1) + \sigma^2(M_2)]/(M_2 - M_1)^2. \qquad (3.1.59)$$

This suggests a simple graphical method of finding the optimum M_2: if we pick a value of M_1 and plot the function $\sigma^2(M_2)$ as a function of $(M_2 - M_1)^2$, then the tangent to this curve from the point $[-\sigma^2(M_1), M_1]$ has the least slope $(\sigma_A{}^2)$ of all lines from that point to the curve (see Fig.

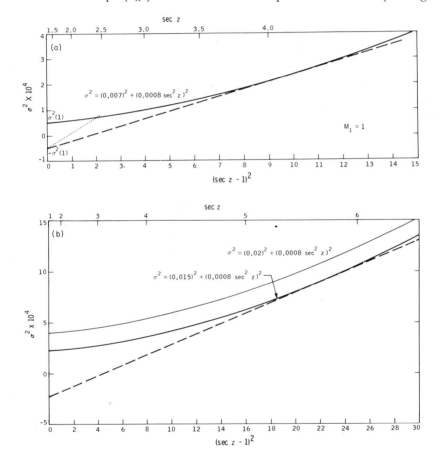

FIG. 18. Graphical determination of optimum air mass for extinction determination. (a) Error law (solid curve) taken from curve of Fig. 17b. Dashed line: determination of optimum M_2. Dotted line: determination of half-optimum M_2. (b) Same scintillation noise as (a), but with constant errors of 0.015 and 0.020 mag. (The dashed line picks out $M_2 \approx 6$ for the former.)

18). Therefore the value of M_2 at the point of tangency is the optimum value. Furthermore, any chord drawn from the specified point intersects the curve in two points, such that their values of M_2 give equally good determinations of the extinction coefficient (apart from systematic errors, of course).

For example, Fig. 18a shows the error law found for the March, 1961, observations of Fig. 17. If we take $M_1 = 1$, the zenith variance is about 5×10^{-5} (mag)^2. The (dashed) tangent to the curve from the point $-\sigma^2 = -5 \times 10^{-5}$, on the vertical axis, meets the curve at about $M_2 = 4.1$. If we draw a line from the same point, but with twice the slope (i.e., twice the value of $\sigma_A{}^2$), we find it meets the curve at $M_2 \approx 2.4$; thus the extinction is determined with half the optimum weight, or $2^{1/2}$ times the minimum error, if we use $M_2 = 2.4$.

What would be the situation if we had used the UBV standard magnitudes, linearly transformed to the instrumental system? In fact, the transformation between V and the instrumental magnitude left a root mean square residual of 0.015 mag per star, which is clearly much larger than the internal errors of the March photometry (especially considering that the standard stars were observed an average of 4 times apiece in this run). Whether this 0.015-mag error is due to the errors in the tabulated values, or to transformation errors, is immaterial for our present purposes; in either case, it acts like a constant error if the published values are used to determine the extinction. The effect of increasing the constant error from 0.007 to 0.015 mag is shown by Fig. 18b; the optimum air mass M_2 has increased to about 6. In general, the situation will be still worse, as the March observations use only the brightest and best-determined of the UBV standards. If the constant error is raised to the more typical 0.02 mag for run-of-the-mill UBV standards, M_2 becomes very large indeed—in agreement with Hardie's statement, and in spite of the rapid increase in scintillation with air mass.

In case we do not spend equal amounts of time at the high and low air masses, Eq. (3.1.12) shows us how to modify the graph. If we multiply through by f, we obtain

$$f\sigma_A{}^2 = \frac{1}{n(M_2 - M_1)^2} \left[\frac{f}{(1-f)} \, \sigma^2(M_1) + \sigma^2(M_2) \right]. \quad (3.1.60)$$

Notice that the term in $\sigma^2(M_1)$ contains a factor $[f/(1-f)]$, which is the ratio of the observing time spent at M_2 to that spent at M_1. For example, if we observe a star at M_2, both rising and setting, as well as at M_1, on the meridian, the factor is 2. Then to find the optimum M_2, we must draw the tangent from the point $-2\sigma^2(M_1)$ instead of just $-\sigma^2(M_1)$. In general, we simply move the pivot point down by a factor equal to the ratio of the observing times. As is evident from the shape of the curves in Fig. 18, increasing this ratio also increases the optimum value of M_2. (This can also be seen from the shapes of the contours in Figs. 3 and 4.)

3.1.7. How Much Time to Spend on Extinction

3.1.7.1. General Principles. Because the errors in the extinction observations propagate through the extinction corrections into the final results, enough effort must be spent in determining the extinction to guarantee adequate precision and accuracy. However, how much is "enough"? The answer obviously depends on the type of work being done.

In general, the variance in a magnitude or color, after correction for extinction, will be

$$\sigma^2 = \sigma^2_{obs} + \sigma^2_{ext} + \sigma_z^2, \qquad (3.1.61)$$

where σ^2_{obs} is the variance due to observational errors of the program star itself, σ^2_{ext} is the variance propagated from extinction errors, and σ_z^2 is the zero-point variance. If each night is reduced separately, or if the zero point is allowed to be a free parameter for each night, the first two terms on the right correspond roughly to the "internal error," and the whole expression is the "external error." The separation of σ_{ext} and σ_z is somewhat artificial, as they usually depend on the same observations, so these errors are not always independent, as assumed in Eq. (3.1.61).

In order to keep σ^2_{obs} small, we usually observe the program star near its minimum zenith distance M^*, and reobserve it several times; then

$$\sigma^2_{obs} \approx \sigma^2(M^*)/n, \qquad (3.1.62)$$

if we have n observations altogether. In good conditions, small values of n (say 2 or 3) suffice to reduce σ_{obs} to a few thousandths of a magnitude.

The contribution of the extinction errors is more complicated to assess, because extinction stars are usually standard (zero-point) stars as well. Thus we must consider the techniques used to collect and reduce the data, as well as the ultimate use to which the results are put.

3.1.7.2. Absolute Photometry. If we are doing absolute photometry (comparing a star to a standard lamp), then

$$\sigma_{\text{ext}} = M^* \cdot \sigma_A; \qquad (3.1.63)$$

and as M^* must exceed unity, the error σ_A in the extinction coefficient appears with full force. It is not always realized that this same large error appears whenever we assume that some instrumental zero point remains constant from night to night, as is often done in the case of colors.[4,7a,20] The assumption of a fixed zero point (which can be ensured only by using a reliable standard source) greatly strengthens the precision of extinction determination when data from different nights are combined; however, if it is not strictly true, it can introduce considerable systematic errors. Thus, we achieve precision at the risk of accuracy.

3.1.7.3. Relative Photometry

3.1.7.3.1. PRINCIPLES. To avoid this problem, it is common practice to let the nightly zero points be free parameters. The determination of the zero point then depends on observations of certain (instrumental) standard stars. It is often claimed that this reduces the problem to relative photometry, so that the extinction errors cancel out. This implication is only partially true, however; if either the program or the standard stars are distributed over the sky, there will be an error of the form

$$\sigma_{\text{ext}}^2 = \overline{(\overline{M}_{\text{std}} - M^*)^2} \cdot \sigma_A{}^2 = \sigma_M{}^2 \cdot \sigma_A{}^2, \qquad (3.1.64)$$

where the rms air-mass difference σ_M can hardly be less than several tenths.[†] Thus, compared to absolute photometry, the extinction error σ_{ext} in relative photometry is reduced only by a factor $(M_{\text{std}} - M^*)_{\text{rms}}/M^*$, which is usually rather modest (say about $\frac{1}{3}$), and which may well be compensated by the increased uncertainty σ_A in the extinction itself, owing to the sparser distribution of observational degrees of freedom over the set of unknown parameters. Furthermore, it is clear that errors in the deduced nightly zero points appear in the results as systematic errors. Once again, the importance of matching the mathematical model to the physical situation is evident.

[†] Note that this error results from the *variance* in air mass, even if the *mean* air masses of extinction and program stars are the same.

3.1.7.3.2. DIFFERENTIAL PHOTOMETRY. Of the two factors on the right of Eq. (3.1.64), σ_A^2 can be found from the slope of a straight line in a diagram such as Fig. 18. The other factor (air-mass variance) deserves further attention. As is well known, it can be made very small by making all observations at nearly the same air mass, usually that of the Pole. A few reference stars near the Pole can then be used to measure and remove temporal changes in both extinction and instrumental zero point.

This practice is now less popular than it once was. Some telescopes are difficult (or even impossible) to use near the Pole. At high latitudes, the polar-altitude condition restricts observations to a small range of declination; at low latitudes, the polar altitude is so low that the observational errors (both σ_{obs} and σ_z) are offensively large. If some other reference altitude is used, the reference stars traverse an appreciable range of air mass during a night, so the extinction error becomes appreciable once again.

Another case in which the air-mass factor is very small is differential photometry of variable stars. Since a comparison star bright enough to be dominated by scintillation noise can usually be found within one or two tenths of a degree of the variable, the air-mass difference can usually be kept below 0.01, up to 2 or 3 air masses. Thus the extinction coefficient needs to be known only to about 0.1 mag/air mass in order to keep magnitude errors below 0.001 mag. However, color-dependent terms in the extinction can cause much larger errors, especially during eclipses of binaries having dissimilar components.

These cases, in which the σ_{ext} term in Eq. (3.1.61) can be made small, require us to look at the σ_z term to determine the proper balance between program and reference stars. If both are bright enough that photon noise is negligible, $\sigma_z^2 \approx \sigma^2(M)/n_{ref}$, where n_{ref} is the number of reference-star observations that can be combined. For example, if the extinction coefficient and the instrumental zero point are constant, widely separated reference observations can be combined; but if there are variations, we may have to compare each program-star deflection with just one adjacent reference observation. In the former (constant) case, we should observe the reference star the *same* number of times as each (constant) program star. For a variable star, each observation provides an independent point on the light curve, so $n_{ref} = 1$ will suffice, unless we are concerned about the systematic accuracy of placing the entire light curve on a standard system. In most cases, however, we have to worry about instrumental or extinction variations on the order of 0.01 mag/hr. Then $\sigma_z \sim 0.01 \, \Delta t$, where Δt is the time difference (in hours) between reference observations.

This is generally larger than σ_{obs} if Δt exceeds a few tenths of an hour. Thus, differential photometry usually requires several reference observations per hour, to achieve maximum precision.

If photon noise is important, the fainter star should receive most of the observing time. An argument like the derivation of Eq. (3.1.15) shows that, if the comparison star receives a fraction F of the total observing time, the program star is best determined if the ratio of the times devoted to the two stars is also the ratio of their standard errors for unit time, σ_c and σ_p, respectively:

$$F/(1 - F) = \sigma_c/\sigma_p, \tag{3.1.65}$$

or

$$F = \sigma_c/(\sigma_c + \sigma_p). \tag{3.1.66}$$

[Here, c and p denote "comparison" and "program" stars, respectively, and we have set $\sigma_{obs}^2 = \sigma_p^2(1 - F)^{-1}$, and $\sigma_z^2 = \sigma_c^2 F^{-1}$.]

In the case of pure photon noise, Eq. (3.1.65) is equal to $10^{0.2(m_c - m_p)}$; if the comparison star is so bright that photon noise is negligible, m_c should be replaced by the magnitude at which photon noise equals scintillation noise. Thus, a 15th mag program star should be observed ten times as long (or ten times as often, if each observation is of fixed length) as a 10th mag comparison star, to achieve the best measurement of their magnitude difference in a given amount of telescope time.

3.1.7.3.3. ALL-SKY PHOTOMETRY. Now let us consider observations over a considerable range of air mass, so that σ_{ext}^2 is appreciable. We must evaluate the air-mass variance factor in Eq. (3.1.64) to calculate this extinction error. Generally, each extinction star is observed at both large and small air mass, but each program star is observed only at its minimum air mass M^*. Also, the zero-point standard stars are usually used as extinction stars; this avoids the need for separate zero-point standards. Thus, the reference air mass $\bar{M}_{std}$ in Eq. (3.1.64) is really $\bar{M}_{ext}$, the mean air mass of the extinction stars. This should be calculated using the appropriate (air-mass-dependent) weights, $1/\sigma^2(M)$, for lower and upper air masses:

$$\begin{aligned}
\bar{M}_{ext} &= \left[\frac{(1 - f)M_1}{\sigma^2(M_1)} + \frac{fM_2}{\sigma^2(M_2)} \right]\left[\frac{(1 - f)}{\sigma^2(M_1)} + \frac{f}{\sigma^2(M_2)} \right]^{-1} \\
&= \frac{\sigma^2(M_2)(1 - f)M_1 + \sigma^2(M_1)fM_2}{\sigma^2(M_2)(1 - f) + \sigma^2(M_1)f}.
\end{aligned} \tag{3.1.67}$$

Thus $\bar{M}_{ext} = (M_1 + M_2)/2$ only if $\sigma(M) = $ constant, and $f = \frac{1}{2}$. If we can assume $\sigma(M) \sim M^p$, Eq. (3.1.67) can be written as

$$\bar{M}_{\text{ext}} = M_1\left[\frac{M_r^{2p} + f_r M_r}{M_r^{2p} + f_r}\right] = M_1[1 + f_r(M_r - 1)(M_r^{2p} + f_r)^{-1}], \quad (3.1.68)$$

where

$$M_r = M_2/M_1 \qquad (3.1.69)$$

is the ratio of the two air masses, and

$$f_r = f/(1 - f) \qquad (3.1.70)$$

is the corresponding ratio of observing times.

For example, the common case of $M_1 = 1$, $f = \frac{2}{3}$, and $p = 2$ gives $M_2 \approx 2.2$ for optimum results; then $\bar{M}_{\text{ext}} = 1.1$. If $p = 1$ and $M_2 = 4$, we have $\bar{M}_{\text{ext}} = 1.33$. Evidently the higher weight of one zenith observation generally overcomes the greater number (f_r) of large-air-mass data, giving a mean air mass near M_1.

On the other hand, even if program stars are observed only at culmination, many of them will be at considerably larger air masses. Their mean air mass can be roughly estimated by the following argument: suppose that program stars are uniformly distributed over the sky, and that they are observed only at culmination. Let us further suppose that only those stars culminating at $M^* \leq M_c$ are used. The uniform distribution means that the number of stars at declination δ is proportional to $\cos \delta$. If the observer's latitude is ϕ, stars culminate at $M^* = 1/\cos|\delta - \phi|$, assuming a flat earth. The cutoff M_c occurs at a declination

$$\delta_c = \phi - \cos^{-1}(1/M_c). \qquad (3.1.71)$$

If we assume all stars poleward of δ_c are observed, including the small region near the pole where $M > M_c$ at low latitudes, we have

$$\overline{M^*} = \int_{\delta_c}^{\pi/2} \cos \delta \sec(\delta - \phi) \, d\delta \Big/ \int_{\delta_c}^{\pi/2} \cos \delta \, d\delta$$

$$= \frac{\left(\dfrac{\pi}{2} - \delta_c\right)\cos \phi - \sin \phi\left[\ln \dfrac{\cos(\phi - \delta_c)}{\sin \phi}\right]}{1 - \sin \delta_c}. \qquad (3.1.72)$$

Figure 19 shows the run of Eq. (3.1.72) with ϕ for several values of M_c. For moderate latitudes and M_c, $\bar{M} \approx 1.5$, which is a few tenths larger than typical values of $\bar{M}_{\text{ext}}$. In fact, the same argument can be used to

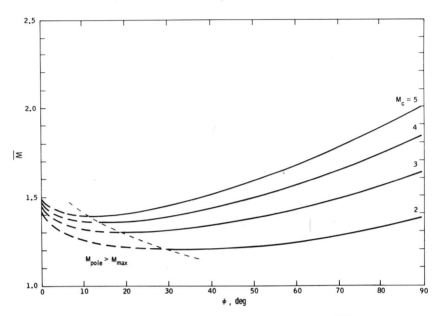

FIG. 19. Mean program-star air mass at culmination. Values of $\overline{M^*}$, estimated according to Eq. (3.1.72) for several values of M_c, as functions of the observer's latitude are shown. The dashed portions, to the left of the dotted curve, indicate the regions where $M > M_c$ near the Pole.

estimate the factor $\overline{(\overline{M}_{ext} - M^*)^2}$ that appears in Eq. (3.1.64). For,

$$\sigma_M{}^2 = \overline{(\overline{M}_{ext} - M^*)^2} = (\overline{M}_{ext})^2 - 2\overline{M}_{ext}\overline{M^*} + \overline{M^{*2}}, \quad (3.1.73)$$

where Eq. (3.1.72) gives $\overline{M^*}$, and

$$\overline{M^{*2}} = \int_{\delta_c}^{\pi/2} \cos \delta \sec^2(\delta - \phi) \, d\delta \Big/ \int_{\delta_c}^{\pi/2} \cos \delta \, d\delta$$

$$= \frac{\left\{ \cos \phi \ln\left[\dfrac{\sec(\tfrac{1}{2}\pi - \phi) + \tan(\tfrac{1}{2}\pi - \phi)}{\sec(\delta_c - \phi) + \tan(\delta_c - \phi)} \right] \atop +\sin \phi[\sec(\delta_c - \phi) - \sec(\tfrac{1}{2}\pi - \phi)] \right\}}{1 - \sin \delta_c}. \quad (3.1.74)$$

Figure 20 shows the course of Eq. (3.1.74) for the same values of M_c used in Fig. 19. These results allow the estimation of the air-mass variance factor by means of Eq. (3.1.73) if $\overline{M}_{ext}$ is specified. Figure 21 shows the results for $\overline{M}_{ext} = 1.1$ and 1.3, the typical values derived above.

As the values shown in Fig. 21 are several tenths at least, we can expect our program stars to have rms extinction-produced errors that are likewise several tenths of the error in the extinction coefficient, according to Eq.

(3.1.64). For example, a typical value of σ_M is about 0.4, if we are at a moderate latitude and avoid stars that transit within about $25°$ of the horizon. Then if σ_A, the standard error in the extinction coefficient, is 0.01 mag/air mass, the resulting rms program-star error will be 0.004 mag. This may be acceptable in some kinds of work; on the other hand, it is certainly larger than the random observational errors of the brighter program stars if an aperture over 40 cm (16 in.) is used (see Part 2, Table I). If a 90-cm (36-in.) telescope is used, this rms extinction error exceeds the photon noise for stars brighter than eleventh magnitude, for a 10-sec integration with B or V filter. Thus, if the final error caused

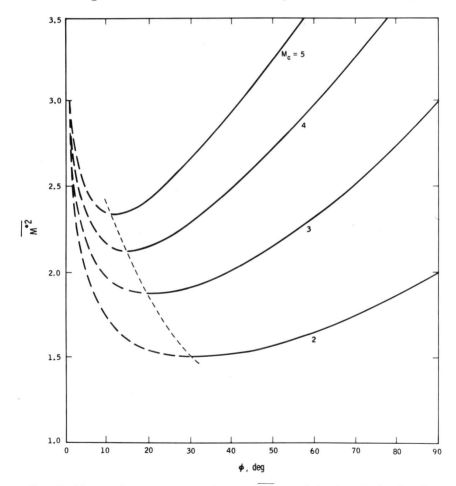

FIG. 20. Mean square program-star air mass $\overline{M^{*2}}$ at culmination, displayed as in Fig. 19.

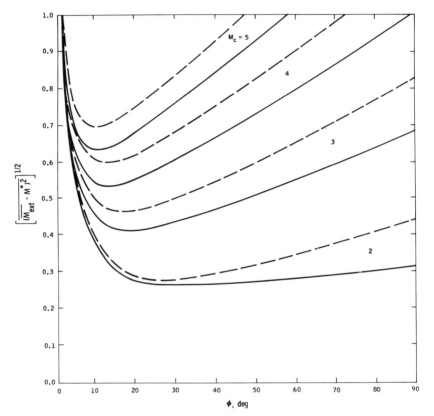

FIG. 21. The rms air-mass difference σ_M between program and extinction stars, for $\bar{M}_{\text{ext}} = 1.1$ (dashed) and 1.3 (solid curves), calculated from Eq. (3.1.73) in the text, with the help of Eqs. (3.1.72) and (3.1.74). As in Figs. 19 and 20, the large values at low latitudes are due to the polar region where $M > M_c$.

by uncertainty in the extinction coefficient is to be kept smaller than the observational errors, we must generally know the extinction coefficient somewhat more precisely than 0.01 mag/air mass.

Let us again ask the question: what fraction F of the observing time should be devoted to extinction, to make optimum use of telescope time? (Again, suppose initially that only one program star is used.) Having determined our observational error law as a function of air mass, we can compute the error σ_A to be expected from n_{ext} observations (suitably divided between M_1 and M_2) of an extinction star; for example, by using Eq. (3.1.12), or its graphical equivalent. Let us summarize this result as

$$\sigma_A{}^2 = \sigma_1{}^2/n_{\text{ext}}. \tag{3.1.75}$$

Here σ_1 may be thought of as the rms extinction error for one extinction observation.

Combining Eqs. (3.1.61) and (3.1.64), we have

$$\sigma^2 = \sigma_{\text{obs}}^2 + \sigma_z{}^2 + \sigma_M{}^2 \cdot \sigma_A{}^2; \tag{3.1.76}$$

σ_{obs}^2 and $\sigma_z{}^2$ may be approximated in terms of the air-mass-dependent error law:

$$\sigma_z{}^2 \approx \sigma^2(\overline{M}_{\text{ext}})/n_{\text{ext}}, \tag{3.1.77}$$

assuming the extinction stars are used to set the zero point, and

$$\sigma_{\text{obs}}^2 \approx \sigma^2(\overline{M^*})/n_{\text{prog}}. \tag{3.1.78}$$

Now n_{ext} and n_{prog}, the number of observations of the extinction and program stars, respectively, are in the ratio

$$n_{\text{ext}}/n_{\text{prog}} = F/(1 - F). \tag{3.1.79}$$

Thus, Eq. (3.1.76) becomes

$$\sigma^2 = \frac{\sigma^2(\overline{M^*})}{N(1 - F)} + \frac{\sigma^2(\overline{M}_{\text{ext}})}{NF} + \frac{\sigma_M{}^2\sigma_1{}^2}{NF}, \tag{3.1.80}$$

where N is the total number of observations. To find the optimum value of F, we set $d\sigma^2/dF = 0$: we find

$$F/(1 - F) = [\sigma^2(\overline{M}_{\text{ext}}) + \sigma_M{}^2\sigma_1{}^2]^{1/2}/\sigma(\overline{M^*}), \tag{3.1.81}$$

or

$$F = \frac{[\sigma^2(\overline{M}_{\text{ext}}) + \sigma_M{}^2\sigma_1{}^2]^{1/2}}{[\sigma^2(\overline{M}_{\text{ext}}) + \sigma_M{}^2\sigma_1{}^2]^{1/2} + \sigma(\overline{M^*})}. \tag{3.1.82}$$

There are too many independent variables involved in Eq. (3.1.82) to allow any universally applicable conclusions to be drawn (note that the air-mass dependence of the random error law is involved in both σ_1 and $\overline{M}_{\text{ext}}$, and that M_c is involved in $\overline{M^*}$). However, the two examples used previously may be useful: pure scintillation noise; and scintillation plus a large fixed error, such as occurs in the UBV standard stars.

In the first example, Eq. (3.1.12) applies, with $p \approx 2$. We shall assume $f = \frac{2}{3}$, in accord with common practice, and then have $M_2 \approx 2.2$ for best extinction measurement, if $M_1 = 1$. Equation (3.1.12) then gives

$\sigma_1^2 = 26.5\,\varepsilon_0^2$; we already have $\overline{M}_{\text{ext}} = 1.1$; and $M_c \approx 2.5$ gives $\overline{M}^*$ ≈ 1.3 for moderate latitudes, and also $\sigma_M \approx 0.4$. Thus $\sigma(\overline{M}_{\text{ext}}) = 1.21\,\varepsilon_0$, and $\sigma(\overline{M}^*) = 1.69\,\varepsilon_0$. These values give $F/(1 - F) = 1.41$, or $F = 0.59$. Thus, in this case, each extinction star should receive 1.41 times as much attention as each program star. As each extinction star would normally be observed 3 times per night (rising, setting, and at culmination), each program star should be observed about twice, on the average. To indicate the actual precision involved, the values of ε_0 given in Table I of Part 2 can be inserted into Eq. (3.1.12) with $n = 3$: we find $\sigma_A \approx 0.003$ mag/air mass for a 40-cm (16-in.) telescope, and about 0.001 mag/air mass with a 150-cm (60-in.). Thus, the extinction *can* be determined very precisely from just a few observations; the major practical requirement is to provide enough extinction measurements per hour to monitor the temporal changes in the extinction coefficient with adequate accuracy. Finally, we note that a final precision of a few thousandths of a magnitude should be reached.

Now consider the case where large random errors of fixed size occur, as in Hardie's method. We suppose that a fixed random error of 0.02 mag (rms) is present in the standard (extinction) stars, but that only scintillation noise [say, for a 40-cm (16-in.) aperture] affects the program stars. In order to avoid the large *systematic* errors at large zenith distances, we take $M_2 = 4$. The large standard-star error forces us to adopt $f \approx \frac{1}{2}$; so from Eq. (3.1.60) we find $\sigma_A = 0.0104$ mag/air mass, for $n = 2$. Thus $\sigma_1 = 0.0073$ mag/air mass. Because the standard stars are dominated by the large constant error, $\sigma(\overline{M}_{\text{ext}})$ is very nearly 0.02 mag; however, we may take $\sigma(\overline{M}^*) \approx 0.0014$ mag, because program stars are affected only by scintillation noise. If we again adopt $\sigma_M \approx 0.4$, we have $F \approx 0.94$ —in other words, between 14 and 15 standard stars should be observed for one program star! The inefficiency of using the *UBV* "standards" for extinction determination is manifest; in this case, it *is* impractical to determine the extinction thoroughly and accomplish anything else. (It should be borne in mind that the situation is even worse for larger telescopes.)

Thus, the use of the relatively imprecise *UBV* standard stars for measuring the extinction causes the observer to throw away most of the statistical weight in his raw data. Evidently, much higher precision should be attained by directly measuring the extinction in the natural system of the photometer. Of course, improved precision of the standard stars (i.e., a revised standard system) might help; but it has not yet been demonstrated that transformations between such broadband systems can be made with the requisite accuracy (a few thousandths of a magnitude) to allow even a revised set of standard values to replace direct instrumental measurements of extinction.

3.1.8. Concluding Remarks on Extinction

If the inherent precision of modern photoelectric observations is to be realized in the final published results, accurate measurements of (and corrections for) atmospheric extinction are necessary. Such measurements can readily be made, with relatively few observations, if proper attention is paid to both instrumental and observational technique, and if the reduction methods are properly matched to the observations. In particular, one should have:

(1) temperature regulation of *all* spectrally selective components (filters, receiver, and standard source) to a few tenths of a degree;

(2) a large enough focal-plane diaphragm to include essentially *all* the light of a star, at *all* zenith distances, to avoid seeing-dependent effects;

(3) no observations at altitudes below about 20°;

(4) frequent-enough observations of extinction stars to monitor changes in extinction with adequate accuracy; and

(5) a realistic reduction and weighting scheme, preferably based on actual, measured characteristics of the instrument used.

If the extinction is determined from *UBV* "standard" values, it is difficult to achieve adequate precision. However, if the extinction is measured directly in the instrumental system, only enough observations are needed to measure the time dependence of the extinction. For example, if 15 stars/hr can be observed, about 5 of these should be extinction stars. If each extinction star is observed 3 times (once each rising, setting, and at transit), adequate coverage is achieved by selecting one extinction star for each 35 min of right ascension from the declination zone that passes within 20° of the observer's zenith. If the observer knows from past experience that changes in extinction are unlikely to occur, and/or is willing to discard nights on which significant changes exist, the number of extinction stars might be reduced to 1/hr of R.A. (three observations per hour of time). It does not appear wise to reduce the density of extinction stars below this level, which is already quite low (only 20% of all observations).

Finally, the accuracy of the results depends on an accurate modeling of instrumental and atmospheric characteristics in the reductions. This means the use of an accurate air-mass formula (bearing in mind that *no* formula is accurate beyond two or three air masses); a physically based (rather than empirical) correction for bandwidth (color-dependent) effects; a proper choice between "colors" and "magnitudes" as observational variables; reduction of several nights together, if instrumental stability permits; and a realistic weighting scheme.

It appears that if the work is carefully done, ground-based photometry with a real accuracy of a few thousandths of a magnitude should be possible.

3.2. Transformation to a Standard System

3.2.1. Introduction

The previous sections show how to obtain reproducible photometric data and reduce them to outside the earth's atmosphere. These reduced data, however, are still on the natural system of the photometer, that is, they are measurements made with the spectral response function of a *particular* instrument. In most cases, one needs to compare and combine observations made with *different* instruments. Thus we must "transform" the instrumental magnitudes and colors to the values that would have been measured with a *standard* instrument, having a different spectral response.

Usually in astronomy, this is a real instrument which has measured magnitudes and colors for a rather small list of "standard" stars. This is a practical procedure because (a) many stars are in fact quite constant over long periods of time, and (b) by using these measured objects as standard lamps, there is no need to know the actual spectral response of the instrument to perform the transformation. The latter is a very important point; for (as explained in Section 2.2.1.1) absolute comparisons with laboratory standards are much more difficult and less precise than comparisons between stars.

On the other hand, laboratory photometry is based on the tabulated spectral response function of a mythical "standard observer," so that, in principle, one must know the instrumental spectral response (and match it[23] to that of the standard observer). This does not lead to a precisely reproducible system, both because of the difficulties in making the

---

[23] H. Wright, C. L. Sanders and D. Gignac, *Appl. Opt.* **8**, 2449 (1969).

necessary absolute measurements of spectral response, and because the standard response function is not well defined between tabulated wavelengths (different interpolation schemes give different intermediate values).

At first, these two approaches seem quite different, but they turn out to have many features in common. First, it is still highly desirable to know the spectral response functions of standard astronomical systems, so that model-atmosphere fluxes can be compared directly with observed colors. This is closely related to the problems of establishing the stellar effective-temperature scale, and of determining bolometric corrections. Second, it turns out that matching the instrumental response to that of the standard system is as important astronomically as in laboratory photometry, because of difficulties in transformation. And finally, both the laboratory photometric units and "visual" stellar magnitudes[23a] are historically based on measurements made with the human eye; the major difference is that the laboratory units are based on the eye's bright-adapted (photopic) spectral response, but the astronomical systems are based on the mesopic or scotopic (dark-adapted) response, which is shifted to shorter wavelengths. Thus the effective wavelength of laboratory photometry is near 5550 Å, while that of the V magnitude[24,25] is near 5400 Å.

3.2.2. Transformations for Blackbodies

3.2.2.1. Temperature Reddening. If we assume that the major difference between instrumental and standard systems is one of effective wavelength, the transformation can be done accurately for blackbodies. Because many stellar and laboratory sources have nearly blackbody spectra, this is a useful and instructive result.

The Planck formula for the blackbody flux per unit frequency interval is

$$F_\nu = 2\pi h\nu^3 c^{-2}/[\exp(h\nu/kT) - 1], \qquad (3.2.1)$$

so the monochromatic magnitude of a blackbody is

$$m_\nu = -2.5 \log_{10} F_\nu(T) + \text{const}$$
$$= -7.5 \log \nu + 2.5 \log[\exp(h\nu/kT) - 1] + \text{const}. \qquad (3.2.2)$$

[23a] H. F. Weaver, *Popular Astron.* **54**, 211, 287, 339, 389, 451, 504 (1946).

[24] R. V. Willstrop, *Mon. Notices Roy. Astron. Soc.* **121**, 17 (1960).

[25] A. Ažusienis and V. Straižys, *Bull. Vilnius Astron. Obs.* No. 16, p. 3; No. 17, p. 3 (1966). See also *Sov. Astron.-AJ* **13**, 316 (1969).

At the high-frequency side of the curve $\exp(h\nu/kT) \gg 1$, so we can adopt the Wien approximation [drop the 1 in Eq. (3.2.2)] and write

$$m_\nu \approx -7.5 \log \nu + 1.08 h\nu/kT + \text{const.} \qquad (3.2.3)$$

Over a modest frequency interval, the curvature of $\log \nu$ is small, so m_ν is nearly a linear function of ν ($= c/\lambda$). Thus if we have observed monochromatic magnitudes m_1 and m_2 of a blackbody at frequencies ν_1 and ν_2, the magnitude m_* at some nearby frequency ν_* can be found by linear interpolation or extrapolation:

$$m_* = m_1 + \left(\frac{\nu_* - \nu_1}{\nu_2 - \nu_1}\right)(m_2 - m_1)$$

$$= m_1 + \alpha(m_2 - m_1). \qquad (3.2.4a)$$

If we know the frequencies ν_1, ν_2, and ν_*, we can calculate the transformation coefficient α; however, if we do not, it can be found empirically from measurements of the same blackbody at all three frequencies.

We note that $(m_2 - m_1)$ in Eq. (3.2.4a) is a color index; thus the last term is usually called the *color term* in the transformation. If the magnitude scales are defined with arbitrary (e.g., instrument-dependent) zero-point constants [on the right sides of Eqs. (3.2.2) and (3.2.3)], we will also have a *zero-point term* in Eq. (3.2.4a) which becomes

$$m_* = m_1 + \alpha(m_2 - m_1) + \beta. \qquad (3.2.4b)$$

There are now *two* transformation coefficients (α and β) to be determined. If this is done empirically, we must observe at least two blackbodies of different colors (i.e., temperatures) to find α and β.

Finally, if we measure in two narrow frequency bands ν_{1*} and ν_{2*}, we can write down two transformations similar to Eq. (3.2.4b), and by subtraction find the linear transformation between the starred and the unstarred color indices:

$$(m_{2*} - m_{1*}) = \gamma(m_2 - m_1) + \delta. \qquad (3.2.5)$$

Again, the transformation involves only a linear color term and a zero-point term, which may be either computed from the definitions of the four magnitude systems, or found empirically. The coefficient γ, like α, is just the ratio of frequency differences (or "color baselines") for the starred and unstarred systems; in fact, Eq. (3.2.4b) is a special case of Eq. (3.2.5) with $m_{1*} = m_1$, and $m_{2*} = m_*$.

3.2.2.2. Atmospheric and Interstellar Reddening. In the approximations above, the monochromatic magnitudes of blackbodies are linear functions of $1/\lambda$. Any selective extinction or *reddening* that is also (in stellar magnitudes) a linear function of $1/\lambda$ will be indistinguishable from a change in blackbody temperature. In particular, the interstellar reddening is nearly proportional to $1/\lambda$ in the visible spectrum. Thus linear color transformations are as applicable to reddened stars as to unreddened ones.

Furthermore, the aerosol component of atmospheric extinction is also approximately proportional to $1/\lambda$; and, over a limited wavelength interval, the Rayleigh extinction in magnitudes (proportional to $1/\lambda^4$) can be approximated by a linear function of $1/\lambda$. [For example, a Taylor series expansion about λ_0^{-1} gives

$$\lambda^{-4} \approx \lambda_0^{-4} + 4\lambda_0^{-3}\left(\frac{1}{\lambda} - \frac{1}{\lambda_0}\right) = (4\lambda_0^{-3})\left(\frac{1}{\lambda}\right) - 3\lambda_0^{-4}.]$$

Thus we expect the monochromatic extinction correction—regarded as a transformation between intra- and extra-atmospheric systems, at a fixed air mass—to be approximately a linear function of $1/\lambda$. In this approximation, atmospheric reddening (like interstellar reddening) is indistinguishable from temperature reddening; only our ability to calculate the air mass and its effect allows us to correct for the atmosphere.

If all reddening is equivalent to temperature reddening, each object can be assigned a unique color temperature T_c. Then Eq. (3.2.2)–(3.2.5) show how measurements at different spectral frequencies are related. Since only one parameter (a color, or a color temperature) is involved, the *same* transformation applies to all objects, reddened and unreddened. To the extent that *broad-band* magnitudes can be regarded as shifted in effective wavelength by the addition of a colored filter (a fixed mass of reddening atmosphere), their transformation to outside the atmosphere should contain a linear color term; Eq. (3.1.57) shows that it does (namely, the term in ΔX).

How valid are these linear one-color transformations? To derive them, we have assumed (a) blackbody sources; (b) the Wien approximation; (c) monochromatic measurements; and (d) $|\nu_1 - \nu_2| = \Delta \nu \ll \nu$, so that curvature in $\log \nu$ can be neglected. However, if $\Delta \nu \ll \nu$, we can assume m_ν is a linear function of ν without the Wien approximation. Furthermore, if our sources have sufficiently smooth spectra, we can assume linearity of m_ν even if the sources are not blackbodies. Thus linear transformations should be valid for any such sources, provided that $\Delta \nu \ll \nu$.

However, the spectra of real astronomical sources are not perfectly smooth, but have absorption and/or emission features. Furthermore, most astronomical color systems have $(\Delta\nu/\nu) \approx 0.1$ or 0.2, i.e., not very small. Consequently, such transformations are not sufficiently accurate to preserve the inherent precision (<0.01 mag) of good photoelectric photometry.

We can improve the transformation empirically by using additional data. For example, if we have measurements in 3 bands instead of 2, we can use a formula like

$$m_* = m_1 + \alpha(m_2 - m_1) + \beta(m_3 - m_2) + \gamma. \tag{3.2.6}$$

This has worked fairly well in some cases, e.g., where m_* is the red-leak of the U filter, and the numbered bands are U, B, and V; however, it fails for cool[25a] or reddened[25] stars. As a second example, both Schmidt–Kaler[26] and Fernie and Marlborough[27] have found systematic transformation errors that are proportional to interstellar reddening. For O and B stars, the reddening is a linear combination of $(U - B)$ and $(B - V)$, so Eq. (3.2.6) applies; but for later spectral types, a different, nonlinear relation in required. As a final example, Argue[28] used equations like (3.2.6) to transform his observations of late-type stars, but found that systematic differences between luminosity classes remained.

Clearly no linear, single-valued transformation is accurate enough to preserve the full weight of good data. Every phenomenon we wish to measure—temperature, luminosity, reddening, and probably also metallicity, rotation, and other peculiarities—seems to require a different photometric transformation. Of course, a careful spectral analysis of each star would provide this information, but this is self-defeating: one of the main goals of multicolor photometry is to provide such data *without* requiring spectra.

3.2.3. Transformations in General

3.2.3.1. The Problem.   Must we then abandon hope of preserving observational accuracy through the transformation to a standard system? To answer this, we must look more closely at the general transformation problem. We saw that blackbody data could be transformed accurately

[25a] C.-Y. Shao and A. T. Young, *Astron. J.* **70**, 726 (1965).
[26] T. Schmidt-Kaler, *Observatory* **81**, 246 (1961).
[27] J. D. Fernie and J. M. Marlborough, *Observatory* **84**, 33 (1964).
[28] A. N. Argue, *Mon. Notices Roy. Astron. Soc.* **125**, 557 (1963).

because they form a one-parameter family of spectral distributions. If interstellar reddening, for example, were identical to temperature reddening, they would be indistinguishable, and the same transformation would apply to both reddened and unreddened stars. In fact, they are not: the spectral features (Balmer decrement and interstellar slope change near 4500 Å) that permit us to separate interstellar reddening from temperature effects in early-type stars are the very features that produce different transformation relations for the two groups. Each additional effect, such as metallicity, luminosity, or rotation, must leave its signature in the stellar spectrum, and hence in the transformation from one response curve to another. The *ad hoc* treatment of each effect separately is not very satisfactory, not only because it requires a huge number of standard stars of different luminosity, reddening, etc., but also because it leaves unsolved the problem of transforming peculiar objects (pulsars, quasars, galaxies, planets, emission-line stars, nebulae,...) to a common basis. Furthermore, these *ad hoc* treatments are in principle unable to cope with the general problem of transforming observations of arbitrary spectral distributions, because the set of all possible spectra (one-valued functions) is larger than a countable infinity, so that even a (countably) infinite number of individual treatments is inadequate.

Another way[29] of viewing the transformation problem is to regard a spectrum as a point or vector in an infinite-dimensional space: the coordinates of the point, or components of the vector, are the spectral power densities at successive wavelengths. A photometric measurement in m different bands projects or maps the infinite-dimensional vector into an m-dimensional subspace. Two such mappings can be mapped into each other in a one-to-one way (i.e., two photometric systems are related by a single-valued transformation) if the subspaces are linearly dependent; the transformation then amounts to a rotation of axes in m-dimensions. However, if one of the subspaces contains an appreciable component orthogonal to the other subspace, this represents spectral information not contained in the other, which is excluded by any transformation between them. Thus two photometric systems are transformable if and only if the response functions of one are a linear combination of the response functions of the other.

The concepts of linear dependence and information content suggest an information-theory approach, as follows. Suppose we treat the general problem of photometric transformations in a manner analogous to

[29] G. H. Conant, Jr., Private communication (1959).

King's treatment[16] of the extinction correction (i.e., the transformation of broad-band data from inside to outside the atmosphere). King showed that transformations between different systems depend on derivatives of the energy distribution $S(\lambda)$ reaching the photometer, and on the second (and higher) moments of the instrumental response function $\phi(\lambda)$ about some effective wavelength λ_1. The leading terms in the Taylor series expansion about λ_1 depend on $[d(\ln S)/d(\ln \lambda)]$ and the mean square bandwidth w^2. In practice, we must approximate this unknown derivative by a color index [cf., Eq. (3.1.51)].

However, different reddening mechanisms generally produce different relations between $[d(\ln S)/d(\ln \lambda)]$ and a color index which depends on a second band centered at λ_2. We can write the color index in stellar magnitudes as

$$C_{12} = 1.086[\ln S(\lambda_2) - \ln S(\lambda_1)] \tag{3.2.7}$$

and, expanding $(\ln S)$ in a Taylor series at λ_1, we have

$$\ln S(\lambda_2) - \ln S(\lambda_1) = \frac{d(\ln S)}{d(\ln \lambda_1)} \cdot (\ln \lambda_2 - \ln \lambda_1)$$

$$+ \frac{d^2(\ln S)}{d(\ln \lambda_1)^2} \cdot (\ln \lambda_2 - \ln \lambda_1)^2 + \cdots. \tag{3.2.8}$$

The first term in Eq. (3.2.8) is the theoretical justification for replacing the logarithmic derivative by a color index. However, in fact, the second (and higher-order) terms are appreciable in accurate photometry, and differ for different reddening mechanisms. These terms would be small if $S(\lambda)$ were a sufficiently "smooth" function. However, these terms are quite large if $S(\lambda)$ has a kink near λ_1—such as the Balmer jump in U, or the interstellar reddening break in B. These higher-order terms spoil the uniqueness of the transformation in terms of a color index, and explain why different relations are required for stars of different reddening and luminosity.

Making the bands narrower does not solve the problem, because then individual spectral features play a larger part in proportion to the band width, and small instrumental wavelength shifts due to temperature variations and manufacturing tolerances become more important. Placing the bands closer together helps, because of the powers of $[\ln(\lambda_2/\lambda_1)]$ which appears in the higher terms of Eq. (3.2.8). However, discrete bands cannot be placed close enough together to solve the problem, even if they are adjacent; for we can always encounter spectral distributions which give the same *response* in each band, but have very different gradients *across* each band. Thus, cutting the spectrum up into adjacent rectangular passbands, no matter how small, does not provide enough information to solve the transformation problem.

At first glance, it may appear that since adjacent rectangular passbands measure all the power in a spectrum, no more information can be obtained, and the problem is insoluble. This is not the case, and we shall show that a completely satisfactory solution is readily attainable. To do this, we adopt a more fundamental point of view: we regard multicolor photometry as low-resolution spectroscopy.

3.2.3.2. The Solution: A Spectroscopic Approach.

Suppose we want to determine the energy distribution in a star's spectrum at some low resolution—say, 500 Å, which would yield a function $S^*(\lambda')$. Here λ' is the varying center wavelength of the spectral window, and the asterisk indicates the effect of smearing the spectrum out by the slit function $W(\lambda - \lambda')$.

We could equally well measure the spectrum by using a Michelson interferometer.[30] The resulting interferogram (output intensity as a function of path difference) is the Fourier transform of the spectrum; if we only need low resolution, we need only measure the central part of the interferogram (small path differences). In fact, the interferogram (Fourier transform) of S^* is just the product of the Fourier transforms of S and W.

We must realize that, as we actually only measure a part of the spectrum, and a part of the interferogram, the measured finite parts are not exact Fourier transforms of each other. However, we can choose W so that a reasonably short piece of the interferogram transforms into S^* as accurately as we wish. In particular, we can determine the values of S^* and its derivatives at every point in the spectrum, well enough to transform our measurements to any photometric system which does not exceed the spectral resolution of our data.

In fact we can do this without using either a spectrum scanner or an interferometer, by making regular photometric measurements through a series of filters having $W(\lambda)$ as the passband shape. To prove this assertion, we regard $S^*(\lambda)$ as a real function on some interval (λ_1, λ_2). Let its interferogram (Fourier transform) be

$$s^*(\omega) = (2\pi)^{-1/2} \int_{\lambda_1}^{\lambda_2} S^*(\lambda) \exp(i\omega\lambda) \, d\lambda, \qquad (3.2.9)$$

[30] L. Mertz, "Transformations in Optics." Wiley, New York, 1965.

where the interferogram "frequency" ω is not to be confused with optical frequency c/λ. Now if we have chosen $W(\lambda)$ so that its transform $w(\omega)$ is negligible for $|\omega| > \omega_{max}$, we guarantee that $s^*(\omega)$ is also negligible outside this range, even if $S(\lambda)$ consists of delta functions (emission lines!). Then the sampling theorem[31] tells us that S^* is completely specified by its values at λ_1, $\lambda_1 + \pi/\omega_{max}$, $\lambda_1 + 2\pi/\omega_{max}$, ... These sampled values are simply photometric observations with filters spaced $\Delta\lambda = \pi/\omega_{max}$ apart in wavelength, as stated above.

In other words, we can render even an emission-line spectrum transformable to a standard system if we smear it out adequately by choosing a suitable passband shape, Also, if we can handle emission nebulae, we can certainly handle any star, no matter how peculiar. A special advantage of such a system is that the *same* transformation applies to all objects, regardless of spectral peculiarities. Thus a small number of ordinary stars can be used as standards.

The requirement that $s^*(\omega)$ be negligible beyond ω_{max} is necessary to achieve accuracy in the reconstruction of S^* from the samples. Any harmonics beyond ω_{max} will appear in the sampling as though they were at image frequencies less than ω_{max} such as $|2\omega_{max} - \omega|$, $|4\omega_{max} - \omega|$, etc., a phenomenon known as *aliasing*. That is why we must choose $W(\lambda)$, and hence $w(\omega)$, so as to prevent these images from appearing in $s^*(\omega)$. We cannot make them vanish, but we can make them very small beyond some point; this is a standard problem in one-dimensional apodization.[32] For the *UBV* bands, frequencies ω far beyond the sampled limit $\omega = \pi/\Delta\lambda$ are still important, and we may regard the *UBV* difficulties as due to the aliasing of these frequencies into the sampled frequency range.

To ensure that all transformation errors are less than some fraction f, it suffices to require that the sum of the amplitudes of all Fourier components with $\omega > \omega_{max}$ be less than f, that is,

$$\int_{\omega_{max}}^{\infty} |w(\omega)| \, d\omega \Big/ \int_0^{\infty} |w(\omega)| \, d\omega \leq f. \qquad (3.2.10)$$

This condition cannot be met for a rectangular passband, whose amplitudes fall off only as ω^{-1}. [We should have seen this from the Gibbs

[31] R. W. Ditchburn, "Light," 2nd ed., chapter 20. Wiley (Interscience), New York, 1963.

[32] P. Jacquinot and B. Roizen-Dossier, *in Progr. Opt.* **3**, 31–184. See also A. Papoulis, *J. Opt. Soc. Amer.* **62**, 1423 (1972).

phenomenon, which results from truncating $w(\omega)$ if $W(\lambda)$ is discontinuous. Another example is the ringed diffraction pattern of a telescope with sharpedged pupil.] Rectangular-passband photometry is clearly nontransformable, for if a spectral line falls into one band and is excluded from the next, there is no way to interpolate the result for an intermediate band: the line is either in or out, but we cannot say which.

However, Eq. (3.2.10) can be satisfied by continuous band profiles; the smoother the profile, the more rapidly $w(\omega)$ becomes negligible, and the farther apart in wavelength the bands may be placed. For example, the sinusoidally modulated or "channelled" spectrum used by Walraven[33] allows bands to be spaced approximately $\Delta\lambda = 1.11 f^{1/2}\lambda_\omega$ apart, where f is the maximum transformation error and λ_ω is the full width between minima of the spectral channel. Thus for $f = 0.01$, $\Delta\lambda \approx \lambda_\omega/9$, which is just under $\frac{1}{4}$ of the full width of a channel at half response.

This may seem a close spacing; but the problem is considerably worse with most glass or interference filters, which give steep-sided or asymmetric passbands with higher harmonic content (bigger Fourier components at large ω). The deeply overlapping bands required for accurate transformation may look redundant at first glance, but the sampling theorem shows that they are not—in fact, they provide just enough pieces of independent information to allow accurate transformations. One can show, for example, that steep-sided filters like those of the UBV·system must be spaced about 100 Å apart to allow accurate transformation. Hence the observed transformation errors represent severe aliasing, due to undersampling by about a factor of 10. The same problems must also occur with the numerous narrower-band systems that have been introduced, as they also have poor overlap between bands.[†]

3.2.4. Matching Response Functions

Until inherently transformable systems are in general use, the photometrist's best hope is to measure his response functions, and, by choosing appropriate filters and detectors, match them as closely as possible to the standard response functions (if these are known). The results of trial-and-error matching are reported by Hardie;[4] methods of designing

[33] T. Walraven, *Bull. Astron. Inst. Netherlands* **15**, 67 (1960).

[†] Systematic transformation errors "up to several hundredths of a magnitude" have been found in the Strömgren *uvby* system [J. A. Graham and Arne Slettebak, *Astron. J.* **78**, 295 (1973).]

a close match are given by Wright *et al.*[23] The *UBV* response functions are not accurately retrievable; a most careful reconstruction has been done by the Vilnius group.[25] The authors of newer systems have been somewhat more careful in measuring filter passbands, but other instrumental factors (transmission, and detector response) are often neglected. The narrow-band systems may be more difficult to reproduce, because of steep-sided and ripple-topped interference filters; also, a 10-Å error is a larger fraction of the bandwidth of a 100-Å filter.

To duplicate existing systems with the greatest accuracy, it should prove helpful to use two instrumental bands similar in shape to the standard, but differing in effective wavelength by a small fraction of the band width, such as a pair of Hardie's "visual" filters. Interpolation between these, using the short-baseline color they define, should give a much more accurate transformation to a standard system than does the usual transformation using long-baseline (i.e., equal to or exceeding the band width) colors. Of course, this doubles the labor of measurement. However, halving the duration of each observation would leave the total time fixed. The resulting $2^{1/2}$ increase in random errors may be a small price to pay for a large decrease in systematic transformation errors, if (as is often the case) the latter are the more important.

3.2.5. Mathematical Models

In King's analysis of the extinction transformation,[16] the measured quantity is expanded in a series whose terms are products of (a) the central *moments* [such as w^2 in Eq. (3.1.45)] of the instrumental response function, and (b) the wavelength *derivatives* (such as N and n) of the stellar spectrum and of the atmospheric extinction. High-order terms in this series can be neglected if, with increasing order, the instrumental moments decrease (which means using sufficiently narrow bands) and the atmospheric derivatives remain moderate (which means avoiding regions of molecular absorption).

The major problem with existing systems is that color indices formed from undersampled spectral data provide a poor estimate of the spectral gradient N within each band. If adequately sampled data were used, the linear approximations derived above should be quite accurate, both for the color term in the extinction and for the color terms in transformation from instrumental to standard systems. In the case of the instrumental factors, we note that two types of deviations from the standard response functions occur. The first, contributed primarily by the ratio of response curves of the standard and instrumental photomultipliers, is a smooth function,[3c] like the atmospheric transmission. We already know from King's analysis that a linear color term takes care of this very well. The second problem is the displacement of filter cutoffs from their standard wavelengths, due to manufacturing errors or temperature shifts. This alters the effective wavelength of an instrumental passband; but a linear color transformation is adequate to correct for this if Eq. (3.2.10) is satisfied. However, even if adequate sampling is used, a formula involving several color indices, such as Eq. (3.2.6) may be required; if n adjacent bands overlap, it seems best to use an n-point interpolation formula to reduce them to the standard magnitude m_* at frequency ν_*. Undersampled data will generally require nonlinear terms to allow, in part, for aliasing.

Whatever analytical form is chosen to represent the data, there remains the statistical problem of finding the best coefficients to use in this formula. A blind application of least squares may produce systematic errors, as the following example shows: Suppose we make two different sets of measurements of the same stars with the same photometer, which remains absolutely unchanged between the two series. Owing to experimental errors, the values obtained for the same star will be slightly different in the two series; how do we combine them? Now, we know the two sets are on the same photometric system, so that in Eq. (3.2.5), for example, we must have $\gamma \equiv 1$ and $\delta \equiv 0$. However, we also know from linear regression theory that the expected least squares value of γ will be less than unity (and $\delta > 0$) because of the imperfect correlation produced by the random errors. Thus a least squares fit of one set of data to the other will produce systematic errors, which depend on the relative sizes of the errors in the two sets of data.[†] Other systematic effects arise because of partially correlated errors between terms with a common element [e.g., V and $(B - V)$, or $(B - V)$ and $(U - B)$.] Such problems (and their solutions) are discussed at length by Deeming,[34] who shows that systematic errors of 0.005 mag in $(B - V)$ are readily attained in ordinary photometry.

Such systematic errors are serious enough in themselves, for they are larger than the random errors of good observations. However, they can become multiplied severalfold in some situations. For example,

[34] T. J. Deeming, *Vistas Astron.* **10**, p. 125.

[†] Such a situation exists between "summer" and "winter" UBV standard stars.[25]

photometry of a faint variable star in a globular cluster or nearby galaxy may be done photographically, relative to comparison stars which have also been calibrated photographically, against faint ($\sim$17 mag) photo-electric standards, which in turn are related to brighter "secondary standards" ($\sim$10 mag), which have been transformed to match some bright "standard stars" that were originally tied to a few "primary standards." Often a big telescope is used for the faint stars, but the brighter "secondary standards" have been set up using a smaller telescope. Thus the faint stars on which the distance scale hangs are separated from bright nearby standards by 3 or 4 transformations, with cumulative systematic errors at each step. As the systematic effects depend on the *squares* of the random errors, they can be quite large at the faint end of the scale. These transformation errors, of course, are in addition to any scale errors due to nonlinearity over a large dynamic range.

Thus, as in correcting for extinction, the photometrist must carefully match his mathematical techniques to the actual situation at hand. A poor choice of model, or misuse of least squares, can produce large systematic errors, even with small residuals.

4. RESHAPING AND STABILIZATION OF ASTRONOMICAL IMAGES*

4.1. Reshaping of Images

4.1.1. Definitions

The terms used to describe pencils of radiation differ between astronomers and optical physicists. Table I shows the equivalences, and illustrates the potential confusion due to different uses of the same word. The astronomical terms will be adopted here. The task of the astronomical spectroscopist is to gather the maximum possible radiant power from a

TABLE I. Comparison of Terminologies Used in Astronomy and Optics[a]

Astronomy	Optics	Units
1. (Specific) intensity	Radiance, brightness	erg cm^{-2} sr^{-1} sec^{-1}
2. Flux	(Illumination, emittance)	erg cm^{-2} sec^{-1}
3. (Radiant power)	Flux	erg sec^{-1}
4. (Luminosity/4π)	Intensity	erg sr^{-1} sec^{-1}

[a] Parentheses indicate that the correspondence is not exact, or that the term is not standard.

given source, and to disperse and detect it appropriately. A useful parameter of a telescope or spectroscope is its *throughput*, the product of area A and solid angle Ω that the instrument accepts. (Other terms in use are *étendue*, *light-gathering power*, and *luminosity*; the last seems especially inappropriate because of its other meanings.) The solid angle for a sizable telescope can usefully be taken as that of a typical seeing disk under moderately good conditions, that is, a circle of 1-arcsec radius, or 7.4×10^{-11} sr. Refined work on extended sources, such as the sun and

---

* Part 4 is by Donald M. Hunten.

planets, tends to use a similar field. Our examples will use a 2.5-m (100-in.) telescope, and therefore a throughput of

$$A\Omega(\text{telescope}) = 3.6 \times 10^{-6} \quad \text{cm}^2 \text{ sr}.$$

It should be noted that this is the *geometrical* throughput, since the problem is one of matching optical systems. To calculate the power transmitted by a system, one must include the transmission T, which should include such factors as central obscuration. The product of this *physical* throughput $A\Omega T$ with the specific intensity of the source gives the power output.

4.1.2. Throughput of Spectroscopes

A systematic treatment of the throughputs of various spectroscopic instruments was first given by Jacquinot.[1] By far the largest solid angle is passed by the Fabry–Perot spectrometer (and the Michelson interferometer), which require only a circular hole instead of a slit. We return briefly to these instruments below, after a detailed discussion of the grating. Though somewhat restricted in its throughput, the grating has so many virtues that it is still the spectroscope of choice for a vast number of purposes. A general discussion of astronomical spectrographs is given by Bowen.[2]

The equation of a grating with spacing d in the mth order is

$$m\lambda = d(\sin \alpha + \sin \beta).$$

We shall assume the angles of incidence α and diffraction β to be equal (except when calculating dispersion). A correction for their inequality is readily made, but is too small to concern us here. Differentiation with respect to β, and substitution of the original equation, gives

$$d\beta/d\lambda = (2/\lambda) \tan \beta.$$

This form emphasizes that the only important factor controlling the angular dispersion is the tilt angle. A different choice of grating space merely specifies the order of diffraction, but does not by itself influence the dispersion. Let the slit have dimensions $(b \times l)$, at distance f_c (collimator focal length). The L-number is $L = f_c/l$. The *resolution* is

---

[1] P. Jacquinot, *J. Opt. Soc. Amer.* **44**, 761 (1954).

[2] I. S. Bowen, *in* "Astronomical Techniques" (W. A. Hiltner, ed.), p. 34. Univ. of Chicago Press, Chicago, Illinois, 1962.

$R = \lambda/\Delta\lambda$, where $\Delta\lambda$ is the width (at half maximum) of the instrumental function. Then the geometrical throughput, with A_g the area of the grating, is

$$A\Omega = A_g \cos\beta \frac{lb}{f_c^2} = A_g \frac{2\sin\beta}{LR},$$

which is inversely proportional to R, if other things remain equal. For astronomical purposes the slit may better be described by its aspect ratio $S = l/b$, rather than its L-number. The throughput is then

$$A\Omega = A_g \cos\beta \left(\frac{2\tan\beta}{R}\right)^2 S.$$

Though convenient for computation, this form can be misleading, because the aspect ratio is seldom the same for any two instruments. The purpose of an image slicer is to convert a square or circular field to a long thin one (to increase S from its original value of 1).

The domain of image slicers is best illustrated by the examples in Table II. Here are compared two instruments, a relatively small one that would probably be mounted on the telescope, and a large one representing roughly the 1971 state of the art. The "Cassegrain" design shows that no efforts are required if a resolution of 2500 is enough—the throughput of the system is limited by the telescope, not the grating. With more effort, a resolution of perhaps 10,000 could be attained in this type of design. The factors that could still be varied are the size and angle of the grating. Also, one could choose to reduce the slit width and overfill the grating, with relatively small light losses. Finally, it must be remembered that this maximum resolution scales inversely with the telescope diameter, as well as proportionally with the grating diameter.

TABLE II. Typical Grating Spectroscopes for Use with a 2.5-m Telescope

Instrument	Cassegrain	Coudé
Grating dimensions (cm)	10×12.5	12.5×25
Angle β	15°	62°
tan β	0.27	1.9
$A_g \cos\beta$ (cm²)	78	123
Aspect ratio S	1	100
Maximum resolution	2500	220,000

The "coudé" spectroscope in Table II assumes the use of an image slicer to obtain an aspect ratio of 100. The throughput is then telescope-limited for resolutions less than 220,000. Apart from the image slicer, most of the gain comes from the factor $\tan \beta$; at $62°$ the grating falls in the "échelle" class, though the rulings need not be as coarse as many échelles. An excellent grating is available with 316 lines/mm; the 12th order falls at 4660 Å, and the free spectral range of ± 400 Å is enough for most applications that need the corresponding high resolution.

The discussion so far has concentrated on matching of telescope and collimator, but the final destination of the light must also be considered. A large aspect ratio for the slit is not particularly helpful with photographic detection; the spectrum may be considerably wider than necessary, with no shortening of the exposure time. For photoelectric detection, the full advantage of image slicing is readily realized. These questions are discussed below.

For the Fabry–Perot etalon, the interference filter, and the Michelson interferometer, the solid angle is limited only by the obliquity of the rays at the edge of the pencil; as Jacquinot[1] shows, the relation is

$$\Omega = 2\pi n^2/R,$$

where n is the refractive index of the spacing medium. The throughput and resolution of the coudé instrument of Table II can be matched by an etalon of diameter only 4 mm. However, the use of an etalon is limited by its very small free spectral range. It can be used for emission spectra with the aid of a narrow-band interference filter. For absorption spectra, the most successful arrangement is the PEPSIOS,[3] which requires three etalons in series, as well as the filter. Though the geometrical throughput remains large, the transmission suffers considerably.

4.1.3. Image Slicers

As the preceding discussion has shown, there is often a great advantage if a square field can be cut into slices, which are then arranged end-to-end: the aspect ratio can be increased to as much as 100. Devices to perform this operation are known as *image slicers*, though it should be remembered that sometimes it is preferable to slice the pupil rather than the image. Image slicers have seen remarkably little use, perhaps because of a widespread impression that they are more of a curiosity than a practical device.

[3] J. E. Mack, D. P. McNutt, F. L. Roesler, and R. Chabbal, *Appl. Opt.* **2**, 873 (1963).

Properly used, they can give a large advantage, especially in the coudé context discussed above, and with photoelectric detection.

One of the chief difficulties with an image slicer is the minute scale. If the coudé focal ratio of our 2.5-m telescope is $f/30$, the angular scale is 2.75 arcsec/mm; the field to be sliced is only 0.72 mm in diameter. Thus, it is frequently desirable to work at even longer focal ratios, which imply additional optics and additional light losses. Different limitations become apparent if a much larger throughput is available, as from an extended source. Of most importance is that the Bowen slicer is not satisfactory at focal ratios much shorter than $f/30$. Again, it may be desirable to slice at a large scale and then refocus on the slit. Benesch and Strong,[4] who did not use this remedy, describe some of the problems of a Bowen slicer in an $f/10$ beam.

4.1.4. Bowen Image Slicer

The Bowen slicer takes advantage of 90° reflections to align end-to-end the slices from a square field. Its operation is illustrated in the perspective sketch of Fig. 1, where the active parts of the mirrors are emphasized by heavier outlines. By means of a slight skewing of the rays, the ends of

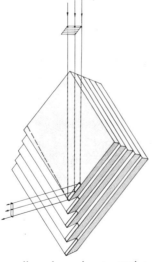

FIG. 1. A Bowen image slicer shown in perspective. The grid at the top represents the square field to be sliced. Typical rays for one slice are shown. The dashed line represents a possible design modification discussed in the text.

[4] W. Benesch and J. Strong, *J. Opt. Soc. Amer.* **41**, 252 (1951).

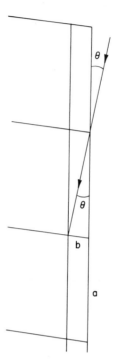

FIG. 2. The output of a Bowen slicer, showing the input ray crossing the slit at angle θ.

the slices can be neatly lined up, as shown in Fig. 2. However, they are not in the same plane: they are arranged in space like the risers of a staircase. This is a principal reason for limiting the f-ratio to values greater than 30. The skewing of the input and output rays is illustrated in Figs. 2 and 3. The slicing mirrors are formed by the corners of much larger glass plates, whose dimensions and angles have been chosen to make alignment and mounting easy. The sharp edge of the first mirror defines the inner edge of the second slice, and so on. The face of each plate must be undercut slightly, because the input beam proceeds slightly away from the observer in Fig. 1. A 45° undercut of the bottom edge is necessary to avoid interference with the output beams. This feature is illustrated further in Fig. 3.

The "slit's-eye" view in Fig. 2 shows that the input beam, which is in the plane of the slit, must cross it at an angle $\theta = b/a$ (radians), where b is the slit width and a the slice length. Normally the area to be sliced is square; if the number of slices is n, the slit length is na and $\theta = 1/n$. Though the illustrations are for $n = 5$, $n = 10$ is more common and values as great as 20 have been used. Thus, negligible error is introduced by the approximation tan $\theta \approx \theta$.

The original description by Bowen[5] said little about construction and alignment, except to point out that the active elements could be the corners of much larger plates. More recently, Pierce[6] outlined an elegant method which makes the job astonishingly easy. If the plates are stacked

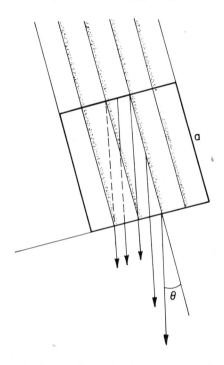

FIG. 3. Input view of a Bowen slicer, showing output rays from one slice. Some of them pass under the corner of the next slice above.

in reverse order, it is merely necessary to form right angles on two edges, and angles of 45° and $\chi = \theta/2^{1/2}$ on the other two, as shown in Fig. 4. These angles each serve two functions: they graduate the dimensions of the plates so that the square faces need merely be lined up, and they provide exactly the right amount of undercutting discussed above. The prescription given here differs from that of Pierce in two respects. First, he gives $\chi = \theta$, not $\theta/2^{1/2}$. Apparently both angles produce a satisfactory

[5] I. S. Bowen, *Astrophys. J.* **88**, 113 (1938).
[6] A. K. Pierce, *Publ. Astron. Soc. Pacific* **77**, 216 (1965).

slicer, perhaps because additional skewnesses can be introduced during alignment; but $\chi = \theta/2^{1/2}$ is believed to be strictly correct. Second, Pierce suggests tapering the 45° edge at $\theta/2^{1/2}$, as indicated by the dashed line on the top mirror in Fig. 1. Unless the slicer is to be used in close contact with a slit, this taper is completely unnecessary; its only optical function is to square off the ends of the slices as viewed from the input. The compound angle is distinctly more difficult for the optician than the simple 45°.

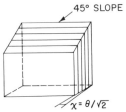

FIG. 4. The Pierce construction of a Bowen slicer. The slices are stacked in the reverse order from Fig. 1 for the optical work. (The corresponding diagram in A. K. Pierce's article [*Publ. Astron. Soc. Pacific* **77**, 216 (1965)] would produce a slicer of opposite handedness.)

The thickness of the plates must be $t = b/2^{1/2}$ to space the surfaces by b along the rays at 45°. Parallelism is important, for nonparallel surfaces will throw different slices in different directions. However, as Bowen[4] suggests, it should be possible to introduce an intentional taper of each plate, just enough to aim each spot at the center of the collimator. In other words, the slicer can also act as a concave field mirror. As far as the writer knows, this suggestion has not been put into practice.

Fabrication of the slicer thus proceeds as follows. A generous number of plates is waxed to a flat, ground down to the specified thickness, and polished. This operation being by far the longest, it is advisable to do as many plates as possible; more slicers can then be produced with little additional labor. The selected plates are waxed into a stack, with an extra one at each end for protection. The angles shown in Fig. 4 are ground and checked with a protractor; extreme accuracy is not necessary. The two sloping faces are usually polished, though this is not obviously necessary. After careful cleaning, the stack is assembled in reverse order and clamped in a jig. Flat, square surfaces should be provided, against which the square edges can be butted. From here on, the slicing corners are extremely vulnerable (and extremely sharp); they should be carefully protected.

Past practice has been to coat the surfaces after assembly, and indeed this may be the only practical procedure, because it would be difficult to avoid scratching coated surfaces during assembly. However, it is not clear how good a coating can be produced on an assembled stack. The faces cannot be perpendicular to the source, and gas must be escaping from the cracks, right beside the faces. Cleaning an assembled stack for recoating offers even greater hazards, and if possible the slicer should be dismantled for this operation. There might be virtue in attempting to glue the plates together with low-vapor-pressure epoxy, perhaps with thin spacers to keep the glue from running over the slicing faces. The principal difficulty with any such scheme is to keep it from interfering with the proper stacking of the plates.

4.1.5. Installation

The beam must be incident on a Bowen slicer at right angles to the axis of the collimator, and at an angle θ with the slit (Fig. 2). In some cases, the fore optics can be designed with this result in mind. Most frequently, however, an extra small mirror is included with the slicer to render the input and output beams parallel, with a small sidewise displacement. This mirror must be parallel to the plates of the slicer. Alignment and testing are easy in the beam of a small helium–neon laser. The individual slices may be traced for several meters, and their widths may be judged by the widths of the diffraction patterns.

4.1.6. Richardson Image Slicer

The Richardson slicer,[7] illustrated in Fig. 5, resembles a confocal cavity or a White cell. The mirrors can be adjusted in two different ways. As illustrated, multiple strips are formed along the slit, in much the same way as a Bowen slicer, and each beam is superposed on the grating. Richardson describes a complementary arrangement, in which a cylindrical lens forms a long, astigmatic image on the slit, and the slicer forms a stack of astigmatic strips on the grating. There seems little to choose between the two arrangements; the first is slightly simpler and its operation is easier to describe.

---

[7] E. H. Richardson, *J. Roy. Astron. Soc. Can.* **62**, 313 (1968); *Proc. ESO/CERN Conf. Auxiliary Instrumentation for Large Telescopes*, Geneva, May 2–5, 1972; *Contr. Dominion Astrophys. Obs.*, no. 202.

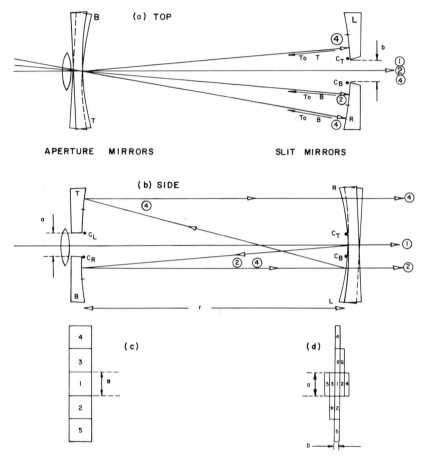

FIG. 5. A Richardson slicer, adjusted for parallel output beams. Parts (c) and (d) show the locations of the various beams on the mirrors. In part (a), the slit width b is greatly exaggerated (see text for discussion).

Each of the two concave mirrors is cut in half; the edges of the *slit mirrors* form the slit jaws. The *aperture mirrors* must have a gap between them to admit the light, but the height of this gap cannot exceed the slice length a. The spacing of the mirrors, equal to the common radius of curvature r, cannot exceed aF, where F is the focal ratio of the spectroscope. This condition, which we shall discuss below, puts some stringent limitations on the practicability of a Richardson slicer in actual situations.

In Fig. 5, the pupil of the system (the telescope primary) is focused between the aperture mirrors, and the field lens produces an image at the

center of the slit. (If more convenient, pupil and image may be interchanged.) Figure 5d shows this image, divided into 5 slices (for this
example). Slice 1 goes through the slit, and the others are reflected back
to the aperture mirrors. Because of the tilt of the slit mirrors, beams 2
and 4 go to the lower mirror (image 2 in Fig. 5c), and 3 and 5 to the
upper one. They next reappear on the slit mirrors, and slices 2 and 3 go
through the slit. On the next round trip, slices 4 and 5 cross the axis
and then fall on the slit. More slices are possible, if the slit is narrower
than $\frac{1}{5}$ of the image, but each additional pair of slices suffers 2 more
reflections.

The arrangement actually used by Richardson has already been briefly
described. A cylindrical lens ahead of the slicer replaces the refocusing
lens shown in Fig. 5. The first astigmatic focus is placed between the
aperture mirrors, and the second on the slit. In Fig. 5b, the centers of
curvature C_T and C_B of the aperture mirrors are placed on axis; in
Fig. 5a, C_L and C_R are displaced by a slit width. A field lens focuses the
aperture mirrors on the grating. The images stacked on these elements
are now rectangles, not the squares shown in Fig. 5(c); and each beam
traverses the same part of the slit. With this adjustment, the light distribution across the spectrogram is a one-dimensional image of the telescope primary, and is therefore rather uniform. Also, a further readjustment can leave a vacancy in the middle of the grating, so that a slice
is not wasted on the back of a plateholder.

In the Richardson slicer, as well as in the Strong image transformer
described below, all slices are in focus. Thus, the focal ratio of the spectroscope can be fairly small; for example, Richardson describes slicers
working at $f/18$. The mirrors may become very small, but can be fabri-
cated on larger pieces of glass. Their diameter should be somewhat
greater than the slit length, and their focal ratio, at the center of curvature, equal to that of the collimator. The cylindrical lens can be modified
to change the focal ratio of the beam from the telescope.

A Richardson slicer for an odd number n of slices requires a mean
number of reflections $(n^2 - 1)/2n$, or approximately $n/2$. With efficient
coatings, the loss of light can be kept small; moreover, for stellar work
the intense central core of the image suffers no reflections. In addition,
there are two lenses which must be coated. A Bowen slicer treats all
slices the same, requiring a single reflection, or two if the input and output beams must be parallel. For photographic stellar spectroscopy the
Richardson slicer is preferred.

4.1.7. Image Transformer

Benesch and Strong,[4] to avoid some of the problems of a Bowen slicer in a fast system, introduced an alternative, the *optical image transformer*. One possible embodiment is shown in Fig. 6. Of the three slices, the middle one goes straight through the slit; the others are diverted to the sides by small plane mirrors, and reflected back by concave mirrors. Only the left-hand beam is shown in full; it returns above the center slice and is deflected through the slit by another small plane mirror. The right-hand beam, similarly, is directed below the center slice. The concave mirrors refocus each slice back on the slit; there is thus no limitation on the f-number of the system. There are, however, disad-

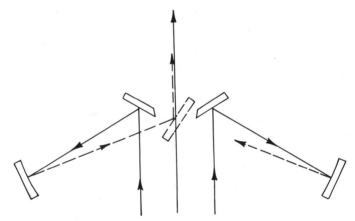

FIG. 6. An image transformer for 3 slices. One mirror has been omitted from the right-hand path.

vantages: each beam except the center one suffers three reflections, and each mirror after the first must be mounted and adjusted individually. In the context of coudé spectroscopy, one mirror per beam would have to be the size of an individual slice. Another, somewhat neater, example shown by Benesch and Strong requires five reflections for each beam. In the infrared these extra reflections absorb very little energy, but at shorter wavelengths they exact a serious penalty. There seem to be few applications that would not be better served by the more recent Richardson image slicer. The specific example discussed by Benesch and Strong involved the illumination at $f/10$ of a 1×75-mm slit with 3 slices. The corresponding Richardson slicer would require mirrors of 75-mm diameter and 750-mm radius of curvature.

4.1.8. Telescope Slicer of Fastie

A scheme proposed by Fastie[8] may be appropriate for a future telescope composed of several independent mirrors. The individual images may be aligned along the slit, and the pencils superposed on the grating by tiny individual prisms.

4.1.9. Fiber Optics

Over a decade ago it was suggested that a fiber-optic "light funnel" would solve the image-slicing problem once and for all. A detailed exposition was given by Kapany,[9] but no description of a working unit has appeared, even in a recent book on fiber optics.[10] The writer constructed one in 1963, and found it to be useless because, even with a collimated input beam, the exit beam filled a cone of some 20°. Even if most of the energy was transmitted (not necessarily true), the intensity was greatly degraded. An excellent quantitative study of this effect has recently been published by Robben and Fraser.[11]

4.1.10. Summary

Most applications should be satisfied by either a Bowen or a Richardson slicer, with the choice depending on details of the installation, the allowable number of reflections, and the available optical facilities. For more than 5 or 7 slices, the Bowen scheme is likely to be preferable.

Addition of a Bowen slicer to an existing coudé spectrograph will shorten the exposure only to the extent that it is no longer necessary to widen the spectrum by trailing the image. The wide spectrum does in principle contain more information than a conventional narrow one, but apparently not enough to interest most observers. Bowen[5] discusses the possibility of introducing a cylindrical lens just ahead of the plate to narrow the spectrum and concentrate the light, but the system is not popular. Conceivably, some other anamorphic optical system might be possible, but no concrete suggestions seem to exist. The recent revival

[8] W. G. Fastie, *Appl. Opt.* **6**, 397 (1967).

[9] N. S. Kapany, *in* "Concepts of Classical Optics" (J. Strong, ed.), p. 553. Freeman, San Francisco, 1958.

[10] N. S. Kapany, "Fiber Optics." Academic Press, New York, 1967.

[11] F. Robben and R. Fraser, *Appl. Opt.* **10**, 1141 (1971).

of interest in image slicing is almost entirely due to photoelectric spec-troscopists, who can gain a large advantage. Photographic observers should perhaps reconsider their past lack of interest.

4.2. Stabilization of Images

4.2.1. Automatic Guiding

We shall distinguish two classes of instrument: (1) *guiders*, in which the detector, nearly always a photographic plate, is shifted to follow the image; (2) *image stabilizers*, in which the image is maintained on a fixed detector. The emphasis will be on stabilizers, which are of interest for spectroscopy, and also for narrow-field photography of objects such as planets. Both guiders and stabilizers require an image tracker to sense the error to be corrected. Tracking is by far the most difficult part of the problem and will be discussed in some detail.

Guiders typically make use of a double-slide plate holder with servo-motors driving the two axes.[12] In some cases the motors are of the stepping type.

The simplest *image stabilizer* is a connection to the slow-motion controls of the main telescope. The early devices of Whitford and Kron[13] and Babcock[14] were of this type. However, there are serious disadvantages, due to the large mass of the telescope. Response to an error signal is slow, and overshoots and backlash tend to cause problems. In practice, only slow drifts can be corrected. With fast guiders of limited range, it is common practice to smooth the error signals and use them to correct the telescope pointing.

Much faster response can be obtained by rotating or translating a small optical element. A refractor plate driven around two axes has been used by Babcock *et al.*[15] For photography of planets, Leighton[16] translated a relay lens which also magnified the image. This scheme would not be

---

[12] W. F. Ball and A. A. Hoag, *Sky and Telescope* **35**, 22 (1968); *Kitt Peak Contr.* no. 278; A. A. Hoag, W. F. Ball, and D. E. Trumbo, *in* "Automation in Optical Astrophysics," p. 71; 11th Colloq. of the Int. Astron. Un., *Publ. Roy. Obs. Edinburgh* **8**, 1971.

[13] A. E. Whitford and G. E. Kron, *Rev. Sci. Instrum.* **8**, 78 (1937).

[14] H. W. Babcock, *Astrophys. J.* **107**, 73 (1948).

[15] H. W. Babcock, B. H. Rule, and J. S. Fassaro, *Publ. Astron. Soc. Pacific* **68**, 256 (1956).

[16] R. B. Leighton, *Sci. Amer.* **194** (6), 157 (1956); reprinted in "The Amateur Scientist" (C. L. Stong, ed.), p. 26. Simon and Schuster, New York, 1960.

satisfactory for illumination of a spectroscope, because while correcting the image position it displaces the axis of the pencil. Rotation of a mirror is attractive; the direction of the pencil can be stabilized, and extremely small rotations are sufficient. This technique has been applied in two quite different ways: control of a Cassegrain secondary, and control of a small mirror in the fore optics.

The "wobbly secondary" technique is popular for rocket-borne telescopes; it was developed independently at Kitt Peak National Observatory[17] and Johns Hopkins University.[18] An important advantage in the ultraviolet is that no extra reflections or refractions are required. There is a small tilt of the pencil, but only of the order of the angular errors being corrected. In principle, control of the secondary could be applied to large ground-based telescopes for fine guiding over a considerable field of view. The effect on image quality has been investigated by J. Simmons at Kitt Peak and by Bottema and Woodruff.[19] Simmons found that, for conventional mirror curves, it is best to rotate the secondary about its first focus (the focal point of the primary). The quality of an image returned to the axis is then comparable to its quality in the original off-axis position. Bottema and Woodruff imposed the requirement that the center of rotation coincide with the vertex of the mirror, certainly preferable from the mechanical standpoint, and varied the form of the mirrors. They found an excellent solution, the "tilted aplanatic," for which both mirrors are hyperboloidal; corrections of up to 10 arcmin would still give excellent images in the center of the field. In practice, only small corrections are likely to be needed, and these complications may be mainly of academic interest.

Stabilization by a mirror at the focus has seen considerable development in the last few years. The optics required are small and inexpensive, and being small can have very rapid response, of the order of milliseconds. Thus, image motion can be completely eliminated. Brief descriptions have been given by Connes and Connes[20] and by Belton et al.[21] Two typical arrangements are shown in Fig. 7. The sky is focused on a field element, which can be either a lens or a mirror. At typical coudé scales

[17] R. C. Anderson, J. G. Pipes, A. L. Broadfoot, and L. Wallace, *J. Atmos. Sci.* **26**, 874 (1969).

[18] M. Bottema, W. G. Fastie, and H. W. Moos, *Appl. Opt.* **8**, 1821 (1969).

[19] M. Bottema and R. A. Woodruff, *Appl. Opt.* **10**, 300 (1971).

[20] J. and P. Connes, *J. Opt. Soc. Amer.* **56**, 896 (1966).

[21] M. J. S. Belton, A. L. Broadfoot, and D. M. Hunten, *J. Geophys. Res.* **73**, 4795 (1968).

of 2 to 4 arcsec/mm, a 30-mm field corresponds to 1 to 2 arcmin. At a Cassegrain focus the element can be correspondingly smaller. The field element forms an image of the telescope mirror on the concave servo mirror, which is typically 10–15 mm in diameter. Two transducers (usually small loudspeakers) control the orientation of the mirror and stabilize the final image. The magnification can be unity, or can be chosen at some other value if required. Though the mirrors are used slightly off axis, the aberrations thus produced are small enough to be tolerable for most purposes.

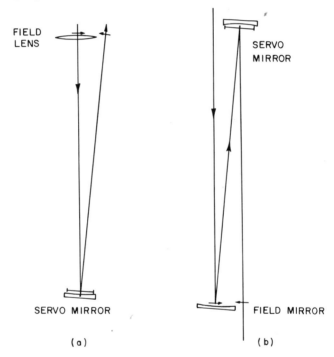

FIG. 7. Recommended arrangements for a concave servo mirror. With a telescope pupil focused on this mirror, the direction of the output beam is fixed.

An important property of this optical system is that it fixes the direction of the output pencil as well as the position of the image. If desirable, it can also take up a substantial misalignment of telescope and spectroscope.

Figure 8 shows the drive unit used at Kitt Peak. The loudspeakers are replacement units for a transistor radio, 5-cm diameter. The drive rods are bent from brass shim stock, 75 μm (0.003 in.) thick, and have

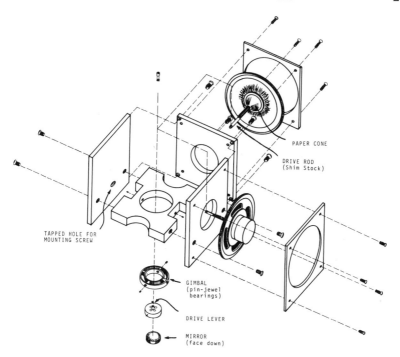

PAPER CONE

DRIVE ROD
(Shim Stock)

TAPPED HOLE FOR
MOUNTING SCREW

GIMBAL
(pin-jewel
bearings)

DRIVE LEVER

MIRROR
(face down)

FIG. 8. Exploded view of a small mirror-driver unit using 5-cm loudspeakers.

the useful property of being stiff to longitudinal forces, but easy to twist. Thus, the two axes do not interfere with each other. The response time of a few milliseconds is fast enough to eliminate not only seeing motions but also the rapid jumps that occasionally occur in the telescope motion.

4.2.2. Dynamics of a Servo Mirror

This discussion concentrates on the unit shown in Fig. 8, but is applicable to any design that includes a restoring force. A mechanical equivalent is shown in Fig. 9. It is convenient, as shown in Fig. 9a, to define an effective mass m_e for the mirror, which represents the inertia seen by the drive rod. The equation of motion is

$$Fl = I(q/l) = mk^2(a/l),$$

where $I = mk^2$ is the moment of inertia of the mirror, m is its mass, and a is the acceleration of the drive rod. For a disk the radius of gyration k is equal to $r/2$; we thus find

$$m_e = m(k/l)^2 = (m/4)(r/l)^2.$$

For Fig. 9b, the equation of motion for the displacement x is

$$(m_1 + m_e)\ddot{x} = F - kx - b\dot{x}.$$

It is convenient to work with the steady-state frequency response, which is readily measured for an actual system by driving it at various frequencies and observing the resulting motion optically. We thus set $F = F_0 \times \exp(i\omega t)$ and $x = A \exp(i\omega t)$;

$$A = \frac{F_0/k}{1 + i\omega(b/k) - (m_1 + m_e)\omega^2/k}.$$

At low frequencies the amplitude is F_0/k, with the driving and restoring forces in balance. The high-frequency amplitude is $-F_0/[m_1+m_e)\omega^2]$; there is a phase reversal and an inverse-square frequency dependence. A resonant peak occurs near the angular frequency given by ω_0^2

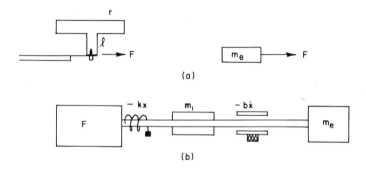

(a)

(b)

FIG. 9. Equivalent mechanical circuits of a servomirror drive (see text for discussion).

$= k/(m_1 + m_e)$. If the damping constant b happens to be large enough, this peak may not be present. Normally, however, the damping has to be supplied; probably the easiest way is to do it electronically, as discussed below. When critically damped $(b = 2k/\omega_0)$, the response to a step function is

$$x/A = 1 - (1 + \omega_0 t) \exp(-\omega_0 t).$$

The response time is of the order of $\pi/\omega_0 = 1/2f_0$, depending on the definition adopted.

The optimum length for the lever arm l makes the reflected mass m_e equal to the mass m_1 of the driver and arm. This choice gives the greatest angular acceleration: shorter arms load the driver too heavily, and longer ones do not use the available linear acceleration as efficiently. No theoretical optimum exists for the mirror size: smaller ones are always faster. However, there is little point in reducing the diameter below 10 or 12 mm, because of other parts, like the gimbal, because the required lever arm may be too short to be practical, and because millisecond response is easy to obtain with mirrors this size. As Fig. 7 suggests, a small mirror of short focus may have to be used further off axis than a large one of long focus. Thus, optical requirements may dictate a minimum size. Under certain assumptions, the attainable image acceleration can be shown to scale as $r^{-3/2}$ for a given driver.

Table III shows the characteristics of two drivers: the small loudspeaker illustrated in Fig. 8, and a much more powerful unit intended as a vibration generator. For the latter, a mirror 50 mm in diameter has been assumed. In the small unit, the lever arm is too short to match the mass of the speaker and drive rod ($m_1 = 0.18$ gm). Thus, a slightly faster response could be obtained at the expense of angular amplitude.

TABLE III. Characteristics of Typical Mirror-Drive Systems

Driver[a]	50-mm Speaker	Vibration generator[b]
Force factor, dyn/mA	20	360
Max. steady force, dyn	3200	540,000
Max. stroke ($2A$), mm	0.037	1.25
Restoring force const (k), dyn/cm	10^6	4×10^6
Moving mass, gm	0.08	6.5
Resonant frequency, Hz	560	120
Drive rod mass, gm	0.10	1
Mirror diameter, mm	12	50
Mirror thickness, mm	2.5	10
Mirror mass (m), gm	0.75	50
Lever arm (l), mm	5	35
Effective mass (m_e), gm	0.27	6.5
Resonant frequency (f_0), Hz	250	85
Response time ($1/2f_0$), msec	2	6

[a] The first 6 lines are for a bare driver.

[b] Type V47, Pye-Ling, Ltd. (Ling Electronics, Manchester Ave., Anaheim, California).

As discussed below in connection with the electronics, it is easy to derive a velocity signal from the voltage induced in the voice coil. This signal can then be used to damp the resonance completely and thus to produce a well-behaved mechanical system. This property makes electrodynamic drivers particularly attractive, in spite of the somewhat involved mechanical couplings they require.

4.2.3. Other Drivers

Figure 10 shows an electromagnetic drive that has been used for "wobbly secondaries" in rocket-borne telescopes.[17] Another arrangement is described by Leighton.[16] Suitable coils are readily obtainable by dismantling relays. Electromagnetic drivers are simple, compact, and powerful—driving forces are similar to that of the vibration generator in Table III. Because the coupling to the mirror mount is magnetic,

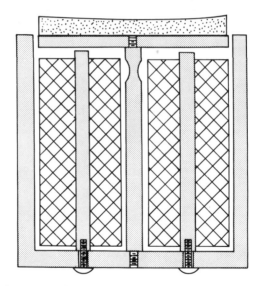

FIG. 10. An electromagnetic mirror drive, with 4 coils attracting an iron disk.

x and y motions do not interfere with each other. However, there are important disadvantages also. The inherent nonlinearity due to variation of the air gap is not serious if the initial gap is large enough. The real difficulty is to damp the resonance. The variable air gap produces an electrical, as well as mechanical, nonlinearity that frustrates attempts to derive a velocity signal. It is necessary to provide either viscous mechanical damping, or separate velocity pickups for electrical damping. The simplicity of the basic design is thus lost.

An entirely different approach is to use servomotors, as described by Bottema et al.[18] With this system there is no restoring force, and therefore no resonance to damp. If a response time of a tenth of a second is sufficient, this system deserves serious consideration. Bottema et al.[18] describe a linkage that neatly decouples the axes. Alternatively, cables and pulleys may be used, as in many laboratory recorders that use essentially the same drive system.

4.2.4. Image Trackers

As the discussion above suggests, there is no lack of elegant and practical devices for following or stabilizing astronomical images. However, all of them require error signals, and obtaining these signals is usually the hardest part of the problem. It has two parts: (1) provision of an image to track, without too much loss of valuable light, and (2) design of a suitable detector.

For a wide-field guider it is established practice to use a bright star near the edge of the field. A versatile system for relaying the light is illustrated by Ball and Hoag.[12] A response time of a few tenths of a second is amply fast, because the rapid trembling is uncorrelated over fields of several arcminutes.

The image stabilizer for a single object (often, but not necessarily, bright) offers a different problem, for fast response is useful, but at the same time the light for the measurement should not be wasted. The most direct (but wasteful) method is to divert some 10% of the light by a beamsplitter. Alternatively, a small part of the telescope aperture may be used to provide an offset tracking beam. A plug can be drilled from the secondary and cemented back in at a small angle.[18] Some Kitt Peak rockets have used a different version of this scheme, in which a thin prism is mounted beside the secondary to divert a fraction of the field by the required amount. Both methods give an auxiliary image of the main object, displaced a fraction of a degree and with a few percent of the intensity. If desired, this image may be isolated by a diaphragm at an image of the plug or prism formed with a field lens; but in most cases this precaution is unnecessary.

Often the main measurement concerns only a small spectral region; some of the remaining wavelengths can then be used for tracking. For example, planetary spectroscopy is normally done in the infrared, and the entire visible region can be split off by a dichroic mirror. Such mirrors can be made, by multilayer techniques, to have a very high reflectivity for the desired wavelength band. At appreciably different wavelengths the transmission, even if not high, is ample for a tracking beam. Another common situation requires an interference filter to isolate a band of a few to a few hundred angstroms. Any such filter has a high reflectivity for nearby wavelengths, as long as the absorbing glass is placed on the far side. Normally the mirror or filter will be plane, but an interesting alternative exists. In Fig. 7a the stabilized image beside the field lens might well be reflected down again by a concave mirror with a dichroic coating. For the transmitted light, the back surface can be polished and chosen to give a lens with any required power, including zero. A plane mirror or filter can be inserted almost anywhere, depending on the experiment, but normally will not coincide with an image; thus, the two wavelength regions will form images well separated in space.

Such techniques are useful only at wavelengths for which efficient reflectors and transmitters are available. It is therefore not surprising that rocket-borne telescopes use a basically different technology that emphasizes the fewest possible reflections.

4.2.5. Tracking Devices

A large body of technology exists in the contexts of missile guidance and automatic navigation. It is summarized in two books,[22,23] of which the latter is much more useful. Though a survey of this literature is somewhat helpful, few of the techniques are applicable to the problem of guiding a ground-based telescope. Emphasis tends to be on acquisition of a target out of a wide field, and on discrimination against large objects such as clouds ("spatial filtering"). Also, the rotating devices, or reticles, are often complicated by the need to modulate the signal at a fairly high frequency for an infrared detector. One such device was used in an automated telescope at Kitt Peak.[24] The ability to acquire an object out of a wide field was valuable here, as was the ability to reject fainter objects and track on the brightest one only.

[22] Quasius and F. McCanless, "Star Trackers and Systems Design." Spartan Books, Washington, D.C., 1966.

[23] L. M. Biberman, "Reticles in Electro-Optical Devices." Pergamon, Oxford, 1966.

[24] S. P. Maran, in "Stellar Astronomy" (H. Y. Chiu, R. L. Warasila, and J. L. Remo, eds.), vol. 2, p. 337. Gordon and Breach, New York, 1969.

We shall therefore concentrate on devices that are intended specifically for astronomical guiding. They fall into two general classes: those that detect separately in four quadrants, and those that scan. Though quadrant detectors appear simpler and more straightforward, they have their disadvantages, and scanning detectors are probably more widely used. The first tracker to be described[13] was in a sense a hybrid of the two. The beam was divided into two parts by a roof mirror; the beams were chopped and then recombined on a single photomultiplier. It tracked in one dimension only, and therefore had two channels instead of the normal four.

A quadrant detector can be made from a pyramidal mirror and four inexpensive photomultipliers. A system of this kind[20,21] has been used at Kitt Peak for several years. Though it works well, it is unlikely to be duplicated. The detector head is bulky and heavy, and there is no convenient way to adjust the zero point. It is necessary either to translate the head physically, or to adjust one of the mirrors that take the beam to it.

Quadrant photomultipliers are similarly inflexible, though much less bulky. Their use is briefly described by Bottema et al.[18] Quadrant silicon diodes are also available, but the sensitivity is too low for astronomical use without additional complications, such as chopping or cooling. There is no point in using a chopper with a quadrant device: it is better to convert the chopper into a scanner, and use a single detector.

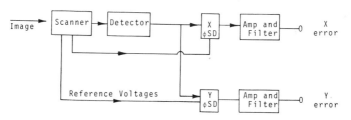

Fig. 11. Block diagram of a conical-scanning system.

"Conical scanning" is widely used in tracking radars and other military devices.[23] Its first use for astronomical tracking was the rotating knife-edge device of Babcock.[14] By this or other means a field of view is made to rotate in a small circle. Unless the image is in the center of the circle, an alternating signal is developed; its amplitude gives the magnitude of the error, and its phase gives the direction. As Fig. 11 indicates, this vector error can be decomposed into its x and y components by two phase-sensitive detectors (Babcock used mechanical commutators rotating with the knife-edge). After suitable filtering to remove carrier-frequency ripple, the x and y error signals are available. If rapid response is required, a high carrier frequency is necessary, and electronic scanning is to be preferred. Mechanical rotation, though slow, would be much more attractive if a hollow-shaft synchronous motor were readily available, with a clear aperture of 20 mm or so. Otherwise, one may be forced to use a belt drive. In any case, the reference voltages should be derived directly from the rotating assembly.

The rotating knife-edge was later replaced by a single ball, magnetically driven around a race.[15] With a hollow shaft, a thin prism can be used to scan the image around a fixed detector, or a "nutating" mirror (with its normal at a small angle to the axis of rotation) can be used with a normal motor.

Electronic scanning is possible with an image dissector, such as the ITT "star-tracker" photomultipliers.[†] These tubes have a cathode some 20 mm in diameter, followed by an electrostatic lens that focuses the photoelectrons on an aperture plate with a small hole. Only those electrons that pass through the hole are measured. A magnetic deflection yoke permits the sensitive spot to be scanned as desired. A carrier frequency of 5 kHz is readily obtained, and could be considerably increased if the need should arise. With the introduction of these tubes, there is no longer much reason to use a rotating reticle. Magnetic scanning is much faster and simpler than a mechanical motion. The only reasons for retaining the latter might be lower cost, the need for a complicated reticle, or the use of infrared. Another revolution may well be in the offing: the use of silicon vidicons, the first really tractable television tubes. It should be possible to use them in exactly the same mode as the star-tracking photomultiplier.

A conical-scanning system was developed, at the writer's suggestion, by T. W. Avery and R. H. Nagel; it has been used on several rocket flights. Figure 12 shows a simplified version suitable for ground-based use. It is considerably more convenient and versatile than the quadrant device mentioned above. In addition to the elements of Fig. 11, it includes power amplifiers and velocity feedback for the mirror unit shown in Fig. 8. The CA 3047, now obsolete, can be replaced by some other type of microcircuit amplifier; the coil drivers will require current

---

† Types FW 118, 129, and 130 with S-1, S-11, and S-20 cathodes; deflection yoke FW 315 (ITT Industrial Laboratories, Fort Wayne, Indiana).

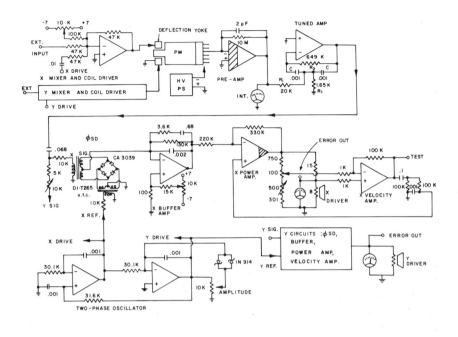

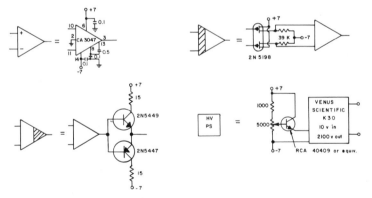

FIG. 12. Circuit of a conical-scanning system. Abbreviated blocks are explained below the main figure.

boosters. The supply voltages of ± 7 V have been chosen to avoid over-driving the miniature speakers. Most of the circuits are applicable to any conical-scanning system.

The 2-phase oscillator solves the equation of simple harmonic motion. The amplitude is limited when the drop across the silicon diodes begins to exceed their effective threshold of about 0.5 V. The frequency was chosen at 5 kHz to match a readily available amplitude to the impedance of the deflecting coils. The mixers at the top of the figure permit the inclusion of a dc bias and an external signal with the basic scan. With the bias, the guide point can be placed anywhere on the cathode. An external square wave, for example, can be used to oscillate or "chop" the image so that two spots on a planet can be compared, or sky subtractions made.

The preamplifier includes an added field-effect transistor (FET) pair for low input current; alternatively, an available FET amplifier can be used. A meter on the output indicates the signal strength, and is used as a guide for adjusting the high voltage to the photomultiplier. A constant signal at this point implies a constant loop gain in the optical servo, and therefore constant performance. In the rocket-borne version, the high voltage is actually controlled by the dc signal. If this option is used, the meter should read the high voltage instead, for a roughly logarithmic indication of the intensity.

The tuned amplifier, due to R. H. Nagel, has the same response as an LC resonant circuit.[25] If R_p represents the resistance of R_i and R_1 in parallel, the important parameters are

$$Q = \tfrac{1}{2}(R_2/R_p)^{1/2} = 10,$$

$$2\pi f_0 = (R_2 R_p C^2)^{-1/2} = 2\pi \times 5000 \quad \text{Hz},$$

$$K_0 = R_2/(2R_i) = 16 \text{ (gain at } f_0).$$

As Fig. 11 indicates, the signal is fed, along with reference voltages, to a pair of phase-sensitive detectors, followed by appropriate filters and amplifiers. Only one of these chains is shown explicitly in Fig. 12. The ring-bridge ϕSD[26,27] uses diodes from an integrated-circuit package for convenience; close matching is not necessary. What is necessary is a reference current larger than the signal current, and a low-impedance load provided by the summing point of the operational amplifier. The corresponding input resistor is placed *ahead* of the ϕSD. One of these resistors is made adjustable so that the gains of the two channels can be equalized. The RC feedback network helps to stabilize the servo loop,

[25] See also J. G. Graeme, G. E. Tobey, and L. P. Huelsman, "Operational Amplifiers, Design and Applications," p. 291. McGraw-Hill, New York, 1971.

[26] R. K. Brown, N. F. Moody, P. M. Thompson, R. J. Bibby, C. A. Franklin, J. H. Ganton, and J. Mitchell, *Proc. Inst. Radio Eng.* **47**, 778 (1959).

[27] D. M. Hunten, "Introduction to Electronics," p. 280. Holt, New York, 1964.

by rolling off the gain above 2 Hz to a plateau of lower gain above 70 Hz. At a few hundred hertz the rolloff of the speaker system takes over.

To drive the speakers, a pair of inexpensive silicon transistors act as a complementary emitter follower to boost the power output. The crossover distortion is taken up by the negative feedback.

The velocity signal required for damping the system is isolated from the drive signal by a bridge network. A slight readjustment of the bridge balance allows for the small asymmetry of the differential amplifier. The gain can be adjusted for other drivers by changing the 100-kΩ feedback resistor. The network at the output filters out low and high frequencies. The adjustment of the bridge is critical, but requires no further attention once it has been done. Figure 12 shows coarse and fine adjustments; the former can be omitted if it is preferred to select the fixed resistors. The most convenient way to monitor the bridge balance is to apply a dc offset to the power amplifier, and observe at the "test" terminal. The advantage of a dc signal is that it is blocked from the velocity feedback loop.

Zero-center meters in parallel with each driver give a convenient indication of the telescope errors, and inform the observer when action is required on the slow motions. Only an occasional check at the eyepiece is needed. If desired, the outputs can be filtered of their high-frequency components and connected directly to the telescope, for complete automation. (At a coudé focus, a coordinate rotation may be required.)

The high-voltage supply recommended is an unregulated dc to dc converter, with a simple manual adjustment of the input voltage. Besides being compact, convenient, and inexpensive, it readily lends itself to a closed-loop sensitivity control as suggested above. It is assumed that the 7-V supplies are well regulated, though most of the stages should work equally well on unregulated power. Probably the only critical items are the dc zero adjustments in the coil drivers and buffers; if desired, they could have their own zener regulators at about 5 V.

4.2.6. Pulse-Counting Image Dissector

A different use of the star-tracking photomultiplier has been described by Ball et al.[12] By counting individual photoelectric events, one should be able to obtain a small, but significant, improvement in signal/noise ratio.[27] The penalty that must be paid is a complicated electronic system. Instead of a conical scan, the device uses a cross scan: up and back the X axis, both positive and negative, and then a similar excursion on the

y axis. The required wave forms are easy to produce with digital circuits. The error signals are detected by a reversible counter, counting up and down during the positive and negative deflections respectively.

No sensitivity comparisons have been made between conical-scanning and cross-scanning detectors. There is no obvious reason why either should be better; the choice depends mainly on convenience.

5. DETECTIVE PERFORMANCE OF PHOTOGRAPHIC PLATES*

5.1. Introduction

Over the past century, photographic observations have made profound contributions to astronomy and astrophysics; and for much of this period, photographs were the best light detectors available for a wide range of applications. Recently, a variety of detectors utilizing photocathodes have begun to live up to their anticipated potential, offering detective performance several times better than the best photographic materials. However, it is clear that many astronomers will use photographic techniques for some time to come and for a number of reasons, including lower cost, relative ease of use, and tradition. Also, it should not be overlooked that there is still a large potential for improved utilization of photographic materials, both through the development of new and better materials and also through the improvement of the techniques used with existing materials. Indeed, much progress has been made in both these areas during the past few years, partly as a result of the efforts of the American Astronomical Society's Working Group on Photographic Materials in Astronomy[†] and the Eastman Kodak Company.

Many of the most important advances in astronomy and astrophysics have resulted from visual inspection and measurement of photographs. In recent years, the emphasis has gradually been shifting from visual measurements to pure machine measurements, partly because technological advances have made, for example, precision microphotometers more readily available, and partly because modern astrophysical theories demand more accurate photometry. Most visual measurements are limited to a photometric precision of about 10%, while modern micro-

---

[†] Much of the work of this group has been published in the *AAS Photo-Bulletin*. For information on how to obtain this publication, write to D. W. Latham, Smithsonian Astrophysical Observatory, Cambridge, Massachusetts.

* Part 5 is by D. W. Latham.

photometers can measure to a precision of better than 1% over a much larger range of densities.

The main thrust of this part will be to describe the detective properties of the three types of Kodak Spectroscopic Plates used most often for photometry in astronomy: types 103a-O, IIa-O, and IIIa-J. We will make no attempt to review the many photographic effects on photometry, since a good introduction to this topic is available from Kodak,[1] nor will we review the many techniques and materials used by astronomers, since this topic is covered extremely well in a recent review article by Hoag and Miller,[2] which includes an excellent bibliography. We will consider only those detective properties that are valid for signals that vary slowly across a photograph, that is, we will not consider adjacency effects or modulation transfer properties. Thus, our discussion and results should be relevant for most observations involving spectrophotometry and wide-area photometry, but they will have less obvious relevance for direct stellar photometry. Our basic point is that an experiment or observation involving photographic photometry can be properly designed only if the detective properties of the photographic materials to be used are well known. The many specific results that are quoted in this part may be taken as illustrations of this principle.

5.2. Photographic Photometry

If a photograph is exposed to a signal whose energy as a function of position (x, y) is given by $E(x, y)$, photometry can be defined as the process by which $E(x, y)$ is determined from the photograph.

One of the first steps in a photometric determination is the measurement of the degree of blackening of various positions on the photograph. This is usually accomplished with an instrument that measures the relative fraction of light $T(x, y)$ that is transmitted by a small area centered on (x, y). In many of these instruments, each transmission reading is converted electronically to a density scale

$$D = -\log(T).$$ (5.2.1)

This has lead to the name "microdensitometer," which is an unfortunate

---

[1] Eastman Kodak Company, "Kodak Plates and Films for Science and Industry," Publ. P-315. Rochester, New York.

[2] A. A. Hoag and W. C. Miller, *Appl. Opt.* **8**, 2417 (1969).

choice for the important reason that if the density varies in the area being measured, the output density corresponds to the mean transmission of the area, not to the mean density.[3] Thus, we prefer the name "microphotometer.[4]"

Because of the many variables in the photographic process that can affect the relation between $E(x, y)$ and $D(x, y)$, accurate photometry requires that each photograph be calibrated by exposing an area of it with a set of known exposures E_i.[5] Then the relation between exposure and density,

$$D = \mathscr{D}(E), \qquad (5.2.2)$$

can be inferred from the set of values (E_i, D_i). The fundamental assumption of photographic photometry is that the function $\mathscr{D}$ and its inverse

$$E = \mathscr{E}(D) \qquad (5.2.3)$$

are valid at any position on the photograph, so that the signal exposure that caused density D can be determined by entering D into $\mathscr{E}(D)$.

5.3. Signal-to-Noise and Detective Quantum Efficiency

If a photograph is to be evaluated visually, then it may turn out that the best result is obtained when the photograph is exposed and processed to optimize some photographic parameter. For example, it may be advantageous to develop for high contrast or to expose and process for optimum density of the image above background fog. Indeed, in some cases it may even be useful to introduce photographic effects that would normally make photometry more difficult. For example, it may be advantageous to use a developer with large adjacency effects in order to improve the visibility of fine features.

However, if a photograph is to be measured for the purposes of photometry, then it should be exposed and processed to give the minimum uncertainty in the value derived for the exposure of each image element—

---

[3] R. J. Zinn and E. B. Newell, *AAS Photo-Bull. No. 2,* 6 (1972).

[4] Unfortunately there is no good review paper available on the design and use of microphotometers. Some useful general comments are given by E. W. Dennison [*AAS Photo-Bull. No. 2,* 7 (1970)]. Detailed information about specific microphotometers available commercially can be obtained from the manufacturers listed by D. W. Latham [*AAS Photo-Bull. No. 1,* 28 (1971)].

[5] A. A. Hoag, ed., *Bull. Amer. Astron. Soc.* **1,** 141 (1969).

for example, in the value derived for the total number of photons that caused the image element. It is best for one to think of this uncertainty in the value by imagining that the experiment was performed over and over, that many photographs were exposed, processed, and measured identically. If systematic errors are disregarded, the uncertainty can be defined as the standard deviation of a single value from the mean of all the values derived for the exposure of the image element. The ratio of the value derived for the exposure to the uncertainty is called the *output signal-to-noise*.

It is possible for one to devise ways to evaluate the signal-to-noise without taking many photographs. For example, let us assume that the uncertainty in a photometric measurement is due entirely to the granularity, that is, to the uncertainty in the blackening by the developed silver grains in an image element, and thereby assume that systematic effects such as large-scale nonuniformities, adjacency effects, erroneous calibration exposures, etc. are negligible; then we can evaluate the noise on a uniformly exposed photograph by measuring and deriving the exposure for many independent image elements and then calculating the standard deviation.

Let us now consider in more detail the case where the exposing signal $E(x, y)$ describes a uniform exposure. Of course, a uniform exposure does not imply that every image element received the same total number of photons. If the mean of the total number of photons arriving at many different areas of the same size a is $N(a)$, then the standard deviation of the number arriving in a single area is $[N(a)]^{1/2}$.[6] Thus, if the photograph were a perfect detector, we would find that

$$\mathscr{E}/\sigma_{\mathscr{E}}(a) = N(a)/[N(a)]^{1/2} = E/\sigma_E(a), \qquad (5.3.1)$$

where $\sigma_{\mathscr{E}}$ is the standard deviation of the value inferred for the exposure of a single element of area a.

For a real photograph, the output signal-to-noise [the left-hand side of Eq. (5.3.1)] is always less than the input signal-to-noise (the right-hand side). The detective quantum efficiency[7]

$$DQE = \left[\frac{\mathscr{E}/\sigma_{\mathscr{E}}(a)}{E/\sigma_E(a)} \right]^2 \qquad (5.3.2)$$

[6] L. G. Parratt, "Probability and Experimental Errors in Science," p. 215. Wiley, New York, 1961.

[7] For example, see J. C. Marchant, *J. Opt. Soc. Amer.* **54**, 798 (1964).

is a measure of how badly the actual detector deviates from a perfect detector. It is easier to understand the rationale behind DQE if Eqs. (5.3.1) and (5.3.2) are combined to obtain

$$\mathscr{E}/\sigma_{\mathscr{E}}(a) = DQE \cdot N(a)/[DQE \cdot N(a)]^{1/2}. \tag{5.3.3}$$

In other words, the actual detector performs as if it were an ideal detector with the input signal decreased by the factor DQE.

If one wants to evaluate the performance of a photographic material as a light detector, then a plot of DQE vs. E is perhaps the single most important piece of information. However, if one wants to evaluate the suitability of a photographic material for a specific photometric problem, then DQE vs. E can be a bit clumsy, and it is usually easier to use the output noise-to-signal $\sigma_{\mathscr{E}}(a)/\mathscr{E}$ expressed as a percentage. For example, a plot of $\sigma_{\mathscr{E}}(a)/\mathscr{E}$ vs. E allows an experimenter to determine immediately what percent photometric precision he can expect for the value determined for the signal that exposed a single image element, at any exposure level. In many practical situations, a plot of $\sigma_{\mathscr{E}}(a)/\mathscr{E}$ vs. D is even more useful because it allows an experimenter to estimate what kind of photometric precision he can expect at various density levels without having to know a priori the absolute exposure needed to generate the densities.

Quantitative information about the detective performance [DQE or $\sigma_{\mathscr{E}}(a)/\mathscr{E}$] of the photographic materials used most often in astronomy is extremely meager. What few data there are show the following two important features: First, the maximum DQE of the most sensitive spectroscopic plates made by Eastman Kodak Company is on the order of 1%.[8] Second, for at least one of these types of plates, the DQE is largest at the rather low specular density of about 0.3 to 0.4.[9]

Especially lacking are data on detective performances at high densities and on the effects the many and varied hypersensitization techniques have on detective performance. Moreover, most of the scanty published data have been derived by using shortcuts to avoid the rather large computational problem of doing the statistics "properly" for the determination of $\sigma_{\mathscr{E}}$. For several years, we have wanted to take advantage of the digitized microphotometer and computerized reduction procedures available at the Harvard College and Smithsonian Astrophysical Observatories in order to evaluate the detective performance of the photographic materials used most often in astronomy. Such a program was initiated

[8] J. C. Marchant and A. G. Millikan, J. Opt. Soc. Amer. **55**, 907 (1965).
[9] H. D. Ables, A. V. Hewitt, and K. A. Janes, AAS Photo-Bull. No. 1, 18 (1971).

recently, and some of the preliminary results are presented in the next section. However, an enormous amount of work remains to be done in this field, especially for the evaluation of the different techniques of hypersensitization.

5.4. Detective Performance of Kodak Spectroscopic Plates, Types IIa-O, 103a-O, and IIIa-J

In this section, we present new DQE and noise-to-signal data for three of the most popular photographic materials used in astronomy: Kodak Spectroscopic Plates, types IIa-O, 103a-O, and IIIa-J. All the density measurements reported here were made with the David Mann microphotometer at the Harvard College Observatory, and the data were reduced on the CDC 6400 at the Smithsonian Astrophysical Observatory with a package of programs called MICRO.[10]

5.4.1. Accuracy and Precision of the Detective Performance Determinations

All the measurements of detective performance were made with a slit nominally 7×200 μm at the emulsion. Although the effective area of the slit may have been somewhat different from the nominal value of 1400 μm² because of various instrumental effects such as uncertainties in the actual slit settings and the usual image spreading introduced by the optics, we expect that this value is accurate within 25%. For example, Fig. 1 indicates that when the optics are in good focus, they typically introduce an image spread of about 1 μm. The response function in Fig. 1 was derived by numerical differentiation of a transmission tracing of a knife edge, with nominal slit widths of 1.5 and 0.5 μm for the illuminating and pickup slits, respectively. We conclude that the uncertainty in the value for the effective area of the measuring slit has not introduced a systematic error much larger than 10% into our DQE and noise-to-signal determinations.

The precision to expect in our determinations of detective performance values is indicated in Fig. 2, which is a plot of output noise-to-signal on an inferred exposure scale $\sigma_{\mathscr{E}}(a)/\mathscr{E}$ vs. the inverse root of the area a of the microphotometer measuring slit. The same uniformly exposed spot on a IIa-O plate was measured with 17 different slit configurations

---

[10] D. W. Latham, *Proc. Colloq. Int. Astron. Un., 11th, August 1970* p. 183; *Publ. Roy. Obs. Edinburgh* **8** (1971).

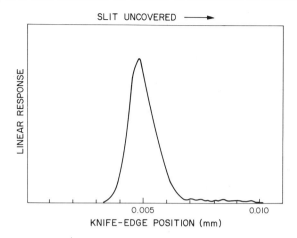

Fig. 1. Image spread of the microphotometer optics determined by numerical differentiation of a transmission tracing of a knife-edge.

covering a range of 50 in the slit area. Different slit heights and different slit widths were used, so that Fig. 2 includes nearly square slits as well as long skinny ones. For each slit configuration, the uniform spot was traced, and a few hundred density readings for positions separated by the slit width were recorded on magnetic tape. Tube sensitometer spots on the same plate were traced with the same slit settings to provide a photometric calibration. In the computerized reductions of these digitized

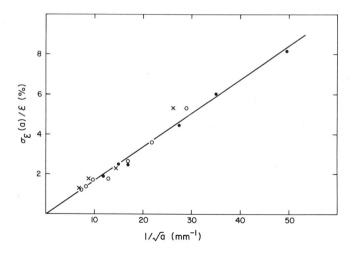

Fig. 2. Dependence of noise-to-signal on the area of the microphotometer pickup slit with differing widths of 7 (●), 18 (○), and 25 (×) μm for the IIa-O plate.

data, each density reading was converted to an inferred exposure by use of the characteristic curve derived from the tube sensitometer calibration spots; then the standard deviation $\sigma_{\mathscr{E}}(a)$ of a single exposure from the mean inferred exposure $\mathscr{E}$ was calculated. This same basic procedure was used to determine all the detective performance values reported in this section. We conclude from Fig. 2 that our noise-to-signal determination have a precision considerably better than 1% in the value of $\sigma_{\mathscr{E}}(a)/\mathscr{E}$.

Figure 2 demonstrates something much more fundamental and useful than just the precision to expect in our noise-to-signal results. Equation (5.3.1) predicts that, for an ideal detector, $\sigma_{\mathscr{E}}(a)/\mathscr{E}$ should vary directly as $1/a^{1/2}$, since for a uniform exposure the total number of registered photons should vary directly as the sampling area. This is confirmed experimentally by the fact that a line passing through the origin of Fig. 2 fits the data. This suggests that the noise-to-signal data presented in this section can be adjusted to different sampling areas by simply scaling by the square root of the area. Of course, some caution must be exercised when scaling the noise-to-signal data to very large slit areas, because effects such as large-scale nonuniformities may limit the detective performance. For example, published data indicate that even with good methods of agitation during development, large-scale nonuniformities limit the attainable noise-to-signal to about 1%.[11]

5.4.2. Detective Performance of IIa-O and 103a-O Plates for Different Developers and Developing Times

In 1965, we produced a set of 40 test plates in order to compare the performance of IIa-O and of 103a-O with four different developers and five different developing times appropriate to each developer. Each 10×12.5-cm plate was exposed identically to blue light in a tube sensitometer and then developed with nitrogen-burst agitation. Recently, these plates were all remeasured and the detective performances determined.

The characteristic curves for five IIa-O plates developed in D-76 are plotted in Fig. 3. The main reason we show these curves is to illustrate how the classical photographic speed, defined as the reciprocal of the exposure required to produce a density of 0.6 above fog, depends strongly on the developing time. The longer a plate is developed, the higher the speed. Indeed, the plate developed 30 min was 3 times faster than the one developed only 5 min.

---

[11] W. F. van Altena, *AAS Photo-Bull. No. 1*, 3 (1971).

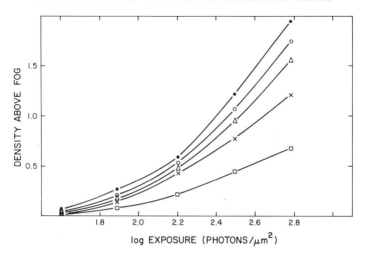

FIG. 3. Characteristic curves for identical exposures on IIa-O plates developed different lengths of time in D-76 with nitrogen-burst agitation: 30 (●), 20 (○), 15 (△), 10 (×), and 5 (□) min.

In contrast to the strong dependence of speed on developing time, the data plotted in Fig. 4 demonstrate the striking result that the output noise-to-signal for the same five plates depends only on the density above fog and has essentially no dependence on developing time at a given density. Moreover, this conclusion does not seem to be limited to

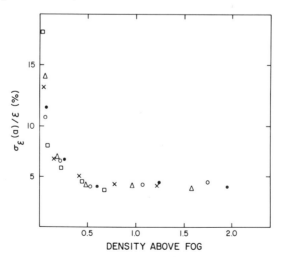

FIG. 4. Noise-to-signal as a function of density above background fog for IIa-O plates developed in D-76: 30 (●), 20 (○), 15 (△), 10 (×), and 5 (□) min.

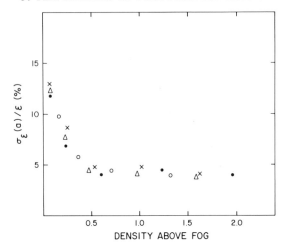

FIG. 5. Noise-to-signal vs. density for IIa-O plates processed with different developers D-76, 30 min (●); D-19, 8 min (○); HC-110, 5.5 min (×); Acufine, 10 min (△).

just differences in developing time. Figure 5 shows that at a given density above fog, the output noise-to-signal has essentially no dependence on developer type.

One of the most striking features of Fig. 5 is that the noise-to-signal performance of IIa-O is virtually constant for densities greater than about 0.4 above fog. This means that it is simple to predict the photometric accuracy of a well-exposed spectrum on IIa-O. If the slit width at the plate is 7 μm and the widening is 200 μm, then each resolution element will have a photometric accuracy of about 4%. To calculate the accuracy for other slit widths and widenings, it is only necessary to scale this number by the square root of the area. For example, for a slit width of 30 μm and a widening of 800 μm, the photometric accuracy of each resolution element should approach 1%.

If IIa-O plates are being used for photometry of very faint objects, then Fig. 5 shows that the developer type and developing time can be chosen to bring up the density of the faintest images. In this case, speed is an excellent indicator of the best choice.

If plenty of light is available, but the scene to be photographed covers a wide range of intensities, then Fig. 5 suggests that the developer type and developing time can be chosen for low contrast in order to increase the dynamic range of the plate.

The noise-to-signal performance of 103a-O is quite similar to that of IIa-O at the same density above fog. This is demonstrated by Fig. 6,

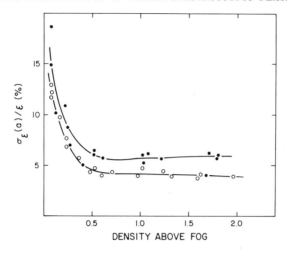

FIG. 6. Noise-to-signal vs. density for IIa-O plates (○) and various developers compared with 103a-O plates (●).

where we have replotted the IIa-O data from Fig. 5, along with the exactly corresponding data for 103a-O. Figure 6 suggests that a well-exposed IIa-O plate provides a photometric accuracy about one-third better than a well-exposed 103a-O plate.

If the noise-to-signal performance of a plate depends only on the density above fog, then any change in photographic technique that affects the shape of the characteristic curve must also affect the $D\underset{\sim}{Q}E$. For example, Fig. 7 shows the $D\underset{\sim}{Q}E$ for IIa-O developed different amounts in D-76. Although the $D\underset{\sim}{Q}E$ data in Fig. 7 should be quite accurate relative to each other, there is a large uncertainty in the absolute $D\underset{\sim}{Q}E$ scale. Since we did not perform an absolute calibration of the exposures provided by our tube sensitometer, it was necessary to adopt the very approximate value of 500 photons/μm^2 as the exposure needed to produce density 1.0 on IIa-O when developed for 6 min in D-19.[12] A new calibration of the spectral sensitivity of IIa-O plates was received from Kodak too late to be incorporated into the calculations. The new data indicate that a value somewhat less than 400 photons/μm^2 would have been more appropriate. This suggests that our absolute $D\underset{\sim}{Q}E$ scale is systematically in error by a factor of 1.1 or 1.2. Indeed, it would disappoint but not surprise us if the absolute $D\underset{\sim}{Q}E$ values presented here prove to be in error by as much as a factor of 1.5. The strong dependence of the peak

[12] J. H. Altman, private communication (1972).

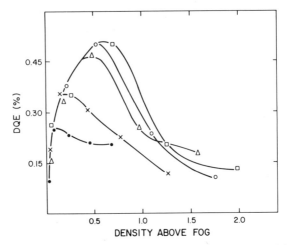

FIG. 7. *DQE* vs. density for IIa-O plates developed in D-76 at 5 (●), 10 (×), 15 (△), 20 (○), 30 (□) min.

DQE on the developing time shown in Fig. 7 is typical of the other three developers tested. Furthermore, there are significant differences in the peak *DQE* provided by the different developers. Figure 8 shows the *DQE* for the four developers plotted at the best developing time for each developer. D-76 comes out on top and is about twice as good as D-19 for densities less than 1. The similar *DQE* results for 103a-O are plotted in Fig. 9. The *DQE* of 103a-O is systematically less than that of IIa-O by about 50%.

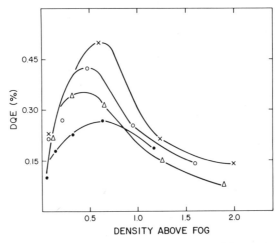

FIG. 8. *DQE* vs. density for IIa-O plates processed with different developers: D-19, 5.5 min (●); D-76, 30 min (×); HC 110, 8 min (△); Acufine, 10 min (○).

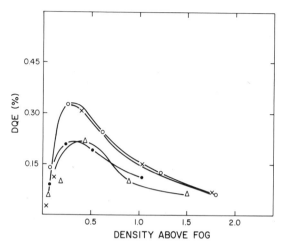

FIG. 9. *DQE* vs. density for 103a-O plates processed with different developers: D-19, 8 min (●); D-76, 30 min (×); HC 110, 4 min (△); Acufine, 10 min (○).

5.4.3. Detective Performance of IIIa-J and IIa-O

A recent investigation by Smith *et al.*[13] shows that the speed of Kodak Spectroscopic Plate type IIIa-J can be increased as much as a factor of 5 by its being baked in an atmosphere of dry nitrogen for approximately 16 hr at 65°C. Several of the original test plates from these experiments

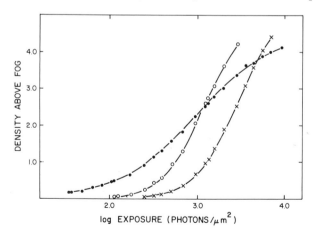

FIG. 10. Characteristic curves for baked IIIa-J plates (○) compared with untreated IIIa-J (×) and IIa-O (●).

[13] A. G. Smith, H. W. Schrader, and W. W. Richardson, *Appl. Opt.* **10**, 1597 (1971).

were lent to us so that we could measure and compare the detective performance of baked and unbaked IIIa-J plates with that of unbaked IIa-O. The characteristic curves for three of these plates are plotted in Fig. 10. Once again, an absolute calibration of the tube sensitometer exposures was not available, and only the relative exposures are accurate. The slope of the IIa-O characteristic curve is not so steep as that of the IIIa-J, because the photographic grains in IIa-O cover a fairly wide range of sizes while those in IIIa-J are nearly all the same size. Although baking increases the speed of IIIa-J dramatically, the baked plate is still slower than IIa-O for the range of densities normally used by astronomers for photometry. However, the output noise-to-signal performance of IIIa-J is about 4 times better than that of IIa-O, as shown by Fig. 11. The most important conclusion to be drawn from Fig. 11

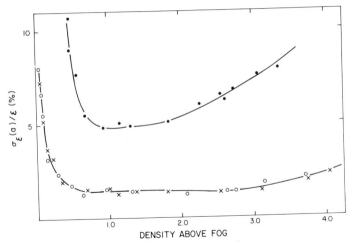

FIG. 11. Noise-to-signal vs. density for IIIa-J untreated ($\times$) and baked ($\bigcirc$) and IIa-O ($\bullet$).

is that the noise-to-signal characteristics of IIIa-J are not changed at all by baking, and depend only on the density above fog. Thus, the increase in speed afforded by baking reflects directly as an increase in DQE, as shown by Fig. 12. Another interesting feature of Fig. 11 is that the noise-to-signal performance of IIIa-J is excellent out to extremely large densities. This means that if the microphotometer used to measure the plate can be assumed to handle densities as high as 4 and 5, then IIIa-J can be used for accurate photometry over a large dynamic range of intensities even though the slope of the characteristic curve is very steep.

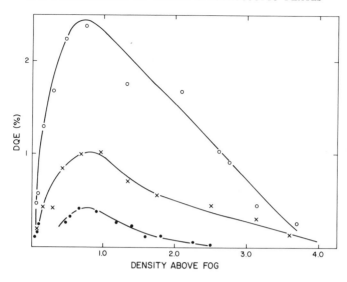

FIG. 12. *DQE* vs. density for baked IIIa-J (○) compared with untreated IIIa-J (×) and IIa-O (●).

Astronomers using photography usually find themselves waiting for photons to arrive and be detected. The waiting time can be minimized by designing the experiment so that the photographic plate is used at its maximum *DQE*. We have shown that the proper choice of developer and developing time can shorten the waiting time by a factor of 2. This assumes, of course, that the exposure time and plate scale can be chosen so that each resolution element receives the number of photons needed to produce the density at which *DQE* is a maximum. The characteristic curves and *DQE* data presented above are sufficient to allow approximate determinations of these parameters for specific experimental situations.

6. TWO-DIMENSIONAL ELECTRONIC RECORDING

6.1. Phosphor Output Image Tubes*

6.1.1. Introduction

Over the past decade the astronomical community has become increasingly receptive to the use of image tubes for astronomical research. No longer is their use confined to the few technically oriented astronomers; today they are considered a part of the standard observatory instrumentation and are used, even by technically unsophisticated researchers, for a wide variety of astronomical problems. This change resulted from the introduction of high-gain, high-resolution, tubes with phosphor output screens. Such tubes are comparatively rugged, have large output formats, and require no special knowledge of high-vacuum techniques for their use or maintenance.

By 1966 two groups were using phosphor output image tubes to dominate the observational side of a young and extremely exciting branch of astrophysics: the study of quasi-stellar objects. Livingston and co-workers at the Kitt Peak National Observatory[1] were using a six-stage high-gain transmission secondary emission (TSE) image tube manufactured by the English Electric Valve Co., and Ford, together with the Carnegie Image Tube Committee,[2] had successfully guided an RCA effort that resulted in a two-stage tube that used a phosphor screen/photocathode sandwich to produce photoelectron amplification. An excellent review article by Livingston[3] covers image tube developments to 1967. There have been some major advances in the intervening years,

[1] W. C. Livingston, C. R. Lynds, and L. A. Doe, in "Advances in Electronics and Electron Physics," Vol. 22, p. 705. Academic Press, New York, 1966.

[2] W. K. Ford, Jr., in "Advances in Electronics and Electron Physics," Vol. 22, p. 697. Academic Press, New York, 1966.

[3] W. C. Livingston, in "Advances in Electronics and Electron Physics," Vol. 23, p. 347. Academic Press, New York, 1967.

* Chapter 6.1 is by E. J. Wampler.

but the Livingston article describes many of the fundamental processes and limitations of these devices.

There exists a certain lack of critical assessment of the performance of image tubes by manufacturers eager to market their products. Photocathode sensitivity, dark emission, response uniformity, and life often do not meet the manufacturer's specifications, or the specifications are too vague to be useful. All image tubes are expensive and manufacturers are sometimes reluctant to ship tubes on consignment to prospective purchasers for evaluation. There are large variations in the performance of tubes constructed by different (or even the same) manufacturer. This increases the difficulty of writing a review article. Some of the consequences of the internal physical processes of the tube can be described in a general way and difficulties that we or others have encountered may be noted. Personal pronouns have been used to indicate those statements that are subjective judgements. In the final analysis, performances quoted here were obtained with selected tubes having characteristics that may (or may not) be available later. The prospective user is warned to evaluate tubes before purchase, even if the evaluation requires a trip to the place of manufacture.

6.1.2. Principles of Operation

Phosphor screens can convert the kinetic energy of an incident electron into light with an efficiency of about 10%.[4] Since the emitted light has a photon energy of a few electron volts, an efficient phosphor bombarded by electrons will emit photons with an efficiency of 0.03 to 0.05 photons/eV. Because it is possible to accelerate electrons released by a photocathode to energies of many thousand electron volts, it is possible to construct practical single stage tubes with absolute photon gains of 20 to 50 when operated with an accelerating potential of 15 kV. The major losses in these tubes are in the low yield of the photocathode (about 10% of the incident photons are converted to photoelectrons) and in the several kilovolts required for the photoelectrons to penetrate the aluminum film deposited on the back of the phosphor screen that prevents light feedback to the photocathode. In order to achieve very high photon gain, several tubes must be cascaded or electron-multiplying sandwiches must be inserted between the photocathode and the phosphor screen.

---

[4] P. C. Ruggles and N. A. Slark, *Electron. Components* **5**, 294 (1964).

The simplest tube is one that uses proximity focusing. The phosphor screen is located very close to the photocathode. Photoelectrons released by the photocathode are accelerated to the phosphor screen by a uniform axial electric field. If the tangential velocity of the photoelectron when it is released by the photocathode is sufficiently low, there will be little transverse migration of the photoelectron before it reaches the phosphor screen. Because the transverse velocity increases with increased energy of the input photon, the resolution of the tube is a function of the color of the illuminating source, decreasing as the photon energy increases. The disadvantage of proximity focusing is that the gap between the photocathode and the phosphor screen must be made very small to achieve good resolution. The field gradients then become large, and dark emission from the photocathode and field emission is increased. Commercial tubes can achieve a limiting resolution (the point at which the modulation transfer function has dropped to about 5%) of 25 line pairs/ mm with photon gains of approximately 10–20.[5]

A variation of this tube uses a bundle of hollow tubes (called a *channel plate*) between the photocathode and the phosphor screen to achieve electron gain.[6] Photoelectrons from the photocathode enter the micron-sized tubes and strike the walls to produce secondary electrons that are accelerated down the tube and produce additional secondaries as they in turn impinge on the tube walls. The electron gain of such bundles can exceed several thousand. Tubes incorporating this principle have been developed by the Army Night Vision Laboratory at Fort Belvoir, Virginia. Photon gains can exceed 10^5 with a limiting resolution of the order of 25 line pairs/mm. This family of tubes (called *Gen II image tubes* by the Army) has been recently declassified and can now be purchased by civilian users. They suffer from gain nonuniformities introduced by the channel plate, high background produced by ion feedback through the channel plate, and vignetting of input photoelectrons by the geometrical properties of the channel plate. As many as 50% of the input photoelectrons can fail to produce an output scintillation.[6] This is equivalent to reducing the quantum efficiency of the photocathode and is important if the major source of noise is the input shot noise of the signal. The advantages of the tube are that it is compact, extremely rugged, and

[5] M. J. Needham and R. F. Thumwood, *in* "Advances in Electronics and Electron Physics," Vol. 28, p. 129. Academic Press, New York, 1969.

[6] B. W. Manley, A. Guest and R. T. Holmshaw, *in* "Advances in Electronics and Electron Physics, Vol. 28, p. 471. Academic Press, New York, 1969.

produces high gain with comparatively low accelerating voltages. Bendix is developing an improved channel plate that should reduce the ion feedback problem.[7]

By focusing the photoelectrons, the resolution of image tubes can be greatly improved. The resolution of a well designed tube seems to be limited by the resolution of the high-efficiency phosphor screen used in the tubes. Tubes can be made with a resolution exceeding 100 line pairs/mm, (E. H. Eberhardt, private communication), although more typically the resolution is 50–70 line pairs/mm.

The most convenient focused tube is one that uses electrostatic focusing. A family of electrostatically focused tubes has been developed to a high state of perfection by the Army Night Vision Laboratory. These tubes (called *Gen I* by the Army) have been produced in enormous quantities by several firms. This fact ensures the scientific user that he will be able to select good tubes at a reasonable cost. Fiber optics faceplates are used to match the curved surfaces required by the electron lens to flat input and output formats. The tubes suffer from aberrations in the electron image[8] and vignetting[9] in the fiber optics coupling plates. De Veny[10] has found radial gain variations in excess of a factor of 10 over the faceplate of the Westinghouse WL-30677. Our measurements of the uniformity of selected Varo tubes (Model 8605) show total variations less than a factor of 2. The use of fiber plates allows the coupling of two or three modules to achieve photon gains as high as 10^5.

Image tubes may also be focused magnetically. Both the TSE tube used by Livingston *et al.*[1] and Powell and Lynds[11] at Kitt Peak and the "Carnegie" tube developed by Ford[2] use magnetic focusing. Magnetically focused tubes use flat input and output plates, an advantage when ultraviolet transmission is required, because high-quality image tubes using fiber optics plates with good ultraviolet transmission are not yet available although there are some attempts at developing fiber plates with good transmission to about 3000 Å. Magnetically focused tubes can be constructed with good imaging qualities over a very large area. Single-stage

[7] C. E. Catchpole and C. B. Johnson, *Publ. A.S.P.* **84**, 134 (1972).

[8] W. M. Wreathall, *in* "Advances in Electronics and Electron Physics, Vol. 22, p. 583. Academic Press, New York, 1966.

[9] D. L. Emberson and B. E. Long, *in* "Advances in Electronics and Electron Physics," Vol. 28, p. 119. Academic Press, New York, 1969.

[10] J. B. De Veny, *Publ. A.S.P.* **82**, 142 (1970).

[11] J. R. Powell and R. Lynds, *in* "Advances in Electronics and Electron Physics," Vol. 28, p. 745. Academic Press, New York, 1969.

tubes for direct photography with clear apertures as high as 144 mm have been developed by ITT (ITT F4094). The primary disadvantage of these tubes, particularly the larger versions, is the large magnet required for their operation. The photocathode must be in a region of relatively uniform magnetic field. This requirement increases the difficulty of designing cooling systems, and, if the tube is to be used for spectroscopy, the design of the spectrograph camera is complicated.[12] High light gain may be achieved either by coupling diode tubes with fiber optics plates, or by incorporating one or more electron multiplying sections within the tube itself. The advantages of the modular approach are that the increased production line yield of the simpler tubes decreases their price and the experimenter can select modules with characteristics suited to his needs.

6.1.3. Astronomical Applications

6.1.3.1. Tubes with Conventional Faceplates. For the purposes of this discussion image tubes will be divided into two categories, those that use fiber optics plates and those that use conventional homogeneous faceplates. As noted above the first tubes to be extensively used for astronomical research were the P 829D TSE image tubes manufactured by the English Electric Valve Co., and the RCA C33011 tube developed by RCA for the Carnegie committee. During this early period high-resolution fiber plates were still classified by the US Army and consequently were not generally available to the astronomical community. The early magnetically focused tubes use conventional glass faceplates and a transfer lens is required to image the output screen on a photographic emulsion. The collection efficiencies of such lenses are low, typically of the order of 3%. Even to achieve this efficiency, lens aberrations substantially reduce the resolution of the system. For instance, the resolution on the output screen of the C33011 is approximately 40 line pairs/mm, but this is reduced to about 25 line pairs/mm by the transfer lens and the photographic emulsion. Quite high photon gains are required of the image tube in order to compensate for the losses introduced by the transfer lens.

The P 829D uses 5 KCl transmission dynodes on alumina support films to provide an average electron gain of about 3000. The total photon

[12] C. G. Wynne and M. J. Kidger, *in* "Advances in Electronics and Electron Physics," Vol. 28, p. 759. Academic Press, New York, 1969.

gain is then about 150,000. When used with fast emulsions this is sufficient to expose a number of photographic grains for each event, even following the light losses introduced by the transfer lens. The limiting resolution of selected tubes has been claimed to be between 40 and 70 line pairs/mm, but in actual use the degradation introduced by the transfer lens and the photographic plate reduce the resolution of the tube to about 30 line pairs/mm.

An additional disadvantage of the tube is that a large number of electrons are apparently scattered in the multiplication process and produce a background that is dependent on the input light level.[1,13] This, together with the light that is transmitted by the photocathode and subsequently scattered back to the photosurface by the internal structures of the tube, adversely effects the contrast performance of the tube. If the tube is used to obtain spectra of stellar objects with absorption features, the depths of the absorption features are reduced by the light-dependent background. An alternate way of describing this phenomenon is to quote the zero frequency modulation transfer function. Unfortunately, tube manufacturers indicate that at very low frequencies the modulation transfer function approaches 100%; they usually do not mention the scattered light problem.

The information gain of the P 829D is further reduced by the fact that the intensity distribution of output scintillations is exponential in form rather than the characteristic peaked distribution expected from a multiplication process obeying Poisson statistics.[14,15] The cause of this deviation from the expected peaked distribution is not understood.[16] In addition Reynolds[15] reports that the tubes suffer from the complete loss of a large fraction of the input photoelectrons in the multiplication process. Some tests[17] indicate that only 30% of the input photoelectrons actually produce a detectable event. The resulting output scintillations have a wide variation in amplitude. This increases the noise-in-signal if ordinary microphotometry is used to determine the intensity of spectral features. Living-

[13] P. H. Batey and N. A. Slark, in "Advances in Electronics and Electron Physics," Vol. 22, p. 63. Academic Press, New York, 1966.

[14] F. J. Lombard and F. Martin, Rev. Sci. Instrum. 32, 200 (1961).

[15] G. T. Reynolds, in "Advances in Electronics and Electron Physics," Vol. 22, p. 71. Academic Press, New York, 1966.

[16] W. L. Wilcock and D. E. Miller, in "Advances in Electronics and Electron Physics," Vol. 28, p. 513. Academic Press, New York, 1969.

[17] P. Iredale, G. W. Hinder, and D. J. Ryden, IEEE Trans. Nucl. Sci. NS-11, 139 (1964).

ston has attempted to count individual events in order to improve the accuracy of the reduction process, but he reports[3] that gain nonuniformities severely limit the precision achievable by this technique. Finally the tubes are handicapped by severe spiral distortion and a rather small (2.5-cm) photocathode.

Some of the defects of the P 829D are largely overcome by newer tubes. The RCA C70021 is a 38-mm, three-stage, magnetically focused tube with phosphor screen/photocathode sandwiches used to obtain radiant light gain in excess of 2×10^5. Lynds (private communication) reports that the C70021 has less spiral distortion than the P 829D and an output pulse height distribution that is more characteristic of a Poisson multiplication process. The disadvantages of the C70021 are that the tube has higher background, slightly poorer (25 line pairs/mm) resolution, poorer contrast performance (the degradation of the modulation transfer function by scattered light electrons), and higher dark current than the P 829D.

The C70021 is one of a family of multiple stage, magnetically focused tubes developed by RCA. The two-stage "Carnegie" tube (C33011) is a member of this family. The light gain of the C33011 when used with a relay lens seems to be insufficient to produce an exposed grain for each photoelectron event. The use of lower gain produces photographic images that appear similar to those obtained by conventional photography. The resolution of the image tube, relay lens, and photographic plate combination is about 25 line pairs/mm, although the image tube by itself is capable of resolving approximately 40 line pairs/mm. After taking into account the resolution loss introduced by the image tube system, Ford estimates a blue information gain of approximately 10. Tests of this system at the 300-cm (120-in.) coudé spectrograph at Lick Observatory indicate that for the tube we used, the information gain in the blue was hardly more than unity. Because the tube had an S-20 photocathode, the information gain is much higher in the red. Burbidge and Kinman[18] have incorporated the C33011 in a low-dispersion spectrograph used at the prime focus of the 300-cm (120-in.) telescope. This instrument proved very effective for identifying quasi-stellar sources, a task that requires extended wavelength sensitivity.

The poor contrast performance of the C33011 that Lick Observatory owns is very striking. Livingston[3] has also commented on this problem and gives a quantitative comparison (in his Fig. 6) of spectra obtained

[18] E. M. Burbidge and T. D. Kinman, *Astrophys. J.* **145**, 654 (1966).

by direct photography and those obtained using the image tube. This deficiency of the C33011 severely limits its usefulness in many areas of astronomical spectroscopy.

6.1.3.2. Tubes Using Fiber Optics Plates. In 1969 tubes with high-resolution (6 μm) fiber optics plates became available for use in unclassified research. This seems to represent a major breakthrough in the development of image tubes for astronomical applications. Good light transfer with high resolution can be achieved by pressing a photographic plate into direct contact with the output fiber of the faceplate. The resolution in readily available tubes is limited to about 70 line pairs/mm primarily by the grain in the phosphor screen. Because the efficiency of light transfer is high, fewer stages are needed to achieve high-speed gains. A single-stage tube is comparable in effective photon gain to the C33011; a two-stage system can record single-photon events on a photographic emulsion. The use of high-efficiency phosphor screens results in a peaked distribution in the brightness of the output scintillations.[19-21]

The fiber optics tubes have much better contrast performance than tubes using conventional faceplates. In part this is due to the inability of photons trapped in the faceplate to diffuse along the photocathode. In part it results from the use of thick photocathodes that decrease the amount of transmitted light that can scatter back to the cathode and to the careful attention to ensuring that the internal surfaces of the tube are black. Finally, it results from the decrease in the number of stages required to produce the required information gain.

However, fiber plates can suffer from a large number of deficiencies. Among the most important of these are: shear dislocations between adjacent parts of the filter plate, large blemishes, random spatial variations of the transmission, and pronounced "chicken wire" or "honey comb" patterns produced by bonding interfaces. In addition, some early plates were radioactive, a condition that resulted in bright scintillations from a dark cathode. By careful tube selection these difficulties can be reduced or eliminated. A tube with a uniformly illuminated photocathode should produce an image that appears smooth and "creamy" when examined either with the naked eye or with a 2.5-cm (1-in.) focal length eyepiece. A Ronchi ruling in optical contact with the entrance faceplate produces a pattern that is useful in checking for shear dislocation in either faceplate.

[19] G. T. Reynolds, *IEEE Trans. Nucl. Sci.* **NS-13**, 81 (1966).
[20] J. McNall, L. Robinson, and E. J. Wampler, *Publ. A.S.P.* **82**, 837 (1970).
[21] S. R. Smith and J. L. Lowrance, *Publ. A.S.P.* **84**, 154 (1972).

Tubes from different manufacturers, or even different tubes from the same manufacturer, vary widely in their ability to pass these simple tests. The situation is reminiscent of the early days of photoelectric photometry when it was necessary to test a large number of 1P21 photo-multipliers in order to obtain useful tubes.

A great advantage of the fiber faceplate tubes is that they are modular in design. High-gain image tubes may be constructed by optically contacting a number of single-stage tubes together. By carefully selecting the single-stage tubes it is possible to obtain high-quality multistage units at a modest cost.

6.1.3.2.1. MAGNETICALLY FOCUSED TUBES. The Hale Observatories have used an ITT F4708 at the prime focus of the 500-cm (200-in.) telescope to decrease the exposure time required to obtain sky-limited direct photographs. According to Oke (private communication) the 40-mm F4708 has performed very satisfactorily. Sky-limited exposures can be obtained in approximately 10 min, a speed gain of 6 to 10. Because the input flux level is high the dark current of the tube is not a serious problem, even without cooling. Gunn[22] has published a photograph of a faint cluster of galaxies that was obtained using the image tube. His photograph may be compared with one taken by Arp[23] using conventional photography.

Magnetically focused image intensifiers can be made with very large formats. ITT produces a tube, available with a fiber output plate, with a useable area 144 mm in diameter (F4094). A 90-mm tube is available both in single- (F4092) and two- (F4093) stage versions. Distortion correction in these tubes is achieved through the use of internal electrodes. Typical figures for linear and spiral distortion, in the 144-mm tube, are $<0.5\%$ and <200 μm, respectively. Similar figures for the 38-mm RCA C70021 are 4.0% and 125 μm, respectively. The paraxial limiting resolution of large single-stage tubes is very good, 80 line pairs/mm are typical and for some tubes the resolution may exceed 100 line pairs/mm (E. H. Eberhardt, private communication). Because the faceplates of the tubes are flat, ultraviolet transmitting input windows may be used in the first stage. Also, image deterioration caused by vignetting and transmission irregularities in the comparatively long fibers near the edge of the faceplate in electrostatically focused tubes is avoided.

[22] J. E. Gunn, *Astrophys. J.* (*Lett.*) **164**, L113 (1971).
[23] H. Arp, *Astrophys. J.* **162**, 811 (1970).

For tubes with apertures below 50 mm the advantages of the magnetically focused tubes must be balanced by the disadvantage of the magnetic focusing assembly and by the greater experience in tube manufacture that has resulted from the US Army demand for electrostatically focused tubes. For large aperture tubes the advantages of magnetic focusing probably overcome the convenience of electrostatic focusing. With either type of focusing the demand for large aperture tubes is small and the manufacturing experience is limited. Without Federal development funds the outlook for future improvement, or even the continued manufacturing, of high-quality large-aperture tubes is bleak.

For direct photography with large telescopes, the large-scale format requires image tubes with large apertures. For spectroscopy, high-resolution tubes allow the use of fast camera systems. Because exposures on faint objects at a dark site may exceed several hours, it may be necessary to cool the image tube to reduce the dark current. The magnetic focusing designs of Baum[24] show that the focusing structure must extend an appreciable distance in front of the photocathode and any magnetic shielding that is required must be situated well away from this structure. For high-resolution image tubes the shielding and focusing requirements are even more stringent than those of earlier tubes. These considerations severely constrain the design of spectrograph cameras. Cassegrain cameras for image tubes have been designed by Wynne and Kidger,[12] and although they are very fast they suffer from large central obscuration ratios. This is equivalent to reducing the aperture of the telescope. An $f/1.4$ semisolid Cassegrain Schmidt camera with 25% central obscuration has been designed by I. S. Bowen and is being manufactured by Boller and Chivens. Some of these cameras may be sufficiently small that they could fit inside the magnetic focusing structure for this image tube. Because the aberrations of fast cameras are angular aberrations, there are advantages to limiting the size of the collimator bundle and exploiting the good resolution of the newer tube.

6.1.3.2.2. ELECTROSTATICALLY FOCUSED TUBES. The Westinghouse WL 30677 was the first high-resolution tube with fiber optics faceplates to gain the attention of astronomers. It was a 40-mm input photocathode and a 25-mm output screen. The paraxial magnification is 0.65 and the tube is a "low-distortion" design, the variation of the magnification from center to edge is less than 2%. The resolution (referred to the input

[24] W. A. Baum, in "Advances in Electronics and Electron Physics," Vol. 22, p. 617. Academic Press, New York, 1966.

faceplate) is about 25 line pairs/mm; comparable to the resolution of the "Carnegie" tube.

The WL 30677 has very good contrast performance, much better than that obtained by the "Carnegie" tube. Beardsley *et al.*[25] have used the WL 30677 to obtain spectra at the Allegheny Observatory. Lynds[26] has obtained spectacular direct plates of NGC 1275 in Hα light with the device. As noted above, De Veny[10] found appreciable vignetting in the WL 30677. The severe vignetting, presumably in the fiber optics bundles, may be caused by the low distortion design (see Fig. 6 of Emberson and Long[9]).

At Lick Observatory a Varo 40-mm Gen I tube (the 8605) is in use at the coudé focus of the 300-cm (120-in.) telescope. Zappala[27] has reported on the early performance of the tube. More recent plates, obtained with better cathode cooling, show lower background. This is important because the thermal background exhibits considerable spatial structure that degrades the quality of the spectrum if it is allowed to contribute significantly to the density of the exposed spectrogram. The structure in the dark emission is much more pronounced than any structure seen when the tube is exposed to light. It appears as a mottled pattern with typical dimensions that are characteristic of the pattern in the fiber bundle, although the fiber structure is not clearly seen. Presumably the fiber pattern influences the thermal emission properties of the photocathode.

The Varo tubes have very high infrared sensitivity; selected tubes have appreciable response to 9000 Å. Typical luminous sensitivities exceed 250 µA/lm and it is possible to obtain tubes with luminous sensitivities approaching 400 µA/lm. The sensitivity in the near ultraviolet is not as striking. Below 4500 Å the sensitivity decreases rapidly, partially as a result of the decreased transmission of the input fiber plate and partially because the photocathode is very thick in order to obtain high red sensitivity. Tubes that we have used can detect λ3727 with an efficiency about 1/10 of that at λ4500. Below λ3600 the tubes are not useful. It will be important to develop tubes with higher uv sensitivity, but the limitations of the fiber plates may require a return to magnetic tubes for applications requiring far uv sensitivity.

[25] W. R. Beardsley, J. K. de Jonge, D. J. Haring and J. R. Hansen, *Publ. Allegheny Obs.* **11**, 1 (1969).

[26] C. R. Lynds, *Astrophys. J.* (*Lett.*) **159**, L151 (1970).

[27] R. R. Zappala, Kitt Peak Nat. Obs., Cont. No. 554, 1 (1971).

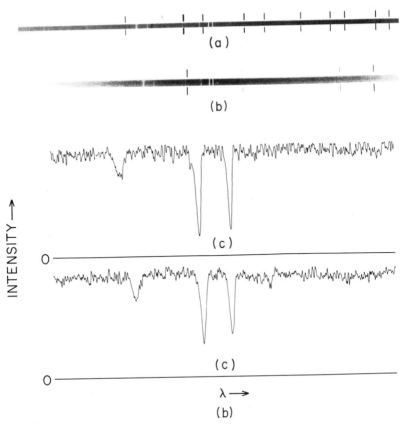

FIG. 1. Spectra of HD 168607 taken by G. Herbig. (a) 126-min exposure on pre-flashed 103a-F; (b) 5-min exposure using a Varo image tube; (c) direct intensity tracings of the two spectra in the region of the sodium D lines. When using the image tube the central intensity of D1 rose from 18 to 29% of the continuum intensity but the equivalent width of the line was unaffected. The original dispersion for both spectra was 16 Å/mm.

Not surprisingly, tubes with such high ir sensitivity have moderately high dark current. At 20°C the dark current may exceed 10^4 electrons/cm²/sec. Fortunately the tubes respond well to cooling. The high room temperature dark current may be reduced by factors of 10^3 by cooling to $-20°$ or cooler. Many tubes show ion scintillations and persistent bright spots when operated at room temperature. Both these defects decrease with time after the high voltage is applied and may be nearly eliminated by cooling the tube.

It is possible to obtain 40-mm Gen I tubes with limiting paraxial resolution of the order of 72 line pairs/mm. Near the edge of the 40-mm format the resolution is much poorer and the pincushion distortion ($\sim$10%) is severe. We have found that only the central 30 mm is of high quality. If tubes are stacked to achieve high gain, the useful area decreases. For a three-tube chain only the central 20–25 mm is useful.

The 8605 has much less vignetting than the WL 30677. Tubes may be selected that show less than a 2:1 gain variation over the central 30 mm. The contrast performance of the 8605 is very good. Figure 1 shows a comparison between coudé spectra obtained with the 50-cm (20-in.) camera using either the image tube or conventional photography. There is some degradation of the spectrum when the image tube is used, but for most applications the degradation is negligibly small. Information gains of 20 to 50 are achieved in the red, but in the blue below 5000 Å the distortions and limited format size decrease the information gain to unity.

6.1.3.3. Other Applications. Zappala[27] described a three-stage "snooperscope" system used at the 300-cm (120-in.) coudé to view faint red stars on the slit jaws. It was necessary to use a three-stage system to overcome the light losses in the transfer optics between the output phosphor screen and the eye. Because such systems do not integrate they achieve gain primarily by converting radiation to which the eye is insensitive into green light. For blue stars the system achieves little or no gain over the unaided eye. This is probably because the peak quantum efficiency of the dark-adapted eye is comparable to the quantum efficiency of photocathodes.

With integrating systems very much fainter objects can be seen. A Westinghouse WL 30677 coupled to a SEC camera tube is used at the Cassegrain focus of the Lick 300-cm (120-in.) telescope to provide a remote display for field acquisition and guiding.[28] Using integration, objects fainter than 22nd magnitude can be detected. This is between 10 and 20 times fainter than can be seen visually. An image tube is necessary for reducing the integration time to an acceptable value. In the Lick application the use of a demagnifying image tube is valuable for matching the telescope scale at the Cassegrain focus to the resolution of the SEC camera tube.

Tubes with electronic magnification can be used to match the optical scale of a telescope or spectrograph camera to the resolving capability of the detector. Tubes with magnification less than unity (such as the

---

[28] E. J. Wampler, *Proc. Amer. Astronaut. Soc.* **28**, 337 (1972).

Westinghouse WL 30677) are effective in increasing the "speed" of a spectrograph. Tubes, such as those manufactured by Varian,[29] with ×5 magnification have very high resolution at the input and would be useful either for direct viewing in "snooperscope" systems or in matching the rather coarse resolution of electronic readout systems to high-resolution optical devices. The use of a small photocathode area that is implied by the high resolution decreases the amount of dark current and can be important if faint objects are to be detected.

The phosphor screens commonly used in image tubes are linear over a very wide range.[30] This property, together with the persistence of the screen, has been exploited in the design of sensitive spectrophotometers.[31-33] The persistence of the phosphor screen of an image tube coupled to an image dissector provides the necessary short-term storage to enable the dissector to scan a large number of information elements without attendant aperture loss. In practice, devices have been constructed[30] that use simple modular construction and simple circuits to obtain photon-limited performance in over 10^3 simultaneous channels. A Lick Observatory system is now being used[34] to obtain spectra of faint galaxies and quasi-stellar sources.

The linearity and stability of the phosphor screen are very important. Using the Lick spectrophotometer, spectra have been obtained of objects as faint as magnitude 22. For such faint objects, accurate sky subtraction requires a stable system. Eleventh-magnitude stars are used as standards. Their use reduces the amount of observing time required to calibrate the system, but the wide dynamic range imposed by this observing program requires the system to be linear. Similarly the determination of the intensity ratios of emission lines requires that the characteristics of the device be insensitive to counting rate.

6.1.4. Future Developments

Much work is needed before the image tubes that use phosphor screens can be considered perfected. Unfortunately their development to date has

[29] R. S. Enck and J. P. Sakinger, *Electro-Opt. Syst. Design Conf.—1970 Proc.* p. 293. Ind. and Sci. Conf. Management, Chicago, 1970.

[30] E. H. Eberhardt and R. J. Hertel, *Appl. Opt.* **10**, 1972 (1971).

[31] L. B. Robinson and E. J. Wampler, *Publ. A.S.P.* **84**, 161 (1972).

[32] J. F. McNall, D. E. Michalski and T. L. Miedaner, *Publ. A.S.P.* (1972) (in press).

[33] W. K. Ford, Jr., L. Brown, and V. C. Rubin, *Carnegie Inst. Year Book* **70**, 327 (1971).

[34] L. B. Robinson and E. J. Wampler, *Astrophys. J.* (*Lett.*) **171**, L83 (1972).

been supported primarily by military interest. Military design goals are sufficiently different from scientific ones to compromise the tube's usefulness in scientific applications. Tube development programs designed to meet scientific goals would be able to build on the technology developed by military contracts. Such development, if adequately supported, should result in simple, easy to use image tubes that are superior to those currently available.

The (so-called) III–V photocathodes may be cited as an example of this contention. The initial development of opaque III–V photocathodes received substantial Department of Defense support and has produced a family of photocathodes that represent a technological breakthrough.[35] However, the development of image tubes using these cathodes is not now receiving similar support. One can only hope that these funding problems are a temporary phenomenon and that tubes using these cathodes, which have such high promise, will be constructed.

ACKNOWLEDGMENTS

This work was supported, in part, by the National Science Foundation (GP-29684), and the National Aeronautics and Space Administration (NGR 05-061-008 and NGR 05-061-006).

---

[35] W. E. Spicer and R. L. Bell, *Publ. A.S.P.* **84**, 110 (1972).

6.2. Electrographic Tubes*

6.2.1. Introduction

Of all of the electronic methods of making pictures (two-dimensional photometric recordings), the electrographic method is the simplest in principle. The optical scene to be recorded is focused upon a photoemissive cathode; the electrons released from this cathode are then accelerated toward and electronically focused onto the surface of a recording medium, usually a silver halide emulsion. The impact of electrons renders a silver halide emulsion developable, as with exposure to light. Thus a permanent picture that looks like a photograph is obtained. No scanning is involved, and neither light nor photography enters into the process after the formation at the cathode of the primary electronic image. It is convenient to think of the electrographic process as being an analog of photography, with the role of light sensitivity being assigned to the photocathode, and that of picture storage to the emulsion.

The simplicity of the process ensures a high signal-to-noise ratio in comparison with other electronic picture-making devices. This is because the introduction of extraneous noise will be limited to the statistical noise from the conversion of electrons into developable plate grains. The separation of the roles of sensitivity and storage allows great flexibility in selecting or designing emulsions, as the light sensitivity of the emulsion is of no importance to the picture recording process. The high sensitivity of the photoemissive cathode, the enormous data storage capability of ultrafine grained emulsions, and the linearity of the electrographic recording are all realized by the electrographic tube. There results a picture-taking device unrivaled for recordings of high information content or for maximum use of faint, hard-won light, such as that collected by large, expensive, astronomical telescopes. With existing equipment, electrographic recordings may be measured so as to recover quantitative exposure values for each elementary point of the original scene with systematic errors of the order of a few percent.

---

* Chapter 6.2 is by **Gerald E. Kron**.

The actual construction and operation of an electrographic tube, though simple in principle, is difficult in practice. Photoemissive cathodes are chemically delicate and easily damaged or destroyed by minute quantities of contaminants. Decker[1] has shown that of all of the common contaminants found in a high vacuum, water vapor is the most harmful to photocathodes. A partial pressure of only 5×10^{-8} Torr of water vapor causes serious loss in the sensitivity of an S-20 photocathode within a few hours. The difficulty of preserving a photocathode in the presence of a silver halide emulsion can be appreciated when one considers that the introduction of 100 cm^2 of thin emulsion into a 1-liter volume will raise the pressure from the ultrahigh vacuum range to between 0.001 and 0.01 Torr from released water vapor alone.

The problem is compounded by the necessity for removing the emulsion for development and substituting fresh emulsion prior to the next exposure without appreciable injury to the photocathode.

Another serious, but more subtle, problem lies in the suppression of the electronic background despite the presence of cesium vapor in tubes with small clearances and applied accelerating voltages of the order of several tens of kilovolts.

Almost all of the time and effort that has gone into the development of functional electrographic tubes has been spent on the solution or partial solution to one or more of these three problems.

The technological obstacles to the practical application of electrography have been overcome by three distinct designs: the classic Lallemand electronic camera, the US Navy electronic camera, and the McGee Lenard window spectracon. Sections 6.2.3, 6.2.4, and 6.2.5 of this chapter will be devoted to descriptions of them.

6.2.2. Electronic Focusing

The content of the sections on practical electronic cameras can perhaps be better appreciated after a brief, qualitative discussion of the two methods for focusing an electron beam from an object surface to an image surface: (1) the electrostatic method and (2) the magnetic method.

Electrostatic focusing depends upon the properties of the electrostatic field formed at the gap between two circularly symmetrical electrodes at different electrical potentials (Fig. 1). Electrons entering the field from

---

[1] R. W. Decker, "Advances in Electronics and Electron Physics," Vol. 28A, p. 364. Academic Press, New York, 1969.

the right are accelerated and guided to form an image on an electronic focal surface. A requirement of the system is that the two surfaces are curved toward each other. Fortunately, the depth of the electronic focus is so great that a plane receiver can be used without appreciable loss in definition, otherwise electrostatic focusing would be all but impracticable.

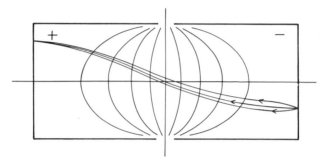

Fig. 1. Electrostatic focusing of accelerated electrons.

The chief on-axis aberration of an electrostatic lens is "chromatic" aberration caused by dispersion in the velocities of the electrons at release. For a photocathode, the mean speed of release is equivalent to an energy of about 0.75 eV. An analysis of the US Navy electronic camera made by Vine of the Westinghouse Electric Corporation[2] indicates that chromatic aberration at 30-kV accelerating potential would limit the definition to 190 line pairs/mm referred to the cathode. The same analysis indicates that definition at a 13 mm radius from the center would be limited by all aberrations to approximately 100 line pairs/mm.

TABLE I. Electrostatic Focusing

Advantages	Disadvantages
Tube is light in weight, simple.	Aberrations and distortions fairly serious.
Permits scale change.	Size of photocathode having tolerable
Tube parts accessible (not buried in magnet).	edge aberrations is severely limited in
Magnetic shielding simple.	tubes of practicable size.
	Focal plane at cathode must be appreciably curved, thus complicating presentation of optical image.

[2] J. Vine, private communication (1968).

The chief distortion of electrostatic electron optics is pin-cushion distortion, which may conveniently be nullified by means of a special barrel distorting optical corrector[3] placed close to the photocathode. Table I gives the chief advantages and disadvantages of electrostatic focusing as applied to the electronic camera.

Magnetic focusing is accomplished by accelerating the electrons from the photocathode by means of a uniform electrostatic field in the presence of a uniform coaxial magnetic field (Fig. 2). Accelerating potentials used in practice range from about 20 to 40 kV, with uniform magnetic fields of the order of several hundred gauss. The electrons are focused in the planes where they have executed an integral number of complete loops on their way from the cathode to the anode focal surfaces. Chromatic aberration is only about a tenth as much as with electrostatic focusing,[4] and so is negligible, and no other serious aberrations exist. The chief source of distortion comes from nonuniformity of the magnetic field,

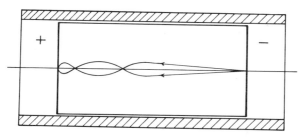

FIG. 2. Magnetic focusing: The electrons are accelerated by a uniform electric field in the presence of a solenoidal magnetic field.

which results in a rotation of the outer portions of the image with respect to the inner portions, a type known as "S" distortion. This distortion is not inherent, and can be controlled by careful attention to the design of the magnet. Table II gives the chief advantages and disadvantages of electromagnetic focusing. Both focal surfaces are planes, which, in principle, can be made as large as desired, all else being held constant, provided that the magnetic field can be kept uniform. Inherent resolution can be many hundreds of line pairs per millimeter.[5]

[3] A. V. Hewitt, G. E. Kron, H. D. Ables, "Advances in Electronics and Electron Physics," Vol. 33B, p. 737. Academic Press, New York, 1972.

[4] P. Grivet, "Electron Optics," p. 424. Pergamon, Oxford, 1965.

[5] W. R. Decker and H. Mestwerdt, "Advances in Electronics and Electron Physics," Vol. 28A, p. 22. Academic Press, New York, 1969.

TABLE II. Electromagnetic Focusing

Advantages	Disadvantages
Object and image surfaces are planes. Aberrations and distortions inherent in the method are very small. No inherent maximum size restriction placed on cathode.	Scale is limited to unity. The magnet causes: Considerable weight. Generation of heat, with need for cooling, of electromagnet. Burial of ends of tube within magnet. Difficulty with magnetic shielding. Requirement that shape of tube be essentially cylindrical.

6.2.3. Magnetic Shielding

An electron that is accelerated through a 30-cm distance by a uniform electrostatic field of 30-kV potential difference will be deflected by 20 μm if subjected to a 1-mG magnetic field at right angles to the motion. An experiment with the US Navy electronic camera indicated that, with magnetic shielding removed, 10 mG applied with Helmholtz coils were required to cause a 20-μm deflection, referred to the photocathode. The relative insensitivity of the real camera to a variation in ambient magnetic field is caused by partial shielding from iron electrodes, and by the focusing action of the tube. Even so, because an electronic camera will resolve dimensions of the order of 20 μm and the magnetic field of the earth is 500 mG, it is clear that the performance of an electronic camera will be affected by motion through the earth's or any other comparable magnetic field unless the camera is shielded well enough to reduce variations in the magnetic field penetrating the camera to less than, say, about 5 mG.

6.2.4. The Lallemand Electronic Camera

The electronic camera was developed and the term *electronic camera* was originated by Prof. André Lallemand of the Paris Observatory. Lallemand's ideas were first published in 1936.[6,7] In 1951, Lallemand and Duschesne published their first results[8] with a functional camera, and

[6] A. Lallemand, *C. R. Acad. Sci. Paris* **203**, 243 (1936).
[7] A. Lallemand, *C. R. Acad. Sci. Paris* **203**, 990 (1936).
[8] A. Lallemand and M. Duchesne, *C. R. Acad. Sci. Paris* **233**, 305 (1951).

proposed that it would be useful for making recordings with astronomical telescopes.

Lallemand avoided the difficulty of preserving the photocathode during plate change by simply abandoning the cathode at the end of a duty cycle and placing a new, preprepared cathode each time the tube was to be used. Preservation of the photocathode during use in the presence of contaminants from the emulsion was accomplished by cooling the plates with liquid air, thus reducing the vapor pressure of the contaminants. The vapor pressure of water, for example, at the temperature of liquid air is only 10^{-13} Torr, a negligible value. Tube background is not a serious problem in the Lallemand camera because the photocathode is prepared outside of the tube, thus eliminating appreciable contamination of the tube with cesium.

Figure 3 is a diagram showing a cross section of a Lallemand camera without its magnetic shield. The structure of the camera is mainly of glass, except for such parts as must conduct either electricity or heat.

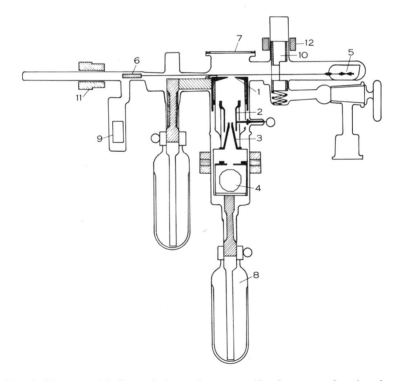

FIG. 3. Diagram of Lallemand electronic camera. Numbers are referred to in the text.

The tube is electrostatically focused, and is a triode, consisting of (1) the cathode surface at negative relative potential, (2) a focusing electrode at intermediate potential, and (3) an anode, relatively positive, followed by (4) a field-free region terminating at the surface of the plate in its holder. The scale of the tube is about 0.7:1, 12 plates are held in the plate-holder, and the diameter of the photocathode is 18 mm. In operation, the focusing electrode is at ground potential, and cathode and anode are at appropriate potentials below and above ground. The overall potential usually used is 27.5 kV. A photograph of the Lallemand camera is shown in Fig. 4.

FIG. 4. Photograph of the Lallemand electronic camera.

Preparation of the tube is a lengthy and rather difficult process.[9] All parts must first be cleaned with a solvent such as acetone, with the aid of an ultrasonic cleaner. After cleaning, the various parts may be handled only with tongs or with gloved hands. The plateholder is loaded with fresh plates, and all internal parts of the camera are assembled and lowered

---

[9] A. Lallemand, M. Duchesne, and M. F. Walker, *Publ. Astron. Soc. Pacific* **72**, 268 (1960).

into the glass envelope which has been standing-by mounted on the preparation table. A new cathode in its capsule (5) is placed in the right side-arm (Fig. 3), and the tow device (6) is attached to the eyelet of the cathode wire. Next, the closure window (7) is placed, and the mechanical vacuum pump on the preparation table is started for the initial pump-down. Meanwhile, an oven, with which the table is also fitted, is arranged around the tube, and the bake, which takes several hours, is started. During baking, the plates, which would otherwise be ruined by the heat, are kept cool by means of ice water in the dewar (8). The new cathode is outside the oven in its side-arm where it does not reach a destructive temperature. A diffusion pump is started shortly after initial scavenging by the mechanical pump, and after several hours of baking the oven is turned off, the tube is valved off from the diffusion pump, the appendage ion pump (9) is started, and the window is sealed with a low vapor pressure vacuum wax. After the tube is cool, the dewar is filled with liquid air, the pressure falls rapidly as the plates and internal parts become cold, and the tube is ready for activation. The tube is activated by towing the cathode in its capsule into registration with the breaker (10), by means of the magnet (11); the capsule is broken with an internal tool operated by another magnet (12), and is then towed into its service position where it is held by a pair of spring clips. The sensitivity of the cathode can now be tested, and if this and the pressure are normal, the tube is ready for use. If the dewar is kept filled with liquid air, the tube can stand-by for more than a week with negligible loss in cathode sensitivity. Thus, the tube, if used for astronomical recording, can survive a normal number of cloudy nights without need for repreparation.

The tube is placed into service by carrying it from the preparation room to its mounting at the telescope. Warm dry air is circulated inside the magnetic shield to reduce surface electrical leakage and to eliminate light-generating events that would fog the recordings. The optical focus is established, and the high voltage is applied. High voltage is generated by two independent electrostatic generators of the Van de Graaf type, one for the negative potential from the grounded focus electrode to the cathode, the other for the positive potential to the anode, which is maintained above ground. The voltages are monitored and set by means of two electrostatic voltmeters of high quality and high precision. The voltages can, however, be maintained only within ranges of 25 V, an accuracy insufficient for the purpose, and as a result the tube is not always in the best electronic focus. The dark slide is opened by means of an electromagnet and the exposure is made. After an exposure, the dark slide is closed, and the plate exchanged for an unexposed one. The plates are mounted on a multisided prism, so plates are changed by rotating the prism to a new facet, accomplished by means of a pawl mechanism actuated by still another electromagnet. Thus, the tube is used until the plates have been exhausted, a process that may take more than one night at the telescope.

After the plates are used, the tube is returned to its preparation site and disassembled after the admission of air. The plates are removed and developed.

The Lallemand electronic camera has an advantage over any other in that the observer can select a photocathode that is adapted to the work he is doing. He can, for example, select a cathode of unusually high sensitivity when such is needed, or he can use an S-1 cathode in case he needs sensitivity in the red and near infrared spectral regions. The provision for cooling the cathode takes care of the high thermal dark current of the S-1 photosurface; thus, the Lallemand camera is the only one presently in use that can give the observer sensitivity in the infrared.

Two disadvantages have prevented large-scale acceptance of the Lallemand electronic camera. First, there is the tedious preparation that represents considerable expense because of the large amount of time required of a highly trained person of advanced technical capability. The photocathodes, when purchased from a commercial supplier, are also expensive, as they cost several hundred dollars each. Second, the use of a new photocathode each time the tube is used restricts its value as a photometric device, because the departure of the cathode from uniform sensitivity must be measured each time the tube is used. This can be done only by using one or more of the precious plates for calibration purposes.

Development work continues on the Lallemand electronic camera, directed by Prof. Lallemand at his laboratory at the Paris Observatory. A camera having a cathode diameter of 10 cm exists there in an advanced state of development. Pictures taken in the laboratory with this camera during an interval of more than a year indicate that it gives excellent definition[10] and that it is a practical device.

The new camera is magnetically focused; hence, it has a 1:1 scale and is fitted with an ingenious plateholder that contains eight round plates of ten centimeter diameter. These plates can be exchanged, one for another at the electronic focal plane until all are used, by means of a

---

[10] A. Lallemand, private communication (1972).

magnetically operated mechanism. Photocathodes are prepared and stored in large dome-shaped capsules about 15 cm in diameter and 10 cm long. The cathodes are denuded, placed into the camera, and the camera is sealed prior to use, all in a large steel vacuum chamber fitted with the necessary controls and viewing ports for accomplishing its rather complicated mission. All of the preparation apparatus remains in the laboratory; only the cylindrically shaped camera itself, about 50 cm long by 20 cm in diameter, along with its focusing magnet, goes on the telescope.

Thoroughout the world there are about nine installations of the classic Lallemand camera. Most of these are for astronomical use, but several are in use for other purposes, mainly by physicists.

The Lallemand camera has now had a useful career of just 20 years duration, the best possible tribute to its designer and to the device itself.

6.2.5. The US Navy Electronic Camera

This camera is in principle identical to the Lallemand camera; in fact, in its early stages, the development of this camera was helped by advice from Prof. Lallemand and Dr. Duchesne, and by a study of the Lallemand camera at Mt. Hamilton, then the site of the Lick Observatory.

The Navy camera,[11] however, incorporates an "air lock" principle that permits the plate to be changed without sacrificing the photocathode. Experience with the tube indicates that photocathodes can be preserved for at least two years, with only minor degradation in sensitivity, even though the tube is used regularly.

A diagram of the camera is given in Fig. 5 and a photograph of the camera on the telescope is shown in Fig. 6. The picture enters through a correcting lens (1) and is focused upon the spherical cathode surface (2) inside the tube. The correcting lens curves the field to the proper radius and corrects for the pincushion distortion. The surfaces of the correcting lens are given low-reflectance coatings.

Glass (3) is used as an electrical insulating material between the cathode and the focus electrode (4), and between the focus electrode and the anode (5), which is the grounded portion of the tube. The voltages that appear across these two insulators are approximately 5 and 25 kV, respectively. The electrons from the photocathode enter the hole in the end of the anode electrode and are brought to a focus on the emulsion (6).

---

[11] G. E. Kron, H. D. Ables, and A. V. Hewitt, "Advances in Electronics and Electron Physics," Vol. 28A, p. 1. Academic Press, New York, 1969.

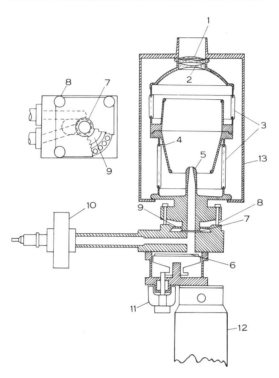

FIG. 5. Diagram of the US Navy electronic camera. Numbers are referred to in the text.

The protective valve (7) utilizes a coin-shaped piece of copper that can be rolled on its edge by gravity into or out of registration with the axis of the tube. To seal the valve for plate changing, the coin is rolled into the tube axis and then squeezed between the two stainless annular seats shown; the force for sealing is applied by four jack screws (8), and the compliance needed to allow motion of one of the seats is supplied by a steel diaphragm (9) of annular shape, as shown in the inset of Fig. 5. Two small ion pumps (10) are mounted on side tubes about 25 cm long to reduce the field from their magnets at the camera, and to attenuate the light that they generate.

The plateholder with its circular plate (6) magnetic turning tool (11) and cooling dewar (12) is near the bottom of Fig. 5. It incorporates a dark slide that protects the plate from the bright ion pump light emitted during heavy pumping shortly after a plate change, and from the blast of electrons that occurs momentarily when the accelerating voltage is applied after stand-by. The tube must be provided with at least two plateholders, as the fresh plateholder must be mounted immediately upon removal of the used one.

The magnetic shield (13) also acts as a light shield, and as armor for the fragile part of the tube. The magnetic shield is lined with an electrical insulator of black silicone rubber. The base end of the shield is firmly coupled to the tube near its center of gravity so that the tube can conveniently be mounted on a telescope by clamping to a flange at the front of the shield, without relative motion of the photocathode caused by flexure when the telescope is moved.

Accelerating and focus potentials are applied by means of a resistive divider, with the high voltage provided by a carefully filtered electronic

FIG. 6. Photograph of the US Navy electronic camera mounted on the Navy 61-in. astrometric reflector at Flagstaff, Arizona.

supply. The focus voltage can be adjusted and held to ± 4 V by means of taps on the voltage divider. The plateholder is loaded with one circular plate of diameter 7 cm; this plate is mounted so as to arrange the recorded images around the circumference of the plate. Six recordings of the full field at 2:1 scale can be accomodated without mutual interference. Eight can be fitted on the plate with an overlap equivalent to only 7% of the photometrically useful area. If the images to be recorded are small, more can be fitted on the plate. For example, as many as 48 low-dispersion spectra can be put on one plate without mutual interference if they are arranged radially.

The Navy camera stands by with the valve protecting the cathode in the closed position, and with the two small ion pumps operating constantly to maintain a low pressure. When the tube is to be used, the plate is cooled with liquid nitrogen; this greatly reduces the emission of water vapor and other contaminants from the emulsion, and the pressure in the part of the tube containing the plate can be reduced to about 10^{-9} Torr within approximately 30 min. The protective valve can now be opened, the high voltage applied, and the pictures taken. The pressure remains low and the tube may be kept in service as long as the plate is kept cold, and until the plate is fully used. The protective valve is then closed and the plate allowed to warm up; the plate can then be exchanged for a new one at any time, a process that requires about 30 min.

The photocathode is formed *in situ*. The parts of the tube are first carefully cleaned according to ultrahigh vacuum practice with fine abrasives and solvents. They are then assembled on a clean bench or in a clean room. A probe device containing the needed supplies of cesium and antimony (the cathode ingredients) is mounted in place of the plateholder. The tube is attached to a suitable ion-pumped vacuum system, and is baked to a temperature of 430°C. After cooling, the anode end and central portion of the tube are reheated selectively to 150°C to reduce occlusion of cesium which would cause the tube to glow when the high voltage is applied. Antimony and cesium are now deposited on the cathode surface by reaching down the axis of the tube with the probe, which is moved into position with a magnet. After the formation of the cathode, the protective valve is closed for the first time to permit removal of the probe device to make way for the first plate.

After considerable experience has been gathered by the technician, he will be able to make photocathodes having a quantum efficiency at 4200 Å of between 10 and 20%. The most important requirement for the production of highly sensitive photocathodes is cleanliness; if proper cleanliness can be achieved, the production of a good photocathode is practically inevitable. The sapphire substrate used in the Navy tube seems to offer an unusually favorable surface on which to form a photocathode.

If for any reason a new cathode must be formed, this can be done by disassembling the tube, recleaning it, and reprocessing. The only replacement parts needed are the gold wire seals, which cannot be reused after having been flattened during previous service.

The plateholder is sealed to the camera with a lead wire gasket, and the tube is sealed off from the parent vacuum systems by means of an all-metal commercially available valve. Thus, the camera structure includes no synthetic or organic materials.

6.2.6. The McGee Spectracon

The spectracon[12] differs from all other electronic cameras in that the recording emulsion is out in the air, where it can be changed without resort to high-vacuum technique. The practical application of this tube therefore approaches closely the convenience of photography. The chief feature of the spectracon is a window, known as a *Lenard window*, made of a material that is permeable to high-velocity electrons, but not permeable to thermal molecules of gas or vapor. The material itself is perfect natural mica, cleft to a thickness of 3 to 4 μm. This material, because it is a single crystal, is extremely strong; a round window with a diameter of 2 cm, properly mounted, will withstand a pressure differential of more than 2 atm.[13]

The spectracon is a magnetically focused electronic camera, and therefore has a scale of 1:1. It is shown diagrammatically in Fig. 7, where the photocathode (1) is shown to the left, and the 7×30-mm Lenard window (2) to the right. The entire assembly weighs about 11 kg, and is a cylinder 12.7 cm in diameter and 38 cm long. The field of 160 G furnished by the solenoid (3) requires a power input of 60 W, which is removed by liquid cooling (4) in order not to heat the photocathode and increase the background from thermal emission. The tube is operated at an accelerating potential of 40 kV and the electrons retain an energy of approximately 20 kV after penetrating the mica. The cathode end of

[12] J. D. McGee, A. Khogali, A. Ganson, W. A. Baum, "Advances in Electronics and Electron Physics," Vol. 22A, p. 11. Academic Press, New York, 1966.

[13] J. D. McGee, D. McMullan, H. Bacik, and M. Oliver, "Advances in Electronics and Electron Physics," Vol. 28A, p. 61. Academic Press, New York, 1969.

the tube is operated at ground potential, thus eliminating corona near the cathode as a serious source of background. About 25% of the electrons are lost by absorption in and reflection from the mica.[14]

The photocathodes are formed in an auxiliary system[12] to which the tube is temporarily attached. The substrate, a rectangular piece of glass 10×30 mm in size, is subsequently transferred from the external system to the tube through a flat glass pipe, which is then sealed off. Any kind

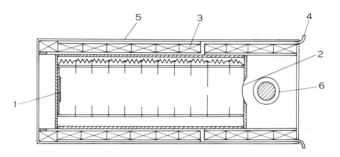

FIG. 7. Diagram of the McGee spectracon. Numbers are referred to in the text.

of photocathode can be formed and introduced into a tube during manufacture, though the useful S-20, or trialkali cathodes slowly lose sensitivity after the transfer.

The solenoid of the spectracon has been carefully designed and empirically adjusted to give an extremely uniform magnetic field that varies no more than $\pm 1\%$ along the significant part of the length of the tube. No figures have been published for evaluating the residual distortion; however, it is so small that it is not appreciable to the eye. The tube is furnished with a double layer mu-metal magnetic shield (5). This shield is not saturated from the effect of the solenoid, and therefore is effective in screening the tube from variations in the ambient magnetic field so efficiently that the Spectracon can be subjected to an entire reversal of the magnetic field of the earth without appreciable shift in the position of the image.

The cathode of the Spectracon may be cooled with the same liquid that serves the magnet. Some models of the spectracon have thermoelectric booster coolers built into the cathode end of the magnet. These coolers, which operate from the magnet current, can keep the cathode at a temperature below $0°C$. This feature raises the possibility of providing the

[14] J. D. McGee, A. Khogali, and A. Ganson, "Advances in Electronics and Electron Physics," Vol. 22A, p. 31. Academic Press, New York, 1966.

Spectracon with the infrared sensitive S-1 cathodes thereby making it competitive with the Lallemand camera in this respect.

The Lenard window, thin as it is, nevertheless represents many molecular layers and scatters the photoelectrons through large angles. The excellent definition of 80 line pairs/mm of which the tube is capable can therefore be achieved only if the emulsion is brought into very close contact with the mica. Since atmospheric pressure causes the window to take a cylindrical form of small radius, the emulsion must have a very flexible substrate in order to allow gentle pressure to form it to the proper curvature and press it into close contact with the mica; 50-μm Formvar, a strong, flexible plastic with good dimensional stability is generally used. The film is wrapped around a foam rubber roller (6) of the proper radius of curvature and the unit pressed against the mica with a force of about 200 gm by means of spring loading. The entire film applicator is made of electrically insulating material so that films may be changed without removing the accelerating potential from the tube. In spite of the remarkable ability of the mica to resist rupture from atmospheric pressure, it is not unknown for the mica to break when the film is pressed against it. Experiments are continuing in an effort to reduce the frequency of such catastrophes, even though they are already sufficiently rare that the spectracon is in practice a useful and effective electronic camera.

The chief disadvantage of the spectracon lies in the slit-shaped window, which decreases the value of the instrument for recording anything but spectra or other slit-shaped scenes. Round mica windows of 2-cm diameter will, as mentioned above, easily withstand atmospheric pressure. However, a round window loaded with atmospheric pressure conforms to a sphere and so far, no practicable method has been found for pressing an emulsion into close contact with a spherical mica surface without a prohibitively large breakage rate.

Experiments with the use of round Lenard windows are in progress at the Royal Greenwich Observatory under the direction of Dr. Dennis McMullan.[15] McMullan has furnished an air lock into which the film can be passed and be presented to a mica window unloaded by atmospheric pressure, and therefore flat. The mica serves as a divider between the ultrahigh, ultraclean vacuum of the tube itself, and the low, heavily contaminated vacuum of the air lock. The round film is held by clamping it around its edge, and is pressed against the mica by inflating it from behind with air at a pressure of about 15 Torr. Contact is excellent,

[15] D. McMullan, *The Observatory* **91**, 199 (1971).

and the pressure is very gentle. An operating model has been built by adding the mica window fitting to a US Navy electronic camera, whose scale reduction of a factor of two permits recording the entire 4-cm cathode through the 2-cm mica window.

The experiments at the Royal Greenwich Observatory are aimed at the production of a Lenard window magnetically focused electronic camera having an 8-cm cathode, with readout through a window of equal size. The ultimate size of a tube of this type should be limited only by the size of available perfect mica of the proper thickness.

Scores of spectracons have been built in the McGee laboratory at the Physics Department of the Imperial College, University of London. There are probably more of them in use throughout the world than any other electronic camera.

6.2.7. Other Developments

In the three previous sections descriptions have been given of the three kinds of electronic cameras that are in routine operation and that are producing recordings of astronomical value from terrestrially mounted optical telescopes. This section will be devoted to a brief description of other developments in the field.

An electronic camera intended for space research has been developed at the Observatory of Meudon.[16] This camera, which was derived from the Lallemand concept, is very unusual in that it is cryopumped by cooling it with liquid helium. It is magnetically focused, with the magnetic field established by means of a superconducting solenoid. A very strong magnetic field of 7 kG is used; this results in many small loops being executed by the electrons on their route. A property of such a system is that resolutions as high as 250 line pairs/mm can be obtained in theory at any image plane; i.e., the depth of focus is infinite. The development is for the purpose of furnishing a remotely operated electronic camera for recording with balloon-borne telescopic equipment.

Development of a large electronic camera with a cathode 15 cm in diameter and very high resolution has been undertaken at the Westinghouse Aerospace Laboratories.[17] The camera design calls for magnetic focusing, with cathode protection by a valve after the US Navy camera

---

[16] M. Combes, P. Felenbok, J. Guerin, and J. P. Picat, "Advances in Electronics and Electron Physics," Vol. 28A, p. 39. Academic Press, New York, 1969.

[17] R. W. Decker and H. Mesterwerdt, "Advances in Electronics and Electron Physics," Vol. 28A, p. 19. Academic Press, New York, 1969.

concept. The intention was to record many pictures very rapidly on roll film, and as a result the development included considerable attention to details of the film holder. Unfortunately, the federal funds for this project were withdrawn and a working model was never built.

At the University of Texas, under the direction of Dr. Paul Griboval, an electronic camera is being developed for use with terrestrially mounted telescopes.[18] This camera is magnetically focused with a cathode having a diameter of 50 mm. The plates are introduced through an air-lock system analogous to that used by McMullan. The high vacuum is separated from the low vacuum by means of an extremely thin film of alumina. Contact is obtained between the emulsion and the alumina by means of electrostatic forces; thus, the alumina is held against the emulsion, and, hopefully, the emulsion can be on a rigid substrate, i.e., glass.

G. R. Carruthers of the Naval Research Laboratory, Washington, has developed an unusual electronic camera[19] for use as a rocket-borne detector in the 900–1600-Å ultraviolet spectral range. This camera is practically an evacuated Schmidt camera with a metallic photocathode coated on a substrate at the optical focal plane. The photoelectrons are then focused magnetically through the hole in the primary mirror onto the emulsion. The photocathodes used are not damaged by exposure to dry air, so installation of the cathode and recovery of the emulsion without destruction of the cathode are simple matters. The device has been used for recording ultraviolet spectra and star images with the Aerobee 150 rocket vehicle. Experience has shown that the camera represents a great advance over the use of photography.

6.2.8. Emulsions and Development

Any silver halide emulsion is sensitive to the impacts of high-voltage electrons, and therefore any available emulsion can be used to produce pictures in an electronic camera. However, one of the valuable advantages of the electronic camera over photography is that it can produce recordings in which the density is practically a linear function of the exposure over a wide range of densities. Not all emulsions, however, are suitable for linear recording. A comparison between the density growth for the Eastman spectroscopic emulsion V-0 and the Ilford nuclear-track emul-

[18] P. Griboval, "Advances in Electronics and Electron Physics," Vol. 33A, p. 67. Academic Press, New York, 1972.

[19] G. R. Carruthers, *Appl. Opt.* **8**, 633 (1969).

sion K.5 exposed in the electronic camera to 30-kV electrons is shown in Fig. 8. Both emulsions were developed for 5 min in D-19 developer at 68°F with continual agitation. Of six fine-grained photographic emulsions tested in the electronic camera, all showed a linearity failure similar to Eastman V-0; hence, the use of nuclear-track emulsion is a necessity.

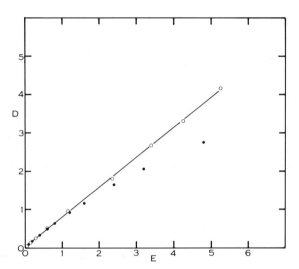

FIG. 8. Density growth from exposure to 30-kV electrons for the nuclear track emulsion Ilford K.5 (○) and for the photographic emulsion Eastman V-0 (●). The line is a straight line drawn through the observed points for K.5.

The chief difference between nuclear-track and photographic emulsions lies in the amount of silver compared with gelatin: nuclear-track emulsions have about five times more. The emulsions almost universally used in the electronic camera are made by Ilford; the types are designated G.5, K.5, and L.4 in order of decreasing grain size.

As shown in Fig. 9, three different kinds of developer employed with K.5 emulsion exposed to 25-kV electrons give as many different density response curves. Measurements on signal-to-noise ratios indicated that the three developers were equivalent, a result to be expected if one assumes that each of the developers developed all the exposed grains. However, the results displayed in Fig. 9 indicate that for high densities, only D-19 development gave a practically linear relation between density and exposure.

Figure 10 shows the resulting density growth in changing the develop-

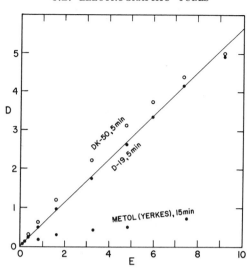

FIG. 9. Effect of developing K.5 electrographs at 25 kV with three different kinds of developers.

ment time in D-19. The data were obtained from a single plate by cutting it into four pieces after exposure. Underdevelopment apparently results in limitation of the linear portion of the response curve, whereas more than necessary development simply leads to useless grain growth. The 5-min development commonly used appears to be a good compromise.

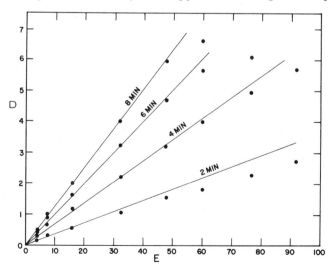

FIG. 10. Effect of development time on K.5 electrographs in D-19 at 25 kV.

6.2.9. Characteristics of the Electronic Camera

A silver halide emulsion employed in an electronic camera is an information storage device. Each photoelectron from the cathode will cause not just one, but several[20] developable grains; thus, transfer of all the information gathered by the cathode, with only slight loss from statistical variation in grain size is assured. Upon development of the plate, this information is made available as variations in density in proportion to the original exposure. The number of grains developed does not depend upon the characteristics of the emulsion. The density, however, does, in the sense that greater density will be obtained with coarser grained emulsions, all other conditions being equal. Thus, the use of a coarse-grained emulsion will result in a growth of high density, but, as there must be fewer of the coarse grains available in a given volume of emulsion, the supply will sooner become exhausted, i.e. the relatively linear portion of the response curve will have been used. The use of a finer grained emulsion will, on the other hand, provide more storage for information within its linear response, but the density level will, other conditions being equal, be less.

Latham[21] has pointed out an interesting consequence of the ability to select grain size in the recording medium:

...When one sets out to make a photometric observation electrographically, he can decide what accuracy is required in the statistics of the photons registered in each resolution element, and can then choose the nuclear emulsion with the appropriate blackening rate to give an easily measured density for the desired number of registered photons.... In classical photography an emulsion is usually chosen to have good detective quantum efficiency in the range of input light levels encountered, and there is little additional choice in the blackening rate. Thus, it is normal to end up building a selection of different (spectrograph) cameras with different light-gathering powers and different focal ratios in order to accomodate the rather narrow range of good detective quantum efficiency of the photographic emulsion and to avoid severe reciprocity failure problems. Thus if an electrographic camera is used as the detector for a spectrograph, one can arrange to choose different nuclear emulsions instead of choosing different cameras, and the spectrograph need have only one camera.... .

The negligible fog level of the nuclear track emulsions frees the valuable low-density end of the characteristic curve for recording low-level phenomena with limitations set only by electron statistics and by

[20] R. C. Valentine, The Response of Photographic Emulsions to Electrons, *Microscopy* 1, Academic Press, London and New York, 1966.

[21] D. W. Latham, *Amer. Astron. Soc. Photo-Bull.* (1972) (in press).

the background level of the electronic camera, a level that can be taken at about 1 electron/sec/mm² of cathode area at room temperature. The high-storage capacity of fine-grained emulsions assures that grain exhaustion will not seriously affect the efficiency of the electrographic process at high densities. The common Cs–Sb photocathode has a quantum efficiency at λ4200 typically of about 10%, i.e., ten photoelectrons are emitted for each 100 photons incident. The maximum efficiency for fast photographic emulsions is about 1%,[22] but the average efficiency over a useful range in density is considerably less. Thus, the information gain over photography has been determined experimentally[23] to be about 30 at photographic density up to one and about 50 up to density 2.5, in comparison with the two photographic emulsions most commonly used in astronomy, Eastman 103a-O and IIa-O. The information gain, based upon signal-to-noise measurements, represents an advantage fully equivalent to an increase in telescope aperture. A 100-cm (40-in.) telescope equipped with an electronic camera becomes the equivalent of a 500-cm (220-in.) telescope and a photographic camera, in its rate of recording photometric data.

Electrographic recordings on very fine-grained emulsions, such as Ilford L.4, tend to be unspectacular in appearance because their information content, present at modest densities but high signal-to-noise ratio, is not readily accessible to print-out, or to visual examination. This is illustrated in Figs. 11a,b; 11a shows a print of a field of galaxies containing the peculiar object Arp 174.[24] The long faint arm reaching up and to the left is barely visible on the original, and is entirely lost in the halftone. However, an isodensity tracing made with a Joyce–Loebl–Tech–Ops isodensitometer, shown in Fig. 11b, shows the arm so well (in spite of its level of density only 0.03 D above a sky background of 0.4 D) that it could be evaluated quantitatively with considerable precision. The original was recorded in 30 min on L.4 emulsion with a US Navy electronic camera having a 12% Q.E. cathode, and the Navy 100-cm (40-in.) Ritchey–Chretien reflector at Flagstaff.

The high storage capacity emulsions such as L.4 make possible the full exploitation of the electrographic method. If only coarse-grained emulsions were available, the chief virtue of the information gain over photography would be a saving in time. The fine-grained emulsions

[22] P. Fellgett, *Mon. Notices Roy. Astron. Soc.* **118**, 230 (1958).

[23] H. D. Ables and G. E. Kron, *Publ. Astron. Soc. Pacific* **79**, 424 (1967).

[24] H. Arp, Atlas of Peculiar Galaxies. California Inst. of Technol. (1966).

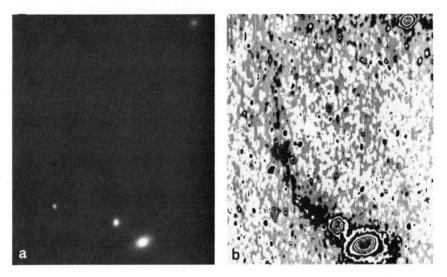

FIG. 11. Retrieval by means of microisodensitometry of photometric information stored in an extremely fine-grained electrograph. The emulsion was Ilford L.4. (a) US Navy electrograph; (q) US Navy photograph.

enable the observer to store a large amount of information during relatively long exposures, thus making possible the recording of difficult objects with small telescopes. This feature is illustrated in Figs. 12a–c. These figures are recordings of a test pattern devised by Baum.[25] In the lower left of the pattern can be seen a rectangular frame containing five round spots in a field of increasing background intensity. Thus, the ability of the recording method to reveal the spots will depend upon its ability to gather and store enough information to bring the signal-to-noise

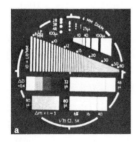 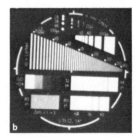 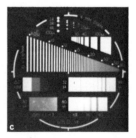

FIG. 12. (a) A photograph and (b) and (c) two electrographs of a Baum test pattern illustrating the effect of increasing storage capacity as described in text.

[25] W. A. Baum, "Advances in Electronics and Electron Physics," Vol. 16, p. 391. Academic Press, New York, 1962.

ratio of the spots to a high enough level for visibility. Figure 12a was recorded photographically on Eastman 103a-O emulsion; 12b was recorded electronically on Ilford K.5, whereas 12c was recorded likewise on Ilford L.4 emulsion. The number of spots visible on the originals were, respectively, one, three, and four. Any reader with sufficient interest to refer to the Baum reference for further details of the rather complex test pattern will find other interesting facts in Fig. 12a–c.

The penetration of 30-kV electrons into a nuclear track emulsion is only a few micrometers;[20] thus, the electronic image is quite shallow and does not go entirely through even the thin 10-μm emulsions commonly used in the electronic camera. The surface uniformity of sensitivity therefore does not depend on the thickness of the emulsion, as it does in photography, but only upon uniformity of grain size, a more easily controlled parameter. The limited range of the electrons also ensures that lateral diffusion will be limited to only a few micrometers, a distance

FIG. 13. Electrograph of a star field (the cluster M67) on L.4 emulsion. North is at top, east to the left. Original was 2 cm in diameter, reduced electronically by half from a 4-cm photocathode.

negligible compared with the circle of confusion of practicable electronic cameras now in use. A further advantage of the thin image is its excellent accessibility to the developer, which results in practically complete absence of Eberhard and border effects.

The reciprocity failure characteristic of photography does not exist[26] in the electrographic recording method. Also absent is the threshold effect, thus rendering useful the low-density end of the characteristic curve.

Figure 13 is a recording on L.4 emulsion of the open cluster M67 made with the 100-cm (40-in.) Ritchey–Chretien reflector and the electronic camera of the US Naval Observatory Station at Flagstaff, Arizona. The focal ratio of the telescope is 7.2, the diameter of the recorded field from the 40-mm photocathode is 16 arcmin. The limiting magnitude of this 5-min exposure in blue light, made during mediocre observing conditions is 19.5. Under better observing conditions we have successfully recorded, with 30-min exposures, the globular clusters associated with the elliptical galaxy M87, indicating a limiting magnitude of the order of 22. The sky background density is about 0.4, whereas background from all other sources is negligible. The resolving power of the electronic camera is safely better than that of the telescope–earth's atmosphere system. Thus, this combination of telescope, electronic camera and emulsion is a well-matched, almost ideal astronomical tool. It is a very inexpensive, and in several important respects, superior substitute for a gigantic telescope condemned to be used with photography.

[26] M. Duchesne, C. R. Acad. Sci. Paris 271, 142 (1970).

6.3. Television Systems for Astronomical Applications*

6.3.1. Introduction

The possibilities of television have intrigued astronomical observers for many years. Early attempts to apply television sensors to astronomical measurement programs yielded marginal results. The advent of planetary space probes and orbiting observatories made it necessary to develop television systems that were acceptable photometric instruments. The demonstrated success in this area was greatly dependent on the complementary development of digital computers that were necessary for processing the raw television data[1]. Recent emphasis on space astronomy applications has led to the development of sensors that are a significant advantage in both space- and ground-based observational astronomy. This section reviews the trade-offs between various types of television sensors and discusses the system considerations involved in the design and application of these television systems to astronomical observations. Examples of observations with the SEC vidicon are represented along with examples of computer processed video data.

Television affords several advantages in astronomical observation. These include:

(a)   high quantum efficiency,

(b)   quantum noise limited signal-to-noise ratio over most of the dynamic range,

(c)   broad spectral sensitivity,

(e)   no reciprocity loss in long exposures,

(f)   data is in convenient form for processing by digital computer.

Integrating television appears particularly advantageous for measurements where the flux is very low, such as high-dispersion spectroscopy and imagery of extended objects in narrow spectral bands.

---

[1] T. C. Rindfleisch, Digital Image Processing for the Rectification of Television Camera Distortions, *Astronomical Use Television Type Sensors, Symp. Proc.* Princeton Observatory, May 20–21, 1970, NASA SP-256, pp. 145–166 (1971).

---

* Chapter 6.3 is by John L. Lowrance and Paul Zucchino.

The fact that the data are available as an electrical signal easily adapted for computer processing is an operational advantage, and so is the immediate availability of pictures and intensity traces from TV display devices.

The selection of television image sensors for astronomy[†] should be based on the following criteria:

(a)   long exposure, with good reciprocity,

(b)   internal tube background sufficiently low so that the S/N ratio is quantum noise limited;

(c)   high effective quantum efficiency;

(d)   maximum number of picture elements per image, i.e., good MTF in single-exposure readout[‡];

(e)   large elemental storage capacity in photoelectrons per picture element;

(f)   flat image plane for ease of matching optical systems and broad spectral coverage.

As discussed later, there are trade-offs to be made among these various criteria. Before proceeding into the details of these trade-offs it may be helpful to review the various generic types of television systems and sensors.

6.3.2. Review of TV Sensor Types

Table I shows the primary types of television systems. The period of signal integration is the distinguishing characteristic in each case.

6.3.2.1. Nonintegrating Type, Image Dissector. The image dissector consists of a photocathode and an electron multiplier. The photoelectron image from the photocathode is focused in the plane of the small entrance aperture of the electron multiplier. The photoelectrons from the projection of the aperture on the photocathode enter the electron multiplier and are amplified. The electron image is deflected magnetically to scan the image past the small aperture. Scanning of the image produces the video signal.

[†] This excludes solar astronomy.

[‡] The modulation transfer function (MTF) is the amplitude response to a sine wave spatial frequency input.

TABLE I. Types of Television Systems

Nonintegrating example	Line integration example	Frame integration	
		Continuous readout, photon counting	Single–Exposure readout
Image dissector	Image intensifier + image dissector	Requirement: External memory for integration	Requirement: Long integrating capability
		Example: Image intensifier +EBS vidicon +digital memory	Examples: SEC vidicon EBS vidicon (cooled) Silicon vidicon (cooled)

The image dissector is simple and relatively accurate photometrically. However, it suffers from not being able to integrate the photoelectron image during the period when it is not entering the electron multiplier. The fraction

$$\frac{\text{aperture area of electron multiplier}}{\text{scanned area on photocathode}}$$

represents the efficiency of the system. This is of course quite small if the resolution is to be significant.

6.3.2.2. Line Integration—Image Dissector + Image Intensifier. This is a sensor in which the image dissector is preceded by an image intensifier. The phosphor of the image intensifier produces photons for some periods of time Δt after being excited by an accelerated photoelectron. If the dissector scans the image in this time Δt, to a first approximation the output signal is the integral of the incoming photoelectrons during the same Δt. Unfortunately the persistence of the phosphor is exponential. This means that the signal seen by the electron multiplier is strongly dependent on when the phosphor was excited relative to when it is scanned by the aperture. Its integration time is also limited to less than a second due to the short time constant of the phosphor. This type of sensor is discussed in detail in Chapter 6.1.

Both the image dissector and this combination of an image dissector and intensifier are not considered competitive with the television sensors which have the capability of frame integration and are not limited by phosphor time constants.

6.3.2.3. Frame Integration.   Nearly all television systems in use today employ sensors which integrate the optical image signal as an electrical image for a period of time. The integrated electrical image is then scanned by means of an electron beam to convert the electrical image to a video signal. Figure 1 shows schematically the various types of sensors considered for astronomical image sensors in space.

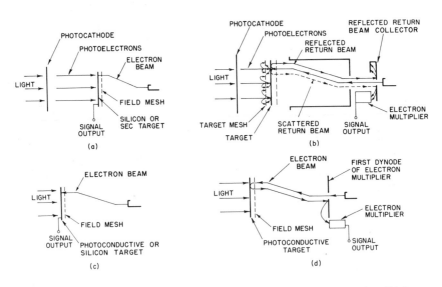

Fig. 1. Types of frame-integrating TV sensors. (a) EBS or SEC tube; (b) image isocon; (c) vidicon; (d) return beam vidicon.

The vidicon utilizes an electron gun to scan the photoconductive target. A continuous electrical contact is provided on the front side of the target by a transparent conductor which serves as the signal electrode. In operation, the scanning beam is used to adjust the charge on the back side of the target to a negative potential relative to the signal electrode. During exposure, the light pattern is focused on the photoconductor. The photoinduced conductivity in the illuminated areas causes the back side of the target to charge to a more positive potential in the illuminated areas. During the readout, the electron beam deposits charge on the positive areas, resulting in a capacitively coupled charge signal appearing at the signal electrode.

The image orthicon utilizes a photocathode as the photon sensor. The photoelectron image pattern developed at the photocathode is focused by an axial magnetic field onto a thin, moderately insulating, target surface. Secondary emission from the target results in a net positive charge deposition on the target. The electron beam from the gun scans the target, landing electrons on the more positively charged areas, and the remainder of the beam returns to an electron multiplier surrounding the electron gun. The signal is the current output from the anode of the electron multiplier.

The image isocon is a modification of the image orthicon in which an improved electron-optic system has been introduced and in which the inherent beam-noise problem of the image orthicon has been largely overcome.[1a,2] When the scanning electron beam approaches the target three possibilities exist: An electron may strike the target and neutralize positive charge; an electron may fail to land and be electrostatically reflected; an electron may land on the target and be scattered back at various angles. These latter scattered electrons constitute the signal return beam in the isocon tube. The reflected electrons are prevented from entering the multiplier by means of a carefully aligned baffle.

The SEC-vidicon tube also utilizes a photocathode as the light detector. Photoelectrons are accelerated to 8 kV and focused onto a thin target made of specially deposited KCl which has the property of providing gain through secondary electron emission within the target. Typical electron gain is 100. The signal is read out at the target signal electrode as in a vidicon.

The EBS-vidicon tube operates in much the same way as the SEC tube. The target, however, is a very thin silicon wafer upon which a tightly spaced matrix of p–n junction diodes has been formed. The spacing of the diodes is of the order of 12 μm. Primary photoelectrons, accelerated to 10-keV energy, impinge upon the target and cause multiple dissociation of electron–hole pairs. The electron gain is typically 2000. The holes are collected at the p-side of the diode where the charge is neutralized by the scanning beam during readout. Signal is read out on the conductive backplate of the target.

[1a] A. D. Cope, and E. Luedicke, The development of camera tubes for recording astronomical images, in "Advances in Electronics and Electron Physics," Vol. 22A, p. 175. Academic Press, New York, 1966.

[2] G. A. H. Walker, J. R. Auman, V. L. Buchholz, B. A. Goldberg, A. C. Gower, B. C. Isherwood, R. Knight, and D. Wright, Application of an image isocon and computer to direct digitization of astronomical spectra (to be published).

6.3.2.4. TV Sensor Summary. Effective quantum efficiency is of primary importance since it determines the exposure time. The image orthicon and image isocon require a mesh between the photocathode and the target. This mesh blocks transmission of $\frac{1}{3}$ to $\frac{1}{2}$ of the photoelectrons, reducing the effective quantum efficiency of the photocathode by a corresponding amount. In the vidicon there is no mesh prior to signal integration and the quantum efficiency of the photoconductor is comparable to or exceeds that of the best photocathodes. The EBS or SEC vidicons also do not have a mesh between photocathode and integrating target. Therefore, on the important point of quantum efficiency, the image isocon and the image orthicon are decidedly handicapped.

Signal-to-noise ratio is another important characteristic. The electron beam returning from the target of an image orthicon is fed into an electron multiplier. The larger the signal on the target the more beam electrons land on the target and correspondingly less return beam electrons enter the electron multiplier. The total noise consists of the statistical noise in the image, the statistical or shot noise in the electron beam entering the multiplier, and the noise due to the electron multiplier itself. The beam shot noise is maximum when the signal is low because the return beam is then at maximum intensity and therefore maximum beam shot noise. This problem is overcome in the image isocon where the fraction of the return beam entering the electron multiplier is the electrons which have been laterally deflected or scattered by the charge pattern on the target. Thus the isocon signal increases as the signal on the storage target increases, and the corresponding beam shot noise is proportional to the square root of the signal. This is a significant improvement over the noise characteristics of the image orthicon. A detailed analysis of the isocon signal-to-noise properties is beyond the scope of this discussion, but in summary the signal-to-noise ratio achieved over the entire dynamic range is still considerably below the statistical signal-to-noise ratio of the image stored on the target.[3]

In the case of the vidicon (which does not have prestorage gain), the signal is due to the beam electrons landing on the target and this signal is normally fed to a preamplifier rather than an electron multiplier. The noise level of the best preamplifiers is about 300 electrons rms per half

---

[3] P. Zucchino and J. L. Lowrance, Progress Report on Development of the SEC-vidicon for Astronomy, *Astronomical Use Television Type Sensors, Symp. Proc.* Princeton Observatory, May 20–21, 1970, NASA SP-256, pp. 27–53 (1971).

cycle of bandwidth. This means that the noise due to the readout process is comparable to statistical noise in 100,000 photoelectrons per picture element. Return beam versions of the vidicon suffer from the same readout limitations as the image orthicon and isocon.

The EBS or SEC vidicons employing the same preamplifier as a vidicon have the advantage of prestorage target gain. Thus the preamplifier noise is reduced relative to the statistical noise in the photoelectron image by the target gain. With an SEC gain of 80 the readout threshold is $400/80$ or 5 photoelectrons rms, and the gain of 2000 of the EBS tube's silicon target makes the preamplifier noise negligible. *So, on the question of achieving a quantum-noise limited S/N ratio, tubes with high prestorage gain have a clear advantage.*

Another important point is resolution in cycles per unit length. The image orthicon is at a disadvantage due to the fact signal charge is stored between the target and the target mesh. When sensed by the electron beam the charge image appears somewhat defocused due to the finite spacing between target and mesh of 12 to 25 μm. With the SEC or EBS vidicon there is no mesh and the target thickness can be considerably thinner with correspondingly higher resolution.[3] The vidicon has the highest intrinsic resolution due to the very thin photoconductive target which can be made less than 1 μm thick.

The remaining very important characteristic is long exposure capability. The image orthicon can be made with a high-resistivity glass target capable of integrating for hours. The EBS silicon target is limited by the dark current of the silicon diodes. At room temperature the integration period is in the order of seconds. When cooled to $-60°C$ the integration period is many minutes. The photoconductive target is also limited by dark current. Cooling reduces the dark current and allows longer integration periods. The SEC-vidicon target is operationally usable for integration periods of many hours.

The following tabulation summarizes the reasons for preferring the SEC or EBS vidicons and the primary reasons for rejecting the other types of TV sensors for use in integrating television systems for astronomical observing.

Image isocon and image orthicon are rejected for the following reasons:

(a)   the target mesh collects a large fraction of the photoelectrons, lowering the net quantum efficiency of the photocathode; and

(b)   the target-mesh arrangement creates a charge image that seriously degrades the MTF in single-exposure readout.

Vidicon and return beam vidicon are rejected[†4] for the following reasons:

(a) the absence of prestorage gain in the target makes it impossible to achieve a quantum noise limited signal due to preamplifier noise or beam shot noise with return beam readout; and

(b) the dark current of the photoconductor target requires cryogenic cooling for long integration.

Reasons for choice of SEC-vidicon or EBS-vidicon type tube are:

(a) the absence of a target mesh results in no photoelectron attenuation;

(b) the target gain is sufficient to make the preamplifier noise negligible when referred to photocathode;

(c) the target capacitance is large enough to allow the collection of several thousand photoelectrons per picture element;

(d) the modulation transfer function in single exposure readout is acceptable; and

(e) in the case of the SEC vidicon the integration time is hours, and with the silicon EBS vidicon integration periods of several minutes are practical with modest cooling, and periods of several hours are feasible with cryogenic cooling.

6.3.2.5. Photon Counting Image Systems.   There is considerable interest in television systems that can be used to detect a single photoelectron and its location in a two-dimensional image.[5,6] In such a system the image must be integrated in an external digital memory. The advantages of these systems are unlimited integration time and wide dynamic range. They may also afford a low system noise threshold for very weak exposures of only a few photoelectrons per picture element. Also one can watch the image accumulate during the exposure.

[4] T. B. McCord and J. A. Westphal, *Publ. A.S.P.*, Feb. 1972.

[5] A. Boksenberg, An image photon counting system for optical astronomy, *Astronomical Use Television Type Sensors, Symp. Proc.* Princeton Observatory, May 20–21, 1970, NASA SP-256, pp. 77–84 (1971).

[6] S. B. Mende, Single photoelectron recording by an image intensifier TV camera system, *Astronomical Use Television Type Sensors, Symp. Proc.* Princeton Observatory, May 20–21, 1970, NASA SP-256, pp. 85–105 (1971).

[†] The silicon target vidicon exhibits exceptionally high quantum efficiency in the near infrared out to 1.1 μm and may be useful for photometry of relatively bright objects of low contrast where a large number of photoelectrons must be collected (Ref. 4).

The basic approach is to amplify the photoelectron image using a multistage image intensifier to a level where a single photoelectron can be detected above the readout noise of a television camera used to view the phosphor output screen of the image intensifier. The frame rate of the television system is made high enough so that the probability of more than one photoelectron arriving at a given location during a frame period is low. A threshold discriminator detects video pulses that exceed the readout noise and sends this information to a digital memory that is scanned in synchronism with the scanning of the television raster.

In such a system, one must consider: the "false alarm rate" due to background within the image intensifier, multiple detection of the same event on successive scan lines, amplifier noise that exceeds the detector threshold, and the range of incoming data rates. The large digital memory required is also a consideration. To duplicate the capability of the 25-mm square target SEC vidicon, a memory of 10^8 bits that can be scanned in a fraction of a second is required.

These types of systems are in the early stages of development. They appear attractive for some applications if the effective quantum efficiency is found to be comparable to systems that integrate the photoelectrons within the tube in an analog fashion. A more detailed discussion is beyond the scope of this section.

6.3.3. System Design Considerations

It is clear that the SEC and EBS vidicons best meet the requirements for an astronomical television image sensor. The argument for television sensors as opposed to photographic film will not be dealt with here except to note that the quantum efficiency of photocathodes is more than an order of magnitude higher than that of photographic film. The following discussion addresses several important television image sensor system considerations.

6.3.3.1. Focal Ratio Considerations. For extended objects the common equation for the flux density in the image plane of an optical system is given by

$$E^* = E/4f^2, \qquad (6.3.1)$$

where E is the scene luminance (power per unit area), E^* is the image illuminance (power per unit area), and f is the diameter of entrance aperture per focal length $(= D/F)$. For most applications of photographic film this expresses the "speed" of the system very well. However, with certain television type sensors it is more useful to express the image illuminance in terms of power per picture element and power per solid angle. Let h be the height of the picture element in the image plane, h^2 be the area of a picture element, and $E*h^2$ be the power per picture element. Rewriting Eq. (6.3.1),

$$E*h^2 = ED^2h^2/4F^2. \qquad (6.3.2)$$

The solid angle subtended by the picture element is $[h/F]^2$. Thus the image illuminance in power per solid angle is $ED^2/4$. Therefore, for a given *angular* resolution, the flux per picture element is only a function of the area of the entrance aperture D^2.

The ratio h/F is adjusted to match the desired angular resolution. The value F should be as large as practical to maximize the MTF (modulation transfer function) of the image sensor and in some cases that of the optics. The total angular field of view θ is

$$\theta = (h/F) \times \text{number of picture elements across the image sensor,}$$

or for a fixed image sensor size L

$$L/h = \text{number of elements} \quad \text{and} \quad \theta = L/F$$

There is then a straightforward system trade-off between the field of view and MTF for a given image sensor size.

The luminosity of astronomical objects is often expressed in stellar magnitude m_v with units of photons per second per angstrom per square centimeter of collecting aperture.

$$\frac{n}{\Delta t} = \frac{2.51^{-m_v}\pi D^2}{\beta^2 4 \times 10^{-3}} \left[\frac{h}{F} \right]^2 \frac{\text{photons/sec/A}^0}{\text{picture element}},$$

where β^2 represents the solid angle subtended by the object.

The optical image of a point source has a finite angular size. This angular size is inserted for β^2 to calculate the average flux per picture element for unresolved objects. The flux rate per individual picture element is a function of the shape of the point spread function of the system.

In cases where the problem is detection, as in the case of stars fainter than the sky background, it is advantageous to contain all the collected signal from the star within one half cycle of the video bandwidth, i.e., within one picture element. In addition, the physical size of the picture element should be consistent with good spatial frequency response of the sensor. This can be done by reducing the f-number or by reducing the video bandwidth and increasing the width of the scan lines by the same factor such that a picture element is physically larger on the target of the tube. In the latter case the spatial frequency of interest is lower and the MTF is higher, an added advantage in detectivity. On the other hand the background noise in the image sensor (Fig. 7) is higher at lower spatial frequencies. The best system design in this case must consider the actual levels of sensor background relative to the image signal, the actual sensors MTF, and the ease of implementing the two approaches.

6.3.3.2. Resolution. The resolution of the television tube, best expressed by its modulation transfer function (MTF), is made up of the product of the MTF's of the image section, the target, and the readout section.

6.3.3.2.1. IMAGE SECTION MTF. Magnetically focused image sections exhibit very high resolution when the voltage gradient between the photocathode and target is high. The theoretical square wave MTF of the image section in the WX31718 SEC-vidicon is shown in Fig. 2 as a function of the photoelectron energy.* There is experimental evidence that it approaches this performance in the actual tube.

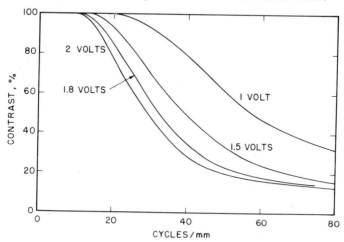

FIG. 2. Magnetically focused image section MTF, SEC vidicon, WX31718 at 8000 V, 80 G, cos α angle distribution, and an initial voltage distribution of $g(V_0) = \sin^2(\pi V_0 / V_{0\max})$.

* J. P. Pietrzyk, private communication, Westinghouse Electric Corporation.

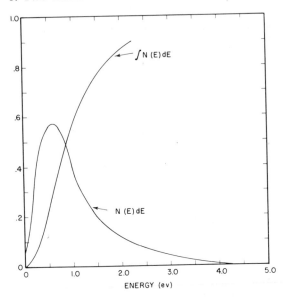

FIG. 3. Photoelectron velocity distribution for a CsNa₂KSb photocathode with λ = 2090 Å at 5.9 eV.

The photoelectron energy or velocity is a function of the photon energy, the band gap, and electron affinity of the photocathode. The bialkali (Na₂KSb) photocathode has an electron affinity of 1.0 V and a band gap of 1.0 V. The resultant velocity distribution is shown in Fig. 3 for various photon energies.[7] Pair production of electrons in the semiconductor photocathode begins at 3 times the band gap. On the average, one of the two electrons resulting from the high-energy photon escapes, but at low velocity. This pair production flattens the quantum yield and limits the high-energy photoelectrons to a small fraction of the total emitted. The curve for 5.9 eV is representative of higher-energy photons as well, with a gradual extension of the high-energy tail of the curve as the photon energy increases. This makes the bialkali photocathode very desirable for those ultraviolet applications where low sensitivity at longer wavelengths is not a requirement. This is because the pair production effect keeps the average energy of the photoelectrons low which maintains a high MTF. One can expect a relative degradation in resolution with solar blind photocathodes such as C₈I due to the higher energy of the emitted photoelectrons.

[7] W. E. Spicer, *J. Phys. Chem. Solids*, Vol. 22, pp. 365–370. Pergamon Press, New York (1961).

6.3.3.2.2. TARGET MTF. The target MTF is a function of the effective thickness of the dielectric layer separating the stored charge from the signal plate as the readout beam recharges the target to cathode potential.[8] This function is plotted in Fig. 4 for various target thicknesses. MTF measurements of the WX31718 SEC vidicon are also shown in Fig. 4. Measurements of the target capacitance indicate a target thickness of about 7 μm. Apparently the effective thickness relevant to the MTF calculation is not directly obtained from the capacitance measurement.

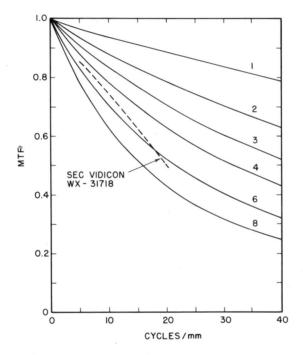

FIG. 4. Target MTF vs. target thickness for MTF = $[1 - \exp(-2\pi kt_2)]/2\pi kt_2$, where k is in half cycles per millimeter and t_2 is thickness in millimeters.

The overall tube resolution is dominated by the target and could be improved by going to a target that is effectively thinner.

The predicted resolution of the EBS (electron bombarded silicon) vidicon is shown in Fig. 5. One notes that the target MTF again dominates the over-all MTF. The target MTF is made up of three factors: the effect

[8] I. M. Krittman, Resolution of Electrostatic storage targets, *IEEE Trans. Electron. Devices* **10** (1963).

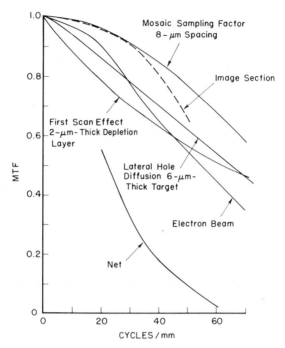

Fig. 5. MTF of EBS vidicon predicted.

of diode sampling, the effects of lateral diffusion, and the electrostatic thickness effect discussed above. The net target MTF is comparable to that of the SEC vidicon.

6.3.3.2.3. READOUT SECTION MTF.   The resolution of the readout section is primarily determined by the aperture response of the electron beam. The analysis of this response is complicated and may be found elsewhere.[9] In general, the magnetically focused beam has higher resolution than what can be obtained with electrostatic focus; and the resolution increases at slow scan due to a lower beam current and a correspondingly smaller beam width. There is also a self-sharpening effect in the readout process due to the leading edge of the beam doing most of the recharging of the charge depleted areas on the exposed portion of the target. The trailing edge of the beam sees a fully charged target resulting in a reading beam that is effectively more narrow than the actual physical width of the electron beam. The electron beam MTF shown in Fig. 5 is representative of the current state of the art.

[9] O. H. Shade, *RCA Rev.* **31**, No. 1, 60–119 (1970).

6.3.3.3. MTF Effects on Signal and Background Noise. The MTF of the image sensor and optical system can be represented by a filter which attenuates the higher spatial frequencies in the image.

Consider the effect of this filter on the quantum noise in the signal and the background. The quantum noise is the statistical variation in the number of quanta measured or counted. This statistical variation has no spatial component except that the number measured is proportional to the size of the picture element. Therefore the MTF of the image section of the sensor does not attenuate the quantum noise, only the spatial information in the signal. The combined MTF of the target and of the electron beam readout of the image sensor does have equal effect on the signal and the quantum noise due to the background during the readout process. Let: MTF_1 represent the optics MTF, MTF_2, the image section MTF, MTF_3, the target and readout section MTF, N_s, the high-frequency signal in photoelectrons per picture element, N_B, the low-frequency background signal in photoelectrons per picture element, and N_r, the rms readout noise expressed in photoelectrons per picture element. The signal-to-noise ratio at the spatial frequency where a picture element is one-half cycle is

$$\text{signal-to-noise ratio} = \frac{N_s(MTF_1)(MTF_2)(MTF_3)}{[(N_B + N_s)(MTF_3)^2 + N_r^2]^{1/2}}.$$

6.3.3.4. Noise As A Function of Spatial Frequency. The signal-to-noise ratio is often expressed as a ratio of the low-frequency or dc signal to the noise at full system bandwidth, i.e., full spatial frequency response. It is more informative to express the signal-to-noise ratio as a function of spatial frequency.

The sensitivity of the television sensor is expressed as readout signal electrons from the target per photoelectron from a photocathode. In the readout of the signal electrons there is a noise threshold due to the pre-amplifier which is expressed in electrons per half cycle of video bandwidth. This noise threshold is independent of the actual dimension on the target that is scanned by the beam during the time period of a video half cycle.

The other noise sources are:

(a) quantum noise in the signal,

(b) quantum noise in the background,

(c) thermal noise in the landing energy of the read beam (beam switching noise), and

(d) random variations in target gain.

The background noise is made of the dc background in the signal itself, plus the background due to dark current emission from the photocathode and other parts of the image section. There is a somewhat different background due to the need to shift the target voltage prior to readout with the electron beam. This shift ensures that the electron beam lands in the lightly exposed area of the image.

An examination of the various noise components as a function of spatial frequency on the target follows.

The quantum noise in the optical background is the square root of the number of photoelectrons in the background of the optical image. The ratio of background to signal is the image contrast and is independent of the optical magnification, i.e., the spatial frequency at the sensor.

The photocathode dark current and other background currents within the image section are strongly influenced by temperature.[10] Figure 6

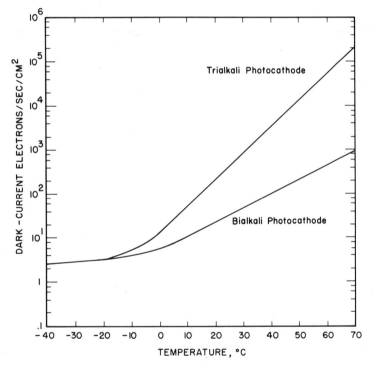

Fig. 6. Photocathode dark current vs. temperature. Full exposure is 2.5×10^6 photoelectrons/mm², target gain is 70, and total capacitance is 5 pF/mm².

[10] M. Rome, in "Photoelectric Imaging Devices" (L. M. Biberman and S. Nudelman, eds.), p. 155. Plenum Press, New York, 1971.

shows the photocathode dark current vs. temperature for bialkali and tri-alkali photocathodes. Based on overall measurements of the tube background, one concludes that the total background obeys the same trend. There is then a clear argument for keeping the image section of the television sensor cool, preferably below $-20°C$. The silicon target tube, the EBS vidicon, requires even lower temperatures to limit the target dark current. The total thermal and field emission induced dark current background is proportional to the actual area of the picture element. Therefore, for a given tube this background decreases as the square of the sensor spatial frequency, and the noise due to this background is inversely proportional to the spatial frequency.

The readout target bias voltage shift can be expressed in photoelectrons. The SEC-vidicon target capacitance is about 5×10^{-12} F/mm². The nominal bias voltage shift is 0.2 V; then

$$Q = CV = 10^{-12} \quad \text{C/mm}^2 \quad \text{or} \quad 6 \times 10^6 \quad \text{electrons/mm}^2.$$

At 10 cycles/mm this becomes

$$[1/20]^2 \times 6 \times 10^6 = 15 \times 10^3 \quad \text{electrons/half cycle.}$$

The noise would be $(1.5 \times 10^4)^{1/2} = 122$ electrons/half cycle rms. At a target gain of 70 this is equal to 1.7 photoelectrons/half cycle rms. This target bias voltage noise is also inversely proportional to the sensor spatial frequency, but is relatively small in any case.

The read beam switching noise is a function of the electron beam cathode temperature and elemental target capacitance

$$\overline{q_n^2} = 2kTC_s$$

At 10 cycles/mm this yields

$$q_n = 2 \times 10^{-17} \quad \text{C} = 122 \quad \text{electrons,}$$

divided by a target gain of 70 to express in photoelectrons

$$1.7 \text{ photoelectrons/half cycle rms.}$$

The dominant readout noise for the SEC vidicon is the preamplifier noise or the noise in the optical input signal; both of which are independent of the actual spatial frequency at the photocathode or storage target of the sensor. This is shown graphically in Fig. 7.

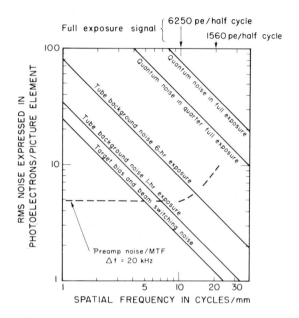

FIG. 7. SEC-vidicon noise versus spatial frequency.

Only in the case of several hour exposures does the system noise become spatial-frequency-dependent because of the internal background. Even this is not relevant in the region of interest, i.e., at 20 cycles/mm where the tubes MTF approaches 50%. In most applications the dominant noise will be the quantum noise in the optical image.

In space applications, background due to energetic particles must also be considered. The cosmic ray background is only 0.5 particles/cm²/sr above 40 MeV. The high-energy proton density above 40 MeV is a strong function of altitude and longitude. At 300 nautical miles altitude the proton density exceeds the cosmic ray background only in the south Atlantic anomaly.[11,12] Presuming a total dose of 1.5×10^5 protons/cm² due to a pass through the south Atlantic anomaly, the Cerenkov radiation should not result in more than 3×10^6 photoelectrons/cm². At 20 cycle/mm resolution this is approximately 20 photoelectrons/picture element/orbit. This is tolerable and could be decreased for very long exposures by shutting off the system during the transit through the high flux region.

[11] F. S. Johnson (ed.), "Satellite Environment Handbook." Stanford Univ. Press, Stanford, California, 1961.

[12] E. G. Stassinopoulos, World Maps of Constant B, L, and Flux Contours. NASA-SP-3054 (1970).

6.3.3.5. Dynamic Range and Contrast Considerations. The modulation transfer function of a diffraction-limited telescope is shown in Fig. 8. The MTF due to atmospheric seeing is also shown for ground-based telescopes.[13] Note that the wide variation in attenuation over the spatial frequency range sets a lower limit on the required dynamic range for the image sensor. At 0.8 lsf (limiting spatial frequency) the response is 10% of the low-frequency response. The scene contrast at high-spatial frequencies will in most cases be low, decreasing the ratio of high-frequency to low-frequency signal in the image even further. Mariner IV pictures of Mars show a typical surface contrast range of 15 to 30%.

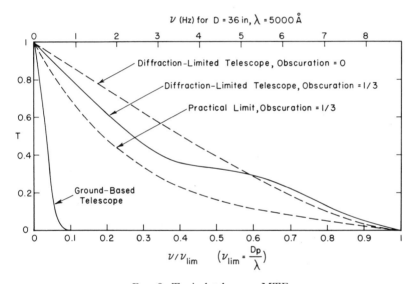

FIG. 8. Typical telescope MTFs.

The generally accepted threshold of detectability is a peak-to-peak signal approximately six times the rms noise.[14,15] Let N be the low-frequency signal or background, $(M_0)(2N)$ be the high-frequency signal in scene, and $(\text{MTF})(M_0)(2N)$ be the high-frequency signal in image, where, for modulation,

$$M_0 = \frac{I_{max} - I_{min}}{I_{max} + I_{min}} \quad \text{and} \quad N = \frac{I_{max} + I_{min}}{2}.$$

[13] I. R. King, The Profile of a Star Image, *Publ. A.S.P.*, **83**, 199–201 (1971).
[14] O. H. Shade, *RCA Rev.* **28**, No. 3, 460–535 (1967).
[15] A. Rose, *Advan. Electron.* **1** (1948).

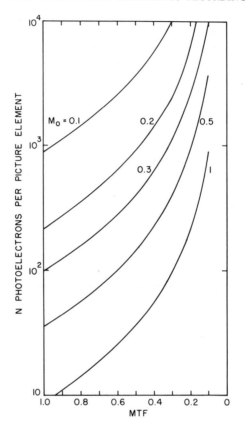

FIG. 9. Number of photoelectrons per picture element that must be accumulated to allow unambiguous detection of a modulation M_0 plotted as a function of the system MTF with $N = (6/2M_0 \times \text{MTF})^2$.

Therefore, for detectability,

$$6N^{1/2} \geq (2M_0)(\text{MTF})(N),$$
$$N \geq [6/(2M_0)(\text{MTF})]^2$$

Figure 9 shows the minimum number of background photoelectrons that must be collected in order for a signal of a given contrast and spatial frequency to be detectable above the statistical fluxations in the background. These data are plotted as a function of MTF for various modulation levels. These curves for a noiseless image sensor present the limitations under the most favorable circumstances. The same data is plotted as a function of spatial frequency in Fig. 10 presuming the "practical"

MTF curve shown in Fig. 8. Also shown is the case where the high-frequency signal-to-noise ratio is 20, a condition that would allow some enhancement or peaking of the high frequencies to rebalance the spatial frequency relationships in the original scene. Objects larger than a picture element can, of course, be detected at lower flux levels. For example, the bars of the standard resolving-power test pattern have a width $\omega_0 = 1/N$ mm and a length $L_0 = 5\omega = 5/N$, where $N = 2f =$ television line number/mm. The bar is then five picture elements long and one picture element wide. The necessary flux per picture element is then reduced by a factor of 5. These curves also imply the long exposure time required for high resolution imagery of faint objects.

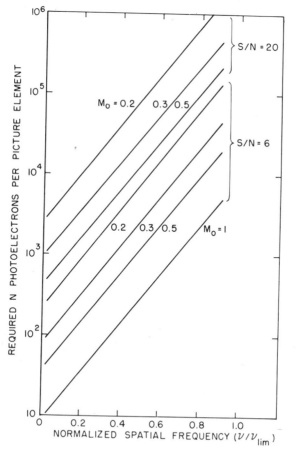

FIG. 10. Figure 9 replotted using MTF of "practical limit" telescope as shown in Fig. 8.

The storage capacity of the image sensor in photoelectrons per unit area must allow these levels of exposure, or some provision must be made for addition of separate exposures by computer.

The storage capacity in photoelectrons per square millimeter of the SEC vidicon is noted in Fig. 7 and may be calculated on a per image element basis as

$$Ne_{max} = \frac{C_T \, \Delta V_{max}}{G_T q_e (2\nu)^2},$$

where Ne_{max} is the maximum number of photoelectrons that can be stored per pixel or image element, C_T is the target capacitance in farads per millimeter squared, ΔV_{max} is the maximum practical change in the target scanned surface potential, usually about 5 V, G_T is the target gain, i.e., stored target electrons per photoelectrons, q_e is the charge of an electron, 1.6×10^{-19} Q, and ν is the maximum spatial frequency in cycles per millimeter. (One half cycle of ν corresponds to the linear size of the pixel.)

In recording spectra there is the important advantage of being able to increase the effective capacity of the sensor in terms of photoelectrons per unit dispersion by trailing the spectra normal to the dispersion as is often done in photographic recording of spectra. The earlier discussion of dynamic range and contrast is equally relevant to spectra. In measuring absorption or emission against a continuum, there is an advantage in knowing roughly where in the dynamic range the high-frequency detail of interest is located, e.g., on top of a continuum that may vary by a factor of two or three over the spectral range being recorded. In order to measure an absorption of 1% such that it is three times the mean square deviation in the continuum, i.e., 3 sigma confidence, the quantum signal-to-noise ratio of the continuum must be at least 300. This in turn requires that 9×10^4 photoelectrons per spectral resolution element be collected.

6.3.3.6. Image System Data Rate.

The optimum rate of reading the data out of the image sensor is dependent on the particular type of sensor. If the signal is taken at a target signal plate, the optimum rate is a function of the total capacitance to ground at the preamplifier input. With the 25-mm square target SEC vidicon, the bandwidth corresponding to optimum readout rate is around 20 kHz but may vary depending on the detailed design of the preamplifier. If return beam readout is used, the readout noise is primarily the beam shot noise and this is independent of the readout rate.

In most cases the analog data will be digitized before processing. The number of bits per picture element must be large enough to make the quantizing noise small compared to the noise in the analog data. The peak-to-peak signal to rms noise ratio due to quantizing is[16]

$$S/N = (12)^{1/2}M,$$

where M is the number of levels. This relationship between the signal-to-quantizing noise ratio and the number of digitizing bits is shown in Table II. Another quantizing consideration is the digital sampling rate relative to the limiting spatial frequency and video bandwidth. The Nyquist criteria states that the minimum sampling rate is once per half cycle of the highest frequency in the data.[17] However, this

TABLE II. Digitizing Bits, Levels, and Signal to Quantizing Noise Ratio

Number of bits	M	S/N
5	32	111
6	64	222
7	128	444
8	256	888

absolute minimum sampling rate results in low-pass filtering of the signal with 63% response at the high-frequency limit. Therefore, in most cases, the sampling rate should be greater than the Nyquist minimum. If the rate is 3 times per cycle at the maximum frequency, the filter response is 83% and for 4 times it is 90%.

6.3.3.7. Preamplifier. For sensors which are read out via a signal plate with low-velocity electron beam scanning of the target, the noise consists mainly of the preamplifier noise and the quantum noise in the image. The preamplifier noise expressed in photoelectrons per pixel (image or picture element) is N_a/G, where N_a is the preamplifier noise in target electrons per pixel and G is the target gain. N_a consists of the shot noise in the channel current of the field effect transistor, the thermal (Johnson)

[16] M. Schwartz, "Information Transmission, Modulation and Noise," pp. 151–154. McGraw-Hill, New York, 1970.

[17] H. S. Black, "Modulation Theory," Chapter 4. Van Nostrand Reinhold, Princeton, New Jersey, 1953.

noise in the resistors of the input network, and excess noise associated with gate leakage of the input transistor.

Figure 11 schematically illustrates the preamplifier employed at Princeton with the SEC vidicon. Its function is to measure the stored charge on each pixel. The first stage is a selected low-noise junction field effect transistor (FET). Since the input FET is intrinsically a voltage-sensitive device the ratio of the signal charge to total shunt capacitance is a key parameter. Negative feedback via C_{fb} is employed to stabilize the performance against device parameter drifts, but the design is such that C_{fb} is kept small in relation to C_{total} so as not to degrade the channel shot noise performance. The bias and feedback resistors are of high value ($\sim 10^8\ \Omega$) to minimize their thermal noise contribution. Upper limits on resistor values are determined by the FET leakage current (especially if operation in a radiation environment is planned) and by the need to avoid low-frequency shifts in the target bias potential.

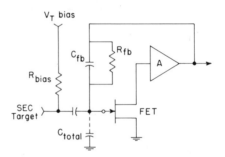

FIG. 11. Preamplifier employed with SEC vidicon. The shot noise figure of merit for input FET is $g_m = C_T I^{1/2}$. See the text for discussion and explanation of symbols used.

Cooling of the preamplifier to $-10°C$ improves the noise performance by reducing the thermal component of the FET leakage current. Further cooling does not help, presumably because surface leakage becomes the dominant FET leakage noise contributor. To date, insulated gate type field effect transistors have proven to be excessively noisy for application to preamplifiers of this type, so we have been unable to take advantage of their extremely low gate leakages.

The channel current shot noise expressed in target electrons per dwell time or half cycle at the input of the preamplifier is calculated as

$$N_c = \frac{C}{g_m}\frac{N}{\Delta t} \quad \text{or} \quad N_c = \frac{C}{g_m} N(2\ \Delta f),$$

where C is the total shunt capacitance to ground at the preamplifier input, g_m is the transconductance of the FET, N is the number of channel current electrons per element dwell time, Δt is the readout dwell time per element, and Δf is the corresponding readout bandwidth. This may be expressed in terms of the FET channel or drain current

$$N_c = \frac{C}{g_m} \left(\frac{I}{q_e \, \Delta t} \right)^{1/2} \quad \text{or} \quad \frac{C}{g_m} \left(\frac{2 \, \Delta f \, I}{q_e} \right)^{1/2},$$

where q_e is the charge of one electron.

The thermal noise current of a resistor is

$$i_n = (4kT \, \Delta f / R)^{1/2},$$

where k is Boltzmann's constant, 1.372×10^{-23} J/K, and T is temperature, K.

A half cycle of bandwidth corresponds to $0.5 \, \Delta f^{-1}$ sec. The number of thermal noise electrons per half cycle is

$$N_r = (kT / \Delta f \, R q_e^2)^{1/2}.$$

The combined bias resistor thermal noise and channel current shot noise in electrons per half cycle (pixel dwell time) is

$$N_a = \left[\frac{kT}{\Delta f \, R(q_e)^2} + \left(\frac{C}{g_m} \right)^2 \frac{\Delta f \, 2I}{q_e} \right]^{1/2}.$$

From this one notes that the input resistance should be as high as practical, the total input capacitance should be low as possible, the g_m should be high, the channel current should be low, and that there is an optimum frequency for minimum total noise.

For a large SEC vidicon and a typical slow scan preamplifier $G = 80$, $f = 20$ kHz, $C = 50 \times 10^{-12}$ F, $g_m = 13.4 \times 10^{-3}$ A/V, $I = 7 \times 10^{-3}$ A, $R = 2 \times 10^8 \, \Omega$, $T = 260$ K, $k = 1.362 \times 10^{-23}$ J/K, $q_e = 1.6 \times 10^{-19}$ C/electron. Solution of the preceding equation for these values yields $N = 243$ electrons/half cycle. This compares with the measured preamplifier noise of 390 electrons per half cycle. The additional measured noise is mostly attributed to the gate leakage current of the FET in the preamplifier.

The dependence of preamplifier performance upon readout rate or element dwell time is shown in Fig. 12. The FET channel shot noise rises with the one-half power of frequency or inversely as the one-half

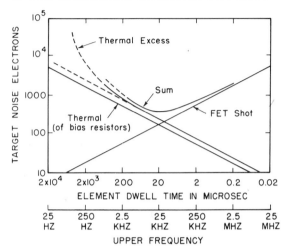

FIG. 12. Preamplifier rms noise vs. readout rate.

power of dwell time, while the other noise contributions behave in a reciprocal manner. This leads to an optimum readout rate with an upper video frequency of 20 kHz corresponding to a dwell time of 25 μsec/image element. The minimum noise point is not a sharp minimum because of the one-half power dependence of the contributions. The minimum can be shifted somewhat higher in frequency by selecting FETs with low shot noise, while sacrificing leakage or "excess" noise characteristics; and conversely can be shifted downward in frequency by using opposite device selection criteria. Standard broadcast readout rates (∼4 mHz) are entirely too fast for optimum threshold performance with sensors that are readout via a preamplifier connected to the target.

6.3.3.8. Photometric Transfer Function. The SEC target consists of a "fluffy" layer of dielectric. The aluminum layer on one side of the target and the scanning electron beam on the other side complete the electrical circuit with the target being capacitive. The exposure proceeds with photoelectrons penetrating the dielectric layer causing selective discharge of an initial charge across the capacitor. As the voltage across the capacitor drops, the discharge per photoelectron decreases. This is somewhat analogous to the exponential response of RC electrical networks. The net effect is a decrease in the target gain expressed in target electrons per photoelectron. This characteristic, shown in Fig. 13, appears to be uniform in shape throughout the target. The gain of the target and the efficiency of the photocathode do vary in different areas of these variations must be corrected in the data reduction.

In the case of the silicon diode target, the target gain is constant as a function of exposure. In both targets the gain is approximately proportional to the accelerating potential between photocathode and target.

Figure 13 shows the photometric transfer function for the 25-mm square target SEC vidicon. The maximum exposure is somewhat arbitrary but is normally around 6250 photoelectrons per picture element. The readout noise is about 5 photoelectrons rms per picture element. Accordingly the signal-to-noise ratio is essentially quantum noise limited over virtually all of the dynamic range. The spectral sensitivity is determined by the response of the photocathode and the transmission of the window. Magnesium fluoride (MgF_2) windows have been developed that allow broad-band transmission from 1150 Å into the infrared. Bialkali, S-20, and S-20 ER photocathodes are also available which will cover the range from 1150 to about 8500 Å. The new III–V photocathodes promise to extend the sensitivity beyond 15,000 Å.

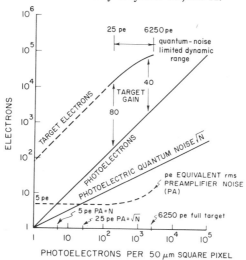

FIG. 13. SEC-vidicon photoelectric transfer function.

6.3.4. Integrating TV System Operation.

Effective use of television sensors for astronomical work requires changes in the operating parameters and in the manufacturing details of TV tubes as contrasted to their other applications. To illustrate this point, and as a guide to prospective users, the major operating and manufacturing differences associated with the astronomical application of a large SEC vidicon will be discussed.

A major construction alteration in SEC vidicons used for long integration astronomy is that they do not have a suppressor mesh between the field mesh and the SEC target. The suppressor mesh is not necessary for the astronomy application because of the sequential operation of the tube. The absence of the suppressor mesh greatly improves the performance of the SEC vidicon.

A recent modification of the SEC vidicon for astronomical use was the introduction of the high storage capacitance target. This target improves the dynamic range of the sensor by raising the number of photoelectrons that can be recorded per unit target area, in a single exposure. Figures 7 and 13 are based on the use of the high storage capacitance target which has a photoelectron storage capacity about 2.5 times greater than that of standard SEC targets. The high storage capacitance target is often not suited for conventional television applications because it exhibits beam discharge lag during fast scan opration with moving scenes. However, that characteristic is not relevant for integrating TV operation with slow scan readout.

Another construction detail of SEC vidicons used for astronomy is the addition of a black aluminum coating to the photocathode side of the target.[18] This coating absorbs light that has passed through the semitransparent photocathode and might otherwise reflect from the target surface back to the photocathode. Also the wall of the image section is coated with chrome oxide to reduce the internal background of the tube caused by surface charging of the image section wall.

6.3.4.1. Sequential Operation of SEC Vidicon.   For the astronomy application, where exposure times of the order of minutes and hours are common, the SEC vidicon is operated in a manner quite different from that employed in broadcast television operation. In normal TV operation the image section of the SEC vidicon is continuously writing an image into the SEC storage target while the reading electron gun is continuously reading out the image that has been integrated in the brief frame or field interval since the beam last scanned a given location.

In the sequential operating modes used in the astronomy application the SEC vidicon is cycled through four modes. They are: PREPARE, EXPOSE, STORE, and READOUT.

In the PREPARE mode the SEC storage target is normalized to make it suitable for exposure. During the EXPOSE mode the photocathode and

[18] P. Zucchino, J. L. Lowrance, Recent developments and applications of the SEC-vidicon for astronomy, in "Advances in Electronics and Electron Physics" (in press).

image section are operating, integrating the incoming image in the SEC storage target. After an exposure the tube may be put into the STORE mode where the photocathode and image section are off and the reading electron gun is also off. The integrated image stored in the SEC target will not degrade perceptibly even after many hours of storage. During the READOUT mode the stored image is scanned out by the reading section's electron gun. During READOUT the photocathode and image section of the tube are off. Since the image section is off, it is safe to have a high field mesh potential during READOUT. (It is also necessary to have had the mesh voltage low during the EXPOSE mode for complete safety during the subsequent READOUT especially if an uncoated SEC target is used.) The high reading beam decelerating field so obtained contributes greatly to the high MTF and low beam pulling characteristics of the suppressor meshless tube.

The PREPARE mode consists of first flooding the photocathode with diffuse illumination while operating the reading beam with low field mesh potential. This is followed by operating the reading gun *only*, first with low mesh voltage and, second, with the high readout mode mesh potential applied while adjusting the target bias upward to the desired level for the EXPOSE mode.

An active PREPARE cycle of this sort is required to obliterate image charge patterns within the SEC target layer that are not neutralized by the readout beam which can only normalize the surface potential of the target. McMullan and Tower have previously reported in detail on the internal charge retention of the SEC target.[19]

Figure 14 illustrates the requirement for an active erase phase in the PREPARE cycle. The lower portion of the figure is a normal readout of a test pattern image. In slow scan operation there is no residual image even on the second scan, that is, the reading beam completely recharges the target surface in one frame, the readout frame. However, if the target is then exposed to flat white illumination, a severe negative image of the previous pattern appears as shown in the middle portion of Fig. 14. The upper portion of the figure shows the image obtained from flat illumination after a PREPARE cycle has followed the test pattern image. Nothing of the test pattern can be seen in this case. (The shading and white marks in the upper portion of Fig. 14 are photocathode and target irregularities of the developmental tube used for this test).

---

[19] D. McMullan and G. O. Towler, *in* "Advances in Electronics and Electron Physics," Vol. 28A, p. 173. Academic Press, New York, 1969.

Another requirement of sequential tube operation is that of "target pulsing." This is a procedure in which the target bias is raised approximately 0.2 V above that used for the final PREPARE step and for the EXPOSE mode. If this were not done, the scanning electron beam would fail to land properly on those areas of the target that have received little or no exposure. The "target pulsing" bias ensures that the scanning beam will land even in those portions of the target where the exposure was zero. Although "target pulsing" ensures that threshold level signals will not be compressed or lost in the readout process, it does give the video signal a rather uneven black or "zero exposure" level.

FIG. 14. Erase illustration. See text for discussion of figure parts.

Figure 15 shows the type of video signals obtained from the SEC tube during sequential operation. The figure consists of oscilloscope traces of single lines of video signal scanned through the middle of the target area. The bottom trace of Fig. 15 is the video signal with the scanning beam turned off. It is the electronic zero level and consists only of preamplifier noise. The middle trace is the signal obtained while reading out a zero exposure. This is the uneven zero level associated with the target pulsing needed to insure proper beam landing. The upper trace is the video obtained while reading out a uniform "full target" exposure.

The increased noise due to the shot noise in the light flux of the exposure can be seen. The unevenness that appears as left to right shading is caused primarily by photocathode nonuniformity in the developmental tube used for these tests.

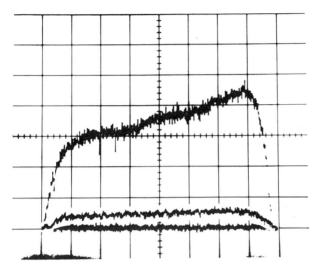

FIG. 15. Video signals. See text for a description of the tracings.

6.3.5. Television Data Processing

The analog video signal can yield useful information directly, but for quantitative photometric measurements the raw data must be rectified to correct for variations in the geometric and photometric transfer function over the format of the image. A major effort has resulted in digital computer programs that can efficiently process the large amounts of data contained in a video image. While it is not practical to discuss these programs in detail it will be helpful to consider the steps necessary to reduce the raw data to calibrated data.

The first step is to calibrate the tube. This is done by illuminating the tube photocathode with uniform illumination at various points along the photometric transfer function. The video is recorded at each level and adjacent picture element are averaged to reduce the noise. This catalog of calibration data is then used for photometric correction of the data obtained in an observation. It is important that the target biasing be the same at the time of observation as during the calibration otherwise a zero exposure offset will be introduced.

The stacking or addition of successive exposures of the same object in order to extend the photometric range of the sensor is quite practical as demonstrated by Crane.[20] Since stacking of frames raises the total photoelectrons observed in each image element, the statistical precision improves as the square root of the number of frames stacked.

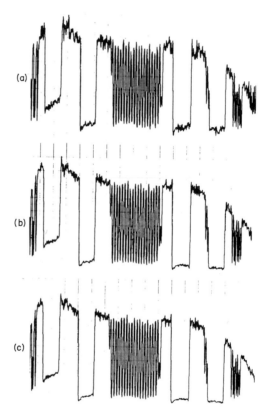

FIG. 16. Video for (a) single, (b) 11, and (c) 36 frames averaged.

Figure 16 shows the same single scan line of digitized video data taken from a single frame, the average of 11 frames, and the average of 36 frames. The noise in the lowest intensity (nominally zero intensity) portions of each line is primarily preamplifier readout noise and it is seen to decrease as the square root of the number of frames averaged. The noise in the highest intensity (nominally 1200 photoelectrons per image

[20] P. Crane, *Bull. of AAS* **3**, 399 (1971).

element) portions of each line does not decrease as the square root of the number of frames stacked. Between one frame and eleven frames the noise decreased roughly 2 to 1 instead of 3.3 to 1. Comparing the first broad pulse of full intensity signal from the left in the one frame and in the 11-frame cases, one sees some structure in the pulse top. Extending the average to 36 frames, the full intensity noise does not seem to reduce at all in peak-to-peak amplitude, and one can find several regions of correlation in the structure of the full intensity traces between the eleven frame and the 36-frame data. It is clear that the statistical noise has averaged down to the point where it is no longer visible among the fine target structure revealed by the stacking process.

The simplest, and for many purposes the most practical, way to reduce the fixed pattern noise in the target is to intentionally use slightly different registrations between the image and the target structure for each exposure. Then the target structure noise, which in single frames is below the statistical image noise, will be averaged down in the same fashion as the image statistical noise. This requires registering the image data between frames prior to stacking. Crane[20] had done this for the 8-frame stacking of Fig. 19.

It should be pointed out that the requirement to stack frames only arises where more than 2000 photoelectrons are to be gathered on any of the 350,000 picture elements usually available. Many astronomical observations do not even "fill" a one-frame exposure. The 6-hr exposure time with the 500-cm (200-in.) Hale Telescope used on PHL 957 (Fig. 17) could have been extended to 24 hr without filling the tube nor resorting to trailing the object along the slit jaws.

6.3.6. Observational Results with Integrating Television

Two major applications of integrating television have been high dispersion spectroscopy of quasi-stellar objects and galaxies and photometry of galaxies. Photometry of other extended objects such as planets and nebulae is also well suited to this type of instrument. Equally interesting is the detection of faint stars against the sky backgrounds, by taking advantage of the wide dynamic range possible with multiframe integration. Two different observations are briefly discussed below.

6.3.6.1. High-Dispersion Spectroscopy on 500-cm (200-in.) Coudé Spectrograph.   The most notable achievement to date has been the observations of very faint quasi-stellar objects and galaxies at high dispersion on the

500-cm (200-in.) Hale telescope's Coudé spectrograph.[21] Spectra of the radio-quiet quasar PHL 957 ($V = -16.5$ mag) were obtained from 4270 to 4495 Å with 0.75-Å resolution in a 6-hr exposure. Thirty-one absorption lines were recorded and several of the lines were resolved and reach zero central intensity. Figure 17 shows the unwidened Coudé

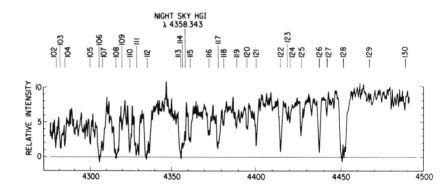

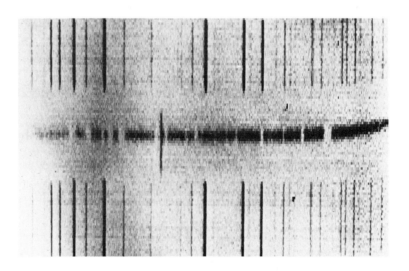

FIG. 17. Spectrum of the quasar PHL-957 obtained with Princeton integrating TV system on Hale 200″ Coudé spectrograph (see text for details).

[21] J. L. Lowrance, D. C. Morton, P. Zucchino, J. B. Oke, and M. Schmidt, *Astrophys.* **171**, 233 (1972).

spectrum of PHL 957 obtained on October 9, 1970 in a 6-hr exposure with an integrating television camera using a SEC-vidicon with a 25-mm square target and a magnetically focused image section. Horizontal lines are the television scan lines, while the smallest rectangles on each line are the digital picture elements whose widths correspond to one third the

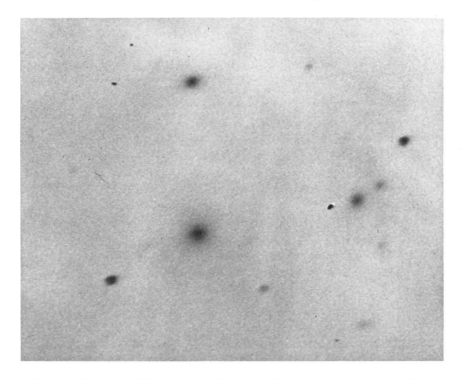

FIG. 18. Galaxies NGC 3841 and 3842 taken with Princeton Integrating TV on Kitt Peak 110-cm (36-in.) telescope, 2.5-min exposure at f/13.5.

net resolution of 0.75 Å. The vertical scale has been magnified 4.5 times relative to the horizontal scale. The comparison spectrum is an iron arc and the emission line crossing the spectrum left of center is the 4358.3-Å line of HgI in the night sky. The intensity trace of the same spectrum is also shown in Fig. 17.

6.3.6.2. Photometry of Galaxies. The same television system has been used at Princeton and Kitt Peak by Crane of Princeton to do photometry of galaxies. Figure 18 shows a digitized television image of the galaxies NGC 3841 and 3842 obtained on the Kitt Peak 80-cm (36-in.) reflector.

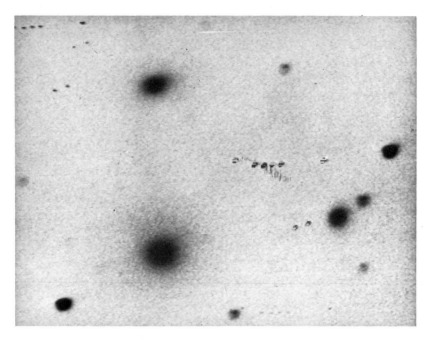

FIG. 19. Digital stacking of eight frames, total exposure of 17 min for NGC 3841 and 3842.

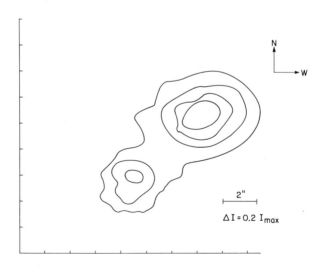

FIG. 20. Plot of equal intensity contours for the central region of the double galaxy NGC 3845 (TV data).

Figure 19 is the result of digitally stacking eight frames of the same objects with a total exposure time of 17 min. An advantage of the television system for this work is the fact that the data is recorded on magnetic tape and can therefore be easily analyzed by computer. The equal brightness contours in Fig. 20 of the doubel galaxy NCG 3845 are an example.

7. X-RAY AND GAMMA-RAY DETECTION BY MEANS OF ATMOSPHERIC INTERACTIONS: FLUORESCENCE AND ČERENKOV RADIATION*

7.1. Introduction

Cosmic X-ray and gamma radiation incident on our planet is absorbed in the upper atmosphere and thus never reaches the surface of the earth. Observations in high-energy astronomy require that the detectors be placed above the atmosphere by balloons, rockets, and satellites. Before 1971, most X-ray astronomy experiments were performed from rockets, with the inherent disadvantages of limited payload and short observation times. In 1971, the first satellite devoted to X-ray astronomy was launched; it circumvented some of the problems in rocket experiments. Balloon-borne X-ray detectors are confined to the hard X-ray spectrum and are hindered by atmospheric background. Despite these difficulties, major discoveries have been made in X-ray astronomy, and progress has been rapid, although experiments are not easy to perform and tend to be expensive.

The difficulties in gamma-ray astronomy are even worse. The fluxes are so low that detectors with large sensitive areas are needed, as well as long observation times. Detectors also tend to be very massive in order to absorb the incident gamma radiation and thus measure its energy spectrum. Since rocket experiments are useless in this field, the observer is limited to balloon flights or satellites. Because there is a continuing need for detectors to become even more sensitive, they tend to become obsolete rapidly. And because new detectors require substantial time to design and construct, there is a long lag between successive observations. Small experiments on satellites have already been flown, and a satellite devoted wholly to gamma-ray astronomy was launched in 1972. Again, gamma-ray astronomy experiments are difficult and expensive to perform.

When these facts are considered, it is interesting to speculate whether,

---

* Part 7 is by G. G. Fazio.

through their interactions in the atmosphere, cosmic X rays and gamma rays can be detected with simple ground-based equipment. The advantages are obvious: (1) extensive regions of the spectrum could be investigated; (2) long observation times would be possible—important because recent results show that sources may be highly variable with time; and (3) the detectors could be simple and inexpensive, and conveniently located.

With these considerations in mind, two techniques have been developed with ground-based optical equipment to perform X-ray and gamma-ray astronomy. The first makes use of the long-known fact that under excitation by X rays, the upper atmosphere fluoresces in the ultraviolet, visible, and infrared spectral regions. This phenomenon has already been applied effectively to the detection of atomic bombs in space, which when detonated give copious yields of X rays.[1-3] Greisen and his colleagues at Cornell University[4,5] have extensively developed the atmospheric fluorescence technique as a method of detecting very large cosmic air showers. In this application, the fluorescence is observed at altitudes much lower than those to which cosmic X rays reach. Greisen's work, however, has stimulated interest in this general field and, particularly, in the applications of this technique to the detection of cosmic X-ray sources. Theoretical predictions by Fichtel and Ögelman[6] indicate that with a relatively simple ground-based detector it should be possible to detect the fluorescent flashes produced in the upper atmosphere by X-ray and gamma-ray bursts created by a supernova explosion. Charman et al.[7] have proposed extension of this technique to the detection of pulsars, variable X-ray sources, and flare stars.

The second ground-based optical technique is used for gamma-ray astronomy at energies above 10^{11} eV. A cosmic gamma-ray photon of this energy can produce in the atmosphere an extensive air shower (EAS), so called owing to the large lateral extent of the many secondary particles (primarily electrons) produced as the shower proceeds through the

[1] S. P. Cunningham and B. C. Murray, Proc. IEEE **53**, 2058 (1965).

[2] D. R. Westervelt and H. Hoerlin, Proc. IEEE **53**, 2067 (1965).

[3] T. M. Donahue, Proc. IEEE **53**, 2072 (1965).

[4] A. N. Bunner, Internal Rep. #62, Cornell-Sydney Univ. Astron. Center (1966).

[5] E. B. Jenkins, Ph.D. thesis, Phys. Dept., Cornell Univ. (1966).

[6] C. E. Fichtel and H. B. Ögelman, NASA Tech. Note No. 4732 (1968).

[7] W. N. Charman, R. W. P. Drever, J. H. Fruin, J. V. Jelley, J. L. Elliot, G. G. Fazio, D. R. Hearn, H. F. Helmken, G. H. Rieke, and T. C. Weekes, IAU Symp. No. 37, "Non-Solar X- and Gamma-ray Astronomy" (L. Gratton, ed.), pp. 41–49, D. Reidel Publishing Co. Dordrecht, Holland (1970).

atmosphere. Since some of the particles in the shower have velocities greater than that of light in air, Čerenkov light is generated. This burst of light, which is partially in the visual spectrum, can be detected with relatively simple ground-based detectors. By nature, Čerenkov light is highly directional, and hence the arrival direction of the initial gamma-ray photon can be determined to within 1°. Another advantage is that, at ground level, the Čerenkov light pool extends over a very large region; hence, the "effective" area of the detector is many orders of magnitude greater than the physical area of the detecting instrument. A disadvantage to this technique for gamma-ray astronomy is that other cosmic-ray particles of comparable energy also produce extensive air showers and provide an isotropic background source of Čerenkov light bursts. The flux of cosmic-ray particles is at least 1000 times greater than the expected gamma-ray flux. Galbraith and Jelley[8] were the first to detect Čerenkov light from the night sky and to search for "point" sources of cosmic rays. Zatsepin and Chudakov[9] first suggested the application of this technique to gamma-ray astronomy.

Efficient ground-based X-ray and gamma-ray astronomy by optical techniques requires a dark observing site with many cloudless nights. Observations will be limited to times when the moon is below the horizon. Atmospheric transparency is also important and hence mountain-altitude sites are preferred, although not necessary.

The subject of this article is limited to optical observations; however, other ground-based techniques can be used to observe cosmic X rays and gamma rays, and we mention two of them here. X rays can cause changes in the properties of the ionosphere (e.g., electrical conductivity of the D region), and these can be detected by radio techniques. Very low-frequency radio waves traveling in the earth–ionosphere cavity would exhibit a phase advance due to a decrease in the effective reflection height caused by absorption of cosmic X rays.[10,11] This technique is particularly appealing because it is simple and because no weather requirements are placed on the observing site.

Extensive air showers above 10^{14} eV can be observed at ground level by detecting their electrons, mesons, and nucleons. Gamma-ray-initiated showers involve predominately electromagnetic interactions; hence, these showers should lack mesons and nucleons, which are produced in nuclear

[8] W. Galbraith and J. V. Jelley, *Nature (London)* **171** (4347), 349 (1953).
[9] G. T. Zatsepin and A. E. Chudakov, *Zh. Eksp. Teor. Fiz.* **41**, 655 (1961).
[10] S. Ananthakrishnan and K. R. Ramanathan, *Nature (London)* **223**, 488 (1969).
[11] G. A. Baird and R. J. Francey, *J. Geophys. Res.* **77**, 1966 (1972).

interactions of cosmic-ray particles. The absence of a penetrating mu-meson component in the shower has been used to help identify cosmic gamma rays; the arrival direction of the gamma ray is determined by the arrival direction of the shower. This technique, however, requires large-area scintillation detector arrays and large mu-meson detectors, which are expensive to build and operate.[12]

7.2. Detection of Cosmic X Rays by Atmospheric Fluorescence

7.2.1. The Fluorescence Process

Protons, electrons, or X rays interacting with a gas may produce ionization and excitation of its molecules. One method by which electronic deexcitation can occur is by the emission of photons (fluorescence). In air, even to very low pressures, the fluorescence process in the visible and near-infrared regions of the spectrum occurs almost entirely in the nitrogen molecule (N_2) and the nitrogen molecular ion (N_2^+). A schematic energy-level diagram is shown in Fig. 1. At low pressures, the three transition systems most responsible for fluorescence are the following: (1) the second positive system N_2(2P), where the line emission is mostly in the near ultraviolet; (2) the first negative system N_2^+(1N), where the emission is mostly in the blue region of the visible spectrum; and (3) the first positive system N_2(1P), where the emission is mainly in the near infrared.

The upper levels of the N_2^+(1N) system can be excited by direct ionization, with excitation by any high-energy interaction, e.g.,

$$e + N_2(X\,^1\Sigma_g^+) \rightarrow e + N_2^+(B\,^2\Sigma_\mu^+) + e.$$

However, the N_2(1P) and the N_2(2P) systems cannot be excited directly from the ground state in a high-energy interaction. These levels can only be excited by low-energy electrons in collisions that involve an electron exchange with a spin change or by cascading from higher levels. Excitation of these levels by X-ray photons occurs through the transfer of the incident X-ray energy into low-energy electrons.

One of the strongest features in the fluorescence spectrum of N_2 and N_2^+ is the (0, 0) band of the N_2^+(1N) system, the head-band wavelength

[12] Y. Toyoda, K. Suga, K. Murakami, H. Hasegawa, S. Shibata, V. Domingo, I. Escobar, K. Kamata, H. Bradt, G. Clark, and M. LaPointe, *Proc. Int. Conf. Cosmic Rays, 9th, London* **2**, 708–711 (1965).

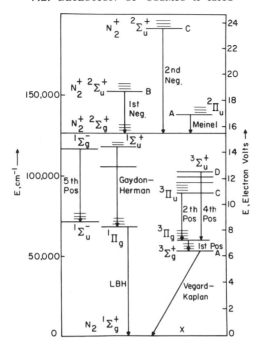

FIG. 1. A schematic energy-level diagram of the N_2 molecule (from A. N. Bunner, Internal Rep. #62, Cornell-Sydney Univ. Astron. Center, 1966).

of which is 3914 Å. A large portion of the energy radiated in this band is emitted within 20 Å of the head-band wavelength. This band is about three times more intense than any other band and accounts for over $\frac{1}{6}$ the total fluorescence energy between 2500 and 12,000 Å for electron input at low energies. The second strongest feature is the (0, 1) band of $N_2^+(1N)$, the head-band wavelength of which is 4278 Å; its emission intensity is 0.34 that of the (0, 0) band. The first negative bands of N_2^+ are particularly important in X-ray and gamma-ray astronomy, as well as in studies of the ionosphere, because their emission intensity is directly proportional to the total energy of the incident particles or photons causing the excitation.

The production efficiency of the 3914-Å band is defined as the total energy emitted in the band divided by the total energy in the incident electrons. At low pressures, the ratio of the cross section of emission of a 3914-Å photon to the formation of an ion pair for electrons above 100 eV is nearly constant and equal to about $\frac{1}{14}$. Approximately 35 eV are required to produce an ion pair, and each 3914-Å photon emitted has an energy of 3.17 eV. Therefore, the efficiency in air at low pressures η_{e0} is approximately

$$\eta_{e0} \approx (\sigma_{3914}/\sigma_{ion})(3.17/35)(0.75) \approx 4.5 \times 10^{-3}, \qquad (7.2.1)$$

where the factor (0.75) is the fraction of the ionization cross section of air due to nitrogen. A more accurate value of η_{e0} determined from an evaluation of present experimental results[13] is

$$\eta_{e0} = 3.4 \times 10^{-3}, \qquad E > 200 \quad \text{eV}, \qquad (7.2.2)$$

or approximately one 3914-Å photon per kiloelectron volt of incident particle energy absorbed. Below an electron energy of 100 eV, the excitation cross section for $N_2^+(1N)$ decreases, causing the decrease in the fluorescence efficiency. Above this energy, the efficiency is almost constant. For electrons above 1 MeV, radiative energy loss (*bremsstrahlung*) becomes important, and unless the radiation is reabsorbed in the gas, the fluorescence efficiency decreases again.

However, the value of η_{e0} is very dependent on pressure because collisional deexcitation is a competing form of energy loss of the excited N_2^+ molecule. From Hartman's data,[13]

$$\eta_e = \eta_{e0}\left[\frac{1}{1 + (p/0.7)} \right], \qquad (7.2.3)$$

where p is the pressure in millimeters of Hg. This pressure dependence of the conversion efficiency is very important in X-ray and gamma-ray astronomy because the atmospheric depth (or pressure) at which a high-energy photon is absorbed is very dependent on its energy. For fluorescence in the upper atmosphere, we shall neglect the temperature-dependent terms in the efficiency equation.

An X-ray photon incident on air transfers almost all its energy to atomic electrons. At X-ray energies to 50 keV, the primary absorption process is the photoelectric effect. The interaction of a photon of energy E_x produces a photoelectron of energy $(E_x - 400)$ eV from the K shell of nitrogen. Almost all the atoms so excited will emit an Auger electron with 322-eV energy. Since there is no method of producing the 3914-Å band by cascading from states above $N_2^+(B\,^2\Sigma_u^+)$, there is a net loss of 78 eV in converting the X-ray energy into electron energy. For oxygen, this loss is 86 eV. Above 50 keV, the Compton process becomes im-

[13] P. L. Hartman, *Planet. Space Sci.* **16**, 1315 (1968).

portant; the incident photon transfers its energy by scattering to one or more electrons before being absorbed by the photoelectron processes. Above 10 MeV, the primary absorption mechanism is electron–positron pair production.

Therefore, for X rays with energy greater than about 750 eV, the fluorescence efficiency in air is constant and approximately equal to the value obtained for excitation by electrons:

$$\eta_x = 3.4 \times 10^{-3}/[1 + (p/0.7)], \qquad E_x > 750 \quad \text{eV.} \qquad (7.2.4)$$

Another important parameter in the fluorescence process is the decay time. Bunner[4] reviews various measurements of this decay time at low pressures and finds a mean value of about 65 nsec. Again, this time is a function of gas pressure, but an important point is that this time is small compared to the time scale of any cosmic X-ray variations or time scales determined by the geometry of the atmospheric volume where the interactions occur. The total observed fluorescence decay time for an X-ray photon incident on the atmosphere is of the order of 1 μsec and is due primarily to the slowing-down time of the electrons.

At high altitudes, atomic oxygen can also be excited into states of O and O^+ at a greater rate than $N_2(1N)$. However, the states in oxygen are metastable, with long decay times, resulting in a lower instantaneous brightness.

7.2.2. X-Ray-Induced Fluorescence Light in the Atmosphere

An X-ray or gamma-ray beam incident on a uniform gas is absorbed according to the relation

$$I(y) = I_0 e^{-\mu \varrho y}, \qquad (7.2.5)$$

where I_0 is the incident intensity, ϱ is the density of the gas (in grams per cubic centimeter), y is the distance traversed in the gas (in centimeters), and μ is the mass absorption coefficient (gm cm^{-2})$^{-1}$, and μ is the sum of the mass-absorption coefficients due to photoelectric effect (μ_τ), Compton absorption (μ_c), and pair production (μ_p). The value of $\mu(E)$ for air as a function of X-ray photon energy E is given in Fig. 2.

The fraction of an X-ray beam that is incident vertically on the atmosphere and that is absorbed in a layer of thickness dh at altitude h is given by

$$dI/I_0 = \mu \exp\left[-\mu \int_h^\infty \varrho(h)\, dh\right] \varrho(h)\, dh; \qquad (7.2.6)$$

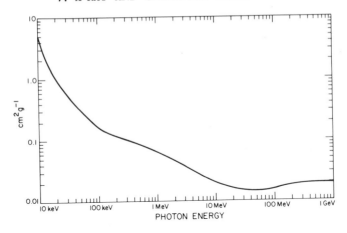

FIG. 2. X- and gamma-ray mass absorption coefficients for air.

therefore, the number of fluorescence photons produced per kiloelectron volt of incident X-ray energy is given by

$$N_{3914}(E) = \left(\frac{10^3}{3.17}\right)\eta_{x0}(E)\mu(E)$$

$$\times \left\{\int_0^\infty \exp\left[-\mu(E)\int_h^\infty \varrho(h)\,dh\right] \cdot \frac{\varrho(h)}{1+[p(h)/0.7]}\,dh\right\}$$

$$= \left(\frac{10^3}{3.17}\right)\int_0^\infty b(h,E)\,dh, \qquad (7.2.7)$$

where $b(h, E)$ is called the *emission distribution function.*

For X rays incident on the atmosphere at an angle θ to the vertical, the above equation is modified, since the X rays are then absorbed at a higher altitude and thus the fluorescence efficiency is improved, particularly for high-energy X rays, but the total incident X-ray flux is reduced by the area factor $\cos\theta$. The emission distribution function then becomes

$$b(h, \theta, E) = \eta_{x0}\mu(E)\cos\theta$$

$$\times\left\{\exp\left[-\mu(E)\sec\theta\int_h^\infty \varrho(h)\,dh\right] \cdot \frac{\varrho(h)}{1+[p(h)/0.7]}\right\}. \qquad (7.2.8)$$

A graph of $N_{3914}(E)$ as a function of incident X-ray photon energy E and incident angle θ is shown in Fig. 3. The sharp decrease in the number of fluorescence photons between 10 and 30 keV reflects the properties of η_x; i.e., higher energy X rays are absorbed lower in the atmosphere,

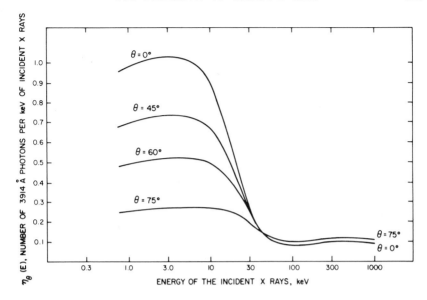

FIG. 3. Production of 3914-Å fluorescence photons, in the atmosphere, per kilo-electronvolt of incident X-ray energy, as a function of X-ray energy and arrival direction θ from the zenith (from J. Elliot, Smithsonian Astrophys. Obs. Spec. Rep. No. 341, 1972).

where the fluorescence efficiency is reduced. Atmospheric fluorescence is most efficient for the detection of cosmic X rays between 750 eV and 20 keV; above energies of 100 keV, the value of N_{3914} is rather constant, but only 10^{-2} of its value for lower energy X rays.

For an X-ray beam of given energy and incident in a normal direction on the atmosphere, the altitude of maximum fluorescence radiation is given in Table I, along with the half-width of the fluorescence layer and the relative efficiency.

TABLE I. The Altitude of Maximum Fluorescence Emission As a Function of X-Ray Energy for Normal Incidence

X-ray photon energy (keV)	Altitude of maximum fluorescence (km)	Half-width of fluorescence layer (km)	$\eta_{x}(p)/\eta_{x0}$
2.3	88	13	0.99
15	55	22	0.64
50	33	25	0.11

Implicit in the above formulas is the assumption that the fluorescence is produced at the same altitude at which the X-ray photon is absorbed. Above 1 MeV, this assumption breaks down owing to the production of the high-energy secondary electrons, which proceed to lower altitudes.

In summary, the upper atmosphere has many excellent properties as a detector for cosmic X rays: (1) fast response time, (2) linear energy response, i.e., over many orders of magnitude the fluorescence light output is linearly proportional to the energy deposited, (3) wide energy bandwidth in the X-ray spectrum, and (4) relatively high efficiency.

7.2.3. Ground-Based Detection of Fluorescence Light

In the previous section, the distribution of fluorescence light in the atmosphere was calculated. We now compute how much of this light reaches a ground-based detector.

Consider a phototube whose photocathode, with area A, is in a horizontal plane, viewing the entire sky. This is the simplest detection system for a parallel beam of cosmic X rays incident on the upper atmosphere at an angle θ to the vertical. If the incident X-ray flux has a differential energy distribution given by $W(E)$, then the volume emissivity in the atmosphere at an altitude h is given by

$$B(\theta, h) = \int_{E_1}^{E_2} b(\theta, h, E) W(E)\, dE \qquad \text{keV (cm}^3 \text{ sec)}^{-1}, \qquad (7.2.9)$$

where $W(E)$ is in units of (keV cm^{-2} sec^{-1} keV^{-1}).

Assuming the fluorescence light is radiated isotropically, the number of photons emitted per unit volume per unit solid angle at altitude h is

$$B(\theta, h)/4\pi\varepsilon \quad \text{photons (cm}^3 \text{ sec sr)}^{-1},$$

where ε is the fluorescence photon energy in kiloelectron volts; for 3914 Å, $\varepsilon = 3.17 \times 10^{-3}$ keV.

Therefore, the total number of photons incident on the photocathode is given by

$$N_\varepsilon = (1/4\pi\varepsilon) \int_v B(\theta, h)\Omega\, dV, \qquad (7.2.10)$$

where Ω is the solid angle subtended by the detector from the source volume element dV.

For the case of the plane-parallel atmosphere with no atmospheric absorption (Fig. 4a), the value of $\Omega\, dV$ is given by

$$\Omega\, dV = 2\pi A \sin\psi\, d\psi\, dh; \qquad (7.2.11)$$

therefore,

$$N_\varepsilon = A/2\varepsilon \int_0^\infty B(\theta, h)\, dh. \qquad (7.2.12)$$

Another experimental situation arises when the sky is viewed with a mirror or through a telescope. For these cases, consider a detector of area A and small solid angle Ω_d, inclined to an angle ψ with respect to

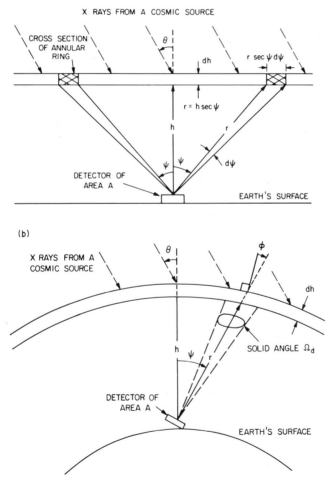

FIG. 4. (a) Fluorescent light from a plane-parallel atmosphere, arriving at a ground-based detector that views a large solid angle; (b) fluorescent light from a spherical shell, arriving at a ground-based detector that views a small solid angle (from J. Elliot, Smithsonian Astrophys. Obs. Spec. Rep. No. 341, 1972).

the vertical. Let us also assume a spherical-shell model for the atmosphere (Fig. 4b), where ϕ is the angle between the normal to the spherical shell and the viewing axis of the detector. Then,

$$N_\varepsilon = A\Omega_d \sec \psi/4\pi\varepsilon \int_0^\infty B(\theta, h) \, dh. \qquad (7.2.13)$$

Absorption and scattering of radiation in the 3500–4000-Å region must also be considered. The most important processes are Rayleigh scattering, aerosol scattering, and absorption by ozone. The attenuation of light may also be time variable. The intensity of a beam of light of wavelength λ after traversing a distance y through a medium is given by

$$I = I_0 \exp\left[- \int_0^y k(y) \, dy \right] = I_0 e^{-\tau}, \qquad (7.2.14)$$

where I_0 is the incident intensity, k is the attenuation coefficient, and τ is the optical depth. The optical depth τ can be calculated from tables given by Elterman.[14] For example, at a mountain site 2.3 km above sea level in southern Arizona, for 3914-Å radiation the value of τ was determined to be 0.41 for a vertical path through the atmosphere.

When absorption is included, Eq. (7.2.10) becomes

$$N_\varepsilon = (1/4\pi\varepsilon) \int_0^\infty B(\theta, h) \exp[-\tau(h) \sec \psi] \, \Omega \, dV. \qquad (7.2.15)$$

Since most of the absorption occurs near the surface of the earth, the absorption term can be removed from the integral as an approximation:

$$N_\varepsilon = [\exp(-\tau \sec \psi)/4\pi\varepsilon] \int_V B(\theta, h)\Omega \, dV. \qquad (7.2.16)$$

Cunningham and Murray[1] have considered the effect of clouds and haze on the fluorescence pulse shape and attenuation. They conclude that the wide-field detection system is little affected up to conditions of heavy overcast or fog. They also conclude that in the narrow field of view, a thin haze will not affect the system, but that a cloud in the field of view would totally obscure the signal and scatter in unwanted light. Hence, the narrow-field system should be operated only under clear skies.

[14] L. Elterman, AFCRL 68-0153, Air Force Cambridge Res. Lab. (1968).

Another important parameter in the detection of atmospheric fluorescence is the time duration of the radiation received at the ground-based detector when an X-ray burst, of the form of a δ function in time, is incident on the atmosphere. Since the absorption and fluorescence processes occur in times of the order of 1 μsec, the width of the detected pulse will depend primarily on the geometry of the experiment. Light from different regions of the fluorescing volume arrives at the detector by different paths; for light from an altitude of 80 km, this effect broadens the pulse to tens or hundreds of microseconds.

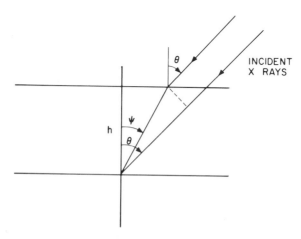

FIG. 5. Geometry of the delay in arrival time at the detector of an atmospheric fluorescence burst.

Figure 5 shows a parallel beam of X rays incident on a plane-parallel atmosphere at an angle θ. The first fluorescent light to reach the detector is from angle θ, followed by light from an ever-increasing angular deviation from the initial direction. The delay in the arrival time of light from an angle ϕ is given by

$$t = (\bar{h}/c) \sec \psi [1 - \cos(\theta - \psi)], \qquad (7.2.17)$$

where c is the velocity of light and $\bar{h}$ is the effective height of the fluorescence layer. The magnitude of the pulse width can be obtained by considering the simple case of an X-ray beam incident in the vertical direction ($\theta = 0$) and of a detector viewing the entire sky (Fig. 4a). In this case,

$$t = (\bar{h}/c)(\sec \phi - 1), \qquad (7.2.18)$$

and therefore the number of photons received as a function of time is given by

$$n(t)\, dt \propto dt/[1 + (ct/\bar{h})]^2. \qquad (7.2.19)$$

From Eq. (7.2.19), we see that: (1) the rise time of the fluorescence pulse is rapid, limited only by the interaction process (~ 1 μsec) and by the rise time of the incident X-ray burst; and (2) the half-width of the pulse in time, for $h \approx 80$ km, is $\Delta t \approx 100$ μsec, and this is the minimum width for an all-sky detector.

For a detector with a limited field of view (Fig. 4b), the pulse width is dependent on the angle ψ that the detector makes with the vertical and on the zenith angle θ of the incident X-ray beam. Cunningham and Murray[1] have computed the pulse width as a function of detector solid angle and incident X-ray angle. For a detector with $\Omega < 10^{-2}$ sr, the pulse half-width is given by

$$\Delta t_{1/2} \approx \frac{2 \sin^2[\tfrac{1}{2}(\theta - \phi)]}{\cos \theta} \left(\frac{\Delta h}{c} \right) + \Delta t_0,$$

where Δh is the thickness of the fluorescing layer and Δt_0 is the pulse width due to the interaction process and the natural width of the incident X-ray pulse. Cunningham and Murray[1] also give, for a ground-based detector oriented toward the zenith, the pulse width as a function of the field of view of the detector and the incident X-ray angle θ.

Table II, taken from Elliot,[15] gives the minimum pulse widths in microseconds as a function of ψ, and θ for a detector with $\Omega = 0.16$ sr, which is the effective solid angle of a 30-cm (12-in.) phototube at the focus of

TABLE II. Minimum Fluorescence Pulse Widths (μsec) As a Function of θ and ψ

ψ (deg)	θ				
	$-60°$	$-30°$	$0°$	$+30°$	$+60°$
0	110	60	8	60	110
30	240	180	90	9	60
45	460	350	200	50	60
60	1100	860	550	240	15

---

[15] J. Elliot, Smithsonian Astrophys. Obs. Spec. Rep. No. 341 (1972).

1.5-m (5-ft) searchlight mirror. The shortest pulse width corresponds again to the case of the incident X-ray beam being in the direction of the detector axis ($\theta = \psi$). Consider three such detectors at the same location, with the same zenith angle ψ, but each toward a different azimuth direction. The pulse widths of an event registered in all three detectors could be used to identify a high-altitude fluorescence pulse and yield some information on arrival direction of the X-ray burst. The arrival direction of the burst can be determined more accurately with the above system by noting the relative arrival times for the three pulses. For $h = 80$ km, $\psi = 45°$, and differential timing accuracy of ± 25 μsec, the direction of the X-ray burst can be located to within an error circle $10°$ in diameter.

Greater accuracy in determination of the direction of the X-ray burst can be obtained by separating the detectors far apart in a triangular pattern and noting the arrival time of the fluorescence pulse at each station. For example, for detectors located 5000 km apart, the arrival direction could be determined to less than $1°$. The disadvantages of this system are that each station must maintain absolute time to within an accuracy of 100 μsec and that darkness and clear weather must be present simultaneously at each.

7.2.4. Sensitivity for Detection of Cosmic X-Ray Sources

To determine what kinds of cosmic X-ray sources can be observed by the atmospheric fluorescence technique, we need to calculate the minimum detectable X-ray flux for a given signal-to-noise ratio.

The signal strength in the receiver photomultiplier tube is proportional to the number of photoelectrons produced at the cathode. From Eq. (7.2.15), we can approximate the signal strength by the formula

$$S = B_s A \Omega_d T \xi t, \qquad (7.2.20)$$

where S is the number of photoelectrons received in time t from the atmospheric layer emitting B_s fluorescence photons $(\text{cm}^2 \text{ sec sr})^{-1}$; A and Ω_d are the detector area and effective solid angle, respectively; T is the atmospheric transmission; and ξ is the product of the photocathode efficiency and the optical transmissivity of the detector. The value of B_s is related to the cosmic X-ray flux by

$$B_s = I_x \eta_x / 4\pi \varepsilon, \qquad (7.2.21)$$

where I_x is the incident X-ray flux (keV cm^{-2} sec^{-1}), and $\varepsilon = 3.17 \times 10^{-3}$ keV for 3914-Å radiation.

The noise signal will be due to fluctuation in the number of photo-electrons produced by the background source, e.g., the night-sky airglow:

$$N = (B_b A \Omega_d T \xi t)^{1/2}, \qquad (7.2.22)$$

where B_b is the brightness of the night-sky light

$$B_b = B_{\lambda b}\, \Delta \lambda \quad \text{photons (cm}^2 \text{ sec sr)}^{-1}. \qquad (7.2.23)$$

At $\lambda = 4000$ Å, the value of B_b is 4.0×10^4 photons (cm² sec sr Å)$^{-1}$.
The signal-to-noise ratio is

$$\frac{S}{N} = B_s \left(\frac{A \Omega_d T \xi t}{B_{\lambda b}\, \Delta \lambda} \right)^{1/2}, \qquad (7.2.24)$$

and the minimum detectable value of B_s is

$$B_{s0} \geq \left(\frac{S}{N} \right) \left(\frac{B_{\lambda b}\, \Delta \lambda}{A \Omega_d T \xi t} \right)^{1/2}. \qquad (7.2.25)$$

Thus, the minimum detectable cosmic X-ray flux is given by

$$I_{x0} = 4 \pi \varepsilon B_{s0}/\eta_x \quad \text{keV (cm}^2 \text{ sec)}^{-1}. \qquad (7.2.26)$$

When the optical bandwidth $\Delta \lambda$ of the detector is being determined, one of two situations arises: (1) the use of broad-band filters permitting large values of $A \Omega_d$, or (2) the use of small interference filters for 3914 Å, with small values of $A \Omega_d$.

In any given optical system, the product $A \Omega_d$ is invariant throughout the system. In an optical detector with an interference filter, the maximum value of Ω_d is given by

$$\Omega_d = 2 \pi \mu^2 (\Delta \lambda)/\lambda, \qquad (7.2.27)$$

where μ is the index of refraction of the filter. Thus, for $\lambda = 20$ Å at 3914, with $A = 20$ cm² (for a phototube of 5-cm diameter), $A \Omega_d$ is only 1.2 cm² sr.

The use of broad-band glass or gelatin filters permits much larger phototubes to be employed; for example, with Corning #5970 or #5850 filters and EMI-9434B, a phototube of 25-cm diameter is possible. One system[15] in operation uses this phototube-filter combination at the focus of a 150-cm (60-in.) mirror ($f/0.43$). In this case, $A \Omega_d = 1150$ cm² sr and $\Delta \lambda \approx 1000$ Å. The value of $\Delta \lambda / A \Omega_d$ is lower, and hence B_{s0} is lower, than in a system using interference filters. Later, we shall discuss the disadvantages of a broad-band system.

To determine the minimum detectable X-ray flux, let us take the broad-band system described above as a typical example:

$$A\Omega_d = 1150 \quad cm^2 \ sr, \qquad T = e^{-\tau} = 0.67,$$
$$B_b = B_\lambda \, \Delta\lambda = 3.4\times10^7 \quad photons \ (cm^2 \ sec \ sr)^{-1}, \qquad \xi = 0.08. \tag{7.2.28}$$

If we require a signal-to-noise ratio of 4, then

$$B_{s0} = 2.9\times10^3(t^{-1/2}) \quad photons \ (cm^2 \ sec \ sr)^{-1}, \tag{7.2.29}$$

$$I_x = 34(t^{-1/2}) \quad MeV \ (cm^2 \ sec)^{-1}$$
$$= 5.4\times10^{-5}(t^{-1/2}) \quad erg \ (cm^2 \ sec)^{-1}. \tag{7.2.30}$$

The strongest continuous source of X rays in the sky is the star-like object Sco X-1. In the energy region between 1.5 and 6 keV, the source strength is maximum and equal to 2.4×10^{-7} erg cm^{-2} sec^{-1}. In principle, to detect this source, the difference in the phototube current when the source is above and when it is below the horizon would have to be measured. The integration time "on" the source is given by Eq. (7.2.30):

$$t_{Sco} = 5.3\times10^4 \quad sec = 15 \quad hr. \tag{7.2.31}$$

If background fluctuations that would mask the effect can be avoided over this integration time, it may be possible to detect this source.

The next class of objects to consider are those sources that emit transient or periodic X-ray bursts of large amplitude. If a burst occurs in a time t_0, the minimum detectable energy at the earth is

$$I_x t_0 \geq 34(t_0)^{1/2} \quad MeV \ (cm^2)^{-1} \tag{7.2.32}$$

$$\geq 5.4\times10^{-5}(t_0)^{1/2} \quad erg \ (cm^2)^{-1}. \tag{7.2.33}$$

Sco X-1 is known to flare, with changes in intensity of the order of 50%; but in the 1–10-keV region, the fraction of emitted power that is periodic is less than 1%. Larger fluctuations are observed at higher energies, but the emitted power is less and the fluorescence efficiency is lower. Cen X-3 and Cyg X-1 show much larger fluctuations in the power emitted in the 1–10-keV region, but the energy received at the earth is approximately 10^{-8} erg cm^{-2} sec^{-1}, 1/20 that received from Sco X-1. If a source has a known periodicity and if the signal can be integrated in phase over n periods, then the minimum detectable flux is lowered by the factor $n^{-1/2}$.

The one X-ray object whose period is well known is the pulsar NP 0532 at the center of the Crab Nebula. It emits radio, optical, and X-ray pulses with a period of approximately 33 msec. In the region 0.5–9 keV, the primary pulse is about 1 msec wide, and the secondary pulse, which follows the primary pulse by 13.2 msec, is about 5 msec wide. X-ray emission at a lower level also occurs between the two pulses. The primary and secondary pulses contain approximately equal amounts of flux in the 0.5–9-keV region; but in the 100–400-keV region, the ratio of the secondary to primary is 2.3 ± 0.2. The pulsar has been detected up to an energy of 1 MeV. The integrated X-ray flux is 1.4 keV cm^{-2} sec^{-1} ($\sim 10^{-9}$ erg cm^{-2} sec^{-1}), corresponding to 47 eV cm^{-2}/period.

Charman et al.[7] first proposed detecting the pulsar by the atmospheric fluorescence technique. Consider the broad-band detector system previously described, which uses three 1.5-m searchlight mirrors, each with a 30-cm (12-in.) photomultiplier tube at the focus. Integration of the pulsar signal can be achieved by cycling a multichannel analyzer, operating in a multiscaler mode, at the pulsar period. To avoid detecting the optical pulsations, the field of view of the mirrors should not include the pulsar. The detector has an area solid-angle factor of 3450 cm^2 sr; the photocathode-filter efficiency factor is 0.081, and the atmospheric transmission is given by $\tau = 0.41$. The number of photoelectrons expected per pulse is 0.7, whereas the noise level due to airglow is 1.36×10^4 photoelectrons per period when the pulse is on. A pulse width of 20 msec for the total width of the signal is used. Therefore, the signal-to-noise ratio for one period is 5.1×10^{-5}. If we expect a signal-to-noise ratio of 4, we must integrate for

$$t = [4/(5.1 \times 10^{-5})]^2(1/30) = 2.1 \times 10^9 \quad \text{sec} = 5.5 \times 10^4 \quad \text{hr}, \quad (7.2.34)$$

which is impractical and nearly impossible. Before the X-ray intensity was well known, Elliot[15] attempted to detect the pulsar with a system like the one above. Integration for 19 hr showed no effect. Randall and Frederick, at Ohio University, are performing a similar experiment.

Grindlay[16] has proposed that flare stars such as UV Cet should emit X rays on a time scale of 1 min, but again the predicted flux of 2×10^{-7} erg cm^{-2} sec^{-1} is too low to detect.

Thus, of X-ray sources now identified, only Sco X-1 may be detected;

[16] J. Grindlay, Astrophys. J. 162, 187 (1970).

the other sources appear too weak to be detected by the atmospheric fluorescence technique. Detection of Sco X-1 by this technique would provide an absolute calibration of the flux sensitivity. However, two other cases where theoretical predictions indicate that sufficient X-ray intensity should exist are supernova explosions and X rays accompanying gravitational radiation bursts. These pulses are very short and occur very infrequently, and hence could easily have been missed by present X-ray detectors. The atmospheric fluorescence technique is an excellent method of searching for such rare energetic events.

7.2.5. A Search for X-Ray Pulses from Supernovae

The total energy released in a supernova outburst is predicted to be as high as 10^{52} ergs. To the present, no X rays have been detected during the initial stages of the explosion, although upper limits of 2×10^{42} erg sec^{-1} have been determined for a supernova in NGC 4254 about 34 days after maximum light emission[17] and 10^{45} erg sec^{-1} 7 days after a supernova in NGC 1275 was reported.[18] Tucker[19] has considered X-ray emission from supernova outbursts as the origin of the diffuse X-ray background and estimates that $\sim 3 \times 10^{50}$ ergs in X rays are emitted in the early phase of such an event. Present isotropy measurements of the X-ray background set a limit on the time width of the X-ray pulse, $t_x \gtrsim 10^{-2}$ years. Colgate[20] has suggested that $\sim 5 \times 10^{47}$ ergs are emitted as high-energy gamma rays in a time of 15 μsec owing to a shock wave associated with the supernova explosion.

Stimulated by Colgate's calculations, Fichtel and Ögelman[6] first proposed searching for fast X-ray pulses from supernovae by means of atmosphere fluorescence. The technique has many advantages for such a search: sensitivity over a large energy interval, long observation times, and the large fraction of the sky that can be explored at the same time. In principle, the arrival direction of the X-ray pulse could also be determined.

The number of supernova events per unit time in a sphere of radius R is given by

$$n = f \varrho_g (\tfrac{4}{3} \pi R^3), \qquad (7.2.35)$$

[17] H. Brandt, S. Naranan, S. Rappaport, F. Zwicky, H. Ögelman, and E. Boldt, *Nature* (*London*) **218**, 856 (1968).

[18] P. Gorenstein, E. Kellogg, and H. Gursky, *Astrophys. J.* **156**, 315 (1969).

[19] W. H. Tucker, *Astrophys. J.* **161**, 1161 (1970).

[20] S. A. Colgate, *Can. J. Phys.* **46**, S476 (1968).

where f is the event rate per galaxy, $3.3 \times 10^{-10} \sec^{-1}$ (one per hundred years) and ϱ_g is the space density of galaxies, 9×10^{-76} cm^{-3}. Radius R is determined by the threshold energy E_t for the detector,

$$E_t = W_{SN}/4\pi R^2, \qquad (7.2.36)$$

where W_{SN} is the supernova energy emitted as X rays or gamma rays. Therefore,

$$n = 3.5 \times 10^{-2} f \varrho_g W_{SN}^{3/2}/E_t \quad \sec^{-1}. \qquad (7.2.37)$$

The numerical factor in the above equation has been multiplied by 0.37 to account for the fact that only the sky within a zenith angle of 75° is included.

For hard X rays and gamma rays, as in Colgate's model, E_t is 10 times the value given in Eq. (7.2.32), i.e., $340(t_0)^{1/2}$ MeV cm^{-2}, owing to the decrease in fluorescence efficiency. The value of t_0 is 10^{-4} sec, as determined by the fluorescence pulse, even if the supernova burst was 15 μsec or less. Using Colgate's value of W_{SN}, we have

$$n = 2.9 \times 10^{-7} \quad \sec^{-1} \qquad (7.2.38)$$

or 1 event every 10^3 hr. Ögelman and Bertsch[21] calculate a rate of about 1 event/18 hr, owing to the different fluorescence efficiency and the different galactic space density that were used.

If the initial X-ray burst of 5×10^{47} ergs from a supernova is in the region 0.7–15 keV, the threshold energy is lower owing to increased fluorescence efficiency. In this energy region, the fluorescence signal is proportional to cos θ, where θ is the zenith angle of the supernova. The net effect of angular dependence is to decrease the number of observable galaxies by a factor of 2. If $t_0 = 10^{-4}$ sec, then $E_t = 0.34$ MeV cm^{-2} and $n = 4.6 \times 10^{-6} \sec^{-1}$, or 1 event/63 hr of observing time.

The first experiments to search for fluorescence pulses generated by prompt X rays and gamma rays from supernova events were by Ögelman and Bertsch.[21] In its original form, their detection station consisted of three 30-cm (12-in.) photomultiplier tubes viewing the entire sky. Two of these tubes had broad-band violet filters (3100–4300 Å), sensitive to 3914 Å and other fluorescence bands; the third tube had a yellow filter with a lower wavelength cutoff of 4300 Å. The signals from the tubes

[21] H. Ögelman and D. Bertsch, *Proc. Int. Conf. Cosmic Rays, 11th, Budapest* (A. Somogyi, ed.), **1**, 35–44 (1970).

with violet filters were integrated on a time scale of 50 μsec and then placed in coincidence. When a coincidence occurred, the delayed signals from all three tubes were displayed on an oscilloscope and photographed. The time of the event was recorded to an accuracy of 10 msec. A signal in the yellow channel was used as an indication that the pulse was not due to fluorescence. The sensitivity of this system is comparable to that of the three-mirror system described above and used by Elliot.[15]

To eliminate any local events that could produce background pulses, Bertsch and Ögelman then operated two identical stations, first over a separation distance of 175 km and then a distance of 3300 km. Accurate arrival times (∼10 msec) were recorded at each station to locate coincident events. The results are given in Table III. The expected number of events is based on the calculations of Bertsch, Fischer, and Ögelman.[22] For 230-hr running time, 4.4 events were expected but none was observed.

Charman and Jelley[23] used two fluorescence detectors, each consisting of a combination of 30-cm (12-in.) photomultiplier tube and broad-band filter mounted at the focus of a 1.5-m (5-ft) mirror. The experiment was operated with the two mirrors viewing the same region of the sky and then with each viewing a different region. The two signals were integrated for 10 μsec and placed in coincidence. When a coincidence occurred, the delayed signals were displayed on an oscilloscope and photographed. In 127 hr of operation on clear nights, no supernova-type events were seen.

At the Smithsonian Astrophysical Observatory's site on Mt. Hopkins, Arizona, Elliot[15] has operated the three-mirror system previously described, with each mirror viewing a different region of the sky. The system was in operation a total of ∼200 hr from March–May, 1971, but no supernova-type events were detected.

The search for coincidences over the 3300-km baseline continues as a collaborative effort between the Goddard Space Flight Center and the Smithsonian Astrophysical Observatory.

The total observing time to date by all groups is of the order of 1000 hr. Depending on the estimates, 1–10 events should have been seen, but none was. Obviously, more observing time is needed.

[22] D. L. Bertsch, A. Fischer, and H. Ögelman, *Proc. Int. Conf. Cosmic Rays, 12th, Hobart, Tasmania* **2**, 811 (1971).

[23] W. N. Charman and J. V. Jelley, AERE Rep. R. 6848 (1971).

TABLE III. Summary of Supernova Running Times

Station location	Operation interval	Threshold (photons cm^{-2} in 50 μsec)	Expected supernova rate (events hr^{-1})	Running time (hr)	2-Station coincidence time (hr)	Expected number coincidences
Goddard, Maryland	September, 1968–September, 1969	200	1/96	450	170	1.7
Fan Mountain, Virginia	June–December, 1969	70	1/21	350		
Mt. Hopkins, Arizona	November, 1969	50	1/13	60	60	2.7
Middle East Technical University, Turkey	June, 1970–January, 1971	100	1/36	376		
Number of events expected in single-station runs	~36	Possible number ≳ 10				
Number of events expected in coincidence	~4.4	Observed number = 0				

7.2.6. A Search for X-Ray Pulses Associated with Gravitational Radiation Pulses

Weber[24] has reported evidence for the detection of gravitational radiation pulses from the region of the galactic center at the rate of ~ 1 per day, with energy $\sim 10^4$ erg cm^{-2} per event.

Jelley[25] recently proposed that the atmospheric fluorescence technique would be an excellent means of searching for X-ray pulses that might accompany these events. Jelley assumes the gravitational pulses arise from collapsing objects and that the time scale is of the order of 0.1 sec for the X-ray pulse. Several models of stellar collapse give times of 0.1–3.0 sec for the bulk of the prompt emission. Using $t_0 = 0.1$ sec, we have from Eq. (7.2.32) a minimum detectable flux of 1.7×10^{-5} erg cm^{-2}. Jelley quotes a higher energy threshold per event of 1.8×10^{-4} erg cm^{-2} owing to the higher signal-to-noise ratio he used and a lower fluorescence efficiency. The lower energy limit, due to absorption, for X rays from the galactic center is about 3 keV. Thus, if only 10^{-8}–10^{-9} of the energy in gravitational radiation is emitted as X rays, the pulse could be detected by a ground-based system.

The fluorescence technique is particularly applicable to this search because of its wide X-ray energy bandwidth and its nondirectional properties. The latter are important, since the beamwidth of Weber's detectors is approximately $70°$; a long-baseline coincidence system in the southern hemisphere or low latitudes in the northern hemisphere would thus be most useful. Arrival times of the fluorescence pulses could be compared with those of the gravitational radiation pulses.

In the past, very little work has been done in the search for fluorescence light pulses in the time scale of 0.1–1 sec; however, Weekes, Smithsonian Astrophysical Observatory, has such a system in operation at the present time.

Recently, Baird and Pomerantz[26] also searched for an X-ray pulse accompanying pulses of gravitational radiation by using a rather simple and small balloon-borne X-ray detector (NaI scintillator). The balloons were launched from the Antarctica as part of a cosmic-ray program and stayed aloft for times up to 6 days. No flux was detected in coincidence with Weber's events and an upper limit to the X-ray flux was 2×10^{-5}

[24] J. Weber, *Phys. Rev. Lett.* **22**, 1320 (1969).

[25] J. V. Jelley, *Nature (London)* **234**, 142 (1971).

[26] G. A. Baird and M. A. Pomerantz, *Phys. Rev. Lett.* **28**, 1337 (1972).

erg cm^{-2} per event. This technique appears to be more sensitive than the fluorescence technique in searching for such X-ray pulses.

7.2.7. Background Sources of Light

Elliot[15] has very thoroughly investigated the sources of background light in atmospheric fluorescence experiments. This chapter is a summary of his results.

7.2.7.1. Night-Sky Airglow. At a good astronomical observing site, the primary source of background light in a fluorescence experiment is the night-sky airglow. The airglow intensity[27] as a function of wavelength, averaged over spectral lines, is given in Table IV.

TABLE IV. Night-Sky Airglow Intensity Averaged over Spectral Lines

Wavelength (Å)	Intensity (10^{-7} erg cm^{-2} sec^{-1} sr^{-1} Å^{-1})
3600	1.1
4000	1.3
4500	2.0
5000	2.0
5500	3

Starlight, when averaged over the sky, contributes about 15% of the total background light at 5300 Å. In the blue region of the spectrum, where fluorescence observations occur, this intensity decreases.

Although the 3914-Å band is prominent in spectra of airglow during aurorae and at twilight, this band is not observed during undisturbed conditions at low magnetic latitudes.[28-30]

The shot noise produced by the night-sky airglow is present on all time scales and is the fundamental limit to the sensitivity of a fluorescence detector. Also, the average level of this background light varies smoothly through the night; on clear nights this variation rarely exceeds ±30%.

[27] C. W. Allen, *in* "Astrophysical Quantities," 2nd ed. Athlone Press, London, 1965.
[28] R. H. Eather and B. J. O'Brien, *J. Atmos. Terr. Phys.* **30**, 1585 (1968).
[29] M. F. Ingham, *Mon. Notices Roy. Astron. Soc.* **124**, 505 (1962).
[30] F. Jacka, R. C. Schaeffer, and J. T. Freund, *Nature (London)* **228**, 984 (1970).

7.2.7.2. Man-Made Light. At almost all observing sites, man-made lights contribute to the general background light. Not only does this light increase the background shot noise, but also, in the search for fast pulses, the 120-Hz component of this light causes the signal baseline to fluctuate, thus changing the discrimination level on the fluorescence pulses. At Mt. Hopkins Observatory, Arizona, the 120-Hz component was barely detectable and so was not a serious problem. Drever, Jelley, and Lawless[31] have reduced the intensity of the 120-Hz component by electronically adding a signal 180° out of phase with the detected signal.

7.2.7.3. Background-Light Effects on the Search for Pulsars. Since an X-ray pulsar produces a periodic narrow burst of radiation at a given fundamental frequency, the frequency spectrum of this radiation contains a significant amount of power in harmonics. For the pulsar NP 0532, the basic frequency is 30 Hz and harmonics can be important up to 300 Hz. In any attempt to detect this pulsar by the fluorescence technique, a major background source is again the continuous airglow; in addition, there are two other important sources of noise in the frequency range of interest: stellar scintillations and man-made 120-Hz light scattered into the detector.

The power spectrum of the scintillation light is greatest at low frequencies, where it can exceed the shot noise due to airglow. The scintillation light decreases with increasing frequency until 500–1000 Hz, where it merges with the airglow noise. The strength of this low-frequency scintillation noise and the shape of its spectrum are determined by (1) meteorological conditions, (2) the zenith angles of bright stars in the field of view, and (3) the diameter of the detector. The sensitivity of a bare photomultiplier viewing the whole sky will be limited by scintillation noise, not shot noise, for frequency measurements less than 300 Hz and pulse widths greater than 500 μsec. This noise level is greatly reduced by using a large-area detector of limited field of view (mirror and photomultiplier tube) aimed away from bright stars.

The man-made light produced by fluorescent and incandescent lamps has a fundamental frequency of 120 Hz and harmonics at 240 and 360 Hz. These frequencies are near the harmonic frequencies of NP 0532 and can add noise to the pulsar signal. This light enters the detector primarily by scattering, and its intensity is dependent on site location and on atmospheric conditions.

[31] R. W. P. Drever, J. V. Jelley, and B. Lawless, preprint (1969).

7.2.7.4. Background-Light Effects on the Search for Pulses of the Order of 1 sec. The primary source of background pulses in the 1-sec time scale is meteors. Here, the intensity and duration of the light can vary over a wide range. Typically, for a bright meteor, the light intensity rises gradually to a maximum brightness ($M_v = -5$) in the order of 1 sec, followed by a decay in tenths of a second.[32] The integral pulse-height spectrum[33] of the light from many meteors follows a power law with spectral index -1.25 ± 0.5. A detector consisting of a 5-cm (2-in.) photomultiplier tube, with a 3914-Å interference filter ($\Delta\lambda = 38$-Å FWHM) and a field of view of 0.59 sr, can observe meteors down to $+3.5$ mag, with a counting rate of about 3 per hr. Near 3914 Å, a fast bright meteor exhibits strong Fraunhofer H and K lines of calcium (3969 and 3933 Å), as well as lines of singly ionized iron, silicon, and strontium. A strong continuum radiation is also observed.

Therefore, in looking for cosmic X-ray bursts with duration of ~ 1 sec, one can easily confuse meteors with high-altitude fluorescence light pulses. This background is serious because the number of meteor pulses detected is large compared with the expected rate of X-ray events. In addition to spectral confirmation to identify meteors, a method of reducing this background is to use a three-detector system operating in coincidence, with each detector viewing a different region of the sky.

7.2.7.5. Background-Light Effects on the Search for Fast Pulses. In the time domain of 10^{-3} to 10^{-5} sec, there are three known sources of light pulses that could be confused with atmospheric fluorescence pulses: lightning, xenon flash lamps, and atmospheric Čerenkov light. Two other types of pulses, of unknown origin, have been identified by Ögelman et al.[34] and observed by other groups[15,23]: (1) pulses with rise times of 2×10^{-4} sec and fall times of 400 μsec (called A-type pulses) and (2) fast atmospheric pulses (called FAP), of 1-msec duration, characterized by a 10-kHz damped oscillation in the yellow region of the spectrum.

Since lightning is a spark discharge in air, its spectrum contains not only the 3914-Å band of N_2^+, but also the much stronger line Hα (6562 Å). Therefore, a phototube with a yellow filter, viewing the entire sky, can act as an anticoincidence signal to omit these pulses. The shapes of

[32] G. S. Hawkins, in "The Physics and Astronomy of Meteors, Comets and Meteorites," McGraw-Hill, New York, 1964.

[33] R. E. McCrosky and A. Posen, Smithsonian Contr. Astrophys. 4, 15 (1961).

[34] H. Ögelman, A. Fischer, and D. Bertsch, Proc. Int. Conf. Cosmic Rays, 12th, Hobart, Tasmania 2, 805 (1971).

the lightning pulses also lack a definite pattern, being highly variable in amplitude and shape. Under certain conditions, distant lightning is more difficult than near lightning to distinguish from a fluorescence event, since the 3914-Å light can be scattered more easily than the Hα light. Lightning can even provide false pulses for stations operating in coincidence over a long baseline (~100 km).

Airplanes carry xenon flash lamps as navigational aids, and even more powerful lamps are used as warning lights atop tall buildings. The flashes have a pulse width of ~250 μsec and occur at a steady rate, typically 1 sec^{-1}. If it were not for their intensity, they could easily be mistaken for a pulsar. The light atop the Prudential Tower of Boston was easily detected, by atmospheric scattering, more than 54 km away.

Čerenkov light bursts are generated by cosmic-ray showers in the atmosphere. These bursts occur in ~10 nsec and have an angular extent of ~1°. The light spectrum is most intense in the blue region, having an energy spectrum, per unit wavelength, proportional to λ^{-3}. A detector used for atmospheric fluorescence can easily detect this light. For the single-mirror—phototube detector viewing in the zenith direction, or for the all-sky detector, the rates were typically of the order of 12 hr^{-1}. The rate falls very rapidly with increasing zenith angle. In the system incorporating three-mirror—phototube detectors, each viewing a different region of the sky, the coincidence rate for Čerenkov light was only 0.5 hr^{-1}. These events are easily distinguished on oscilloscope traces by their very fast rise time and short duration.

The anomalous pulses detected by several groups provide a dilemma as to their origin. Ögelman et al.[34] propose that the A-pulses (200-μsec rise time, 400-μsec decay time) are caused by atmospheric electric-current systems. These A-pulses can reach event rates of 50 hr^{-1} on certain nights, and they show a diurnal variation peaking at 20^h UT. They appear predominantly in clear weather and are distinctly different from lightning.

The FAP pulses were first noticed as coincident pulses in stations separated by 175 km. They are characterized by (1) 1-msec duration, (2) damped 10-kHz oscillation in yellow filter tube and 200-μsec rise time, 400-μsec fall time in pulse in blue filter tubes, and (3) an approximate 2.5-min repetition interval.

The typical energy represented by these pulses is at least 4×10^{-8} erg cm^{-2} in the yellow region and 1.5×10^{-8} erg cm^{-2} in the blue. From the fact that these pulses are observed over a long baseline, one can deduce that if the light is fluorescence, 10^9 erg of energy was deposited in the upper atmosphere. It is known that electrons and protons released from the magnetosphere can cause fluorescence.[35] Ögelman et al.[34] have proposed that these particles generate the FAP pulses, with the injection triggered by solar plasma interacting with the magnetosphere. They note a 27-day repetition rate of an enhancement in the rate of FAP particles and associate this with flares and solar rotation.

Elliot,[15] using the three-mirror detector system, has also detected the FAP pulses but attributes them to some type of low-altitude event, probably man made. The pulses have too much yellow light, and the measured relative pulse delays and widths are not consistent with high-altitude fluorescence. Further study on the nature of these pulses is necessary.

7.2.8. Conclusions and Future Experiments

In the investigation of known cosmic X-ray sources, the atmospheric fluorescence technique cannot compete with rocket and satellite experiments. The most obvious use of the technique in X-ray astronomy is the search for infrequent, strong, cosmic X-ray pulses. Present detection techniques could have easily missed such events. The advantages of the fluorescence technique is all-sky coverage, long observing times, wide X-ray energy range with linear response, and low cost.

X-ray bursts of the necessary magnitude and pulse width have not been detected, but only proposed theoretically as accompanying supernova explosions or gravitational radiation pulses. Table V gives the minimum source energy in an X-ray pulse of 0.7–15 keV needed to detect the

TABLE V. Minimum Source Energy (ergs) for an X-ray Pulse (0.7–15 keV) to be Detectable with the Fluorescence Station Used by Elliot[a]

Origin of the pulse	Duration of the pulse		
	10^{-5} sec	10^{-3} sec	10^{-1} sec
Crab Nebula (2 kpc)	1×10^{38}	1×10^{39}	1×10^{40}
Galactic center (8 kpc)	2×10^{39}	2×10^{40}	2×10^{41}
Virgo cluster of galaxies (11 Mpc)	4×10^{45}	4×10^{46}	4×10^{47}

[a] J. Elliot, Smithsonian Astrophysical Obs. Spec. Rep. No. 341 (1972).

[35] A. Dalgarno, Ann. Géophys. 20, 65 (1964).

event with present experiments. Searches for these pulses should continue. Evidence from experiments in progress is not strong enough to eliminate the existence of such energetic X-ray bursts.

From experience gained through several such stations operated by various groups, future detection systems should have the following properties:

(1) They should consist of two or more stations operated in coincidence and located at least 1000 km apart, preferably on the same longitude to maximize observing time.

(2) Elliot[15] has suggested each station consist of three detectors operated in coincidence and oriented at the same elevation angle, but having azimuth angles 120° apart, to reduce background signals and provide some arrival direction information. A fourth detector, equipped with a yellow filter and viewing the entire sky, should also be used to identify background pulses.

(3) Each individual detector should consist of a 12.7-cm (5-in.) photomultiplier tube (with a quantum efficiency of $\sim$28%) and a 3914-Å interference filter. The phototube should also be collimated to give the advantages of a narrow field of view.

(4) Theoretical estimates of the X-ray pulse widths vary over a large range. Therefore, the station should ideally have the capability of detecting and recording pulses with widths from 10^{-5}–1 sec, and of timing the arrival of these pulses to as high an accuracy as possible.

An attempt should be made to detect the continuous X-ray flux from Sco X-1. Detection of Sco X-1 would permit an absolute flux calibration of the detector system as well as provide a means of developing more sensitive detectors. Finally, much more work should be done on identifying the origin of the anomalous 100-μsec light pulses observed.

7.3. Detection of Cosmic Gamma Rays by Atmospheric Čerenkov Radiation

7.3.1. Čerenkov Radiation in the Atmosphere

When a fast particle traverses a dielectric medium with a velocity v greater than the phase velocity c_n of light in that medium, the particle generates an optical shock wave similar to the shock-wave effect in acoustics. This effect is illustrated in Fig. 6. The emission angle θ of the light with respect to the particle direction is given by

$$\cos \theta = 1/\beta n, \qquad (7.3.1)$$

where β is the velocity of the particle in units of the velocity of light in vacuum and n is the index of refraction of the dielectric medium.

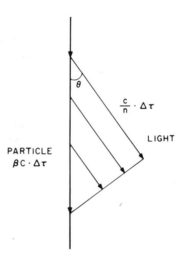

FIG. 6. Geometry of the Čerenkov light wave front emitted by a relativistic particle.

From Eq. (7.3.1) we note two important facts:

(1)   For a given value of n, there exists a threshold velocity for the particle to emit Čerenkov radiation

$$\beta_{\min} = 1/n; \qquad (7.3.2)$$

(2)   For $\beta = 1$, there is a maximum angle of emission of Čerenkov radiation

$$\theta_{\max} = \cos^{-1}(1/n). \qquad (7.3.3)$$

In air, under conditions of STP, $n = 1.0029$; thus, when $\beta = 1$, then $\theta_{\max} = 1.3°$, and the threshold energies are 21 MeV for electrons and positrons, 4.3 GeV for muons, and 39 GeV for protons.

In air, the value of n can be written as

$$n = 1 + \eta, \qquad \eta \ll 1, \qquad (7.3.4)$$

and hence,

$$\theta_{\max} \approx (2\eta)^{1/2} \quad \text{rad} \approx 81\eta^{1/2} \quad \text{deg}, \qquad (7.3.5)$$

and the threshold kinetic energy is

$$KE_{min} = m_0 c^2 [(2\eta)^{-1/2} - 1].$$ (7.3.6)

In the first approximation, $\eta \propto \varrho \propto P$, where ϱ and P are the atmospheric density and pressure, respectively. The variation of η with altitude h is given by

$$\eta = 2.9 \times 10^{-4} \exp(-h/h_0),$$ (7.3.7)

where h_0 is the scale height, 7.1 km.

The rate of production of photon energy per unit path length for a particle of charge Ze is

$$(dW/dl) = (Z^2 e^2/c^2) \int_{\beta n > 1} [1 - 1/(\beta^2 n^2)] \omega \, d\omega,$$ (7.3.8)

where ω is the angular frequency of the radiation.

Hence, the number of photons between λ_1 and λ_2 is

$$dN/dl = 2\pi\alpha(1/\lambda_2 - 1/\lambda_1) \sin^2 \theta,$$ (7.3.9)

where $\alpha = 1/137$. With the approximation of $\sin \theta \approx \theta$, in the wavelength interval between 3500 and 5000 Å

$$dN/dl \approx 780\eta \quad \text{photons cm}^{-1}.$$ (7.3.10)

For air at STP, the value of dN/dl is 30 photons cm^{-1}.

The spectral distribution of the light, i.e., the number of photons per unit path length per unit wavelength, is proportional to $1/\lambda^2$; hence, most of the Čerenkov light emission is in the blue and ultraviolet regions of the spectrum.

Jelley[36] and Boley[37] have reviewed extensively the production of Čerenkov radiation in the atmosphere. In his review Jelley considered a variety of effects that could modify this simple picture, e.g., absorption, dispersion, diffraction, refraction, and Coulomb scattering. The most important of these effects is the last. From Rossi,[38] the mean square angle of scattering is given by

$$\langle \theta_s^2 \rangle = (E_s^2/p^2\beta^2)t,$$ (7.3.11)

[36] J. V. Jelley, *Progr. Elementary Particle Cosmic Ray Phys.* **9**, 39–159 (1967); "Cerenkov Radiation and its Applications," 304 pp. Pergamon, Oxford, 1958.

[37] F. I. Boley, *Rev. Mod. Phys.* **36**, 792 (1964).

[38] B. Rossi, *in* "High Energy Particles," Prentice Hall, Englewood Cliffs, New Jersey, 1952.

where t is the particle path length measured in radiation lengths, which for air is 37.7 gm cm^{-2}, and E_s is a constant equal to 21 MeV. For highly relativistic particles in air, i.e., $\beta \approx 1$ and $E \approx p$,

$$\langle \theta_s \rangle_{rms} \approx 21/E \quad \text{rad.} \tag{7.3.12}$$

Thus for 100-MeV electrons, $\langle \theta_s \rangle_{rms} \approx 12°$, or approximately 5 times the Čerenkov light cone angle.

Since the Čerenkov photons travel with a velocity slower than that of the particle emitting them, the delay time after traversing a distance D is given by

$$\tau_D = \eta D/c.$$

For air, $\tau_D \approx 10^{-8}$ sec. This time is also the typical pulse width for the light.

Blackett[39] first proposed that the Čerenkov radiation emitted by all cosmic rays incident on the earth's atmosphere must contribute to the night-sky light. This contribution, however, amounts to only 10^{-4} of the total brightness of the night sky.

7.3.2. Properties of Čerenkov Radiation Generated by Cosmic-Ray Air Showers

A high-energy gamma-ray photon or electron incident on the atmosphere can generate a cascade of photons and electrons in which the electrons produce gamma-ray photons by *bremsstrahlung*, and these photons in turn produce electrons and positrons by pair production. The number of particles in such a shower builds up exponentially, reaching a maximum number at a depth in the atmosphere given by

$$t_{max} \approx \ln E/E_c, \tag{7.3.13}$$

where t_{max} is the depth measured in radiation lengths. A radiation length X_0 is the length in which an electron loses $1 - (1/e)$ of its energy by radiation. In air,

$$X_0 = 37.7 \quad \text{gm cm}^{-2} = 2.92(1/P_{atm})(T_k/273) \cdot 10^2 \quad \text{m,} \tag{7.3.14}$$

where P is the air pressure in atmospheres, T_k is the absolute tem-

[39] P. M. S. Blackett, *Rep. Conf. Gassoit Comm. Phys. Soc.*, p. 34, Physical Society, London, 1948.

perature of the air in degrees Kelvin; E_c is a constant called the "critical energy," 84.2 MeV, which is the mean energy lost by ionization in traversing a radiation length. The number of electrons at the maximum of the shower is given by

$$N_{e,\text{max}} \approx E_0/10^9. \tag{7.3.15}$$

After the maximum is reached, the number of particles decreases owing to absorption in the atmosphere. Figure 7 shows the number of electrons in a shower as a function of depth in the atmosphere for various initial gamma-ray energies. For example, a shower initiated by a 10^{12}-eV photon would reach a maximum development at $\sim$8.2-km altitude and contain $\sim 10^3$ electrons. Very few of these electrons would reach a depth corresponding to sea level.

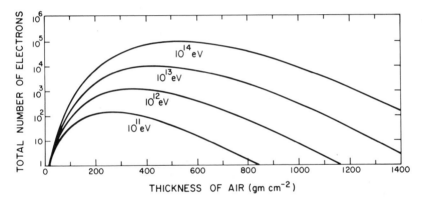

FIG. 7. Number of electrons as a function of thickness of air for an electromagnetic shower of various energies.

However, most of the high-energy particles incident on the atmosphere are nucleons, primarily protons, and the ratio of gamma rays to protons is less than 10^{-3}. The primary result of the interaction of a high-energy proton in the atmosphere is a cascade consisting of high-energy nucleons and mesons. These series of nuclear collisions occur near the axis of the incoming particle at various depths in the atmosphere, and the resulting particles are propagated along the direction of the initial particle. A photon–electron cascade also develops, as a secondary phenomenon, owing to the decay of mesons into two gamma-ray photons. Charged π-mesons $\pi^{\pm}$, also produced in the cascade, can produce $\mu^{\pm}$ mesons, which are relatively long lived ($\sim 10^{-6}$ sec); they interact weakly and hence are able to propagate to sea level.

Thus, a nucleon-initiated shower consists of a core of very high-energy nucleons and mesons, as well as some high-energy electrons and gamma-ray photons, surrounded by an extended distribution of an electromagnetic cascade of electrons and gamma-ray photons, plus some μ-mesons.

A shower initiated by a proton of energy 10^{12} eV reaches its maximum development at 7.7 km ($\sim$350 gm cm^{-2}) and contains $\sim$500 particles. The number of μ-mesons produced is approximately $0.01E_p$ (GeV), or 10 μ-mesons.

The primary contribution to the Čerenkov light from an air shower is from the electromagnetic cascade, since the nuclear particles contribute only a small fraction to the total number. Thus, in the first approximation, a gamma-ray-initiated shower cannot be distinguished from a proton shower of similar energy by ground-based detection of Čerenkov light. The physical properties of the Čerenkov light produced in showers have been studied extensively by Zatsepin and Chudakov,[40] Long,[41] and Rieke.[42] The average properties of a burst of Čerenkov light accompanying an air shower are given in Table VI.

TABLE VI. Average Properties of Čerenkov-Light Burst Accompanying an Air Shower

Photon intensity (at 2000-m altitude, for a 10^{11}-eV gamma-ray photon primary)	$\sim$10 photons m^{-2}
Duration	$\sim$10^{-8} sec
Angular extent	$\sim$1° half angle
Radius of light pool (at 2000-m altitude)	$\sim$100–200 m (10^4–10^5 m^2 area)

The lateral distribution of Čerenkov light from a 10^{11}-eV photon-initiated shower as a function of the detector's field of view, the temporal characteristics of the light pulse, and the light intensity as a function of distance from the shower core, are shown in Figs. 8 and 9, respectively. These data were taken from Rieke's[42] work.

All the above data correspond to detection of Čerenkov light when the detector points to the zenith and the shower axis is near the detector. Data for the size and duration of the light burst for off-axis showers are also contained in the review papers of Zatsepin and Chudakov,[40] Long,[41]

[40] V. I. Zatsepin and A. E. Chudakov, Sov. Phys.—JETP 15, 1126 (1962).
[41] C. D. Long, Ph.D. thesis, Nat. Univ. of Ireland, Dublin (1967).
[42] G. H. Rieke, Smithsonian Astrophys. Obs. Spec. Rep. No. 301 (1969).

and Rieke.[42] As the zenith angle of the shower increases, both the threshold energy for detection of the primary particle and the light-pool area increase.

Although the difference in the Čerenkov light produced by proton-initiated extensive air showers (p-EAS) and gamma-ray showers (G-EAS) is small, the difference becomes important if we wish to distinguish between the two types of primaries. A G-EAS gives more Čerenkov

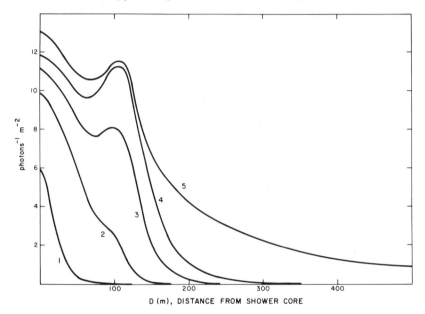

FIG. 8. Lateral distribution of Čerenkov light from a 10^{11}-eV photon-initiated air shower as a function of the half-angle a of the detector field of view. The detector is pointed in the direction of the shower axis (from G. H. Rieke, Smithsonian Astrophys. Obs. Spec. Rep. No. 301, 1969). (1) $a = 1/6°$; (2) $a = 1/2°$; (3) $a = 5/6°$; (4) $a = 7/6°$; (5) a large distribution of total light from shower.

photons at a given level than does a p-EAS with the same energy, or equivalently, the effective primary energy threshold is lower for G-EAS. Another distinction is that the lateral distribution of the Čerenkov photon density at a given atmospheric depth is slightly flatter for G-EAS than for the corresponding p-EAS at $r \sim 100$ m from the core. Rieke[42] has shown that an enhancement exists in the Čerenkov photon density at $r \approx 100$ m for G-EAS. However, one of the most significant distinguishing characteristics is the number of μ-mesons in the shower. For G-EAS, the ratio of μ-mesons to electrons is almost zero.

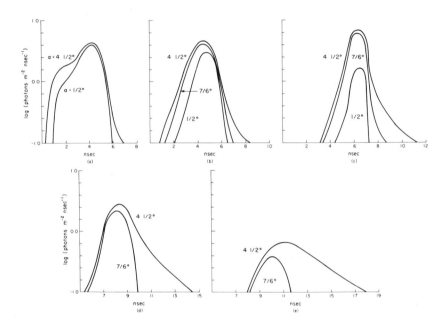

FIG. 9. Temporal characteristics of Čerenkov light pulse from 10^{11}-eV photon-initiated air shower as a function of the half-angle of the detector field of view and distance from the shower core: (a) 0 m, (b) 50 m, (c) 100 m, (d) 150 m, and (e) 200 m. The detector is pointed in the direction of the shower axis.

7.3.3. Detection Techniques for Cosmic Gamma Rays

Čerenkov light bursts produced by cosmic-ray air showers are relatively simple to detect. The basic equipment, which is shown schematically in Fig. 10, consists of a mirror with a photomultiplier tube at the focus, and the associated electronics necessary to amplify the signal and count the number of light pulses above a given amplitude. Typically, war-surplus searchlight mirrors of 1.5-m (5-ft) diameter and 20-arcsec resolution have been used.

The threshold sensitivity in these experiments is limited by the background light; hence, observations must be made on dark, clear, moonless nights. For the most sensitive experiments, e.g., for gamma-ray astronomy, atmospheric transparency becomes important, as well as the number of observing nights. A mountain site at 2300 m, in the Arizona desert, has proved an ideal location for these observations.

A typical integral pulse-height distribution of light pulses from the night sky is shown in Fig. 11. The curve is characterized by a very steep slope due to the shot noise of background light from the night sky (e.g., star light and airglow), superimposed by a much flatter slope due to the Čerenkov light bursts from cosmic-ray air showers. The spectral index in this latter region is -1.6, which is identical to the integral spectral index of the primary cosmic radiation.

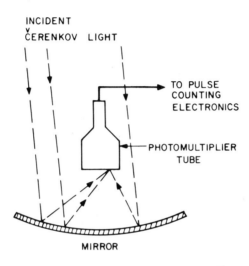

FIG. 10. Simple apparatus for the detection of atmospheric Čerenkov light from an air shower.

The absolute threshold (photons per meter squared) of the receiver for Čerenkov light can be estimated from the noise level detected. In most experiments, this noise level is determined by the fluctuations in the number of photoelectrons emitted at the cathode of the phototube owing to night-sky background light. This source greatly exceeds the noise due to the normal dark current in the phototube. Owing to the limited bandwidth of the phototube and the electronics and to the intensity of the background light, photoelectron "pile-up" exists. The average number of photoelectrons emitted at the cathode is given by

$$\overline{N}_{\mathrm{e}} = \phi_{\mathrm{B}} A \Omega \varepsilon \tau, \tag{7.3.16}$$

where ϕ_{B} is the mean flux of the night-sky background light, A is the collecting area of the receiver, Ω is its field of view, ε is the cathode conversion efficiency, and τ is the resolving time of the system ($\tau = 2f_c$,

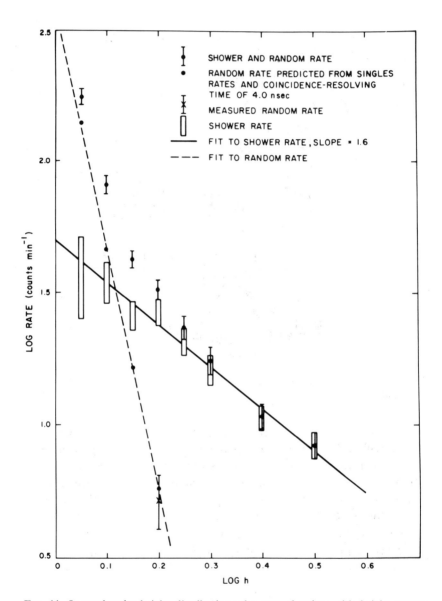

FIG. 11. Integral pulse-height distribution; the rate of pulses with height greater than h (arbitrary units) for a system of two searchlight mirrors operating in coincidence, with 200-MHz electronics, half-angle field of view of 0.5°, at an altitude of 2320 m (from G. H. Rieke, Smithsonian Astrophys. Obs. Spec. Rep. No. 301, 1969).

where f_c is the bandwidth of the system). The noise level is then given by

$$\overline{\Delta N_e} = (N_e)^{1/2} = (\phi_B A \Omega \varepsilon \tau)^{1/2}. \tag{7.3.17}$$

Jelley[36] has given a more accurate formula for $\overline{\Delta N_e}$ that takes into account the effects of the secondary emission ratio of the first dynode. For a given signal-to-noise ratio ξ, the minimum detectable Čerenkov light density is given by

$$\phi_{\check{C}} = \xi(\overline{\Delta N_e})/\varepsilon A = \xi(\phi_B \Omega \tau/\varepsilon A)^{1/2}. \tag{7.3.18}$$

In this equation, we assume the angular field of view of the detector is greater than the extent of the Čerenkov light burst. Typical characteristics for two detector systems in current use are given in Table VII.

The ability to detect a cosmic gamma-ray source by using atmospheric Čerenkov radiation is, in most cases, related to the threshold energy of the detector. For given values of Ω, τ, and A, the integral shower count rate R varies as $E_T^{-1.6}$, where E_T is the threshold cosmic-ray particle energy. The statistical noise varies as $R^{1/2}$ and hence as $E_T^{-0.8}$. If the gamma-ray spectrum of the source is of the form $F_\gamma \propto E_T^{-\gamma}$, then the signal-to-noise ratio is given by

$$\xi = F_\gamma/R^{1/2} \propto E_T^{(0.8-\gamma)}. \tag{7.3.19}$$

Thus, if $\gamma > 0.8$, the maximum signal-to-noise ratio is achieved at the lowest threshold energy. Jelley[36] has shown that, for constant values of Ω and τ, the minimum detectable energy varies as $1/D$, where D is the diameter of the detector mirror. Hence, the larger the detector mirror is, the smaller the threshold energy. The Smithsonian Astrophysical Observatory's 10-m reflector represents a practical limit of 9×10^{10} eV for cosmic gamma rays (Fig. 12).

Although, in any given detector, the pulses due to background light can be eliminated by high discriminator settings, the use of two or more detectors operating in fast-time coincidence permits the threshold level to be closer to the noise level. This results not only in lower threshold energy but also in increased sensitivity due to the greater stability in the shower count rate.

The background light seen by the detector can also vary owing to bright stars drifting in and out of the field of view and to variations in the night-sky transparency. In the search for small variations in the count rate as a function of time, as in gamma-ray astronomy, use is made of

TABLE VII. Characteristics of Instruments Used to Detect Atmospheric Čerenkov Light from Cosmic-Ray Air Showers

Detector	Phototube	Effective area (m²)	Effective solid angle (sr)	Electronic bandwidth (MHz)	Photon-density threshold (photon m⁻²)	Altitude (m)	Rate (counts min⁻¹)	Energy threshold (eV)	Collecting area (m²)
Mt. Hopkins searchlights									
$\alpha = 2°$			1.4×10^{-3}		80–120		165	1.4×10^{12}	1.0×10^{5}
$\alpha = 1°$	RCA 4518	2.5	3.5×10^{-4}	100	40–60	1280	55	1.2×10^{12}	4.4×10^{4}
$\alpha = 0.5°$			1.5×10^{-4}		30–40		20	9×10^{11}	1.6×10^{4}
$\alpha = 0.5°$			2.2×10^{-4}	200	30–40	2320	25	6×10^{11}	2.2×10^{4}
Mt. Hopkins 10-m reflector	RCA 4522	60	1.5×10^{-4}	100	4–6	2320	300	9×10^{10}	1.3×10^{4}

FIG. 12. The 10-m reflector at Mt. Hopkins, Arizona.

a small lamp in the field of view, operated in a servo loop to control the average phototube current to a constant value and hence keep the noise level constant.[43]

Since the detectors have a narrow solid angle, it is necessary to orient them in the direction of the suspected celestial source. With the search-light system, the usual practice has been to operate two or more in co-incidence with the drift-scan mode. In this mode, the detectors are aligned manually on stars 20–30 min of right ascension ahead of the suspected gamma-ray source; the earth's rotation then sweeps the field of view of the detectors over the source. Each drift scan takes about 40–50 min, and even longer for sources of high declination. The source

[43] J. H. Friun and J. V. Jelley, *Can. J. Phys.* **46**, S1118 (1968).

is in the field of view typically for only 8 min of this time. Detection of a source would be indicated by an increase in the shower count rate while the source transits. Many drift scans must be accumulated on each object, since the expected anisotropy is less than 1%.

In the first year of operation, the 10-m reflector was also used in the drift-scan mode. A two-channel coincidence system was achieved by focusing the outer mirror elements to one focus and the inner mirror to another focus $4°2$ apart. Although the drift-scan mode has advantages in terms of stability and ease of operation, it is most inefficient in that less than 20% of the observing time is spent on the source. To increase the time on source, a tracking mode was adopted for the 10-m reflector. In this mode, two phototubes were used at the focus, separated by $4°2$ and centered symmetrically about the reflector axis. The mirrors were aligned to a single focus on the reflector axis. The reflector tracked the suspected gamma-ray source in such a manner that one phototube viewed the source while the second viewed the background shower rate "off" source. Every 10 min, the fields of view of the tubes were interchanged by moving the reflector in azimuth. Approximately 1 min was required for this operation, and thus 90% of the time was spent observing the source. Since a coincidence system was no longer possible, the discriminator thresholds were set well above the noise level.

The major limitation to the sensitivity of these experiments is the isotropic background due to cosmic-ray proton showers. Several groups have sought to distinguish gamma-ray-initiated showers from proton-initiated showers by making use of subtle differences in the light distributions from the two types of showers (cf., Section 7.3.2). Tornabene and Cusimano[44] used an array of spaced detectors to take advantage of the difference in the lateral distribution of the light. O'Mongain et al.[45] used a fast coincidence system to detect a fast annular component of the light in gamma-ray-initiated showers. This ring of light is generated by high-energy electrons in the shower core and focused by the atmosphere.

However, the most successful experiments in distinguishing proton-initiated air showers (p-EAS) from gamma-ray showers (G-EAS) have been performed by Grindlay.[46] He has been able to distinguish the Čerenkov light from the electrons at the maximum of the shower's

[44] H. S. Tornabene and F. J. Cusimano, Can. J. Phys. 46, S81 (1968).

[45] E. P. O'Mongain, N. A. Porter, J. White, D. J. Fegan, D. M. Jennings, and B. G. Lawless, Nature (London) 219, 1348 (1968).

[46] J. E. Grindlay, Nuovo Cimento 2B, 119 (1971); Smithsonian Astrophys. Obs. Spec. Rep. No. 334 (1971).

electromagnetic cascade (at a height h_{max}) from the Čerenkov light due to the unscattered and local shower "core" of predominantly pions, muons, and secondary electrons. These latter particles would be present only in p-EAS. The technique uses two searchlight-mirror detectors operated in the coincidence mode and separated by about 70 m, with each mirror rotated inward from the vertical by an angle θ, so that each is pointed at the shower maximum (see Fig. 13 for $E_\gamma \approx 10^{12}$ eV, $h_{max} \approx 6.2$ km, and $\theta \approx 0.35°$). A third mirror system is used to detect the penetrating shower core ($h \approx 3.5$ km and $\theta = 0.65°$). Because the light

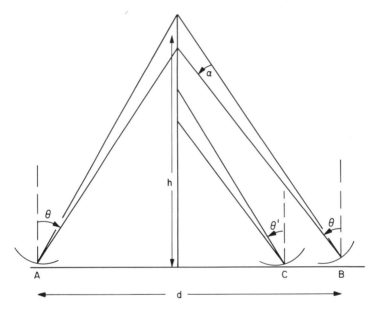

FIG. 13. Geometry of detectors used by Grindlay to distinguish proton-initiated showers from gamma-ray-initiated showers. Points A, B, and C are Čerenkov light detectors separated by the distance d.

from the lower component is relatively nearby, it is rich in the ultraviolet component and hence can be distinguished from light at the shower maximum. A G-EAS is registered only when light from the shower maximum is seen not accompanied by light from the lower level.

With this technique, Grindlay has distinguished the two types of shower with at least 70% probability. For gamma-ray energies greater than or equal to 5×10^{11} eV, the combined effects of p-EAS rejection and increased angular resolution have made possible an order of magnitude increase in sensitivity. The drift-scan mode was used in these experiments, and the complicated pointing geometry permitted only 5% of the operating time to be spent viewing the source. Similar observations are now being performed by tracking the source with the mirrors of the stellar interferometer at Narrabri, Australia.

7.3.4. Results and Future Experiments

In the early 1960s, two groups, one led by Chudakov[47] in the USSR and the other by Jelley in England and Long et al.[48] in Ireland, began using the detection of atmospheric Čerenkov radiation to search for cosmic gamma rays from discrete sources. Among the first investigated were strong cosmic radio sources, such as Cygnus A, Cassiopeia A, Taurus A (Crab Nebula), and Virgo A, as well as several quasars (3C 147, 3C 192, and 3C 273). The threshold energies for these experiments were about 5×10^{12} eV and the sensitivity about 10^{-10} photon cm^{-2} sec^{-1}. No significant enhancements in the shower rate in the direction of these sources were observed. Any enhancements observed were less than 1–2% of the p-EAS shower rate. In 1967, work began at Mt. Hopkins, Arizona, first with searchlight mirrors, and then in 1968 with the 10-m reflector, and observations with the latter instrument have continued to the present (Rieke;[42] Weekes et al.[49]). The group at University College, Dublin, Ireland, in collaboration with the group at UKAERE, Harwell, England, has remained active, with experiments being performed in Ireland and on Malta. Other groups currently active in gamma-ray astronomy include Tornabene (USA), Tata Institute of Fundamental Research (India), the Crimean Observatory (USSR), and recently the University of Sydney (Australia).

The primary source for observations has been Taurus A (Crab Nebula), a supernova remnant, seen to explode in 1064 A.D. At the center of the Nebula is the pulsar NP 0532, with the shortest period known. The Nebula exhibits strong synchrotron radiation in the radio, optical, and X-ray regions of the spectrum and hence is a strong candidate for gamma-ray emission. Several models based on Compton scattering of the high-energy electrons on their own synchrotron radiation predict a detectable

[47] A. E. Chudakov, V. I. Zatsepin, N. M. Nesterova, and V. L. Dadykin, *J. Phys. Soc. Japan* (Suppl. A-III) **17**, 106 (1962).

[48] C. D. Long, B. McBreen, N. A. Porter, and T. C. Weekes, *Proc. Int. Conf. Cosmic Rays, 9th, London* **1**, 318 (1965).

[49] T. C. Weekes, G. G. Fazio, H. F. Helmken, E. O'Mongain, and G. H. Rieke, *Astrophys. J.* **174**, 165 (1972).

gamma-ray flux. Until recently, no significant evidence for a gamma-ray flux could be presented. Grindlay,[50] using the proton-shower rejection technique, has reported evidence for pulsed gamma rays with energy greater than 6×10^{11} eV from the pulsar, occurring at the pulsar period. Fazio et al.[51] have reported detection for a continuous gamma-ray flux with energy greater than 2×10^{11} eV from the Nebula, with evidence that the flux is time variable, reaching its maximum 60–120 days following a pulsar spinup.

Other pulsars have been investigated, but with little success. Chatterjee et al.[52] have shown that the most promising candidates in addition to NP 0532 are PSR 0833-45, the pulsar in the Vela supernova remnant, and CP 0950, the closest pulsar.

Future work will certainly be concerned with monitoring the Crab Nebula and attempting to measure the spectrum and to determine the ratio of pulsed to continuous gamma radiation. The search for other objects will continue. A list of present upper limits on a variety of objects is given by Weekes et al.[49]

It appears impractical to construct a mirror larger than the 10-m reflector, and efforts has been recently devoted to lowering the background effects, particularly by Grindlay's method. Preliminary experiments now being performed at Mt. Hopkins and at Narrabri, Australia, will determine to what degree this method will be pursued.

[50] J. E. Grindlay, *Astrophys. J. Lett.* **174**, L9 (1972).

[51] G. G. Fazio, H. F. Helmken, E. O'Mongain, and T. C. Weekes, *Astrophys. J. Lett.* **175**, L117 (1972).

[52] B. K. Chatterjee, G. T. Murthy, P. V. Ramana Murthy, B. V. Sreekantan, and S. C. Tonwar, *Nature (London)* **231**, 127 (1971).

8. POLARIZATION TECHNIQUES*

8.1. Introduction

Most of observational work in astronomy and astrophysics is concerned with only two aspects of radiation received from the observed objects: its *direction* and its *intensity*, considered as a function of wavelength and time. A third aspect of this radiation is its *polarization*, also considered as a function of wavelength and time. The importance of information revealed by studies of this third aspect of radiation from astronomical objects has recently begun to be appreciated.

8.1.1. Physical Mechanisms Producing Polarization of Light of Astronomical Objects

The following optical phenomena are responsible for the observed polarization:

(a) *reflection* from solid surfaces: moon (polarization discovered by Arago[1] in 1811), Mars, Mercury, minor planets;

(b) *scattering* of light by *small grains*: zodiacal light, comets, Venus (mostly scattering on droplets), Jupiter, reflection nebulae, atmospheres of late-type stars, spiral galaxies, interstellar polarization of starlight; among these objects the circular component of polarization has been detected only recently for Jupiter,[2] for late-type stars,[2a] and, following much earlier predictions of theory,[3] for interstellar polarization;[3a]

---

[1] D. F. J. Arago, "Oeuvres," Vol. **10**, p. 564, Paris and Leipzig 1854–1862.

[2] J. C. Kemp, J. B. Swedlund, R. E. Murphy, and R. D. Wolstencroft, *Nature (London)* **231**, 169 (1971); **232**, 165 (1971).

[2a] T. Gehrels, *Astrophys. J.* **173**, L23 (1972); K. Serkowski, *ibid.* **179**, L101 (1973).

[3] H. C. van de Hulst, "Light Scattering by Small Particles." Wiley, New York, 1957; R. Eiden, *Appl. Opt.* **5**, 569 (1966).

[3a] J. C. Kemp, *Astrophys. J.* **175**, L35 (1972); J. C. Kemp and R. D. Wolstencroft, *ibid.* **176**, L111 (1972).

---

* Part 8 is by K. Serkowski.

(c) *scattering* of light by *molecules* (Rayleigh scattering): Jupiter and other outer planets, Venus, possibly late-type stars;

(d) *scattering* of light by *free electrons* (Thomson scattering): solar corona, envelopes of early-type stars;

(e) *Hanle effect*[4] (resonance scattering of bound electrons in magnetic field): linear polarization in emission lines of solar chromosphere and corona;

(f) *Zeeman effect*: sunspots and magnetic stars (circular and linear polarization in spectral lines), radio-frequency emission lines of molecules and of neutral hydrogen in interstellar medium;

(g) *grey-body magnetoemission*:[5] white dwarfs (circular and linear polarization);

(h) *gyro-resonance emission*[4] (magnetic *bremsstrahlung*): solar chromosphere and corona;

(i) *synchrotron emission*[6] (in some cases *inverse Compton scattering* or *electrostatic bremsstrahlung* are also possible): Jupiter's decimetric emission, Crab nebula, pulsars, galactic background radio emission, radio galaxies, quasars.

8.1.2. Stokes Parameters[3,7,8]

In general, light can be regarded as *partially elliptically polarized.* Such light, described by the Stokes parameters I, Q, U, and V may be decomposed into two beams:

(a) *natural*, unpolarized light of intensity $I \cdot (1 - P_E)$, and

(b) *fully elliptically polarized* light of intensity $IP_E = (Q^2 + U^2 + V^2)^{1/2}$, where P_E is the *degree of polarization.*

[4] J. M. Beckers, *in* "Solar Magnetic Fields" (R. Howard, ed.), p. 3. Reidel, Dordrecht, 1971.

[5] J. C. Kemp, J. B. Swedlund, J. D. Landstreet, and J. R. P. Angel, *Astrophys. J.* **161**, L77 (1970); J. C. Kemp, *ibid.* **162**, L69 (1970).

[6] W. H. McMaster, *Rev. Mod. Phys.* **33**, 8 (1961); T. Takakura, *Solar Phys.* **1**, 304 (1967); M. P. C. Legg and K. C. Westfold, *Astrophys. J.* **154**, 499 (1968); R. D. Blandford, *Astron. Astrophys.* **20**, 135 (1972).

[7] W. A. Shurcliff, "Polarized Light." Harvard Univ. Press, Cambridge, Massachusetts, 1962; J. W. Simmons and M. J. Guttman, "States, Waves and Photons—A Modern Introduction to Light." Addison-Wesley, Reading, Massachusetts, 1970; G. V. Rozenberg, *Uspekhi Fiz. Nauk* **56**, 77 (1955).

[8] D. Clarke and J. F. Grainger, "Polarized Light and Optical Measurement," Pergamon, Oxford, 1971.

The vector of a light wave for this latter beam describes an ellipse on the celestial sphere. The angle θ which the long axis of an ellipse makes with the direction l toward the northern celestial pole, measured *counter-clockwise*[†] from this latter direction,[9] is called the *position angle* of the plane of vibrations. If the ratio of the short axis to the long axis of the ellipse is denoted by $\tan \beta$, the Stokes parameters are

$$Q = IP_E \cos 2\beta \cos 2\theta = IP \cos 2\theta, \tag{8.1.1}$$

$$U = IP_E \cos 2\beta \sin 2\theta = IP \sin 2\theta, \tag{8.1.2}$$

$$V = IP_E \sin 2\beta = IP_V, \tag{8.1.3}$$

where $P = P_E \cos 2\beta$ is the *degree of linear polarization*, and $P_V = P_E \sin 2\beta$ is the *degree of ellipticity*, positive for right-handed elliptical polarization, negative for left-handed polarization. Hence

$$P = (Q^2 + U^2)^{1/2}/I, \tag{8.1.4}$$

$$\theta = \tfrac{1}{2} \tan^{-1}(U/Q). \tag{8.1.5}$$

The Stokes parameters describing a mixture of several *incoherent* light beams are the sums of the respective Stokes parameters describing the component beams. From this additiveness of Stokes parameters results the *principle of optical equivalence*[3]: It is impossible by means of any instruments to distinguish between various incoherent superpositions of light beams that may together form a beam with the same Stokes parameters.

8.2. Analyzers for Linearly Polarized Light

8.2.1. Basic Properties of Analyzers

An *analyzer* is a device which subdivides the incident light into two beams, one of them linearly polarized in the plane parallel to the *principal plane* of the analyzer, the other in the perpendicular plane. In some

[9] K. Serkowski, Polarization of Starlight, *Advan. Astron. Astrophys.* **1**, 289 (1962).

[†] The angle θ is counted in opposite direction than usually. This change is made to be in accordance with the direction in which the position angles are counted in astronomy.

analyzers (e.g., Nicol prism, Polaroid, or wire-grid analyzer) one of these beams is extinguished. Such analyzers have a disadvantage for astronomical purposes because in most astronomical polarization measurements in the optical range the accuracy is limited by the number of available photons. Therefore economy of light is of great importance and preference should be given to *beam splitting analyzers*; both perpendicularly polarized beams outgoing from the beam-splitting analyzer can be measured simultaneously.

If the accuracy of measurements is limited by the photon statistics, the amount of information on wavelength dependence of polarization obtained in a given time is proportional to the number of detectors used simultaneously for different spectral regions. Therefore, it is advisable to subdivide the light into several spectral regions using, e.g., thin film *dichroic filters*. In the case of standard *UBVRI* spectral regions the light loss on each of such filters can be made less than 15%.[10]

Denoting by ϕ the position angle of the principal plane of the analyzer, measured from the direction l toward the northern celestial pole, the intensities of the two beams outgoing from the analyzer (the upper sign for one beam, the lower sign for the other), for the incident light described by the Stokes parameters I, Q, U, and V, are[9]

$$I'(\phi) = \left(\frac{T_1 + T_r}{2}\right)^{1/2} I \pm \left(\frac{T_1 - T_r}{2}\right)^{1/2} (Q \cos 2\phi + U \sin 2\phi)$$

$$= I\left[\left(\frac{T_1 + T_r}{2}\right)^{1/2} \pm \left(\frac{T_1 - T_r}{2}\right)^{1/2} P \cos 2(\theta - \phi)\right], \quad (8.2.1)$$

where T_1 is the *transmittance* of unpolarized light by the two identical analyzers oriented parallel, T_r the transmittance of two perpendicularly oriented analyzers. For a perfect analyzer $T_1 = \frac{1}{2}$ and $T_r = 0$, and Eq. (8.2.1) reduces to

$$I'(\phi) = \tfrac{1}{2}(I \pm Q \cos 2\phi \pm U \sin 2\phi)$$

$$= \tfrac{1}{2}I[1 \pm P \cos 2(\theta - \phi)]. \quad (8.2.2)$$

8.2.2. Wire Grid Analyzers and Dipole Antennas

An analyzer which is most efficient in the far infrared is a grid of parallel wires. If the spacing d between the wires is five times larger

---

[10] K. Serkowski, D. S. Mathewson, and V. L. Ford, *Astrophys. J.* (in press).

than the diameter of wires,[11,12] the grid is an efficient polarizer for wavelengths longer than $5d$. Wire grids with about 3000 wires per millimeter, embedded in a silver bromide substrate, are commercially available.[†] They are efficient polarizers for wavelengths 2–30 μm and their transmittance for unpolarized light is 20% at 2.5 μm and about 40% at wavelengths longer than 8 μm. These grids have been used for astronomical polarimetry in the near infrared by Dyck et al.[13] who overcame difficulties in obtaining grids embedded in a sufficiently plane-parallel layer of substrate; when a wedge-shaped analyzer is rotated, the telescope mirror image moves on the detector, diminishing the accuracy of measurements.

An extreme case of a wire grating is the *dipole antenna* used in radio astronomy. Such an antenna is most sensitive to radiation which is linearly polarized with electric vector parallel to the length of the dipole. The linear polarization is observed by repeating the measurements at different orientations of the dipole. A better method for obtaining accurate measurements of linear polarization is to use a radio interferometer with two circularly polarized feeds, one sensitive to right-handed circular polarization, the other to left-handed.[14] For measuring circular polarization both feeds should be first sensitive to right-handed, then both to left-handed polarization.

Another extreme case is a *crystal* used as a grating. Polarization by reflection from a crystal at Bragg angle of 45° was used recently in X-ray polarimeters.[15]

Still another degenerated case of a wire grating is a *spectrograph slit* which polarizes light strongly if its width is not more than an order of magnitude larger than the wavelength of light.[16] To minimize the polariza-

[11] L. Mertz, "Transformations in Optics." Wiley, New York, 1965.

[12] G. R. Bird and M. Parrish, Jr., *J. Opt. Soc. Amer.* **50**, 886 (1960); J. B. Young, H. A. Graham, and E. W. Peterson, *Appl. Opt.* **4**, 1023 (1965); M. Hass and M. O'Hara, *ibid.* **4**, 1027 (1965); D. W. Kerr and C. H. Palmer, *J. Opt. Soc. Amer.* **61**, 450 (1971).

[13] H. M. Dyck, F. F. Forbes, and S. J. Shawl, *Astron. J.* **76**, 901 (1971).

[14] R. G. Conway and P. P. Kronberg, *Mon. Notices Roy. Astron. Soc.* **142**, 11 (1969); **147**, 149 (1970); R. G. Conway, J. A. Gilbert, E. Raimond, and K. W. Weiler, *ibid.* **152**, 1P (1971).

[15] H. W. Schnopper and K. Kalata, *Astron. J.* **74**, 854 (1969); J. R. P. Angel and M. C. Weisskopf, *ibid.* **75**, 231 (1970); M. C. Weisskopf, R. Berthelsdorf, G. Epstein, R. Linke, D. Mitchell, R. Novick, and R. S. Wolff, *Rev. Sci. Instrum.* **43**, 967 (1972).

[16] R. C. Jones and J. C. S. Richards, *Proc. Roy. Soc. (London)* **A225**, 122 (1954); H. R. Wyss and H. H. Günthard, *J. Opt. Soc. Amer.* **56**, 888 (1966).

† Perkin-Elmer Corporation, Instrument Division, Norwalk, Connecticut.

tion effects spectrograph slits and the focal plane diaphragms in polarimeters should be nonmetallic, made, for instance, from ebonite.[17,17a] Thge *diffraction gratings* used in spectrographs polarize light very stronly[18] and the polarization may be completely different for wavelengths differing by only a few angstroms. For these reasons accurate polarimetry with spectrographs is possible if either the spectrograph is rotated together with an analyzer or if a fixed analyzer is placed in front of the spectrograph slit and the polarization is modulated, e.g., by placing a rotating half-wave plate in front of the analyzer (cf., Section 8.5.3).

A fixed orientation of the analyzer relative to the detectors is also desirable because the sensitivity of the detectors depends on the plane of polarization of incident light.[19] One possible reason for this dependence, in the case of end-on photomultipliers, is that part of the light reaches the venetian-blind dynodes which are tilted and therefore their photosensitivity strongly depends on the plane of polarization of incident light. The first stellar polarimeters, in which a rotatable *Glan–Thompson prism*[20] or *Polaroid*[21] was used as an analyzer, had a *Lyot depolarizer* (cf., Section 8.4.1) placed in front of the photocathode to eliminate the polarization sensitivity of the photomultiplier.

8.2.3. Polarizing Beam Splitters

Economy of light requires using all wavelengths over the widest possible spectral range simultaneously. Since most optical components producing spectral resolution change the state of polarization of the incident light, the analyzer used for polarization measurements should be placed in the light beam *in front* of the optical component subdividing the light into

[17] M. M. Pospergelis, *Astron. Zh.* **42**, 398 (1965) (English transl.: *Sov. Astron.-A.J.* **9**, 313); **45**, 645, 1229 (1968) (English transl.: *Sov. Astr.-A.J.* **12**, 512, 973); Y. N. Lipskij and M. M. Pospergelis, *Astron. Zh.* **44**, 410 (1967) (English transl. *Sov. Astron.-A.J.* **11**, 324).

[17a] K. Serkowski, in *"Planets, Stars, and Nebulae Studied With Photopolarimetry"* (T. Gehrels, ed.) Univ. of Arizona Press, Tucson, 1973.

[18] J. B. Breckinridge, *Appl. Opt.* **10**, 286 (1971); E. Brannen and D. G. Rumbold, *ibid.* **8**, 1506 (1969); J. J. Cowan, E. T. Arakawa, and L. R. Painter, *ibid.* **8**, 1734 (1969); J. M. Simon and M. C. Simon, *ibid.* **12**, 153 (1973).

[19] E. P. Clancy, *J. Opt. Soc. Amer.* **42**, 357 (1952); S. A. Hoenig and A. Cutler, *Appl. Opt.* **5**, 1091 (1966).

[20] J. S. Hall, *Publ. U.S. Naval Obs.* **17**, Pt. 1 (1950).

[21] Y. Öhman, *Stockholm Observ. Medd.* No. 54 (1943); W. A. Hiltner, *Astrophys. J. Suppl. Ser.* **2**, 389 (1956).

several wavelength bands; dichroic filters or a diffraction grating can be used as such an optical component. This means that the analyzer should be transparent and efficient over the widest possible spectral range.

The wavelength dependence of transmittance T for the commercially available birefringent crystals transparent in the visual region is plotted in Fig. 1. The ordinate scale is proportional to $\log[\log(1/T)]$ so that

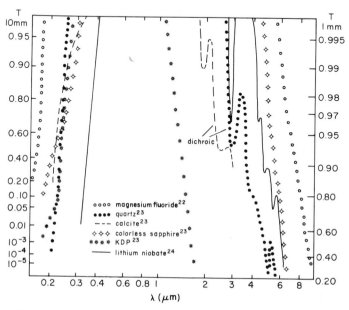

FIG. 1. Wavelength dependence of transmittance of some uniaxial crystals for unpolarized light. The thickness is 10 mm (left scale) or 1 mm (right scale) and KDP is potassium dihydrogen phosphate.

graphs for different crystal thickness are parallel to each other. The light losses by reflection from crystal surface are neglected, i.e., we assume that crystals have a perfect antireflection coating. The wavelength dependence of birefringence is plotted for the same crystals in Fig. 2.

[22] A. Duncanson and R. W. H. Stevenson, *Proc. Phys. Soc. (London)* **72**, 1001 (1958); E. D. Palik, *Appl. Opt.* **7**, 978 (1968); G. C. Morris and A. S. Abramson, *ibid.* **8**, 1249 (1969).

[23] D. E. McCarthy, *Appl. Opt.* **2**, 591 (1963); **4**, 317 (1965); **6**, 1896 (1967); S. F. Pellicori, *ibid.* **3**, 361 (1964).

[24] K. Nassau, H. J. Levenstein, and G. M. Loiacana, *J. Phys. Chem. Solids* **27**, 989 (1966).

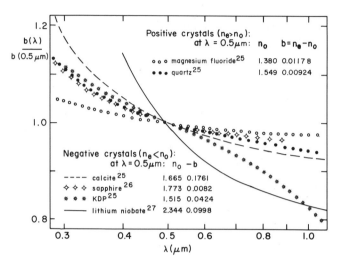

Fig. 2. Wavelength dependence of birefringence for some uniaxial crystals.

The different kinds of polarizing beam splitters are shown in Fig. 3. The simplest one is the *plane-parallel calcite plate* cut parallel to the cleavage plane. When placed in front of the telescope focal plane it gives double images of stars on a photographic plate or on the faceplate of a television camera. The spacing of the images, for $\lambda = 0.55$ µm, equals the thickness of calcite multiplied by 0.109. However, the focal plane for the perpendicularly polarized beams is not the same, which makes accurate polarimetry difficult. Also, the two beams have slightly different intensities even for the unpolarized incident light; the reason is that the light losses by reflection from the calcite surface are different for the two beams. Both these inconvenient features of a single calcite plate are avoided if two cemented plane-parallel calcite plates are used, with their principal planes crossed.[28] Such a pair of calcite plates seems to be the best analyzer for stellar polarimetry by photographic or television techniques. Very faint stars, which are close in the sky to the stars observed, do not affect the results if the observations at two position angles of the

[25] J. M. Beckers, *Appl. Opt.* **10**, 973 (1971); **11**, 681 (1972).

[26] M. A. Jeppesen, *J. Opt. Soc. Amer.* **48**, 628 (1958).

[27] G. D. Boyd, W. L. Bond, and H. L. Carter, *J. Appl. Phys.* **38**, 1941 (1967).

[28] E. Bartl, *Mitt. Univ. Sternw. Jena* No. 39 (1959); K. Serkowski, *Acta Astron.* **10**, 227 (1960); J. S. Hall and K. Serkowski, *in* "Basic Astronomical Data" (K. A. Strand, ed.), p. 293, Univ. of Chicago Press, Chicago, Illinois, 1963; P. W. J. L. Brand, *Opt. Acta* **18**, 403 (1971).

analyzer differing by 180° are averaged; nevertheless, accurate measurements are not possible in crowded regions of the sky.[9]

Similar techniques can be used for *polarimetry of extended objects* (planets, galaxies) if a grid of narrow strips of either opaque material,[29] of thin half-wave plates[30] rotating the plane of polarization by 90°, are placed in the telescope focal plane immediately in front of the double calcite plate or Wollaston prism. This grid, reimaged by a lens on the photographic plate or on the faceplate of a television camera, gives images of the narrow strips of an extended object formed alternately by the light polarized in the two perpendicular planes. When using a television camera this method probably is not better than using a multi-component Wollaston or Rochon prism[30a] giving two images of the field studied, each filling half of the camera's faceplate (cf., Section 8.7).

For photoelectric polarimetry a large separation angle between the two perpendicularly polarized beams is convenient. However, none of the polarizing beam splitters for which such angle is large approximates sufficiently well the perfect analyzer. In the air-spaced *Glan–Foucault* or cemented *Glan–Thompson* prism (Fig. 3) one of the two perpendicularly polarized beams is totally reflected from a tilted calcite surface. However, this beam is contaminated by the beam of perpendicular polarization which is also partially reflected from this surface; such contamination could be avoided only if the reflection were at Brewster's

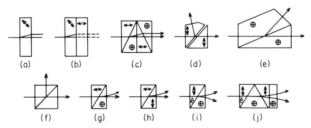

FIG. 3. Beam-splitting analyzers: (a) plane-parallel calcite plate; (b) double calcite plate; (c) double Rochon prism; (d) Glan–Foucault prism (modified by Archard and Taylor); (e) Glan–Thompson (Foster) prism; (f) thin-film polarizing beam-splitter; (g) Rochon prism; (h) Senarmont prism; (i) Wollaston prism; (j) three-wedge Wollaston prism. The directions of light beams and the directions of crystallic optical axes are indicated by arrows; the circled crosses denote the optical axis perpendicular to the plane of drawing.

[29] Y. Öhman, *Mon. Notices Roy. Astron. Soc.* **99**, 624 (1939); *Stockholm Obs. Ann.* **13**, No. 11 (1941); **19**, No. 4 (1956).

[30] A. Dollfus, *Ann. Astrophys. Suppl.* No. 4 (1957).

[30a] D. L. Steinmetz, W. G. Phillips, M. Wirick, and F. F. Forbes, *Appl. Opt.* **6**, 1001 (1967).

angle or if the cement used in the Glan–Thompson prism had an index of refraction equal to that of the extraordinary ray in calcite, which is $n = 1.490$ at $\lambda = 0.5$ μm. *Gedamine*, which has $n = 1.52$ and is transparent to wavelengths as short as 0.25 μm, may be most suitable for this purpose;[7,31] cements whose transmittance goes far into the ultraviolet[†] usually have $n \approx 1.43$ and some contamination of the reflected beam in Glan–Thompson prism will occur. This contamination in the air-spaced Glan–Foucault prism, though diminished in its Archard–Taylor[8] modification, is very serious and, moreover, the convergence (or divergence) of the incident light beam cannot exceed $f/8.5$. Similar difficulties are encountered for *thin film polarizing beam splitters*[32]: the reflected beam has a degree of polarization considerably below unity, particularly if the film is designed for a very wide spectral range or if the incident light beam is appreciably convergent.

The *Wollaston, Rochon,* and *Senarmont* prisms are the only beam-splitting analyzers which give both perpendicularly polarized beams with at least 99.9% polarization. The Wollaston prism is the most convenient one as it gives twice as large an angular separation between the beams than the other two prisms. If a very wide spectral region is required, the magnesium fluoride *optically contacted* Wollaston prism, which is transparent from 0.11 to 7.5 μm, is the most perfect beam-splitting analyzer.[22,30a] Magnesium fluoride is much harder than calcite and can therefore be worked to very high optical precision; optical contacting, almost impossible for calcite, is not difficult with magnesium fluoride.

The Wollaston prism is usually made of two or three wedges of uniaxial crystal with optical axes perpendicular to the direction of incident light. If the *construction angle* of the outer wedges is A (Fig. 3) and the Wollaston prism is made of N wedges, the angle between the two outgoing perpendicularly polarized beams is

$$\alpha \approx 2 \tan^{-1}[(N - 1) \mid n_o - n_e \mid \tan A], \qquad (8.2.3)$$

[31] Y. Bouriau and J. Lenoble, *Rev. Opt.* **36**, 531 (1957); S. F. Pellicori, *Appl. Opt.* **9**, 2581 (1970).

[32] A. F. Turner and P. W. Baumeister, *Appl. Opt.* **5**, 69 (1966); P. B. Clapham, M. J. Downs, and R. J. King, *ibid.* **8**, 1965 (1969).

[†] The extensive tests made by Pellicori[31] indicate that the best ultraviolet transmitting and shock-proof cement for calcite is Dow–Corning R-63-488 optical resin; according to Pellicori, 8% of curing agent supplied with cement and 20% of Dow–Corning Sylgard 1330 thinner should be added. Different samples of calcite vary considerably in their ultraviolet transmittance.

where n_o and n_e are the refractive indices of the crystal for the ordinary and extraordinary rays. The largest practical construction angle is $A \approx 35°$ which gives for a calcite three-component Wollaston prism $\alpha \approx 28°$ for light of wavelength $\lambda = 0.5$ μm. The Wollaston prism approaches very closely the perfect analyzer[33] having $T_r \approx 10^{-4}$.

If the images of a telescope mirror or of a focal plane diaphragm have to be formed by the light beams transmitted by a Wollaston prism, the quality of such images is considerably better for the *three-component Wollaston prism* than for the two-component one.[34] The ordinary and extraordinary beams will make equal angles with the incident light beam if the construction angle for the third wedge is slightly larger than for the first wedge so that the outer faces are not exactly parallel. For the three-component calcite Wollaston prism with $A = 10°$ the angle between the outer faces should equal 38 arcsec.

In the Wollaston prism with an odd number of wedges the reflection losses for the two outgoing beams are not the same. Therefore these two beams do not have the same intensity even for unpolarized incident light. To compensate for this, a thin plane-parallel plate of the same birefringent material as used for the wedges and with optical axis perpendicular to that of the outer wedges should be cemented to one of them.

8.3. Retarders

8.3.1. Properties of Retarders

The retarder[7] is, in the simplest case, a plane-parallel plate of uniaxial[†] crystal cut parallel to its optical axis which makes an angle ψ with the direction l defined in Section 8.1.2. The incident light, described by the

[33] J. Brandmüller and C. Hoffmeister, *Optik* **11**, 1 (1953); R. K. Lee, Jr. and F. Moskowitz, *Appl. Opt.* **3**, 1305 (1964).

[34] W. Bartolomeytchuk, *Z. Instrum.* **68**, 208 (1960); R. A. Soref and D. H. McMahon, *Appl. Opt.* **5**, 425 (1966); U. J. Schmidt, and W. Thust, *Optik* **32**, 570 (1971).

[35] F. Q. Orrall, *in* "Solar Magnetic Fields" (R. Howard, ed.), p. 30. Reidel, Dordrecht, 1971; R. D. Wolstencroft and K. Nandy, *Astrophys. Space Sci.* **12**, 158 (1971).

[35a] R. J. Archer and C. V. Shank, *J. Opt. Soc. Amer.* **57**, 191 (1967); H. Weinberger and J. Harris, *ibid.* **54**, 552 (1964).

[†] Retarders are most easily made of a sheet of biaxial *mica*. These are not recommended for astronomical applications because mica absorbs strongly the near ultraviolet and has appreciable linear dichroism.[35,35a] The mica retarders act as imperfect linear polarizers.

Stokes parameters I, Q, U, V, after passing through the retarder has the parameters I', Q', U', V', described by the *transformation matrix*[9]

$$\begin{bmatrix} I' \\ Q' \\ U' \\ V' \end{bmatrix} = \begin{bmatrix} 1 & 0 & 0 & 0 \\ 0 & \cos^2 2\psi + \sin^2 2\psi \cos \tau & (1-\cos \tau) \cos 2\psi \sin 2\psi & -\sin 2\psi \sin \tau \\ 0 & (1-\cos \tau) \cos 2\psi \sin 2\psi & \sin^2 2\psi + \cos^2 2\psi \cos \tau & \cos 2\psi \sin \tau \\ 0 & \sin 2\psi \sin \tau & -\cos 2\psi \sin \tau & \cos \tau \end{bmatrix} \begin{bmatrix} I \\ Q \\ U \\ V \end{bmatrix} ;$$

$$(8.3.1)$$

the *retardance* τ, which is the phase difference, in radians, introduced by a retarder between the light beams polarized parallel and perpendicular to the optical axis of crystal, equals for normal incidence of light on the retarder

$$\tau = 2\pi(n_e - n_o)s/\lambda. \qquad (8.3.2)$$

Here s is the thickness of the retarder, λ the wavelength, while n_e and n_o are the refractive indices of crystal for the *extraordinary* and *ordinary* rays, i.e., for the vibrations of the electric vector of the light wave which are parallel and perpendicular to the optical axis of crystal, respectively. The quantity $\Delta = \tau\lambda/2\pi$ is called the *path difference*. If the direction of the incident light makes a small angle i with the normal to the surface of the retarder, and the plane of incidence makes an angle ω with the optical axis of crystal, the retardance is[36]

$$\tau \approx 2\pi(n_e - n_o)\frac{s}{\lambda}\left[1 - \frac{i^2}{2n_o}\left(\frac{\cos^2 \omega}{n_o} - \frac{\sin^2 \omega}{n_e}\right)\right]. \qquad (8.3.3)$$

A pair of retarders, one made of positive crystal, the other of negative crystal, with a properly chosen thickness ratio,[†] makes a wide-field retarder for which the retardance τ is almost independent of the angle of incidence unless this angle is very large.[25] Equation (8.3.1) is valid only if the outer faces of the retarder have an antireflection coating. Otherwise the interfering multiple-reflected beams may appreciably change the polarization of the outgoing beam.[8,36a] A computer program for ray

[36] B. Lyot, *Ann. Astroph.* **7**, 31 (1944); M. Françon, *in* "Handbuch der Physik" (S. Flügge, ed.), Vol. 24, p. 444. Springer, Berlin, 1956; M. L. Roblin, *Opt. Acta* **18**, 41 (1971).

[36a] D. A. Holmes, *J. Opt. Soc. Amer.* **54**, 1115 (1964).

[†] This ratio equals $s_2/s_1 = [(n_e - n_o)_1/(n_o - n_e)_2]n_{o2}n_{e2}/n_{o1}^2$, where index 1 refers to positive crystal, 2 to negative.

tracing in uniaxial crystals has been published[37] which facilitates the design of retarders.

The retarder with $\tau = \pi/2$ is called a *quarter-wave plate*; it changes circularly polarized light, for which $I = |V|$, into linearly polarized light for which $I = (Q^2 + U^2)^{1/2}$. The retarder with $\tau = \pi$ is a *half-wave plate*, which changes right-handed circularly polarized light into left-handed and light linearly polarized at position angle θ into light linearly polarized at position angle $2\psi - \theta$. Small deviations of retardance from $\pi/2$ or π radians can be easily detected.[37a]

8.3.2. Achromatic Retarders

The half-wave plates and quarter-wave plates should be placed in the astronomical polarimeters in front of the optical components which separate the light of different spectral regions because the latter usually polarize light. If a wide spectral range must be covered, the retardance of the retarder should change as little as possible over the total range.

Fresnel rhombs approach this ideal quite closely. In the double Fresnel rhomb a quarter-wave path difference is obtained, without shifting the optical axis, by four total reflections in fused silica.[38] Even for an appreciably convergent incident beam the deviations of retardance from $\pi/2$ are very small over the spectral range from 0.3 to 1.0 μm. Unfortunately, for an incoming beam 10 mm in diameter, the rhomb must be approximately 175 mm long; a rhomb giving half-wave path difference would be twice as long. Avoiding stress birefringence in thick Fresnel rhombs is very difficult.[39]

Another method of obtaining achromatic retarders is by combining two plates of different birefringent materials.[40] The conditions for the retardance τ to be the same for the two wavelengths λ_1 and λ_2 are, from Eq. (8.3.2),

$$\lambda_1 = 2\pi[b_1(\lambda_1)s_1 \pm b_2(\lambda_1)s_2]/\tau, \qquad (8.3.4)$$

$$\lambda_2 = 2\pi[b_1(\lambda_2)s_1 \pm b_2(\lambda_2)s_2]/\tau, \qquad (8.3.5)$$

where b_1 equals the difference in refractive indices $n_e - n_o$ for the first

[37] A. Thomescheit, *Optik* **32**, 283, 539 (1970); **33**, 47, 127 (1971).

[37a] A. C. Hall, *Appl. Opt.* **2**, 864 (1963); B. R. Grunstra and H. B. Perkins, *Appl. Opt.* **5**, 585 (1966).

[38] V. A. Kizel, Y. I. Krasilov, and V. N. Shamraev, *Opt. Spectrosc.* **17**, 248 (1964).

[39] J. M. Bennett, *Appl. Opt.* **9**, 2123 (1970).

[40] D. Clarke, *Opt. Acta* **14**, 343 (1967).

material, b_2 for the second material, while s_1 and s_2 are the thicknesses of the two plates; the plus signs should be taken if the sign of b is different for the two plates and their optical axes are parallel, the minus sign if both b have the same sign and the optical axes are perpendicular.

The best choice of materials would be a pair of positive and negative crystals because, in accordance with Eq. (8.3.3), its retardance would depend little on the angle of incidence. In this respect the best combination would be colorless sapphire[†] (negative crystal) and magnesium fluoride (positive crystal).[41] Unfortunately, the refractive indices of these materials differ considerably and incident natural light would become polarized by the reflection from their cemented interface. The pairs involving soft materials, such as calcite[41a] or potassium dihydrogen phosphate,[25] cannot be recommended because the required plane-parallelism and flatness of their surfaces, better than $\lambda/10$, can hardly be achieved in practice.

For these reasons it seems that the most convenient pair is quartz with magnesium fluoride,[25,41] both positive crystals; these materials are hard, easy to polish with high precision, and transparent over a wide spectral range. Solving Eqs. (8.3.4) and (8.3.5) for these materials, with optical axes crossed, assuming $\tau = 180°$ (half-wave) for $\lambda_1 = 0.31$ μm and $\lambda_2 = 0.66$ μm, we find the thickness of the quartz plate $s_1 = 0.304$ mm and of the magnesium fluoride plate $s_2 = 0.262$ mm. The wavelength dependence of retardance for this pair of plates is shown in Fig. 4. It can be seen that the achromatism over that wide a spectral range leaves much to be desired.

The direction of the optical axis for an achromatic retarder having one component made of quartz depends somewhat on wavelength. This is caused by *optical activity (circular birefringence)* of quartz, which for the direction perpendicular to optical axis is about half as large as that along the optical axis, with sign reversed.[42] The joint effect of linear and circular birefringence is best explained by using the *Poincaré sphere*.[7,8,42a] The normalized Stokes parameters Q/I, U/I, and V/I are the coordinates of

[40a] M. P. Wirick, *Appl. Opt.* **5**, 1966 (1966).

[41] V. Chandrasekharan and H. Damany, *Appl. Opt.* **7**, 939 (1968).

[41a] S. B. Ioffe, and T. A. Smirnova, *Opt. Spectrosc.* **16**, 484 (1964).

[42] G. M. Ramachandran and S. Ramaseshan, *in* "Handbuch der Physik" (S. Flügge, ed.), Vol. 25/1, p. 1. Springer, Berlin, 1961.

[42a] J. E. Vos and B. S. Blaisse, *Opt. Acta* **17**, 197 (1970).

† Another negative crystal is lanthanum fluoride, transparent from 0.2 to 9 μm.[40a]

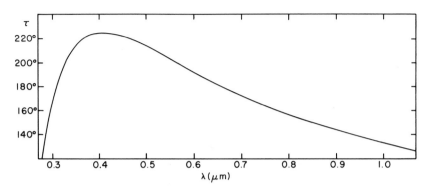

Fig. 4. Wavelength dependence of retardance for a half-wave plate of magnesium fluoride (0.262 mm) and quartz (0.304 mm).

point P describing the state of polarization of light and situated inside or on the surface of the Poincaré sphere of unit radius with center at the origin of the coordinates. The passage of this light through a retarder with optical axis at position angle ψ and with retardance τ changes its state of polarization to one obtained from P by rotation by an angle τ around such an equatorial diameter of the Poincaré sphere which crosses its surface at a point with coordinates $\cos 2\psi$, $\sin 2\psi$, 0. Similarly passage of light through an optically active element rotating the plane of polarization by an angle ϱ changes the state of polarization P into one obtained from P by rotation by an angle 2ϱ around the polar diameter of the Poincaré sphere.

The joint effect of infinitesimally small linear and circular birefringence is described by a composition of the two corresponding rotations of the Poincaré sphere[42] (Fig. 5a). Such composition is equivalent to rotation of the sphere by an angle $(\tau^2 + 4\varrho^2)^{1/2}$ around the diameter crossing the

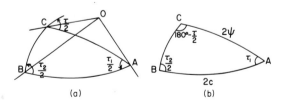

(a)   (b)

Fig. 5. (a) Composition of two rotations. A rotation about AO through twice the internal angle at A followed by a rotation about BO through twice the internal angle at B is equivalent to a rotation about CO through twice the external angle at C. (b) Spherical triangle illustrating the relationships between the parameters of Pancharatnam's achromatic retarder.

surface of the sphere at a point with coordinates $\cos(2\varrho/\tau)\cos 2\psi$, $\cos(2\varrho/\tau)\sin 2\psi$, $\sin(2\varrho/\tau)$. A composition of an arbitrary number of such infinitesimal rotations is equivalent to a rotation around the same diameter by a correspondingly larger angle.

The passage of light through a half-wave plate of quartz and magnesium fluoride may be described as a composition of rotations of the Poincaré sphere around two axes making an angle $2\varrho/\tau$. Such composition is equivalent to a rotation around an axis intersecting the Poincaré sphere somewhere along a small circle of diameter $2\varrho/\tau$ and a center with coordinates $\cos(\varrho/\tau)\cos 2\psi$, $\cos(\varrho/\tau)\sin 2\psi$, $\sin(\varrho/\tau)$. The position of the axis on this circle is determined by the retardance of, e.g., the quartz plate and depends strongly on wavelength. At $\lambda = 0.35$ μm the ratio $2\varrho/\tau$ for quartz equals $1/150$; around that wavelength the position angle ψ of an equivalent optical axis for an achromatic half-wave plate of quartz and magnesium fluoride will change by about $0.2°$ over $\Delta\lambda = 200$ Å. Stronger wavelength dependence of ψ may be expected if the optical axes of the two plates are not exactly perpendicular.

A way of further improving the achromatism of retarders was proposed by Pancharatnam.[43] A combination of three retarders, the first and the last having their corresponding optical axes parallel and identical retardance, will behave as a single retarder with retardance τ and with an equivalent optical axis making an angle ψ with the optical axis of the first plate, where τ and ψ are given by

$$\cos(\tau/2) = \cos \tau_1 \cos(\tau_2/2) - \sin \tau_1 \sin(\tau_2/2) \cos 2c, \quad (8.3.6)$$

$$\cot 2\psi = [\sin \tau_1 \cot(\tau_2/2) + \cos \tau_1 \cos 2c]/\sin 2c. \quad (8.3.7)$$

Here τ_1 is the retardance of each the first and the last retarder, τ_2 that of the central one, and c is the angle between the optical axis of the central retarder and the optical axis of the other two. Equations (8.3.6) and (8.3.7) are just the relationships between the sides and angles of the spherical triangle ABC shown in Fig. 5b.

The retardance τ will be equal to the required retardance τ_0 (usually $\pi/2$ or π) for three wavelengths λ_1, λ_2, and λ_3 for which the retardances of each of the outer plates are $(1-f)\tau_1$, τ_1, and $(1+f)\tau_1$, and the retardances of the central plate are $(1-f)\tau_2$, τ_2, and $(1+f)\tau_2$, respectively; for the constant f a value between 0.15 and 0.3 is usually taken. Assuming $\tau_2 = \pi$ rad, Pancharatnam obtains the following condi-

[43] S. Pancharatnam, *Proc. Indian Acad. Sci.* **A41**, 137 (1955).

tions for the equality of retardance τ at wavelengths λ_1, λ_2, and λ_3:

$$\sin(f\tau_1) = \sin(f\pi/2)\sin\tau_1/\cos(\tau_0/2), \qquad (8.3.8)$$

$$\cos 2c = -\tan(f\pi/2)/\tan(f\tau_1). \qquad (8.3.9)$$

For an achromatic half-wave plate we have $\tau_0 = \pi$ rad; when τ_0 approaches π rad the right side of Eq. (8.3.8) approaches a constant value only if τ_1 approaches π rad. This means that all three retarders of Pancharatnam's combination should be half-wave plates. For better achromatism each of them can be a pair of quartz and magnesium fluoride plates with wavelength dependence of retardance shown in Fig. 4. In this case we may take, e.g., $f = 0.25$ obtaining $c = 57.2°$. For *Pancharatnam's half-wave plate* not only the retardance τ but also the position angle ψ of the equivalent optical axis changes with wavelength, as can be seen from Eq. (8.3.7). For our example of quartz and magnesium fluoride components this wavelength dependence is shown in Fig. 6. The deviations of the retardance from $180°$ are negligible in the spectral range from 0.3 to 1.0 μm, but the changes in ψ are quite considerable. Fortunately, in

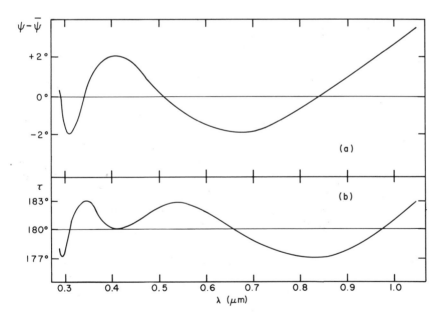

FIG. 6. Wavelength dependence of (a) position angle of the equivalent optical axis and (b) of retardance for the Pancharatnam's achromatic half-wave plate of magnesium fluoride and quartz.

the case of the polarimeter with rotating half-wave plate this wavelength dependence of ψ can be eliminated, as described in Section 8.5.3. The achromatism of Pancharatnam's retarders can be further improved by increasing the number of components.[44]

8.3.3. Variable Retarders

When an electric field is applied along the optical axis of a uniaxial crystal of *ammonium dihydrogen phosphate* (ADP) or *potassium dihydrogen phosphate* (KDP) the crystal becomes biaxial, with a new optical axis perpendicular to the direction of the electric field. The birefringence along this new axis is proportional to the electric field. A plate of ADP or KDP with flat surfaces perpendicular to the original optical axis and with semitransparent or gridlike electrodes deposited on these surfaces is called the *Pockels cell.* The voltages needed to obtain a quarter-wave plate are on the order of 4 kV for KDP and half that much for *potassium dideuterium phosphate* (KD*P), which is commercially available. Usually the applied voltage is alternating with high frequency which causes some heating because the Pockels cell acts as a lossy capacitor.

The field of view of Pockels cells is severely limited by the existence of natural birefringence of the material. If we are viewing not along the natural optical axis but at a small angle i to this axis, the retardance equals[45]

$$\tau \approx \frac{2\pi s}{\lambda} \frac{n_0{}^2 - n_e{}^2}{2n_0 n_e{}^2} \sin^2 i; \qquad (8.3.10)$$

for KDP at wavelength $\lambda = 0.5$ μm we have $\tau = 45°$ at $i \approx 0.045 s^{-1/2}$ rad where the thickness s is in millimeters. The plates must be thicker than about 1.5 mm or they can fracture spontaneously.[45] For some astronomical applications the variable retarders of *lithium niobate* may be attractive since for them the applied electric field is transverse, perpendicular to the direction of the modulated light beam, so that the electrodes do not cause any light losses. Unfortunately lithium niobate is opaque in the near ultraviolet (see Fig. 1).

The disadvantages of Pockels cells—need for high voltage, low transmittance, and small field of view—are avoided in the *piezooptical bi-*

[44] E. O. Amman, *J. Opt. Soc. Amer.* **56**, 952 (1966); C. M. McIntyre and S. E. Harris, *ibid.* **58**, 1575 (1968); A. M. Title, *Solar Phys.* (in press); *Appl. Opt.* (in press).

[45] B. H. Billings, *J. Opt. Soc. Amer.* **39**, 797 (1949); **42**, 12 (1952); G. Hepner, J. Gouzerh, and M. Rousseau, *Nouv. Rev. Opt. Appl.* **2**, 3 (1971).

refringence modulator[46] which recently has become available commercially.[†] Stress birefringence is produced in transparent materials such as fused silica, fluorite, or rocksalt by an acoustic vibration sustained by a transducer of small power, usually less than 1 W. Standing waves are set up in a block of transparent material if the length of this block is equal to half the wavelength of acoustic waves. In practice, frequencies near 50 kHz are used. The useful angular aperture for resulting variable retarder can be very large, on the order of 50°.

By combining stress birefringence plates of different materials it may be possible to produce achromatic variable retarders.[46a] Stress birefringence has particular importance in the *far infrared* studies for lack of naturally birefringent materials at wavelengths longer than 16 μm; up to 16 μm retarders of cadmium sulfide[46b] can be used. The variable retarders of *cesium iodide*[47] could be used for wavelengths up to 60 μm provided sufficiently large stress birefringence can be induced. At 60 to 400 μm natural birefringence of quartz, transparent at these wavelengths, can be utilized.[47a]

8.4. Depolarizers

Depolarizer is a device changing the polarized light in such a way that its *average* degree of polarization becomes zero. Usually the average either over *time* or over *wavelength*[8] is taken. Averaging over time is superior because it can be applied to monochromatic light. Linearly polarized light can be depolarized in this sense by a rapidly *rotating achromatic half-wave plate*.[48] If the retardance of the half-wave plate is $\pi + \varepsilon$ rad, where $\varepsilon \ll 1$, the ratio of the degrees of linear polarization of the transmitted and incident beams, averaged over time, equals $\varepsilon^2/4$. The

[46] A. E. Ennos, *Opt. Acta* **10**, 105 (1963); M. Billardon, and J. Badoz, *C. Acad. Sci. Paris* **262**, 1672 (1966); L. F. Mollenauer, D. Downie, H. Engstrom, and W. B. Grant, *Appl. Opt.* **8**, 661 (1969); J. C. Kemp, *J. Opt. Soc. Amer.* **59**, 950 (1969); J. C. Kemp, J. B. Swedlund, and B. D. Evans, *Phys. Rev. Lett.* **24**, 1211 (1970); J. C. Canit, M. Billardon, and S. Pauthier-Camier, *Nouv. Rev. Opt. Appl.* **1**, 191 (1970); J. C. Kemp, R. D. Wolstencroft, and J. B. Swedlund, *Astrophys. J.* **177**, 177 (1972).

[46a] A. J. Michael, *J. Opt. Soc. Amer.* **58**, 889 (1968).

[46b] T. E. Walsh, *J. Opt. Soc. Amer.* **62**, 81 (1972).

[47] E. K. Plyler and N. Acquista, *J. Opt. Soc. Amer.* **48**, 668 (1958).

[47a] E. D. Palik, *Appl. Opt.* **4**, 1017 (1965).

[48] B. H. Billings, *J. Opt. Soc. Amer.* **41**, 966 (1951).

[†] Morvue Electronic Systems, 10520 S.W. Cascade, Tigard, Oregon.

rotating half-wave plate does not diminish the circular polarization, only changing its sign. Placing in front of it an achromatic quarter-wave plate rotating at half the speed[48] makes a perfect *monochromatic depolarizer* operating on any state of polarization. However, more convenient are depolarizers averaging the polarization over wavelength; the one most widely used is the Lyot[49] depolarizer.

8.4.1. The Lyot Depolarizer

This device[48] consists of two retarders with retardances τ ($\gg 2\pi$) and 2τ while an angle between the optical axes is $45°$. The transformation matrix for the Lyot depolarizer, with the light going first through the thinner[†] plate oriented at $\psi = 0°$, is obtained by multiplying the corresponding matrices (8.3.1) for both retarders and averaging over wavelength; we use *angular brackets* for denoting such averages. For visualizing the effect of inaccurate relative orientation of depolarizer plates we assume $\frac{1}{4}\pi + \varepsilon$ rad, where $\varepsilon \ll 1$, as the angle between the optical axes, obtaining for a depolarizer the transformation matrix

$$
\begin{bmatrix} I' \\ Q' \\ U' \\ V' \end{bmatrix} \approx
\begin{bmatrix}
1 & 0 & 0 & 0 \\
0 & \langle\cos 2\tau\rangle & \frac{1}{2}(\langle\cos\tau\rangle - \langle\cos 3\tau\rangle) & -\frac{1}{2}(\langle\sin\tau\rangle + \langle\sin 3\tau\rangle) \\
0 & 2\varepsilon & \langle\cos\tau\rangle & \langle\sin\tau\rangle \\
0 & \langle\sin 2\tau\rangle & -\frac{1}{2}(\langle\cos\tau\rangle - \langle\cos 3\tau\rangle) & \frac{1}{2}(\langle\cos\tau\rangle + \langle\cos 3\tau\rangle)
\end{bmatrix}
\begin{bmatrix} I \\ Q \\ U \\ V \end{bmatrix}.
$$

(8.4.1)

The largest terms in this expression, other than 1, are $\langle\cos\tau\rangle$ and $\langle\sin\tau\rangle$. If the light transmitted through a Lyot depolarizer has a Gaussian distribution of spectral energy density $I(\lambda^{-1})$ with rms deviation $\sigma(\lambda^{-1})$, then[9]

$$
|\langle\sin\tau\rangle| \leq \exp\{-2[\pi(n_e - n_o)s\sigma]^2\} \geq |\langle\cos\tau\rangle|. \quad (8.4.2)
$$

For a quartz depolarizer with plates 1 mm and 2 mm thick and for the half-width of transmission band 250 Å centered at $\lambda = 4500$ Å, we have $\sigma \approx 500$ cm^{-1} and $|\langle\sin\tau\rangle| \leq 0.013$. The degree of polarization for the light passing through this depolarizer will diminish by a factor of at least 70.

[49] B. Lyot, *Ann. Obs. Paris-Meudon* **8**, Pt. 1 (1929).

[†] The case of light going first through a thicker plate, discussed in an earlier paper[9] [Eq. (24–27)], is slightly less favorable.

The depolarizing action of a Lyot depolarizer can be improved if the retardance not only depends on wavelength but also changes over the surface of the depolarizer. One of its surfaces can be, for example, corrugated and cemented to a plane-parallel plate of isotropic material.[50] The same effect, varying τ by about $\pm\pi$ rad, can be obtained by making the depolarizer plates slightly wedge-shaped,[42] by making one of them slightly convex and the other concave, or by placing the depolarizer in a strongly divergent light beam. Also continuous rotation of depolarizer improves its performance.[17a]

8.4.2. Optimum Orientation of the Lyot Depolarizer[17a]

In polarimeters using a Wollaston prism as an analyzer, the measurements are often made alternately with and without a Lyot depolarizer inserted in front of the analyzer. In this application the *first* plate of the depolarizer should have an optical axis in the *principal plane* of the Wollaston prism, i.e., parallel to the optical axis of one of the component wedges of the Wollaston prism. Such an orientation of a depolarizer relative to the analyzer has two advantages. First, as seen from Eq. (8.4.1) the small deviation ε of the angle between the optical axes of the depolarizer plates from 45° has in this case no influence on the depolarizer efficiency. Secondly, the effects of the polarization of light by reflection from the depolarizer surfaces are eliminated.

The reflectances of the depolarizer surfaces for light waves vibrating parallel and perpendicular to the optical axis are not the same. In the case of calcite the reflection coefficients for the ordinary and for the extraordinary rays at $\lambda = 0.5$ μm are 6.1 and 3.8%, respectively. The depolarized light outgoing from the calcite Lyot depolarizer becomes 1.2% polarized independently of polarization of the incoming beam; for quartz or magnesium fluoride depolarizers the polarization is about 0.07%. This effect can be diminished by coating the outer surface of the depolarizer with a thin film of refractive index intermediate between the refractive indices of the birefringent material for ordinary and extraordinary rays. The effect does not affect the polarimetric measurements at all if the optical axis in the first plate of the depolarizer is in the principal plane of the Wollaston prism.

If, on the contrary, the depolarizer is inserted in front of the Wollaston prism with the optical axis of the *second* plate of the depolarizer in the

[50] C. J. Peters, *Appl. Opt.* **3**, 1502 (1964); **9**, 2403 (1970).

principal plane of the Wollaston prism, the intensities of the two beams outgoing from the Wollaston prism will not be equal even if the depolarizer is illuminated by unpolarized light. Particularly undesirable is an observing procedure such that the depolarizer is inserted always at the same orientation while the orientation of an analyzer is changing. In this case the constant polarization introduced by the depolarizer adds to the instrumental polarization produced by the telescope mirrors. If, in addition, the detectors are rotated together with an analyzer, the shifts of an image of telescope mirror on the detectors, caused by inserting a slightly wedge-shaped depolarizer, are different at each orientation of detectors and their effect is therefore difficult to eliminate.

8.4.3. Corrections for Depolarizer Inefficiency

The performance of a depolarizer can be tested by placing it between a rotatable analyzer with its principal plane at position angle ϕ and a fixed analyzer with its principal plane parallel to the optical axis of the first plate of the depolarizer and oriented at a position angle $0°$. From Eqs. (8.2.2) and (8.4.1) the intensity of the transmitted beam, assuming that the analyzers are perfect, is

$$I' = \tfrac{1}{4}I[1 + P_D \cos 2(\phi - \phi_D)], \tag{8.4.3}$$

where

$$P_D \cos 2\phi_D = \langle \cos 2\tau \rangle, \tag{8.4.4}$$

$$P_D \sin 2\phi_D = \tfrac{1}{2}(\langle \cos \tau \rangle - \langle \cos 3\tau \rangle), \tag{8.4.5}$$

are the normalized Stokes parameters, Q/I and U/I, for the light outgoing from the depolarizer, and I is the intensity of unpolarized light incident on the rotatable analyzer.

To account for the imperfection of the depolarizer, the degrees of polarization of celestial objects, computed on the assumption that the depolarizer is perfect, should be multiplied by the *depolarization correction factors*

$$D \approx 1 + P_D \cos 2\phi_D. \tag{8.4.6}$$

The position angles θ, calculated on the same assumptions, should be corrected[51] by adding

$$\Delta\theta \approx (\tfrac{1}{2}P_D \sin 2\phi_D) \quad \text{rad.} \tag{8.4.7}$$

For the standard UBV spectral regions and a quartz depolarizer 5 mm thick used at Siding Spring Observatory, P_D for most stars does not exceed 0.02, except for late M-type stars for which the values 0.03–0.06 were obtained.[10,51] Large values of P_D are also found, as expected, in the red spectral region for early-type stars that have $H\alpha$ in emission.

8.4.4. Plane-Parallelism of Depolarizer or Retarder

A depolarizer or retarder which is inserted or rotated in front of the analyzer should be very accurately plane-parallel to avoid shifting the image of the *telescope mirror on the detector* (cf., Section 8.5.1). Because of the highly nonuniform distribution of sensitivity over the surfaces of most detectors the practical rule is that the relative shift $\Delta D'$ of an image of the telescope mirror formed by a Fabry lens on the detector should not exceed 0.01% of the diameter D' of this image.

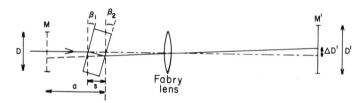

FIG. 7. Shift $\Delta D'$ of an image M' of telescope mirror on the detector caused by inserting a wedge-shaped plate in front of Fabry lens.

Let us suppose that the plate (e.g., depolarizer or retarder) of thickness s and mean refractive index n is inserted into the light beam with its back surface at a distance a behind the image M of the telescope mirror but in front of the Fabry lens which forms another image M' of the telescope mirror on the detector. Let the diameter of image M be equal to D and the diameter of image M' be D'. Then (Fig. 7) the relative shift of the image M' caused by inserting the plate equals

$$\Delta D'/D' \approx (s/D)(1 - 1/n)\beta_1 + (a/D)(n - 1)(\beta_2 - \beta_1), \quad (8.4.8)$$

where β_1 and β_2 are the tilts of the front and the back surfaces of the plate relative to the planes perpendicular to the line joining the centers of the telescope mirror and Fabry lens. If M is not the image of the telescope mirror but the mirror itself, then a/D is the focal ratio of the telescope. In this case, for an $f/10$ telescope, the angle between the

[51] K. Serkowski, *Astrophys. J.* **154**, 115 (1968); **160**, 1083 (1970).

surfaces of the inserted plate should be $|\beta_2 - \beta_1| < 4$ arcsec if the relative shift of the image on the detector has to be less than 0.01%. Tilting the plate has in this case little effect on the results.

There are two solutions giving $\Delta D' = 0$ for a wedge-shaped plate inserted into the beam:

$$\beta_1 = 0, \qquad a = 0; \qquad\qquad\qquad\qquad (8.4.9a)$$

$$\beta_2 = 0, \qquad a = s/n. \qquad\qquad\qquad\qquad (8.4.9b)$$

They correspond to either the first or the second surface of the plate set perpendicular to the line joining the center of the image of the telescope mirror with the center of the Fabry lens. Both these solutions require a lens that forms an image of the telescope mirror on the back surface or inside the inserted plate.

8.5. Optimum Design of an Astronomical Polarimeter

8.5.1. Factors Limiting the Accuracy of Polarimetry

Ideally the precision of polarimetric measurements should be limited only by *quantum noise*, i.e., by the fluctuations in the number of arriving photons. In this case, if n photoelectrons/sec are emitted from the photocathode of a photomultiplier and if the photoelectric current is measured for $t/4$ sec at each of the four position angles of a perfect analyzer, $\phi = 0$, 45, 90, and 135°, the *normalized Stokes parameters*, determined from Eq. (8.2.2), are

$$Q/I = [I'(0°) - I'(90°)]/[I'(0°) + I'(90°)], \qquad (8.5.1a)$$

$$U/I = [I'(45°) - I'(135°)]/[I'(45°) + I'(135°)]. \qquad (8.5.1b)$$

By I' we may understand here the number of counted photons. For small polarization $I'(\phi) \approx \frac{1}{4}nt$. Since the number of counted photons is subject of *Poisson statistics*, the mean error of this number is $\varepsilon(I') = (\frac{1}{4}nt)^{1/2}$ and the *mean error* ε of each of the normalized Stokes parameters Q/I and U/I, as given by Eqs. (8.5.1), is

$$\varepsilon = (2/nt)^{1/2}. \qquad\qquad\qquad\qquad (8.5.2)$$

The degree of polarization P corrected for its systematic increase caused by the lack of precision of the measurements equals,[52] for $P > \varepsilon$,

$$P \approx [(Q^2 + U^2)/I^2 - \varepsilon^2]^{1/2} \qquad\qquad (8.5.3)$$

[52] K. Serkowski, *Acta Actron.* **8**, 135 (1958).

and its mean error equals ε, as given by Eq. (8.5.2). The position angle θ is calculated from Eq. (8.1.5) and its mean error is[9]

$$\varepsilon_\theta \approx \begin{cases} 12^{-1/2}\pi & \text{rad} = 51.96° & \text{for } P \ll \varepsilon, \\ \tfrac{1}{2}\varepsilon/P & \text{rad} = 28.65°\varepsilon/P & \text{for } P \gg \varepsilon. \end{cases} \qquad (8.5.4)$$

The number n of photoelectrons produced per second by a photocathode of *quantum efficiency* q illuminated by the light of a star of magnitude m collected by a telescope of diameter D centimeters equals

$$n = 3.95 \times 10^{18} \lambda \, \Delta\lambda \, F_\lambda T q D^2 10^{-0.4m}, \qquad (8.5.5)$$

where the wavelength λ and the width of the spectral interval $\Delta\lambda$ are expressed in micrometers, T is the joint transmittance of unpolarized light by the earth's atmosphere, the telescope optics (the obstruction of light by the secondary mirror should be taken into account), and the analyzer, while F_λ is the absolute flux density for a 0th mag A0V star, expressed in watts per centimeter squared · micrometers. The values of F_λ are listed for various spectral regions by Johnson[53] while the numerical constant in Eq. (8.5.5) equals $\pi/(4hc)$, where h is the Planck constant in watts · squared second and c is the speed of light in micrometers per second. Assuming $D = 100$ cm, $T = 0.2$, and for the blue spectral region of the *UBV* system $\lambda = 0.44$ μm, $\Delta\lambda = 0.09$ μm, $q = 0.20$, and $F_\lambda = 7.2 \times 10^{-12}$ W/cm² μm, we obtain from Eq. (8.5.5), $n = 45,000$ electrons/sec for a star of blue magnitude $B = 10.0$ mag. Hence, from Eq. (8.5.2), for an observation lasting 10 min ($t = 600$ sec) the mean error of the degree of polarization equals $\varepsilon = \pm0.00027$, i.e., $\pm0.027\%$. This error would be smaller by a factor of $2^{1/2}$ if two photomultipliers were used, placed in the two perpendicularly polarized beams emerging from a Wollaston prism.

The theoretical precision given by Eq. (8.5.2) is not always achieved in practice. In the *conventional Wollaston polarimeter*[54,55] (Fig. 8) the Wollaston prism is rotated in 30 or 45° steps together with two photomultipliers. At every position angle the sensitivity of the photomultipliers is somewhat different because of their different orientation in the earth's

[53] H. L. Johnson, *Ann. Rev. Astron. Astrophys.* **4**, 193 (1966).

[54] W. A. Hiltner, *Observatory* **71**, 234 (1951); *in* "Astronomical Techniques" (W. A. Hiltner, ed.), p. 229. Chicago Univ. Press, Chicago, Illinois, 1962; A. Behr, *Veröff. Göttingen Sternwarte*, No. 114 (1956); No. 126 (1959); J. S. Hall, *Lowell Observ. Bull.* **7**, 61 (1968).

[55] T. Gehrels and T. M. Teska, *Publ. Astron. Soc. Pacific* **72**, 115 (1960); *Appl. Opt.* **2**, 67 (1963).

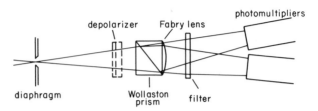

FIG. 8. Conventional Wollaston prism polarimeter.

magnetic field, changing efficiency of cooling the photomultipliers, and because bending of mechanical parts of the instrument results in illuminating somewhat different areas on the photocathodes. To eliminate the effects of this change in sensitivity the measurements are made at each position angle with and without the Lyot depolarizer which should be placed between the focal plane diaphragm[†] and the Wollaston prism. Of course, the depolarizer fulfils its purpose only if it is so accurately plane-parallel that its insertion does not shift the images of the telescope mirror on the photocathodes (cf., Section 8.4.4).

These images, formed by the *Fabry lenses*, should be very accurately focused on the photocathodes as otherwise slightly different *centering of the star* in the focal plane diaphragm during the measurements with and without the depolarizer results in illuminating different areas of the photocathodes. The relative shift of the image on the photocathode resulting from this effect is inversely proportional to the square of the diameter of this image;[56] Fabry lenses of long focal length should therefore be preferred. If a star cannot be centered in a focal plane diaphragm of 100-cm telescope with accuracy better than 1 arcsec and an image of telescope mirror on the cathode is only 7 mm in diameter, this image should not be out of focus by more than 1 mm if its relative shifts caused by inaccurate centering have to be less than 0.01%. So accurate focusing is possible only with achromatic Fabry lenses unless different photomultipliers are used for various spectral regions separated by dichroic filters.

Denoting by $I_1(\phi)$ and $I_2(\phi)$ the signals from the two photomultipliers illuminated by the light beams from the Wollaston prism set at position angle ϕ, and by $I_{d1}(\phi)$ and $I_{d2}(\phi)$ similar signals with the depolarizer in

[56] K. Serkowski and W. Chojnacki, *Astron. Astrophys.* **1**, 442 (1969).

[†] Inserting a depolarizer in front of a diaphragm is less favorable because some light from a star other than the one observed may enter the diaphragm after multiple reflections from depolarizer surfaces.

front of the analyzer, we calculate for each ϕ the parameter [cf., Eq. (8.2.1)]

$$S(\phi) = \left[\frac{I_1(\phi)/I_2(\phi)}{I_{d1}(\phi)/I_{d2}(\phi)} - 1 \right] \Big/ \left[\frac{I_1(\phi)/I_2(\phi)}{I_{d1}(\phi)/I_{d2}(\phi)} + 1 \right]$$

$$= \left(\frac{T_1 - T_r}{T_1 + T_r} \right)^{1/2} P \cos 2(\theta - \phi); \qquad (8.5.6)$$

knowing the parameters $S(\phi)$ we calculate the degree of polarization and the position angle

$$P = [(T_1 + T_r)/(T_1 - T_r)]^{1/2} \{[S(0°)]^2 + [S(45°)]^2\}^{1/2}, \qquad (8.5.7)$$

$$\phi = \tfrac{1}{2} \tan^{-1}[S(45°)/S(0°)]. \qquad (8.5.8)$$

Measurements at two position angles of the Wollaston prism differing by 45° are, in principle, sufficient for determining the degree of linear polarization P and the position angle θ for the incident light. However, to eliminate possible systematic errors caused by deviations of the depolarizer from plane-parallelism it is advisable to make measurements every 45° over the range of analyzer position angle from 0 to 315°.

Unfortunately, half of the observing time t is spent unproductively on the measurements with the depolarizer. Taking in Eq. (8.5.6) the ratios of the results obtained with and without the depolarizer increases the mean error resulting from quantum noise by a factor of $2^{1/2}$. For these two reasons the mean errors resulting from quantum noise for the observations made alternately with and without the depolarizer are $2^{1/2}$ times larger than those given by Eq. (8.5.2), even though two photomultipliers are used.[17a]

For observations with a conventional Wollaston polarimeter the simultaneous use of two photomultipliers for each spectral region is essential for diminishing the effects of *atmospheric scintillation*. Since air is not birefringent, scintillation is the same for both perpendicularly polarized beams of starlight.[57] In a photometer or polarimeter with one photomultiplier the errors caused by atmospheric scintillation are approximately equal to those resulting from the quantum noise for a star of 7th blue magnitude for a 40-cm telescope, 8th blue magnitude for a 150-cm telescope.[58,59]

In the conventional Wollaston polarimeter the atmospheric scintillation

[57] A. H. Mikesell, A. A. Hoag, and J. S. Hall, *J. Opt. Soc. Amer.* **41**, 689 (1951).
[58] A. T. Young, *Astron. J.* **72**, 747 (1967); *Appl. Opt.* **8**, 869 (1969).
[59] K. Serkowski, *Astrophys. J.* **141**, 1340 (1965).

is not eliminated completely because of the *inhomogeneous sensitivity* of the photocathodes (presence of "hotspots"). If K_1 and K_2 are the sensitivities of the corresponding areas of the two photocathodes where an image of a certain point of the telescope mirror is formed, the fraction of atmospheric scintillation that remains not canceled is approximately

$$F = \langle (K_1/\langle K_1 \rangle - K_2/\langle K_2 \rangle)^2 \rangle^{1/2}, \qquad (8.5.9)$$

where the angular brackets denote averaging over the image of the telescope mirror. For several EMI photocathodes this fraction was found to be about 15%.[10,56] This means that atmospheric scintillation, in a conventional Wollaston polarimeter with such photomultipliers, becomes the dominant factor limiting the precision for stars about two magnitudes brighter than for a single-channel photometer, i.e., for stars brighter than the 6th magnitude in case of a 150-cm telescope.

8.5.2. Polarimeter with Wollaston Prism Rotating Relative to Photomultipliers

The most straightforward method of avoiding the loss in polarimetric precision caused by spending half of the observing time on the measurements with depolarizer is to rotate the Wollaston prism with respect to the photomultipliers. The effects of changes in photomultiplier sensitivity with time are eliminated if the measurements are made alternately at position angles of the Wollaston prism differing by 180°, and if the rotation of this prism does not shift the image of the telescope mirror on the photocathodes. The latter may be achieved if a thin three-component Wollaston prism (cf., Section 8.2.3) with a construction angle not exceeding $A = 10°$ is used.

An additional *transfer lens* should form an image of the telescope mirror on the lower surface of the Wollaston prism whose upper surface should be perpendicular to the axis of rotation in accordance with Eqs. (8.4.9). Under these conditions small deviations of the angle of rotation of the Wollaston prism from 180°, due to mechanical imperfection, do not cause shifts of the images of the telescope mirrors on the detectors; the only effect is that slightly different areas of the Fabry lenses are illuminated. The transfer lens should form images of the focal plane diaphragm close[†] to the Fabry lenses; each of the photomultipliers has

---

[†] They should not form exactly on Fabry lens to avoid the possibility of forming a sharp stellar image on a dust particle on lens surface.

its own Fabry lens. These lenses should be large enough to accept the beams of light at position angles of a Wollaston prism 0, 45, 180, and 225°; the measurements at these position angles give Stokes parameters Q and U and no parts of the polarimeter other than the Wollaston prism need to be rotated. Only if the highest precision is sought, the measurements should be repeated at orientations of the entire polarimeter differing by 90°. Then some systematic errors, particularly those caused by imaging on the photocathodes any dust particles which are on the Wollaston prism,[35] are eliminated.

A plane-parallel or slightly convex plate of birefringent material should be cemented to the back surface of the three-component Wollaston prism, with optical axis parallel to its surface and making 45° with the optical axes of the wedges of the Wollaston prism. For wide spectral regions such a retardation plate will depolarize the light beams emerging from the Wollaston prism and different sensitivity of detectors for differently polarized light will not affect the results.

This method of avoiding the need for observing alternately with and without a depolarizer which was outlined here was first used in a polarimeter for spectral range 1–5 μm recently built for the Steward Observatory of the University of Arizona;[2a] Mr. E. H. Roland performed the ray tracing and detailed engineering.

8.5.3. Polarimeter with Rotating Half-Wave Plate[17a]

A disadvantage of the polarimeter described in Section 8.5.2 is that the depolarizing plate is needed and therefore very narrow spectral regions cannot be observed. Also, as with the conventional Wollaston polarimeter, the *inhomogeneous sensitivity* of the detector surface and the *varying position of a star* in the focal plane diaphragm affect the accuracy of polarimetric measurements, limiting it for bright objects to about ±0.01%.

All the disadvantages mentioned here are eliminated if, instead of rotating the Wollaston prism with respect to the photomultipliers, an achromatic half-wave plate is rotated in front of a fixed Wollaston prism.[60] This plate can be either set at discrete orientations, e.g., every 22.5°, or rotated continuously. The best half-wave plate for this purpose appears

[60] B. Lyot, *C. R. Acad. Sci. Paris* **226**, 25 (1948); H. Wille, *Optik* **9**, 84 (1952); I. Appenzeller, *Publ. Astron. Soc. Pacific* **79**, 136 (1967); D. Clarke and R. N. Ibbett, *J. Sci. Instrum. Ser. 2* **1**, 409 (1968); A. Behr, *Eur. Southern Observ. Bull.* No. 5, p. 9 (1968); J. Hämeen-Anttila, *Publ. Astron. Soc. Pacific* **84**, 185 (1972).

to be Pancharatnam's retarder of quartz and magnesium fluoride (cf., Section 8.3.2) or a nine-component Pancharatnam retarder.[44]

Let the angle between the principal plane of the Wollaston prism and the equivalent optical axis of Pancharatnam achromatic half-wave plate be $\psi + \psi_0(\lambda)$, where ψ is independent of wavelength and ψ_0 is a small term dependent on wavelength. The intensities of the two light beams emerging from the Wollaston prism oriented at $\phi = 0°$ are, from Eqs. (8.2.2) and (8.3.1),

$$I' = \tfrac{1}{2}[I \pm Q \cos 4(\psi + \psi_0) \pm U \sin 4(\psi + \psi_0)], \qquad (8.5.10)$$

where the upper sign is for the first beam, the lower for the second, and the Stokes parameters for the beam incident on the rotating half wave plate are I, Q, U, and V. The wavelength-dependent terms ψ_0 can be eliminated if we insert another Pancharatnam half-wave plate between the rotating Pancharatnam plate and the Wollaston prism.[61] This additional plate, which should be exactly identical to the rotating one and should have its equivalent optical axis approximately parallel to the principal plane of the Wollaston prism, can be cemented to the Wollaston prism to diminish the light losses; also, all the outer surfaces of optical components should have a multilayer antireflection coating. If now ψ is the angle, independent of wavelength, between the equivalent optical axes of two Pancharatnam plates the intensities of the two beams emerging from the Wollaston prism are

$$I' = \tfrac{1}{2}(I \pm Q \cos 4\psi \pm U \sin 4\psi)$$
$$= \tfrac{1}{2}I[1 \pm P \cos(2\theta - 4\psi)]. \qquad (8.5.11)$$

If the retardances of the half wave plates are not π rad but $\pi + \varepsilon_1$ for the rotating plate and $\pi + \varepsilon_2$ for the fixed plate, where ε_1 and ε_2 are small, then Eq. (8.5.11) takes the form

$$I' \approx \tfrac{1}{2}I \pm \tfrac{1}{2}(1 - \tfrac{1}{4}\varepsilon_1{}^2)[Q \cos 4\psi + (1 - \tfrac{1}{2}\varepsilon_2{}^2)U \sin 4\psi]$$
$$\pm \tfrac{1}{8}\varepsilon_1{}^2 Q \pm \tfrac{1}{2}\varepsilon_1 V \sin 2\psi. \qquad (8.5.12)$$

For Pancharatnam's plate of crystallic quartz and magnesium fluoride (cf. Section 8.3.2) ε does not exceed 0.04 rad for the spectral range 0.3–1.0 μm and the terms containing ε_1 and ε_2 in Eq. (8.5.12) are negligible.

[61] K. Serkowski, *Proc. 15th I.A.U. Colloq.*, *Veröff. Remeis Sternw. Bamberg* **9**, No. 100, p. 11 (1971).

A method of obtaining identical Pancharatnam plates is first to make each of the six components of the plates in a square shape of size twice as large as necessary and plane-parallel to within 1 arcsec, and then cut each component in four.[†] Out of four Pancharatnam plates obtained in this manner the two which match most closely are selected.

According to unpublished estimates by A. T. Young if a rotating half-wave plate makes N revolutions per second and telescope diameter is D centimeters the photometric error caused by *atmospheric scintillation* is smaller than without modulation by a factor approximately equal $(ND)^{5/6}/100$ if this factor is larger than unity. Hence the scintillation is reduced ten times if $N \approx 4000/D$ and is not reduced at all if $N < 250/D$.[58,62] The remaining scintillation will be reduced by a factor of 5 to 10 if both perpendicularly polarized beams from Wollaston prism are measured simultaneously (cf., Section 8.5.1).

Let each quarter of the revolution of the half-wave plate be subdivided into n equal time intervals, corresponding to n memory locations in an on-line computer or multichannel analyzer. The photon counts obtained from the photomultiplier during each such time interval can be added to a corresponding memory location. Because of the effects of averaging the photon counts over n sections of sine curve, the percentage polarization P, calculated with Eq. (8.5.11) on the basis of such averaged photon counts, is related to the true percentage polarization P_0 by the formula

$$P_0 = [(\pi/n)/\sin(\pi/n)]P \cong [1 + \tfrac{1}{6}(\pi/n)^2]P, \qquad (8.5.13)$$

e.g., for $n = 12$, we have $P_0 = 1.0115P$.

An important advantage of polarimeters with rotating half-wave plate is that the measured linear polarization modulates the signal with frequency $4f$, where f is the frequency of rotating half-wave plate; most of disturbing instrumental effects, in particular those caused by shifts of the image of the telescope mirror on the photocathodes, modulate the signal mainly at frequencies f and $2f$. Nevertheless, these shifts should be minimized and therefore the rotating half-wave plate should be either plane-parallel to within a few arcseconds, which is difficult to achieve,

[62] L. V. Ksanfomaliti and N. M. Shakhovskoy, *Izv. Krim. Astrophys. Obs.* **38**, 264 (1967).

[†] The half-wave plates are being manufactured in this way for the writer by B. Halle Nachfl., Hubertusstr. 11, Berlin-Steglitz.

or the image of the telescope mirror should be formed by an additional lens on the back surface of the half-wave plate, as described in Section 8.4.4. Again, this *transfer lens* will form images of the focal plane diaphragms close to the two *Fabry lenses* which in turn form images of the telescope mirror on the two cathodes (Fig. 9). In addition, the transfer lens will diminish the angles of incidence of light on the half-wave plates. An adverse effect of this lens is that any dust particles on the half-wave plate are sharply imaged on the photocathodes by Fabry lenses. Fortunately, only a small component of resulting signal modulation will have the same frequency 4*f* as that caused by measured polarization. This component can be eliminated by repeating the measurements at orientations of the three-wedge Wollaston prism differing by 180°.

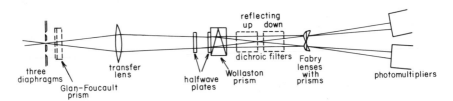

Fig. 9. Polarimeter with an achromatic half-wave plate rotating in front of a Wollaston prism. One of dichroic filters reflects the ultraviolet light to a pair of photomultipliers which are above the plane of drawing, other dichroic filter reflects down the visual light; the near infrared light is transmitted through both dichroic filters.

Dichroic filters[10] or a *grating spectrograph* may be placed between the Wollaston prism and the Fabry lenses to subdivide the light into several beams, corresponding to different spectral regions, directed to different photomultipliers. The reflectivity of dichroic filters[63] and diffraction gratings[18] depends both on the angle of incidence and on polarization of incident light. This reflectivity can be made the same for both perpendicularly linearly polarized beams from the Wollaston prism provided the axes of the beams lie in a plane perpendicular to the plane of incidence on the filters or grating and a Wollaston prism is of somewhat unusual design. The crystalline optical axis in each of the components of the Wollaston prism should make 45° with the *refracting edge* of the prism. The plane of polarization of each of the two outgoing beams will then make 45° with the plane of incidence on the dichroic filters or on the diffraction grating.

63 V. R. Costich, *Appl. Opt.* **9**, 866 (1970).

8.5.4. Eliminating the Polarization of Sky Background

The linear polarization of night sky background at low galactic latitude often exceeds 5%,[59] and may exceed 50% when a large contribution of moonlight or twilight is present.[3,64] Aurora may cause rapid fluctuations in the brightness of the night sky background.

In the conventional Wollaston polarimeter the measurement of stellar polarization is followed by the measurement of polarization and brightness of the sky background close to the star observed. A systematic error may be introduced if an *invisible faint star* is included in the measurement of sky background. The measurement of stellar polarization will not be affected if the degree of polarization and the position angle for such a faint star are the same as for the star observed; this is an advantage when measuring the interstellar polarization, where neighboring stars usually have rather similar polarization. If, however, the difference between the degrees of polarization of bright and faint stars is ΔP and if the intensities of their light are I and I_c, respectively, the measured polarization of the bright star may be in error by

$$\varepsilon(P) = (I_c/I)\,\Delta P. \tag{8.5.14}$$

For the polarimeters described in Sections 8.5.2 and 8.5.3, the polarization of the sky background may be canceled by using *three focal plane diaphragms* of the same diameter. The distance between the centers of adjacent diaphragms should be such that the ordinary image of the central diaphragm on the Fabry lens overlaps with the extraordinary image of one of the side diaphragms. The star should always be centered in the central diaphragm. The polarization of the sky background is canceled because the ordinary image of the sky background in the central diaphragm is superimposed upon the extraordinary image of the sky background in one of the side diaphragms. For a polarimeter with a Wollaston prism rotating relative to the photomultipliers (Section 8.5.2), the line joining the diaphragms should make an angle 22.5° with principal plane of Wollaston prism; the ordinary and extraordinary images of the two diaphragms will not overlap but they still both illuminate the same Fabry lens and the same photomultiplier (see also Piirola[63a]).

Even though the polarization of the sky background is canceled, its

[63a] V. Piirola, *Astron. Astroph.* **27**, 383 (1973).

[64] T. Gehrels, *J. Opt. Soc. Amer.* **52**, 1164 (1962); R. Dumont, *Ann. Astrophys.* **28**, 265 (1965); C. R. Nagaraja Rao and Z. Sekera, *Appl. Opt.* **6**, 221 (1967); I. Ibragimov, *Bull. Inst. Astroph. Dushanbe* No. 54 (1970).

brightness has to be measured separately as it is needed for calculating the polarization of the observed star. There is, however, an attractive possibility of replacing such a sky measurement by a measurement of the star through an additional analyzer, inserted in front or immediately behind the three diaphragms. This analyzer should be thin and should not change the spectral transmittance of the polarimeter; a Glan–Foucault prism may be the best choice for this purpose. With this analyzer inserted, with its principal plane at position angle $\phi = 0°$, the intensities of the two beams from a Wollaston prism in a polarimeter having rotating half-wave plate and three diaphragms are

$$I_G' = \tfrac{1}{2}KI(1 \pm \cos 4\psi) + KI_b, \qquad (8.5.15)$$

where plus sign is for one beam, minus for the other; K is the transmittance of the Glan–Foucault prism for the starlight, while I and I_b are the intensities of starlight and of sky background in one diaphragm, respectively. The intensities of the beams from a Wollaston prism with the Glan–Foucault prism removed are [cf., Eq. (8.5.11)]

$$I' = \tfrac{1}{2}I[1 \pm P \cos(2\theta - 4\psi)] + I_b. \qquad (8.5.16)$$

The observations give us the ratio of the coefficient $\tfrac{1}{2}IP$ at $\cos(2\theta - 4\psi)$ in Eq. (8.5.16) to the average value of I' which is $\tfrac{1}{2}I + I_b$; when this ratio is divided by a similar ratio measured with the Glan–Foucault prism we obtain the degree of polarization P of starlight, free of the effects of sky background,

$$P = \frac{\tfrac{1}{2}IP}{\tfrac{1}{2}I + I_b} \bigg/ \frac{\tfrac{1}{2}KI}{\tfrac{1}{2}KI + KI_b}. \qquad (8.5.17)$$

When observing with the Glan–Foucault prism, the result at each position angle ψ should be taken with a statistical weight inversely proportional to I_G' in order to take into account the changing contribution of photon noise. If the measured polarization does not exceed 10% and the sky brightness $I_b < I/100$, there is no need to spend more than 10% of the total observing time on the measurements with the Glan–Foucault prism.

The procedure described here has the following advantages compared to measuring the sky background separately:

(1)   the principal plane of the Glan–Foucault prism serves as a reference for the observed position angles;

(2)   the effects of deviation of the retardance of half-wave plates from $\lambda/2$ are determined and can be eliminated;

(3)   the effects of the mutual contamination of the two perpendicularly polarized light beams from the Wollaston prism, caused, e.g., by scattering of light on optical surfaces, are eliminated;

(4)   the depolarization of stellar light caused by possible stress bi-refringence in the transfer lens can be evaluated if the measurements with the Glan–Foucault prism are made at four orientations of this lens differing by 22.5°;

(5)   for bright stars, the *nonlinearity* caused by overlapping pulses[†] when photon-counting[65] can be accurately evaluated from the observations with Glan–Foucault prism; and

(6)   the sky background does not need to be measured separately. An invisible faint star whose light may happen to be in one of the side diaphragms will affect the results in accordance with Eq. (8.5.14), in the same way as in the case of the observations with a single diaphragm. However, only the results obtained with one of the two photomultipliers will be affected which will indicate presence of such a star;

(7)   polarimetric accuracy will increase by a factor of about $2^{1/2}$ for the stars giving signals comparable to that of sky background.

8.5.5. Measurements of Circular Polarization

The easiest way of measuring circular polarization with the polarim-eters described in the previous sections is by inserting behind the focal-plane diaphragms a quarter-wave plate which can be set with an optical axis alternately at two position angles $\psi = 45°$ and $\psi = 135°$. The two measurements, reduced in the same way as the measurements of linear polarization without the quarter-wave plate, give the two values of the pseudo-degree of polarization, P_1 and P_2. From these we obtain the degree of ellipticity for the star observed

$$P_V \approx (P_2 - P_1)/(2 \sin \tau). \tag{8.5.18}$$

[64a] F. Gex and J. Tassart, *Nouv. Rev. Opt. Appl.* **2**, 175 (1971).

[65] E. I. Terez, *Izv. Krim. Astrophys. Obs.* **38**, 257 (1969); H. M. Dyck and M. T. Sandford, *Astron. J.* **76**, 43 (1971); A. T. Young, *Appl. Opt.* **8**, 2431 (1969).

[†] The true, corrected pulse rate equals approximately $\delta/[1 - (\delta/\delta_0) - 0.67(\delta/\delta_0)^2]$, where δ is the observed *pulse rate* and δ_0 is the bandwidth of the whole electronic system; δ_0 can usually be kept constant to about 2%. Similar errors are introduced by nonlinearity of electronic system when the direct-current technique of measuring the photoelectric current is used. An advantage of pulse counting is elimination of the effect of photomultiplier *fatigue*[64a] and of fluctuations in dynode amplification.

Here $\tau \approx 90°$ is the retardance of the quarter-wave plate used, found from the same measurements using the formula

$$\cos \tau \approx (P_2 + P_1)/2P, \qquad (8.5.19)$$

where P is the degree of linear polarization for the star observed; the methods of determining the handedness of circular polarization has been described recently.[65a] If the deviation of the angle of rotation of quarter-wave plate from $\pi/2$ is $\Delta\psi$ rad, the spurious circular polarization amounting to $P \, \Delta\psi$ will appear. To eliminate this effect the measurements should be repeated at two perpendicular orientations of the entire polarimeter.[2a,17a]

When the observing procedure described above is used with a conventional Wollaston polarimeter or with one with the Wollaston prism rotatable relative to the photomultipliers the measurements should be made only at Wollaston prism position angles $\phi = k \cdot 90°$, where $k = 0, 1, 2,$ and 3. The circular polarization is then measured with the maximum efficiency.[56] When using the rotating half-wave plate between the quarter-wave plate and Wollaston prism the efficiency of the measurements of circular polarization is twice smaller because the measurements at some orientations of a half-wave plate do not give any information about circular polarization. They give instead the information about linear polarization.

To obtain the highest possible efficiency of the measurements of circular polarization with the rotating half-wave plate the latter may be followed by a Pockels cell[66] or a piezooptical modulator[2,5] (cf., Section 8.3.3) giving alternately the retardance $\tau = -\pi/2$ and $\tau = +\pi/2$ rad and having its optical axis at position angle $\psi = 45°$; this configuration is similar to one used in Pulkovo solar *magnetograph*.[66a] The rotating achromatic half-wave plate can depolarize the linearly polarized component of light very efficiently[48,67] (cf., Section 8.4) while the circularly polarized component changes only its handedness. The high efficiency of measurements is obtained only if variable retarder takes the values of retardance intermediate between $-\pi/2$ and $+\pi/2$ rad for a small fraction of total

[65a] W. Swindell, *J. Opt. Soc. Amer.* **61**, 212 (1971); W. J. Cocke, G. W. Muncaster, and T. Gehrels, *Astrophys. J.* **169**, L119 (1971).

[66] H. W. Babcock, *Astrophys. J.* **118**, 387 (1953); J. R. P. Angel and J. D. Landstreet, *ibid.* **160**, L147 (1970).

[66a] R. N. Ikhsanov and Y. P. Platonov, *Solnechnye Dannye*, No. 11, 78 (1967).

[67] J. Tinbergen, *Astron. Astroph.* **23**, 25 (1973).

observing time; this is not easy to achieve in practice. Also, designing an achromatic variable retarder is difficult.

For these reasons a good method of measuring circular polarization is by inserting a *rotating achromatic quarter-wave plate* between the rotating achromatic half-wave plate and the Wollaston prism. This quarter-wave plate is rotated by a *stepping motor* in such a way that for most of the time its optical axis is consecutively at position angles $\psi = 45° + k \cdot 90°$, where $k = 0, 1, 2,$ and 3. The photon counts made at these orientations are stored in different memory locations. The frequencies of rotation of quarter-wave and half-wave plates should be incommensurable so that for every position angle of quarter-wave plate the half-wave plate has uniform distribution of orientations. If the retardance of half-wave plate is $\pi + \varepsilon_1$, and that of quarter-wave plate $\frac{1}{2}\pi + \varepsilon_2$, difference between the intensities of one beam from Wollaston prism at two orientations of quarter-wave plate is

$$
\begin{aligned}
I'(\psi = 135° &+ \Delta\psi_1) - I'(\psi = 45° + \Delta\psi_2) \\
&\approx \tfrac{1}{4}V \cos \varepsilon_1 \cos \varepsilon_2(\cos 2\, \Delta\psi_1 + \cos 2\, \Delta\psi_2) \\
&- \tfrac{1}{4}U\varepsilon_1{}^2(\Delta\psi_1 - \Delta\psi_2),
\end{aligned} \tag{8.5.20}
$$

where ε_1, ε_2, $\Delta\psi_1$, and $\Delta\psi_2$ are small. If an achromatic half-wave plate is such as shown in Fig. 6, while $\Delta\psi_1$ and $\Delta\psi_2$ do not exceed $3°$, the coefficient of U in Eq. (8.5.20), which is a *coefficient of linear-to-circular conversion*, will not exceed 10^{-5}. The contamination of the measured circular polarization of observed object by its linear polarization can be diminished further by orienting the entire polarimeter so that the principal plane of Wollaston prism is parallel to the plane of vibrations for the observed object.

The linear polarization produced by *reflections* from surfaces of quarter-wave plate can appear as spurious circular polarization. Also, *circular dichroism* (Cotton effect) of quartz[35] in retarders may produce an instrumental circular polarization. These effects, and also the polarization by transfer lens, can be eliminated by placing a rotating half-wave plate in front of transfer lens and making for each object two measurements of circular polarization, one with and one without a second rotating achromatic half-wave plate, inserted between the transfer lens and quarter-wave plate. Inserting this retarder, rotating with frequency incommensurable with that of the other two, changes the sign of the measured circular polarization while that of spurious or instrumental circular polarization remains unchanged. Obviously, it would be preferable to avoid using the transfer lens altogether. Measurements of circular polarization for relatively bright objects by the method here described with precision on the order of $10^{-3}\%$ seem to be possible.

Some of the advantages of the polarimeter with a rotating half-wave plate are incorporated in the *Dollfus polarimeter*.[67,68] In this instrument incident linearly polarized light is changed by an achromatic quarter-wave plate into a circularly polarized light. The light subsequently falls on an achromatic variable retarder with $\tau = 0$ and $180°$, alternately. Another achromatic quarter-wave plate converts the circular polarization back to linear. The Wollaston prism, in a fixed orientation, directs the perpendicularly polarized beams to two photomultipliers.

Each observation of linear polarization consists of two measurements, at orientations of the first quarter-wave plate differing by $45°$. This plate may be followed by a rotating half-wave plate[67] to eliminate the residual linear polarization. The main disadvantage of the Dollfus polarimeter, apart from the difficulty of obtaining a good achromatic variable retarder, is that the result of the measurement is the degree of linear polarization multiplied by $\sin \tau$, where τ is the retardance of the first quarter-wave plate. This plate cannot be of Pancharatnam's type because the wavelength dependent direction of the equivalent optical axis would spoil the accuracy of measurement. Unless a Fresnel rhomb is used as a quarter-wave plate, τ will deviate from $90°$ appreciably and this will depend on the effective wavelength of the incident light and hence on the spectral energy distribution of the observed object. With the first quarter-wave plate removed, Dollfus polarimeter becomes a fine instrument for measuring circular polarization.

For a polarimetric study of transient events, such as solar phenomena or stellar flares, measuring the *linearly* and *circularly* polarized components of light *simultaneously* is essential; a slightly decreased precision is usually the price of this simultaneity. A simplest method of measuring all four Stokes parameters simultaneously is rotating an achromatic quarter-wave plate in front of Wollaston prism.[35] Linear polarization produces modulation with twice higher frequency than circular polarization. The intensity modulation of the beams from Wollaston prism, resulting from linear polarization of the object, has twice smaller amplitude than in the case of rotating half-wave plate. With a similar instrument, having a rotating Polaroid between the rotating quarter-wave

[68] A. Dollfus, *C. R. Acad. Sci. Paris* **256**, 1920 (1963); M. Marin, *Rev. Opt.* **44**, 115 (1965).

plate and fixed analyzer,[17] very accurate circular and linear polarization measurements of the Moon were obtained.

Better separation of linearly and circularly polarized components is obtained in the polarimeters in which two or more retarders, one constant and others variable, are placed in front of an analyzer. Polarimeters with various combinations of such retarders have been built, mainly for solar studies (*vector magnetographs*).[4,69] In some of them each of the Stokes parameters Q, U, and V modulates the signal at different frequency. These instruments often have *no moving components*. They are usually limited to a relatively narrow spectral range. However, even if the variable retarders are not achromatic, polarimeters utilizing them could be designed which work in several spectral ranges separated by dichroic filters. The retardance of retarders at these various spectral ranges should differ by a multiple of 2π or π rad.

8.5.6. Polarimetric Method of Measuring Effective Wavelengths and Radial Velocities

An *effective wavelength* of each spectral region used can be determined by measuring the percentage polarization for this region with a polarizer, followed by a fixed retarder, inserted in front of the polarimeter. A quartz retarder which is quarter-wave at $\lambda = 0.45$ μm, with its optical axis at $45°$ to the principal plane of the polarizer, gives the normalized Stokes parameter Q/I which changes almost linearly with inverse wavelength, from -0.85 to $+0.85$ for $\lambda = 0.3$–1.1 μm. Polarization measured with 0.1% accuracy gives the inverse effective wavelength of wide-band spectral region with an accuracy of about ± 14 cm^{-1}, i.e., ± 3.5 Å at $\lambda = 0.5$ μm.

A polarizer followed by a wide-field retarder several centimeters thick, made, for example, of quartz and sapphire, may provide a *wavelength calibration* for high dispersion photoelectric *spectrometer*. If the optical axis of the retarder makes $45°$ with the principal plane of polarizer, oriented at $\varphi = 0°$, the Stokes parameters Q and V for transmitted light change sinusoidally as a function of wave number k ($= 1/\lambda$) with period $\Delta k = [(n_e - n_o)s]^{-1}$ [cf. Eqs. (8.3.1) and (8.3.2)], where s is the thickness of the retarder. An achromatic quarter-wave plate, with optical axis making $45°$ with that of thick retarder, would change the Stokes parameter

[69] H. Takasaki, *Appl. Opt.* **5**, 759 (1966); J. M. Beckers, *Solar Phys.* **5**, 15 (1968); W. Livingston and J. Harvey, *Kitt. Peak Contr.* No. 558 (1971); A. Cacciani and M. Fofi, *Solar Phys.* **19**, 270 (1971).

V into U, leaving Q unchanged. Hence the position angle of the plane of vibrations for the light transmitted through a polarizer, thick retarder, and quarter-wave plate is a linear function of wave number changing from 0 to 180° with a period Δk. For a quartz and sapphire retarder 3 cm thick the period is $\Delta k = 40$ cm^{-1}, i.e., $\Delta \lambda = 10$ Å at $\lambda = 0.5$ μm. Since the light is fully linearly polarized the error of $\pm 1\%$ in the degree of polarization corresponds to an error of $\pm 0.28°$ in position angle and consequently to an error of ± 0.016 Å in wavelength. Every point across the profile of a spectral line, measured with this accuracy, will give a radial velocity with an error of ± 1 km/sec. The spectral energy distribution across the line profile will remain unchanged after passage of light through the polarizing optics.[17a,69a]

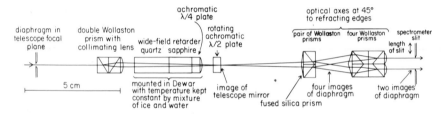

FIG. 10. Radial velocity meter which can be inserted in front of an entrance slit of high-resolution photoelectric spectrometer.

A scheme of *polarimetric radial velocity meter* which can be inserted in front of an entrance slit of photoelectric spectrometer is shown in Fig. 10. Two images of a star are formed along the spectrometer slit, polarized at 45 and 135° relative to the direction of slit; this minimizes the effects of the instrumental polarization in the spectrometer. The light of these two images, after passing through the spectrometer, is measured by two detectors or by two arrays of detectors (cf. Chapter 8.7). Their output is switched to different locations in computer memory in phase with rotation of an achromatic half-wave plate.

The principal advantages of the polarimetric radial velocity meter is that it avoids the loss of starlight on the entrance slit of the spectrometer and that it offers the possibility of observing the extended emission-line objects with wide slit. The result is independent of position of an object within the entrance slit if only the thick retarder is accurately plane-parallel. Spectral resolution needed for obtaining radial velocity with

[69a] K. Serkowski, *Publ. Astron. Soc. Pacific* **84**, 649 (1972).

desired accuracy is at least an order of magnitude lower than with conventional methods. Namely the *resolution* of spectrometer should not be lower than $\frac{1}{6}\,\Delta\lambda$, i.e., in our example 1.6 Å; otherwise the errors resulting from the fact that the averaging is done over Stokes parameters instead of position angle may become appreciable.

The instrument can be calibrated by measuring the position angle of the plane of vibrations for a number of monochromatic light sources of known wavelength. In the case of an echelle spectrometer these monochromatic sources can be observed in different orders of spectrum than that used for stellar measurements; changes in tilt of grating and other shifts and readjustments in the spectrometer during the measurements should have no effect on the measured wavelengths. The only stringent condition is the constancy of temperature of retarder. The change in temperature between the two consecutive calibrations with a monochromatic source should not deviate from linearity by more than about 0.01 °C if accuracy on the order of 1 km/sec is sought. Extremely constant temperature can be achieved if the retarder is cooled with a mixture of ice and water and kept in a Dewar with double windows on each side.

8.5.7. Birefringent Interferometers

The main difficulty with retarders is the dependence of their retardance on wavelength. With a *birefringent interferometer* this wavelength dependence, instead of being an obstacle, allows measurements of the wavelength dependence of polarization and the spectral energy distribution[70] (cf. Chapter 11 by H. W. Schnopper and R. I. Thompson).

The basic part of this instrument is a *compensator*, which is a component equivalent to a retarder with variable thickness s. The compensator, oriented at position angle $\psi = 45°$, is followed by a beam-splitting analyzer with principal plane at position angle $\phi = 0°$. According to Eqs. (8.2.2), (8.3.1), and (8.3.2) the difference in intensity of the two perpendicularly linearly polarized beams emerging from this birefringent interferometer, for incident light having Stokes parameters I, Q, U, and V, is

$$I_1'(s, k) - I_2'(s, k) = Q(k) \cos[2\pi skb(k)] - V(k) \sin[2\pi skb(k)], \quad (8.5.21)$$

where $k = 1/\lambda$ is the *wave number* for the incident light, and $b(k) = n_e - n_o$

[70] A. L. Fymat and K. D. Abhyankar, *Appl. Opt.* **9**, 1075 (1970).

is the *birefringence* of the material of the compensator; the analyzer is assumed to be perfect.

Let us suppose that the incident light covers a wide spectral range with wave numbers from 0 to k_{max}, which can be subdivided into N intervals δk. The difference in intensity of the two beams emerging from the birefringent interferometer is now

$$I_1'(s_j) - I_2'(s_j) = \sum_{i=1}^{N} \{Q(k_i)[\cos 2\pi s_j k_i b(k_i)] - V(k_i) \sin[2\pi s_j k_i b(k_i)]\},$$
$$(8.5.22)$$

where $k_i = i \cdot \delta k$. If we measure this difference in intensity for $2N$ values of the thickness s of the compensator we obtain $2N$ linear Eqs. (8.5.22) which may be solved to obtain the values of $Q(k)$ and $V(k)$ at N values of wave number k. Solving such equations is equivalent to performing the *Fourier transform* of *interferogram* $I_1'(s) - I_2'(s)$; the efficiency of computing the Fourier transforms was enormously increased recently when the Cooley–Tukey[71] algorithm was invented.

We may obtain the wavelength dependence of Stokes parameters Q and U, instead of V, by placing an achromatic quarter-wave plate, oriented at position angle $\psi = 0°$, in front of the compensator. With such a quarter-wave plate the Stokes parameter V in Eq. (8.5.22) should be replaced by $-U$, the equations taking the form

$$I_1'(s_j) - I_2'(s_j) = \sum_{i=1}^{N} \{Q(k_i) \cos[2\pi s_j k_i b(k_i)] + U(k_i) \sin[2\pi s_j k_i b(k_i)]\},$$
$$j = 1, \ldots, 2N. \quad (8.5.23)$$

Such an instrument measures simultaneously both components of the linear polarization.

Another variant of the instrument is obtained by placing an analyzer, e.g., a pair of plane parallel calcite plates with principal planes at $\phi = 0$ and $90°$, in front of the compensator. For the light emerging from such an analyzer $I_a = Q_a = I$ and $U_a = V_a = 0$. Therefore the interferogram takes now the form

$$I_1'(s_j) - I_2'(s_j) = \sum_{i=1}^{N} I(k_i) \cos[2\pi s_j k_i b(k_i)], \quad j = 1, \ldots, N. \quad (8.5.24)$$

The spectral energy distribution $I(k)$ of the incident light is now obtained as a cosine Fourier transform of the interferogram, as for the Michelson

[71] J. W. Cooley and J. W. Tukey, *Math. Comput.* **19**, 296 (1965); J. W. Brandt and O. R. White, *Astron. Astrophys.* **13**, 169 (1971).

interferometer.[72] Linearly polarized light, with plane of vibrations at position angle $0°$, is split upon entering the compensator into two perpendicularly polarized light beams, L_1 and L_2, of equal intensity and with planes of vibration at position angles 45 and $135°$. The path difference between these two perpendicularly polarized light beams L_1 and L_2 is $\varDelta = sb(\lambda)$. The beam-splitting analyzer placed behind the compensator splits each of the light beams L_1 and L_2 into two beams with planes of vibration at 0 and $90°$. The components of L_1 and L_2 with planes of vibration at $0°$ interfere with each other and form the first beam, of intensity I_1', emerging from the beam-splitting analyzer; the components of beams L_1 and L_2 with planes of vibration at $90°$ form the second beam with intensity I_2'.

Let the measurements of the interferogram, given by Eq. (8.5.24), be made for the following N values of path difference $\varDelta$

$$\varDelta_j = j\varDelta_{\max}/N, \qquad j = 1, \ldots, N, \qquad (8.5.25)$$

where $\varDelta_{\max}$ is the maximum value of path difference for which the interferogram is measured. Each of the intensities obtained from Eqs. (8.5.24) is, in general, a sum of intensity $I(k)$ at a wave number k smaller than k_N and of intensities at some wave numbers larger than k_N. Here k_N is the *Nyquist wave number* equal to

$$k_N = N/(2\varDelta_{\max}). \qquad (8.5.26)$$

In order to obtain unambiguous results we should put $k_{\max} = k_N$ and cut off the radiation of wave numbers larger than the Nyquist wave number; in the case of ground-based observations in the optical region we may take $k_N = 33333$ cm^{-1}, corresponding to the atmospheric cutoff wavelength of 3000 Å. Equation (8.5.26) expresses the *sampling theorem* which states[71]: If a function has a Fourier transform that is zero at frequencies greater than or equal to some finite frequency $k_{\max}$, this function is fully specified by values spaced at equal intervals not exceeding $1/(2k_{\max})$.

From Eqs. (8.5.25) and (8.5.26) we obtain the *instrumental line width* of the interferometer

$$\delta k = \delta\lambda/\lambda^2 = k_{\max}/N = 1/(2\varDelta_{\max}). \qquad (8.5.27)$$

---

[72] P. Connes, *Ann. Rev. Astron. Astrophys.* **8**, 209 (1970); W. H. Steel, "Interferometry." Cambridge Univ. Press, London and New York, 1967; R. J. Bell, "Introductory Fourier Transform Spectroscopy." Academic Press, New York, 1972.

This is the equivalent width of the line profile obtained for monochromatic light by solution of Eq. (8.5.24); it has unacceptably strong side lobes which are very confusing. These side lobes can be reduced to below 1% of $I(k_0)$, and the instrumental line profile made similar to that of a diffraction limited slit spectrograph, if we *apodize* the spectrum using, e.g., the convolution formula[73]:

$$I_A(k_i) = 0.54I(k_i)+0.23[I(k_{i-1})+I(k_{i+1})], \quad i = 2, \ldots, N-1, \quad (8.5.28)$$

where $I(k_i)$ are the results of the Fourier transform of the interferogram. Such apodizing has been called *hamming* by Blackman and Tukey.[73] The price of reducing the side lobes by apodizing is the decrease in spectral resolution by a factor close to two. Even though the intensity $I(k)$ is obtained at N values of wave number only every second value is an independent result. Therefore, in practice the instrumental line width for the birefringent interferometer equals

$$\delta k = \delta\lambda/\lambda^2 \approx 1/\Delta_{max}. \quad (8.5.29)$$

Birefringent interferometers have been used for studying the spectral energy distribution of astronomical objects by Mertz[11] in the visual spectral region and by Boyce and Sinton[74] in the infrared around 2 μm. It is particularly attractive to employ the interferometer in the infrared where the precision is limited by detector noise. Since the detector is illuminated by radiation of all wavelengths at the same time, the signal-to-noise ratio is better than for an ordinary one-channel spectrum scanner by a factor close to the number of resolution elements in the spectrum; this is called *multiplexing* or *"the Fellgett[75] advantage."*

In the most practical design of a birefringent interferometer, suggested by Mertz, a compensator[76] is used in which an image of the telescope mirror is formed on the optically contacted interface of a long Wollaston

[73] J. F. James and R. S. Sternberg, "The Design of Optical Spectrometers," p. 128. Chapman and Hall, London, 1969; R. B. Blackman and J. W. Tukey, "Measurement of Power Spectra." Dover, New York, 1959.

[74] P. B. Boyce and W. M. Sinton, *Sky and Telescope* **29**, 78 (1965); W. M. Sinton, *in* "Infrared Astronomy" (P. J. Brancazio and A. G. W. Cameron, eds.), p. 55. Gordon and Breach, New York, 1968.

[75] P. B. Fellgett, *J. Phys.* **19**, 187, 237 (1958).

[76] L. and F. Lenouvel, *Rev. Opt.* **17**, 350 (1938); A. Girard, *Opt. Acta* **7**, 81 (1960).

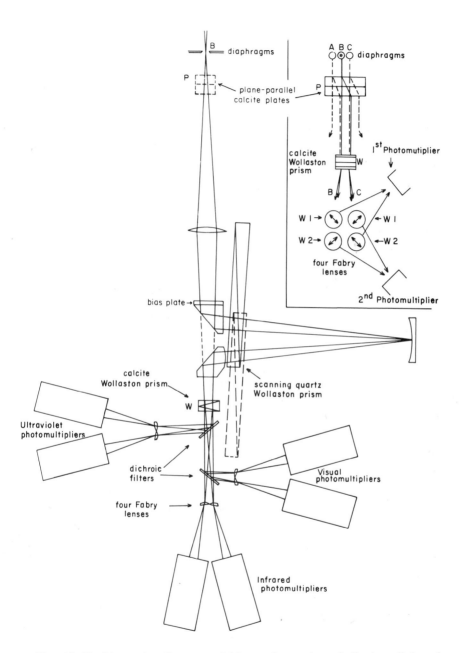

FIG. 11. Birefringent interferometer. An insert shows schematically the splitting of light beams by analyzers in a plane perpendicular to that of the main figure.

prism made of quartz[†] or magnesium fluoride (Fig. 11). The transmitted light is then reimaged again on the same optically contacted interface by a concave ellipsoidal mirror having foci at the primary and secondary images of the telescope mirror. The long Wollaston prism, which has a construction angle A of several degrees, is rapidly moved back and forth in a direction perpendicular to the line joining the center of the mirror surface with its center of curvature. This motion is monitored by a photocell illuminated by an auxiliary light beam going through a *Ronchi grating* attached to the scanning Wollaston prism. The direction of the prism motion makes an angle A/n with the outer surfaces of the Wollaston prism; here n is the refractive index of prism material. The images of the telescope mirror therefore remain focused on the optically contacted interface of the prism during its motion. Three focal plane diaphragms can be used to eliminate the sky background (cf., Section 8.5.4); when the plane-parallel calcite plates are inserted for measuring $I(\lambda)$ both perpendicularly polarized beams emerging from these plates can be used, as explained in an insert to Fig. 11. The calcite Wollaston prism should have the optical axes making $45°$ with the refracting edge (cf., Section 8.5.3).

The birefringent interferometer placed between the focal plane diaphragms and the beam splitting analyzer of the polarimeter appears to be an efficient way to study the wavelength dependence of polarization at moderately high spectral resolution ($\delta k \approx 50$ cm^{-1}). As it can be used with large focal plane diaphragms it is particularly well suited for studying nebulae and galaxies; or, it may replace a cross-dispersion in an *echelle* spectrometer. In the infrared, the interferometer is practically the only choice.

8.6. Instrumental Corrections

8.6.1. Instrumental Polarization

The plane mirrors of the *Newtonian* and *coudé* telescopes or of *heliostats* modify all four Stokes parameters of the incident light. Their effects depend in a complicated way on the state of polarization of the incident light, the angles of incidence, and the time-dependent properties

---

[†] To avoid the effects of the *optical activity* in quartz, it should be cut at 56.2° to the optical axis.[42]

of the aluminum films.[77] Eliminating these effects by compensating plates is a hopeless task. Many discouraging attempts have proven convincingly that precise polarimetry with these types of telescopes is extremely difficult.

In Cassegrain and prime focus telescopes and in refractors, the instrumental linear polarization expected from the Fresnel formulas cancels for the axial beam. However, different regions of the mirror or objective may be taken with unequal weights because of inhomogeneous sensitivity of photocathodes on which they are imaged. Also, polarization irregularly distributed over the mirror surface may appear if the necessary precautions were not taken during *aluminizing*: the glow discharge for final cleaning of the mirror surface before aluminizing should be used sparingly and it should occur everywhere normal to the mirror surface so that the direction of incidence of the charged particles is perpendicular to the mirror.[78] A protective dielectric coating may also polarize or depolarize light.[79]

For a polarimetric telescope an *alt-azimuth mounting* is advantageous because the instrumental polarization can be determined by comparing the results obtained for the same star in the east and in the west. This method is often used in radio astronomy since many radio telescopes have this type of mounting. For the *equatorially* mounted telescope the best method of eliminating polarization introduced by the telescope optics is rotating the entire telescope tube, together with the mirrors. Two such Cassegrain telescopes with *rotatable tube*,[80] each with mirror 61 cm in diameter, are now in use, one at Yerkes Observatory and the other at the Siding Spring Observatory in Australia. The purpose of these telescopes has been fulfilled by observing with them a number of nearby stars of intermediate spectral types for which no linear polarization was found within the limits of error; the best observed of these stars are listed in Table I. Since these stars can be used as unpolarized standards for determining the instrumental linear polarization of other telescopes, there is no need to build more telescopes with rotatable tubes. The mean

[77] V. Bumba, *Bull. Astron. Inst. Czech.* **13**, 95 (1962); F. W. Jäger and L. Oetken, *Publ. Potsdam Sternw.* **31**, No. 103 (1963); No. 107 (1968); R. H. Miller, *Appl. Opt.* **2**, 61 (1963); E. Wiehr, *Solar Phys.* **18**, 226 (1971); M. Makita and K. Nishi, *Ann. Tokyo Astron. Obs. 2nd ser.* **12**, 121 (1970).

[78] G. Thiessen and P. Broglia, *Z. Astrophys.* **48**, 81 (1959).

[79] A. N. Saxena, *J. Opt. Soc. Amer.* **55**, 1061 (1955); D. A. Holmes and D. L. Feucht, *ibid.* **55**, 577 (1965).

[80] W. A. Hiltner and R. Schild, *Sky and Telescope* **30**, 144 (1965).

TABLE I. Unpolarized Nearby Standard Stars Observed with Rotatable Tube Telescopes

HD	Star	α_{1975}	δ_{1975}	m_v	Spectral type (MK)	Distance (pc)	Percentage polarization[a] P	m.e.	Position angle θ (deg)	Reference
432	β Cas	$0^h07.8^m$	$+59°01'$	2.2	F2 IV	14	0.009	±0.009	32	b
10476	107 Psc	$1^h41.2^m$	$+20°10'$	5.2	K1 V	8	0.016	±0.006	175	c
20630	$\varkappa$ Cet	$3^h18.1^m$	$+3°17'$	4.8	G5 V	10	0.006	±0.008	135	d
38393	γ Lep A	$5^h43.5^m$	$-22°27'$	3.6	F6 V	8	0.005	±0.008	130	c, d
39587	χ^1 Ori	$5^h52.9^m$	$+20°16'$	4.4	G0 V	10	0.013	±0.007	20	c
43834	α Men	$6^h11.0^m$	$-74°45'$	5.1	G5 V	9	0.009	±0.010	142	c, d
61421	α CMi	$7^h38.1^m$	$+5°19'$	0.3	F5 IV	4	0.005	±0.009	145	c, d
100623	$-32°8179$	$11^h33.3^m$	$-32°42'$	6.0	K0 V	10	0.016	±0.012	57	c
102870	β Vir	$11^h49.4^m$	$+1°55'$	3.6	F8 V	10	0.017	±0.014	162	c
114710	β Com	$13^h10.8^m$	$+28°00'$	4.3	G0 V	8	0.018	±0.014	116	c
115617	61 Vir	$13^h17.2^m$	$-18°08'$	4.8	G6 V	9	0.010	±0.006	132	c, d
142373	χ Her	$15^h17.9^m$	$+42°30'$	4.6	F9 V	18	0.012	±0.009	31	e
155885/6	36 Oph AB	$17^h13.9^m$	$-26°32'$	4.3	K1 V	5	0.005	±0.007	61	d
165908	99 Her AB	$18^h06.2^m$	$+30°33'$	5.0	F7 V	17	0.002	±0.007	39	e
185395	θ Cyg	$19^h35.9^m$	$+50°09'$	4.5	F4 V	15	0.003	±0.007	139	e
188512	β Aql	$19^h54.1^m$	$+6°20'$	3.7	G8 IV	14	0.012	±0.005	154	e
198149	η Cep	$20^h44.8^m$	$+61°44'$	3.4	K0 IV	14	0.006	±0.005	101	e
209100	ε Ind	$22^h01.4^m$	$-56°53'$	4.7	K5 V	4	0.006	±0.008	88	c, d
210027	ι Peg	$22^h05.9^m$	$+25°13'$	3.8	F5 V	14	0.002	±0.006	45	e
216956	α PsA	$22^h56.0^m$	$-29°45'$	1.2	A3 V	7	0.006	±0.009	89	d

[a] Spectral region 0.4–0.6 μm.

b I. Appenzeller, Z. Astrophys. 64, 269 (1966).

c K. Serkowski, D. S. Mathewson, and V. L. Ford, Astrophys. J. (in press).

d K. Serkowski, Astrophys. J. 154, 115 (1968).

e N. R. Walborn, Publ. Astron. Soc. Pacific 80, 162 (1968).

values of the normalized Stokes parameters Q/I and U/I, observed for the unpolarized standard stars with any telescope, should be considered as describing the instrumental linear polarization and should be subtracted[9] from all other measurements of these parameters with that same telescope. It remains to be established whether the same standard stars are also free of circular polarization.

The objectives of *refractors* or the correcting plates of catadioptric telescopes are likely to depolarize light and to rotate the plane of polarization because of stress birefringence;[81] they produce similar effects as the plane mirrors in Newtonian or coudé telescopes. Taking proper account of these effects is difficult[77] and polarization measurements made with such instruments are likely to be affected by systematic errors. The easiest method of testing the telescope for the presence of retardance introduced by its optics is by measuring the degree of ellipticity for the strongly linearly polarized blue sky, approximately 90° from the sun. For a qualitative test it is enough to inspect the objective visually, when pointed toward the blue sky, through a properly oriented *"tint plate,"* which is a retarder giving path difference $\Delta = 0.57$ μm, followed by a Polaroid. Retardances as small as 5° in different areas of the objective will be visible as different shades of purple. Retardances even 20 times smaller than that can be detected visually with a rotating retarder of $\lambda/20$ path difference.[82]

Another method of testing for depolarization by telescope optics is by measuring standard stars with large and constant interstellar linear polarization. Unfortunately, there are very few strongly polarized stars which have been tested for constancy of polarization with time. Large interstellar polarization is usually observed for distant supergiant stars. These stars often have, in addition to interstellar polarization, an intrinsic polarization changing with time.[51]

Stars that may be suggested as free of intrinsic polarization are early-type stars without infrared excess and without emission lines in their spectra; also the classical Population I cepheids seem to have constant polarization. A few stars suggested as strongly polarized polarimetric standards are listed in Table II. The degrees of polarization $P(\lambda)$, at wavelengths λ other than the wavelength of maximum polarization λ_{max} given in this table[10,83] can be found from the formula describing the

[81] K. Serkowski, *Acta Astron.* **15**, 79 (1965).
[82] M. M. Swann and J. M. Mitchison, *J. Exp. Biol.* **27**, 226 (1950).
[83] G. V. Coyne, T. Gehrels, and K. Serkowski, *Astron. J.* (in press).

TABLE II. Standard Stars with Large Interstellar Polarization[a]

HD	Star	α_{1975}	δ_{1975}	m_v	Spectral type MK	λ_{max} (μm)	$P(\lambda_{max})$ (%)	$\theta(\lambda_{max})$ (deg)
7927	φ Cas	1ʰ18.5ᵐ	+58°05'	5.0	F0 Ia	0.51	3.4	94
14433	+56°568	2ʰ20.1ᵐ	+57°07'	6.4	A1 Ia	0.51	3.9	112
21291	2H Cam	3ʰ27.0ᵐ	+59°51'	4.2	B9 Ia	0.52	3.5	115
23512	+23°524	3ʰ45.1ᵐ	+23°33'	8.1	A0 V	0.60	2.3	30
43384	9 Gem	6ʰ15.4ᵐ	+23°44'	6.2	B3 Ia	0.53	3.0	170
80558	HR 3708	9ʰ17.9ᵐ	−51°27'	5.9	B7 Iab	0.61	3.3	162
84810	l Car	9ʰ44.5ᵐ	−62°23'	3.3–4.0	F0–K0 Ib	0.57	1.6	100
111613	HR 4876	12ʰ49.8ᵐ	−60°12'	5.7	A1 Ia	0.56	3.2	81
147084	o Sco	16ʰ19.1ᵐ	−24°07'	4.6	A5 II–III	0.67	4.3	32
154445	HR 6553	17ʰ04.3ᵐ	−0°51'	5.7	B1 V	0.57	3.7	90
160529	−33°12361	17ʰ40.2ᵐ	−33°29'	6.7	A2 Ia+	0.53	7.2	20
183143	+18°4085	19ʰ26.3ᵐ	+18°13'	6.9	B7 Ia	0.56	6.1	0
187929	η Aql	19ʰ51.2ᵐ	+0°56'	3.5–4.3	F6–G2 Ib	0.56	1.8	93
198478	55 Cyg	20ʰ48.1ᵐ	+46°01'	4.9	B3 Ia	0.53	2.8	3
204827	+58°2272	21ʰ28.3ᵐ	+58°37'	7.9	B0 V	0.46	5.6	60

[a] Errors of $P(\lambda_{max})$ and $\theta(\lambda_{max})$ are likely to be less than ±0.1% and ±1°, respectively.

wavelength dependence of interstellar polarization[61]

$$P(\lambda)/P(\lambda_{max}) = (\lambda_{max}/\lambda)^{-1.15\ln(\lambda_{max}/\lambda)} \qquad (8.6.1)$$

8.6.2. Zero Point of Position Angles

A straightforward method of relating the observed position angles to the equatorial coordinate system was used by Gehrels and Teska.[55,84] The telescope is pointed horizontally *in the meridian* towards a bright star or daylight sky at low altitude. A sheet of *Polaroid* is loosely hung on a plumb line in the light beam in front of the polarimeter. After measuring the linear polarization the Polaroid is rotated around the axis of the plumb line by 180° and the measurement is repeated. The oblique incidence of light on the Polaroid does not affect the results if the point on the Polaroid where the plumb line is attached is chosen so that both measurements give nearly the same position angle θ_0. This position angle (or $\theta_0 + 90°$) should be subsequently subtracted from all the position angles observed without the Polaroid in order to obtain the position angles expressed in the equatorial coordinate system.

E. J. Weber has found, in a laboratory test at the Kitt Peak National Observatory, that Polaroids may rotate the plane of polarization by as much as 0.2°. This rotation can be determined by the method described above if the measurements are repeated using several parallelly oriented identical Polaroids instead of a single one. To avoid using Polaroids a tilted glass plate can serve as a polarizer. Readings of telescope setting circles when this plate is exactly horizontal, as indicated by a spirit level, relate the plane of polarization to equatorial coordinate system.[17a]

For testing the constancy of the zero point of position angles in the course of the observations, the stars listed in Table II may be repeatedly observed. It would be dangerous, however, to rely on any one of these stars as changes in the position angle of their polarization with time or with wavelength are not impossible. When a star is observed at very low altitude the terrestrial atmosphere may slightly rotate the plane of polarization.[49] The depolarization by atmosphere is very small.[85]

In some cases a convenient method of relating the observed position angles to the equatorial coordinate system is to measure the position

---

[84] R. L. Rowell, A. B. Levit, and G. M. Aval, *Appl. Opt.* **8**, 1734 (1969); D. E. Aspnes, *ibid.* **9**, 1708 (1970).

[85] D. H. Höhn, *Appl. Opt.* **8**, 367, 2081 (1969); S. Jorna, *ibid.* **10**, 2661 (1971).

angles of polarization for the entire disc of the planets Mars and Venus. This method, suggested by Gehrels, is based on the observation that the plane of vibrations for the light of these planets remains always perpendicular to the plane through the sun, the earth, and the object.[49,86] The position angle of this plane of vibrations can be calculated[55] from the coordinates of Venus or Mars tabulated in the almanacs. The degree of polarization for these objects amounts often to several percent so the position angle can be measured quite accurately.

8.7. Astronomical Polarimetry in the Future: Television and Image Tube Techniques

The most important advance in astronomical polarimetric techniques to be expected in the near future is the application of television and image tube techniques. The promise of achieving very high photometric accuracy with television techniques, which already is in sight, lies in the *counting of individual photons*.[87] The method seems to be limited to relatively faint sources which would give no more than one photoelectron per resolution element during the scanning of one frame.

The application of television to the polarimetric study of starlight would make it possible to image the *echelle spectrum* of a star, with cross-dispersion produced by a diffraction grating,[88] on the faceplate of an *image intensifier*, followed by a television camera. For one-dimensional scanning an *image-dissector*[89] can be used instead of television camera. Phosphor in image tube can act as a temporary storage device which by its decay keeps information about photon arrival until this information is picked up by the image-dissector. The method is again limited to faint sources giving no more than one photon per resolution element in total duration of single scan. The only method in sight, which permits

[86] T. Gehrels, T. Coffeen, and D. Owings, *Astron. J.* **69**, 826 (1964); A. Dollfus and D. L. Coffeen, *Astron. Astrophys.* **8**, 251 (1970).

[87] A. Boksenberg and D. E. Burgess, in *"Astronomical Observations With Television-Type Sensors"* (J. W. Glaspey and G. A. H. Walker, eds.), p. 21. Institute of Astron. and Space Science, Vancouver, 1973.

[88] D. J. Schroeder and C. M. Anderson, *Publ. Astron. Soc. Pacific* **83**, 438 (1971); J. F. McNall, D. E. Michalski, and T. L. Miedauer, *ibid.* **84**, 145 (1972).

[89] J. McNall, L. Robinson, and E. J. Wampler, *in* "Astronomical Use of Television-Type Image Sensors" (V. R. Boscarino, ed.), p. 117. NASA SP-256, Washington, D.C., 1971; *Publ. Astron. Soc. Pacific* **82**, 837 (1970); W. K. Ford, Jr., and L. Brown, *in* "Carnegie Institution Yearbook 71," p. 221. Washington, D.C., 1972.

achieving high polarimetric accuracy by counting hundreds of thousands of photons per second per resolution element is a *semiconductor diode image tube*[90] (Digicon[†]). The photoelectrons emitted by a conventional photocathode are accelerated to high energy and form an image of photocathode on a long array of PN diodes. Each diode is connected by a separate wire to external amplifier. Every photoelectron produces a strong pulse in diode; these pulses can be counted easily with high speed.

The polarimeter with a rotating half-wave plate (Section 8.5.3) or with piezooptical modulator (Section 8.5.5) in front of the Wollaston prism would be most suitable for these applications. An echelle spectrograph would be placed between the Wollaston prism and an image tube. The polarization at many hundreds of wavelengths would be measured simultaneously. An on-line computer would be needed with the number of memory locations several times larger than the number of resolution elements of the image tube; the photon counts made at different settings of the retarder would be stored in different arrays of memory locations. Such an instrument would be particularly well suited for *radial velocity* measurements (cf., Section 8.5.6) and, when used in the circular polarization mode, for accurate measurements of the *magnetic fields* of faint stars.

In the case of extended objects like planets, nebulae, or galaxies, the polarization of hundreds of resolution elements could be measured simultaneously using a similar instrument but without the spectrograph. The ordinary and extraordinary images of the rectangular diaphragm would be formed on the faceplate of an image intensifier. Another application of television techniques to extended objects would be to use a narrow slit in the focal plane of the telescope and to produce the spectral dispersion perpendicular to the slit. This would give simultaneous polarimetric and photometric area scanning and spectrum scanning. A slit would not be needed if spectrograph were replaced by a polarimetric radial velocity meter and an interference filter separating a single emmission line. A distribution of radial velocities over the surface of nebula or galaxi would be obtained.[17a,69a]

---

[90] E. Beaver and C. McIlwain, *Rev. Sci. Instrum.* **42**, 1321 (1971); E. A. Beaver, C. E. McIlwain, J. P. Choisser, and W. Wysoczanski, *Adv. Electronics and Electron Phys.* **33B**, 863 (1972).

[†] Commercially available at Electronic Vision Corporation, 11661 Sorrento Valley Road, San Diego, California.

Developing these photon counting techniques, which increase *hundreds of times* the efficiency of observing astronomical objects, may have an impact on the progress of astrophysical research as that of telescope with mirror an order of magnitude larger than presently available.

ACKNOWLEDGMENT

The polarization programs at the University of Arizona are supported by the National Science Foundation and the National Aeronautics and Space Administration.

9. THE INSTRUMENTATION AND TECHNIQUES OF INFRARED PHOTOMETRY*

9.1. Introduction

The first infrared observations were made by Sir William Herschel in the year 1800 using simple thermometers to detect the heating effect of the infrared rays from the sun. Almost a century later, this simple yet elegant detection scheme was extended by S. P. Langley with the development of the first bolometer, a device utilizing a resistance thermometer to sense smaller amounts of heat. In the nineteenth century, observations were made of the sun and moon utilizing bolometers and thermocouples operating at ambient temperature. Early in this century, Pettit and Nicholson at the Mt. Wilson Observatory, and W. W. Coblentz at Lowell Observatory, refined these techniques so that accurate radiometric measurements of planetary radiation were accomplished despite the lack of high-performance filters. Highly sensitive PbS photodetectors became available after World War II, and Gerard Kuiper was the first to exploit them in his spectroscopic studies of the planets. William Sinton was able to make 10-μm spectra of the bright planets utilizing the large collecting area of the new 5-m (200-in.) telescope on Palomar Mountain. In addition to spectral resolution, Sinton introduced high spatial resolution scans of the planets. He also showed that the extreme far infrared should not be neglected, using a Golay cell to detect the sun and the moon at about 1000 μm. During this period, John Strong pioneered the use of high-altitude aircraft and balloons to overcome atmospheric absorption and worked steadily to improve spectroscopic techniques. H. L. Johnson extended his work on photoelectric stellar photometry into the near infrared and set the stage for the great advances which followed the introduction of cooled detectors in the early 1960s.

Military interest in infrared techniques stimulated the development of cooled detectors utilizing semiconductor technology borrowed from

---

* Part 9 is by F. J. Low and G. H. Rieke.

the transistor industry. The cryogenic requirements of these new detectors were met economically because of the widespread use of liquid helium, and the need for high-performance filters was satisfied by the development of interference filters made by vacuum deposition of dielectric films. It proved relatively straightforward to apply these devices to multiband photometry utilizing existing optical telescopes. Following Johnson's work at shorter wavelengths, photometric systems were established at the University of Arizona for each of the infrared windows from 1 to 25 μm. At 5, 10, and 22 μm, the helium-cooled germanium bolometer was used.[1] This detector provided four orders of magnitude improvement in sensitivity over uncooled detectors and was utilized at wavelengths out to 1000 μm.

Observations in the spectral region from 25 μm to submillimeter wavelengths had been hindered not only by the lack of sensitive detectors, but also by the high absorption at these wavelengths caused primarily by atmospheric water vapor. Once it was found through ground-based observations that the universe contains many bright infrared sources emitting in the far infrared, efforts were made to overcome the atmospheric limitations by means of aircraft and balloons. It is now possible to carry out routine far infrared observations by these means.

Even at balloon altitudes where the residual absorption becomes almost negligible, the earth's atmosphere still emits appreciable infrared radiation. In order to overcome this problem and the problem of the radiation emitted by the telescope itself, sounding rockets have been used with small infrared telescopes in which the entire apparatus is cooled to cryogenic temperatures. These instruments have proven useful in obtaining information about the brightness of the cosmic background at infrared wavelengths and have carried out rapid surveys of the sky to discover previously undetected sources.

In principle, a large (1-m) cooled telescope accurately pointed in a spacecraft could extend the sensitivity of broad-band observations at 10 μm at least two orders of magnitude beyond present limits of earth-based instruments. It is interesting to note that since the first stars were measured at 10 μm in 1963, the sensitivity of earth-based photometry has been improved by about two orders of magnitude. This was accomplished without increasing the size of the telescope.

The techniques discussed in this chapter are directly applicable to the measurement of absolute fluxes from discrete sources in broad bands

[1] F. J. Low, *J. Opt. Soc. Amer.* **51**, 1300 (1961).

ranging from 1% to octaves in width. The problems of high-resolution infrared spectroscopy are not discussed, since they are dealt with in a separate chapter.

In the following, the current state of the art in infrared detectors will be reviewed with emphasis on the germanium bolometer. The associated apparatus useful for infrared photometry—spectral filters, polarizers, beam splitters, cryostats, and electronics—will be described. These items have general application in infrared photometry. In the remainder of this chapter, ground-based infrared photometry at wavelengths short of 30 μm will be emphasized; such photometry has constituted the bulk of infrared work in astronomy and the techniques for carrying it out are well developed. At wavelengths from about 40 to 300 μm, observations are not possible from the ground. Since techniques used in aircraft, balloons, and rockets are still evolving rapidly, they will not be reviewed here; however, the material presented indicates the direction in which the far infrared instrumentation will move. From about 300 μm to millimeter wavelengths, ground-based observations are again possible, but have been largely neglected. It is not yet clear whether infrared or microwave instrumentation will predominate in this spectral band. The description of ground-based photometry begins with a discussion of the design of telescopes and modulators (spatial filters) for use in the infrared. Atmospheric limitations and the absolute calibration of photometric standards are also considered. Finally, there is a brief description of the operating procedures in ground-based infrared photometry.

9.2. Detectors

Incoherent detectors, such as bolometers and photoconductors, respond only to the incident power level or number of incident quanta, but they provide the highest possible signal/noise when observing continuous sources. Coherent detectors, such as laser-pumped heterodynes, which can measure both the phase and amplitude of the signal, have not yet played a significant role in infrared astronomy. The Heisenberg uncertainty principle limits the sensitivity of coherent detectors severely. Only if the signal is highly monochromatic can the coherent detector compete successfully with the incoherent detector at infrared wavelengths. In practice, the crossover from coherent detection at radio frequencies to incoherent detection in the infrared occurs at about 3×10^{11} Hz ($\lambda = 1$ mm).

There are three basic classes of incoherent detectors which have been used in infrared astronomy: (a) the thermal detectors such as bolometers, (b) the photoconductive detectors, and (c) the photovoltaic detectors. We will discuss each of the three basic classes separately and then compare the levels of performance that can be achieved with presently available detectors. Finally, we will attempt to predict the direction in which future efforts to improve infrared detectors for astronomy will proceed.

The physical mechanism of the incoherent thermal detector is as follows: radiation is absorbed in a nearly black surface and converted efficiently into heat; this heat flows through the sensing element and through a weak thermally conducting path to a large heat reservoir or "heat sink"; the temperature of the heat sink cannot change so the sensing element reaches an equilibrium temperature determined simply by the product of the heat current and the thermal resistance between the heat sink and the sensing element. The time constant governing the approach to thermal equilibrium is the product of the heat capacity of the sensing element and the thermal resistance; thus, there is an inherent compromise between response time and sensitivity. Increasing the thermal conductance to the heat sink decreases the sensitivity, but shortens the response time. In the case of bolometers, the temperature of the sensing element is measured by its electrical resistance. In principle, a wide variety of temperature-dependent parameters could be utilized. In all cases, however, it is important that in measuring the temperature of the sensing element no extraneous noise is introduced. By definition, a "perfect" thermal detector is one in which the predominant noise is produced by the fundamental fluctuations in temperature of the sensing element which occur as energy flows randomly between the heat sink and the sensing element. The quantum efficiency is essentially perfect and independent of wavelength; furthermore, all of the energy of the photon is utilized.

The physical mechanism of the incoherent photodetector is as follows: Photons are absorbed in a semiconducting material by exciting electrons and holes, thus changing the electrical conductivity. The free carriers recombine spontaneously so that the conductivity reaches an equilibrium value determined by the lifetime and mobility of the free carriers. In semiconductors, there is an energy gap between bound and free carriers; thus, a photon must have energy $h\nu$ greater than this energy E in order to produce a change in conductivity. The carriers can also be excited thermally, so it is necessary to cool the semiconductor below the temperature T, where $kT < E$. In an ideal photoconductive detector, the predominant noise is produced by the random generation and recombination of free carriers. Photons with $hv > E$ are detected, but all of their energy is not utilized. In practice, the quantum efficiency may be less than unity and may depend on wavelength leading to less than perfect performance.

The physical mechanism of the incoherent photovoltaic detector is as follows: A large area p–n junction is illuminated by the incident radiation, holes and electrons are generated as in the photoconductor, but are separated by the junction. A photovoltaic potential is produced which is capable of sustaining a current. Excess noise across the junction can be eliminated by reducing the photovoltaic potential to nearly zero by drawing off the charge through a low-input impedance current-measuring amplifier. The current is simply the product of the number of incident photons per second, the charge on the electron, and the quantum efficiency. The random generation of carriers produces noise, but since there is no recombination, the photon noise is less by root two than in photoconductors. At low-light levels, the finite resistance of the junction produces Johnson noise which ultimately limits the sensitivity of these devices. As with photoconductors, the quantum efficiency may be significantly less than unity.

9.2.1. Thermal Detectors

The simplest but least sensitive detectors are the uncooled thermal detectors such as the Golay cell, the thermistor bolometer, and the pyroelectric detectors. The Golay cell is, in effect, a gas thermometer where changes in gas pressure at constant volume are sensed by a thin membrane which deflects a light beam. These devices are inherently microphonic and quite delicate. Thermistor bolometers, like Langley's early metal-strip bolometers, sense the heating effect of infrared radiation by virtue of their temperature-dependent electrical resistance. Although thermistors display a large temperature coefficient of resistance, they are subject to electrical noise, especially at low frequencies. The recently developed pyroelectric detectors sense thermal changes in an interesting fashion. There are dielectric materials, such as triglycine sulfate (TGS), which have large temperature-dependent electrical polarization. When used as the dielectric medium in a capacitor, TGS spontaneously produces a temperature-dependent electric charge which may be sensed by a high-input impedance field-effect amplifier. Because electrical noise in a reactive circuit is quite small, this device has good sensitivity and can be used to relatively high modulation frequencies.

All uncooled thermal detectors are limited by ambient thermal fluctuations and by their large heat capacity. Thus, they no longer play a significant role in infrared astronomy.

Well before 1960 it was fully appreciated that thermal detectors could be vastly improved by cooling to low temperatures. The phenomenon of superconductivity, in which the electrical conductivity of a metal changes abruptly at low temperatures from a finite value to infinity, was actively explored by several experimenters. Unfortunately, superconducting bolometers proved to be limited by extraneous noise associated with instability phenomena during the phase transition from the normal to the superconducting state. Also, it was found that the extremely rapid change of resistance with temperature introduced additional practical difficulties.

The low-temperature germanium bolometer developed by Low[1] in 1961 was the first thermal detector to approach the theoretical limit of performance at liquid helium temperatures. It is the most widely used detector in infrared astronomy and will be discussed here in some detail.

The conduction mechanism in doped germanium at low temperatures may be understood as follows: both p-type and n-type impurities are present in the lattice, but there is always a majority of holes or electrons determined by the major dopant. At low temperatures, no free carriers are excited, but carriers are able to hop between localized impurity sites. The total number of carriers is fixed by the number of impurities and is independent of temperature. Empirically, for the gallium-doped germanium normally used, the resistance R is given by the expression

$$R(T) = R_0(T_0/T)^A, \qquad (9.2.1)$$

where A is a constant equal to about 4, depending weakly on the doping level and degree of compensation.

The standard method of construction is as follows: the gallium-doped germanium is diced and etched to form detector elements of the proper shape. Electrical contacts are made by indium soldering techniques. Metal leads are attached which support the element in a vacuum. One of the leads is usually grounded to a copper substrate which will in turn be bolted to the helium-cooled work surface of the cryostat. The other lead is attached to a sapphire solder land which serves as an excellent thermal conductor, while providing the necessary electrical isolation. Since the germanium element is in a vacuum, the path for conduction of heat between the element and the helium-cooled heat reservoir is solely via the metal leads. Thus, by choosing the composition of the metal leads, a thermal conductivity can be obtained which provides the required thermal conductance. Careful control of the diameter and length of the leads allows precise adjustment of this important parameter. It will be seen in the equations which govern the operation of the detector that the thermal conductance G is the most important single parameter.

It is necessary to blacken the surface of the germanium element to absorb the incident radiation. This can be accomplished in a variety of ways. For wavelengths from the visible to about 200 μm, the blackening is accomplished by painting the front and back surfaces of the detector with a commercially available black paint.[†] This paint has good adhesion properties at low temperatures and an absorption efficiency almost independent of wavelength (0.98 from the visible to 60 μm, 0.40 at 300 μm). It does not add appreciably to the heat capacity of the bolometer. To measure the change of resistance produced by absorbed radiation, a bias current is applied to the detector through a large load resistor R_L. It can be shown that as the bias current is increased, ohmic heating in the detector increases its temperature and ultimately produces saturation. Consequently, there is an optimum value for the bias current given by the relation

$$P = EI = 0.1 T_0 G, \tag{9.2.2}$$

where P is the ohmic heating produced by the bias current I. Figure 1 shows a plot of the current versus the voltage E, which is known as the *load curve*.

The responsivity of a detector is defined as the electric signal produced by a given small change in the incident radiation. For the germanium bolometer, the value of the responsivity at the optimum bias current is given by

$$S = -0.7(R_0/T_0 G)^{1/2}. \tag{9.2.3}$$

It can be seen from this relation that the thermal conductance G controls the responsivity. The higher the thermal conductance, the smaller the signal produced by a given input. This same parameter governs the thermal response time τ according to the relation

$$\tau = C/G, \tag{9.2.4}$$

[†] 3M Black Velvet, No. 101-C10; 3M Company, St. Paul, Minnesota.

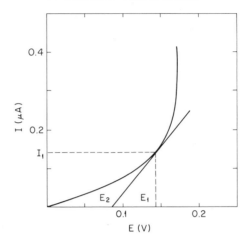

FIG. 1. Load curve for a germanium bolometer at 2 K. Responsivity S is defined as $\Delta E/\Delta Q$. R. C. Jones [*J. Opt. Soc. Amer.* **43**, 1 (1953)] has shown $S = (Z - R)/2E$, where Z is the slope of the load curve at the operating point I_1, E_1. It can be shown that $S = -E_2(2E_1I_1)^{-1}$.

where C is the heat capacity of the germanium element and its associated indium contacts. The thermal conductance enters in two additional important ways; first, when a background power Q is applied, the heat must be conducted away to the heat sink producing a steady temperature drop across the leads just as in the case of the ohmic heating from the bias current. When Q is much greater than P, the bolometer is, in effect, saturated by the background. The responsivity decreases and the thermal time constant increases. The only solution to this problem is to reduce the background power or increase the thermal conductance. Second, the fundamental noise source under zero background conditions is the temperature fluctuation produced by the random flow of heat between the detector element and the heat sink. The amplitude of this so-called *phonon noise* is controlled by the thermal conductance. When operated under low background conditions, and in the absence of extraneous sources of noise, the fundamental limit to the sensitivity of the germanium bolometer is given approximately as

$$\text{noise equivalent power} \equiv \text{NEP} \approx 4T_0(kG)^{1/2}, \qquad (9.2.5)$$

where k is Boltzmann's constant. For further discussion of noise in cryogenically cooled bolometers refer to the work of Low and Hoffman.[2]

[2] F. J. Low and A. R. Hoffman, *Appl. Opt.* **2**, 649 (1963).

In summary, it is seen that the thermal conductance G enters in four ways: (1) it controls the responsivity, (2) it controls the thermal time constant, (3) it controls the ability of the detector to operate under high background loading, and (4) it controls the ultimate noise level under low background conditions. Thus, in order to optimize a germanium bolometer in a specific application, it is necessary to choose the correct value for the thermal conductance. If speed is the only consideration, the G value can be made quite high (of order 10^{-3} W/K) and response times in the microsecond range can be achieved. Usually, however, sensitivity and background considerations predominate. The optimum value for the thermal conductance is often set by the condition for background saturation stated above. If, on the other hand, the background loading is not significant, and a low modulation frequency can be accepted, then the minimum G value is utilized to produce the highest possible sensitivity.

Table I summarizes the parameters of a bolometer with a low thermal conductance, $G = 1 \times 10^{-7}$ W/K. This is near the minimum value of G that can be used at ordinary helium temperatures and limits such a bolometer to low background applications. The table shows how the performance of the detector is degraded as background power is increased.

The dimensions listed are near the minimum values that are practical. Further reduction in the volume of germanium would not reduce the

TABLE I. Measured Performance of a Ge Bolometer under Low Background Conditions at 10 μm

Area A	0.28×0.29 mm^2			
Thickness t	0.25 mm			
Temperature T	1.5 K			
Thermal conductance G	1.1×10^{-7} W/K			
Background power Q	$<1 \times 10^{-8}$	2.7×10^{-8}	5×10^{-8}	W
Resistance R_0	7×10^{6}	3×10^{6}	1.7×10^{6}	Ω
Responsivity $S_{\max}$	4.3×10^{6}	3×10^{6}	1.8×10^{6}	V/W
Time constant τ	8×10^{-3}	9×10^{-3}	11×10^{-3}	sec
Modulation frequency f_0	20	18	15	Hz
Noise E_n	30	40	50	nV/Hz$^{1/2}$
NEP	7×10^{-15}	1.3×10^{-14}	2.5×10^{-14}	W/Hz$^{1/2}$
Photons/sec at 10 μm	$<1 \times 10^{12}$	2×10^{12}	4×10^{12}	

heat capacity because of the contribution from the contacts and leads. When used with cooled field optics, the minimum area is more than adequate for most applications on ground-based telescopes. Only at wavelengths longer than several hundred micrometers is it necessary to increase the volume of the detector. At long wavelengths, absorption is mainly in the germanium rather than at the surface. Methods of enhancing long-wavelength absorption will not be dealt with here.

9.2.2. Photodetectors

PbS photoconductors have played a major role in infrared astronomy and are still widely used in the near infrared. With small areas (0.25×0.25 mm) they can be obtained commercially with NEP values as low as 1×10^{-14} W/Hz$^{1/2}$ under low background conditions. When cooled to liquid nitrogen temperature, spectral response extends to almost 4 μm.

InSb photovoltaic detectors also can be operated at liquid nitrogen temperatures and can provide NEP values as low as 1×10^{-14} or better. InSb has a slightly longer wavelength limit than PbS and is much faster.

Germanium and silicon can be doped with various impurities to provide a wide variety of extrinsic photoconductors with response from a few to several hundred micrometers. Hg:Ge and Cu:Ge have been widely studied and have found use at 10 and 20 μm in certain astronomical applications. Following the discussion of Quist,[3] Cu:Ge and Hg:Ge detectors have been constructed by several groups; they are also commercially available.

9.2.3. Comparison of Available Detectors

Infrared detectors are commonly compared on the basis of their normalized detectivity D^* defined by the relation

$$D^* = A^{1/2}(\text{NEP})^{-1} \quad \text{cm}^{1/2} \text{ Hz}^{1/2} \text{ W}^{-1}, \qquad (9.2.6)$$

where the area A is measured in centimeters and the NEP is defined as the input signal power which produces unity signal/noise when the noise is measured as the rms value in unity bandwidth. The D^* convention may be misleading because the NEP may not be strictly proportional to area over the entire range of interest. Specifically, in astronomy we frequently are able to use the smallest detector area which affords an

[3] T. M. Quist, *IEEE* **56**, 1212 (1968).

improvement in sensitivity. It is a common property of most detectors that the NEP improves as the area is decreased only down to a limiting size of perhaps a few tenths of one millimeter. Further reduction of the size may degrade the performance and is to be avoided.

For astronomical purposes the most relevant basis for comparison is the minimum value of the NEP under identical background conditions. Thus any meaningful comparison of detectors must include a measurement of the incident background power and should be made as close as possible to the actual conditions under which the detectors will be used.

Table I shows that for background power levels greater than 10^{-8} W it is possible to construct germanium bolometers which are background limited in the sense that the noise is dominated by the fluctuations in the incident stream of 10-μm photons. This so called "photon" noise is much lower in the far infrared, but the bolometer would still be saturated, for example, at a slightly lower NEP if the same background power were accepted at $\lambda > 100$ μm. At present, it is not possible to reduce the G value much below the value given in Table I so 1×10^{-14} W/Hz$^{1/2}$ represents the current practical value for the minimum NEP of germanium bolometers.

Doped germanium photodetectors have been compared in performance to germanium bolometers at 10-μm background levels of 10^{-6} to $\sim 10^{-8}$ W. In principle, background-limited performance is achievable with both types of detectors and the noise level should be determined by $N^{1/2}$ (where N is the number of incident photons) for the bolometer and by $(2N)^{1/2}$ for photoconductors. In practice, the performance of the bolometer is more than root two better than for the photoconductor. Apparently the quantum efficiencies are only 15–20% for currently available photodetectors.

At background levels well below 10^{-8} W photoconductors should be more sensitive than bolometers. However, in the near infrared where background levels may be quite low the lowest NEP values still are not much better than 1×10^{-14} W/Hz$^{1/2}$.

9.2.4. Future Developments

A perfect thermal detector, such as the germanium bolometer, yields the highest sensitivity achievable when used under background conditions for which it has been optimized. Thus, efforts will be continued to produce detectors which are optimized at lower background levels than those currently available. Cooling below 1.5 K will be required and temperature-sensitive phenomena in the 0.1–0.3 K temperature range will be explored. For example, paramagnetic salts can be used to make temperature-sensitive inductances at extremely low temperatures. It seems possible that one or two orders of magnitude improvement in sensitivity over existing thermal detectors will be achieved.

As indicated by Quist,[3] the noise in doped germanium photodetectors arises mostly from the generation and recombination of carriers independent of the background level. Thus, in principle, the NEP of a photoconductor decreases indefinitely as the square root of the background. In reality, there are practical difficulties such as the extremely high impedances that are encountered. Helium-cooled field effect transistors can provide the high input impedance, low capacitance, and low noise required by such detectors. Under the extremely low background conditions of a cooled space telescope, these detectors should yield one or two orders of magnitude improvement over existing thermal detectors.

9.3. Associated Apparatus

9.3.1. Infrared Filters

Multilayer dielectric interference filters are commercially available in the United States and in Europe. From 1 to 30 μm, quite acceptable performance can be obtained in both broad-band (~1 octave) and narrow-band ($\Delta\lambda/\lambda \geq 0.01$) filters. Transmission up to 0.90 in the passband and attenuation of $>10^4$ in the stopband can be achieved. At 22 μm, performance is not as good as at shorter wavelengths where control of film thickness is better, but a completely blocked filter should have at least 0.50 transmission over several micrometers of bandwidth.

One of the great advantages of these interference filters is the fact that they can usually be repeatedly cooled to low temperatures without degradation. There is a small but reproducible shift in their parameters. Unfortunately, this statement must be applied with some caution. Experience has shown that some interference filters do not survive rapid cooling to liquid helium temperatures. In certain instances, the problem was one of adhesion of the thin films to the substrate. More subtle and less easily detected changes have occurred; therefore, it seems imperative that when cooling such filters frequent checks of their performance be made both to test for leaks and for changes in transmission. When possible, the rate of cooling should be minimized.

Optical Coating Laboratories, Inc. (OCLI) have developed an important and extremely useful variation of the standard interference filter known as the *circular variable filter*. The films are deposited near the edge of a flat, disk-shaped substrate several centimeters in radius. The film thicknesses vary with position along the circumference of the disk so that the passband changes linearly as the disk is rotated. Resolution for a spot size of a few millimeters can be made as high as $\Delta\lambda/\lambda = 0.01$, and two octaves can be covered in one rotation of the disk. Since these devices can be cooled, compact low-resolution spectrometers can be built with minimum background radiation admitted to the detector.

At wavelengths beyond 30 μm, interference filters are not available commercially and they are difficult to make by vacuum deposition techniques. F. J. Low and K. R. Armstrong have developed the technique of powder deposition used by D. A. Harper to produce thin layers of micrometer-sized diamond particles which are nearly opaque at visible

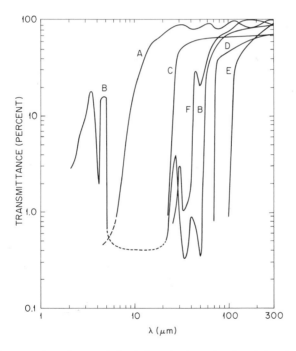

FIG. 2. Transmission of several far infrared filters developed by Armstrong and Low. (A) 5-μm-thick diamond particles 5 μm in diameter on thin plastic film; (B) 1.6-mm-thick Teflon at 300 K; (C) 1-mm-thick sapphire at 4.2 K; (D) 1-mm-thick sapphire at 70 K; (E) 1-μm sapphire particles on thin plastic film at 300 K; (F) 1-mm crystal quartz antireflection coated both surfaces by thermally bonded polyethylene films.

and near infrared wavelengths, but transmit well beyond 30 μm. Figure 2 shows transmission and blocking results for several cooled filters now in use.

Interference filters have been constructed in the far infrared using metal mesh grids in place of semitransparent metal films. Ulrich[4] has described both the theoretical and the practical aspects of these filters.

9.3.2. Windows, Beam Splitters, and Polarizers

9.3.2.1. Windows. We refer the reader to Harshaw Chemical Co. for useful literature on infrared transmitting materials. Eastman Kodak supplies IRTRAN windows. Table II lists the preferred windows for various applications. Note that KBr and CsI are highly hydroscopic and should be protected by a thin film of polyethylene (~12 μm thick) whenever possible. Unfortunately, there is no material which serves as

TABLE II. Window Materials for Various Applications[a]

Material	Passband	Transmission	Index of refraction
Sapphire (Al_2O_3)	0.25–6	0.88	1.7
BaF_2	0.25–13	0.92	1.4
KBr	0.25–25	0.92	1.5
CsI	0.25–60	0.86	1.7
Polyethylene	15–∞	0.90	1.5
Quartz	0.25–4.5	0.91	1.5
Quartz	40–∞	0.80	2.0

[a] The thicknesses are from 0.5 to 4 mm.

the universal window. Pure silicon and diamond approximate the ideal but have high-reflection loss because of their high index of refraction. Silicon is widely used as a lens material. TPX is a useful plastic, since it is clear and has nearly the same index at visual as at far infrared wavelengths.

9.3.2.2. Beam Splitters. Dichroic beam splitters are commercially available which separate the visual band from the infrared. They are

[4] R. Ulrich, *Infrared Phys.* **7**, 37 (1967).

useful in a number of applications; for example, in the guider of a photometer it is possible to use a beam splitter to provide separate optical and infrared foci.

True beam splitters are required in Michelson interferometers and can be made by depositing thin coatings of silicon or germanium on a suitable low index substrate. In the far infrared, mylar plastic films are frequently used.

9.3.2.3. Polarizers. Metal grid polarizers are commercially available from Perkin–Elmer Corporation.

9.3.3. Cryogenics

Cooling the detector, its associated field optics and, where possible, the spectral analyzer or filter is essential if the highest performance is to be achieved. In some cases, the entire telescope is cooled to reduce background radiation emitted by the optical elements of the instrument. The degree of cooling required depends on factors which were discussed in the detector section; here we describe the several most applicable solutions to the cryogenic problem.

Both passive and mechanical cryostats are used to maintain low temperatures in infrared systems. Small, portable, closed-cycle refrigerators have been developed for use in military surveillance applications and are used in aircraft and spacecraft. This type of apparatus is costly, requires considerable maintenance, and may introduce undesirable mechanical vibration. Thus, passive cryostats containing the two most common cryogenic fluids, nitrogen and/or helium, are almost universally used in astronomical applications.

TABLE III. Specifications of Infrared Laboratories Model HD-2 and Model HD-3 Dewars

Specifications	HD-2	HD-3
Liq. N_2 capacity		1.0 liters
Liq. He capacity	1.2 liters	1.0 liters
Liq. N_2 hold time		24 hr
Liq. He hold time (4.2 K)	24 hr	120 hr
Liq. He hold time (2.0 K)	15 hr	75 hr
Weight empty	4.5 kg	6.0 kg

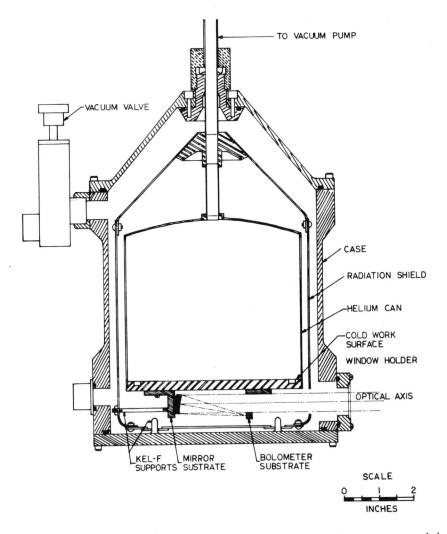

FIG. 3. Model HD-2 dewar uses vapor-cooled radiation shield. Note access to cooled work surface by removable bottom plate.

Table III lists the characteristics of two commercially available dewars[†] widely employed on ground-based, airborne, and balloon-borne telescopes. Figures 3 and 4 show the mechanical details of the dewars and illustrate two convenient geometries.

[†] Infrared Laboratories, Inc., Tucson, Arizona.

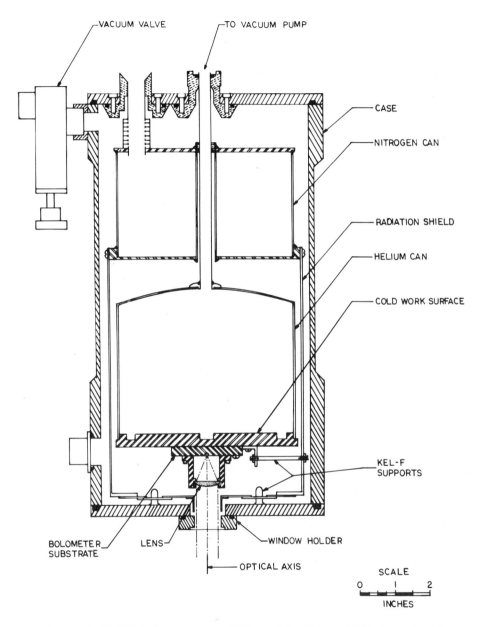

FIG. 4. Model HD-3 dewar uses liquid N_2-cooled radiation shield. As in Model HD-2 Kel-F plastic supports permit accurate alignment of internal structure with respect to case. Supports also greatly reduce microphonics.

Small infrared telescopes have been cooled to liquid helium temperatures and flown successfully in sounding rockets.

The problem of maintaining low temperatures for long periods of time aboard spacecraft is a challenge for cryogenics engineering. On the Mariner 6 mission to Mars, the Ge:Hg detectors were cooled during the encounter by a miniature, two-stage, nitrogen–hydrogen, Joule–Thompson valve device manufactured by Air Products and Chemicals, Inc.

To obtain continuous cooling over much longer periods, solid cryogen systems have been studied. Although unsolved problems remain, it should be possible to design solid nitrogen systems to provide 50–70 K temperatures for 6 months at a weight of about 45 kg. Gillespie and Low have shown that, given a 50 K shield temperature, small quantities of liquid helium can be stored for considerable periods (loss rates of 1 liter/ month were reported). The problem of separating the vapor from the liquid in zero gravity can be avoided by storing the helium as a solid under pressure. Thus, passive systems appear to be attractive in spacecraft applications of the future, as well as in present systems.

9.3.4. Electronics

The most critical section of the electronics apparatus associated with an infrared photometer is the preamplifier. Figure 5 shows the circuit diagram for a solid state amplifier using low-noise field effect transistors which completely satisfies the requirements of germanium bolometers at temperatures above 1.5 K. Selected transistors of the type shown can provide noise levels referred to the input of less than 5 nV/Hz$^{1/2}$ at 20 Hz. The input resistor is high enough that it contributes no noise. Note that the load resistor must either be cooled or made very large compared to the detector impedance. Note that the voltage gain, nominally 10^3, is feedback stabilized.

The block diagram shown in Fig. 6 illustrates the entire electronic system. After phase-sensitive detection, the dc deflections are integrated, digitized, and recorded. The analog signal is also displayed and recorded at the telescope. The system shown here is similar to apparatus developed by H. L. Johnson. Note that all the information required for computer processing of the data is recorded automatically. Alternatively, a small computer could be used to process the data in real time. The automatic telescope "wobbler" is an essential part of the system, since it is necessary to alternately take deflections from positive and negative beams as described in Section 9.8.

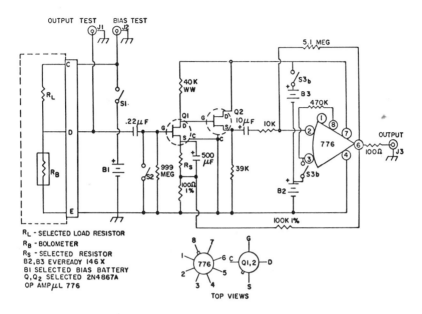

FIG. 5. Circuit diagram for Model LN-6 low-noise preamplifier and bias circuitry for germanium bolometer.

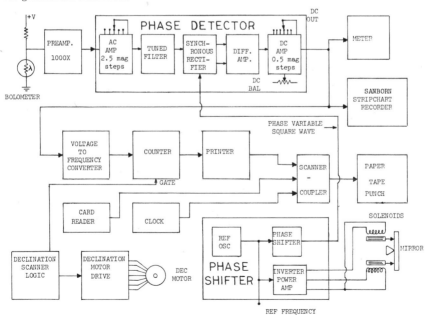

FIG. 6. Block diagram of electronic system used in ground-based photometers.

9.3.5. Field Optics

A lens placed at or behind the focal plane of the telescope which images the primary mirror onto the detector is called a *Fabry*, or *field, lens*. It can be used to perform a number of important functions. Since it spreads the energy from any point within the field stop over the entire detector, it overcomes the problems associated with the nonuniform response which many infrared detectors have over their sensitive areas. A field lens also restricts the view of the detector and therefore reduces the background radiation incident on it. Finally, a field lens permits the use of a detector smaller than the field aperture.

Although refracting optics can be used to perform these functions over narrow passbands, dispersion in the refractive material results in unsatisfactory results over broad passbands. An alternative that has been found useful for telescopes with relatively long focal ratios is to use a small, off-axis mirror to perform the functions of a field lens. On the 155-cm (61-in.) telescope at 10 μm and with a 2-mm (6-arcsec) field stop, a spherical field mirror is used to form a diffraction-limited image of the telescope on a detector 0.25 mm in diameter.

With a small field of view used at long wavelengths, even diffraction-limited field optics may couple the telescope very inefficiently to the detector. In the example of the 155-cm (61-in.) telescope cited above, the coupling is only about 85% efficient at 10 μm and considerably worse at 20 μm. This problem can be alleviated by using larger fields of view or larger detectors at the longer wavelengths.

9.4. Telescope Design

Among the characteristics or design objectives of virtually all large optical telescopes in use today are the following:

(1)   The Cassegrain *f* ratio is relatively small.
(2)   The direct view of the sky from the Cassegrain focus is blocked by baffles. All surfaces within view except the mirrors are blackened in order to reduce the amount of light scattered into the focal plane.
(3)   The unvignetted field of view should be large and well-corrected.
(4)   The mirror surfaces are coated with aluminum.

Conventional telescopes are suitable for infrared observations at wavelengths short of 5 μm. However, beyond 5 μm the characteristics listed above seriously compromise the effectiveness of such instruments.

Beyond this wavelength, specialized design is required to minimize the effects of thermal emission from the telescope and sky, which are usually at temperatures between 260 and 300 K. Unfortunately, a 300 K blackbody has its maximum emission at 10 μm, in the middle of the best atmospheric window in the far infrared. As discussed in Section 9.6., the mean emissivity of the sky in the 10-μm window band is 0.10 or less under good conditions; the effective emissivity of a conventionally designed telescope is of the order of 0.5. Therefore, detectors are background-limited on conventional telescopes at a level of sensitivity less than half as great as the fundamental limit imposed by the sky emission. In the 21-μm window, the sky emissivity is considerably higher; conventional telescopes will come closer to the fundamental sensitivity limit at this wavelength. Although the sky emission is relatively weak, its variability can also limit the sensitivity of instruments beyond 5 μm; this subject is discussed in Sections 9.5 and 9.6. Under good conditions, the photon noise from the background radiation emitted by a conventional 150-cm (60-in.) telescope with a small field of view exceeds the sky noise.

Most of the background from a conventional telescope comes from black surfaces within view of the detector. For example, the sky baffle removes unwanted background from the sky in the visual region, but in the infrared it has the opposite effect. At 10 μm, this baffle has an emissivity very near 1.0, while the sky itself has an emissivity of 0.1 or less. Thus, the background level can be reduced by removing the baffle.

TABLE IV. Characteristics of the Telescopes Used in Calculating the Background Radiation[a]

Characteristics	$f/9$	$f/45$
Primary diameter (cm)	152	152
Sky baffle diameter (cm)	57.2	24.4
Secondary mirror diameter (cm)	45.7	15.0
Detector to secondary (cm)	363	670
Field stop (mm)	0.4	2.0
Field of view (arcsec)	6	6

[a] The dimensions of the $f/9$ telescope are scaled from the Steward Observatory 225-cm (90-in.); those of the $f/45$ telescope are from the NASA 155-cm (61-in.) at the Catalina Observing Station, Lunar and Planetary Laboratory.

In order to provide specific examples, the amount of background radiation from two different telescopes has been computed. The designs of the telescopes have been based respectively on the Steward Observatory 229-cm (90 in.) ($f/9$) and on the NASA 155-cm (61-in.) at the Lunar and Planetary Laboratory ($f/45$). The parameters of the telescopes are

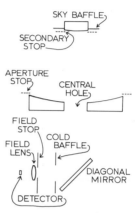

FIG. 7. Optical design of an infrared telescope.

shown in Table IV; Fig. 7 should clarify the terms used below. In computing the background contributions of the various parts of the telescope, it was assumed the telescope was at 280 K and that the field optics were diffraction-limited at 10 μm and restricted the field of view of the detector to a full width at half power equal to the Cassegrain f ratio of the telescope.

TABLE V. The Background Radiation in the N Photometric Band from the Two Telescopes Described in Table IV

Characteristics	$f/9$ (nW)	$f/45$ (nW)
Central hole in primary	5.7	0.9
Sky baffle	4.9	9.2
Edge of secondary	4.5	0.7
Sky	1.9	2.2
Mirror surfaces	1.6	1.9
Secondary supports	1.0	1.4
Total	20	16
Effective emissivity	0.5	0.4

The results of the calculations are shown in Table V. The major sources of emission are related to the telescope characteristics listed at the beginning of this section as follows:

(1) A small Cassegrain f ratio necessitates a large central hole in the primary. Notice the difference in this respect between the $f/9$ and $f/45$ telescopes.

(2) The emission of the sky baffle has already been discussed.

(3) In order to have a large, unvignetted field of view, the secondary mirror must be oversized, i.e., larger than the distance from the secondary to the focal plane divided by the f number. However, an oversized secondary reflects background radiation from the parts of the telescope around the edge of the primary, all of which usually have high emissivities.

(4) Conventional aluminum coatings have an emissivity of about 0.02; ultrahigh vacuum-deposited silver has an emissivity about four times lower and would therefore be more satisfactory for use in the infrared.

Despite the similar background levels obtained with these two telescopes, they differ in suitability for use in the infrared. The small secondary mirror of the $f/45$ telescope permits easy modulation of the signal, as discussed in Section 9.5. The 20% lower background level of the $f/45$ telescope results in 10% less photon noise. In addition, the smaller central obstruction in this instrument combined with the restricted field of view of the detector required to minimize the background received from the sky baffle result in a signal level 20% greater. Thus, the ratio of signal-to-noise with a background-limited detector at 10 μm is 30% greater for the $f/45$ telescope. Put another way, the 152-cm (60-in.) $f/45$ telescope should be equivalent in signal-to-noise at 10 μm to a 200-cm (80-in.) telescope designed like the $f/9$ instrument.

Significant reductions in the background level can be made in telescopes designed for operation in the infrared. From the above discussion, it is clear that such telescopes should have small central obstructions. In addition, highly emissive surfaces in view of the detector, such as sky baffles, should be eliminated so far as possible. Since any telescope must be of finite size, a low-emission aperture stop must be included in the design. Figure 7 shows the four stops that can be used to determine the effective aperture of a Cassegrain telescope. The emission from the defining stop must be kept at a minimum; at the same time, the effective aperture of the telescope should be kept as large as possible in order to maximize the signal received at the focal plane.

For conventional telescopes used in the infrared, the field stop usually also serves as the defining stop in order to minimize the background from the oversized secondary and sky baffle. This procedure has the advantage that the field stop can be placed with the detector inside the cryostat and therefore can be made cold enough that its emission is negligible. However, because of diffraction, the field of view of the detector defined by the field optics and field stop does not have sharp edges. A figure of merit for the sharpness of the edges of a diffraction pattern is

$$\Delta V = a[2(r_1 + r_2)/(\lambda r_1 r_2)]^{1/2}, \qquad (9.4.1)$$

where a is the diameter of the aperture, λ is the wavelength, r_1 is the distance from the source to the aperture, and r_2 is the distance from the aperture to the screen on which the diffraction pattern is observed. For $\Delta V \lesssim 1$, the Fraunhofer approximation holds, i.e., the diffraction pattern is considerably larger than the geometrical shadow of the aperture on the screen. If $\Delta V \gtrsim 10$, nearly all of the energy falls within the geometrical shadow. In the case of a detector system with a relatively small field stop, $\Delta V \sim 2$. For example, the field optics of the $f/45$ telescope system with 6-arcsec field of view have $\Delta V = 1.6$.

On telescopes with small f ratios, a cooled baffle in front of the field stop can be used to reduce the background falling on the detector. Such an arrangement increases in effectiveness the farther the baffle is placed in front of the field stop and the smaller the field stop. In the case of the $f/9$ telescope described above, if the baffle is 100 mm in front of the field stop (about the maximum attainable with conventional cryostats), $\Delta V \approx 50$. In addition to diffraction, the field of view will be blurred by an amount proportional to the ratios of the diameter of the field stop to that of the cooled baffle. For the $f/9$ telescope, this blurring is less than 4% of the $f/9$ beam width for a 0.4-mm (or 6-arcsec) field stop. A cooled baffle provides much less satisfactory results for telescopes with large f ratios. For the $f/45$ telescope described above and a 2-mm (6-arcsec) field stop, $\Delta V \approx 20$ and the blur because of the size of the field stop is nearly as large again as the $f/45$ beam width.

An alternative that can be used on a telescope designed for the infrared is to make the secondary mirror undersized so that the edge of the secondary defines the effective aperture of the telescope. Such a system should be built so that beyond the edge of the secondary only sky is visible to the detector; then, because of the low emissivity of the sky, the stop contributes relatively little to the background. The sky does emit more than the cooled field stop, so this stop should still be kept reasonably tight. Using the edge of the secondary as the defining stop has the advantage that diffraction effects are minimized; for the $f/45$ telescope, $\Delta V \approx 90$, while the $f/9$ telescope has $\Delta V \approx 350$. As a result, the field of view of the detector can be sharply confined, and essentially need include only the telescope mirrors, central obstruction and sky. Of these, only the central obstruction has a high emissivity.

The final alternative is to place the defining stop at the primary mirror of the telescope. This placement is nearly always adopted in the visual region because it provides the maximum possible effective aperture and there is no vignetting. However, in most cases this stop has an effective emissivity near 1.0 and is therefore unsatisfactory for the infrared. The effective emissivity could be reduced by surrounding the primary mirror with a low-emissivity, cylindrical skirt which would reflect the extraneous rays onto the primary and from there onto the sky. Even with such a skirt, the effective emissivity of the telescope will be higher than with an undersized secondary as the defining stop, since the extraneous rays must undergo two additional reflections before reaching the sky.

An $f/45$ Cassegrain telescope with no sky baffle, very thin secondary supports, the minimum central hole in the primary, a slightly undersized secondary, and low-emissivity mirror surfaces would be excellent for work in the infrared. With a 6-arcsec field of view in the 10-μm window, the background radiation reaching the detector from such a telescope would be of the order of 5 nW (1 nW = 10^{-9} W), divided about equally between the sky radiation within the field of view and radiation from the telescope. Thus, the performance in this wavelength region would essentially reach the limit set by the radiation of the sky.

It is of interest to calculate the performance of such a telescope. We shall compute the one-standard-deviation level of significance for an observation consisting of 15 min of integration on the source and an equal time on the background. The rms fluctuation of a monochromatic background level is

$$\overline{\Delta n^2} = \bar{n}(e^{hc/\lambda kT})(e^{hc/\lambda kT} - 1)^{-1}, \qquad (9.4.2)$$

where $\bar{n}$ is the mean rate at which photons arrive at the detector. In a rigorous treatment, the background noise must be computed by integrating over the system band-pass.[2] If $e^{hc/\lambda kT} \gg 1$, this integration can be simplified by introducing the approximation

$$\overline{\Delta n^2} \approx \bar{n}. \qquad (9.4.3)$$

If the band-pass is sufficiently narrow, the integration can be dispensed with altogether. Under these approximations, the rms fluctuation of the 5-nW background in an integration time of 15 min is 3.3×10^{-16} W; since both the source and background measurements are subject to this level of fluctuation and the measurement consists of the difference between the two, the effective one-standard-deviation flux level from the source is $2^{1/2} \times 3.3 \times 10^{-16}$ W $\approx 5 \times 10^{-16}$ W.

Table VI lists the various factors by which a signal reaching the top of the atmosphere is attenuated before it finally reaches the detector. The telescope area factor refers to the percentage of the full telescope aperture which is effective in gathering energy after allowance has been made for losses due to the diffraction effects in the field optics; the

TABLE VI. Attenuation at 10 μm of a Signal Incident at the Top of the Atmosphere before It Reaches the Detector

Factors	Attenuation factor
Atmosphere	0.82
Central obs.	0.97
Telescope area factor	0.83
Modulator	0.95
Mirror surfaces	0.93
Dewar window	0.92
Infrared filter	0.82
Electronic filter	0.81
Total	0.35

electronic filter factor refers to the loss of the higher harmonics in the square-wave-modulated signal, due to the slow response of the detector and the response of the band-pass electronic filter usually used in the lock-in amplifier. The other terms in Table VI should be self-explanatory.

The efficiency estimates in Table VI can be compared with the measured overall system efficiency, which can be calculated from an observation of a star if the detector responsivity, electronic gain, and absolute flux from the star are known. The efficiency measured in this fashion agrees within a factor of 1.1 with the overall efficiency from Table VI.

Assuming a 152-cm (60-in.) telescope with a nominal collecting area of 1.8×10^4 cm² and allowing for the width of 5.3 μm of the 10 μm, or N, photometric band, the flux incident at the top of the atmosphere which would be measurable at the one-standard-deviation level of significance is

$$\frac{5 \times 10^{-16} \text{ W}}{0.35 \times 1.8 \times 10^4 \text{ cm}^2 \times 5.3 \text{ μm}} = 1.5 \times 10^{-20} \quad \frac{\text{W}}{\text{cm}^2 \text{ μm}}.$$

This calculated result can be compared with the measured performance level of a background-limited detector on the NASA 155-cm (61-in.) telescope; under conditions where the calculated background level is 22 NW and therefore the calculated one-standard-deviation level is 3.2×10^{-20} W/cm² μm, the measured one-standard-deviation level, after 15 min of integration each on source and background, is 2.9×10^{-20} W/cm² μm. In view of the uncertainties in the estimates, the agreement between the measured and calculated sensitivities is fortuitous. For example, the telescope emission and overall system efficiency enter in the calculation of the sensitivity, while the absolute value of the measured sensitivity depends upon the absolute calibration; because of the uncertainties in these three quantities, a discrepancy of 30 or 40% between the calculated and measured sensitivities would not have been surprising.

So far this discussion has been confined to 152-cm (60-in.) telescopes with 6-arcsec fields of view. Of course, the background incident on the detector is proportional to the area of the field of view. Therefore, the smallest possible field must be used for maximum sensitivity in the infrared. However, there is a lower limit on the field of view that can be used for accurate photometry. This limit depends upon the size of the image produced by the telescope, the accuracy of the telescope drives and stiffness of the telescope mount, the amount of wind, and the seeing. The most fundamental limit is set by the diffraction limit of the telescope. The diameter of the first null in the diffraction pattern is $\alpha(\text{arcsec}) = 5.03 \times 10^5 \lambda/D$, where D is the telescope aperture and λ is the wavelength; for $D = 150$ cm and $\lambda = 10$ μm, $\alpha = 3.3$ arcsec. With an excellent telescope and under ideal observing conditions, a field of view only slightly larger than α could presumably be used; however, such conditions are only rarely achieved. Experience indicates that a field substantially smaller than 6 arcsec cannot be used to obtain accurate results at 10 μm on a routine basis with a 150-cm telescope. Three arcsec is probably the smallest field that can be used routinely on even the largest telescopes.

The background received by the detector is proportional to $A\Omega$, where A is the telescope area and Ω is the field of view solid angle. The photon noise due to the background is therefore proportional to $(A\Omega)^{1/2} \propto D\theta$, where θ is the diameter of the field of view, and the signal-to-noise is proportional to $A/D\theta \propto D/\theta$. Thus, for a given field of view, the background-limited signal-to-noise is proportional to the telescope aperture. However, the maximum sensitivity will be obtained with the minimum field of view. For *small* telescopes, the minimum useful field is determined by the diffraction limit and is inversely proportional to the aperture, i.e., the signal-to-noise is proportional to $D^2 \propto A$. For *large* telescopes, the minimum useful field of view (about 3 arcsec) is determined by factors other than the diffraction and is independent of D; therefore, the signal-to-noise is proportional to D. The transition between these two cases occurs approximately over the range of apertures where the diameter of the diffraction-limited image ranges from 4 to 2 arcsec. At 10 μm, this transition range is from 127 to 254 cm aperture; at 20 μm, it is from 254 to 508 cm.

In principle, the discussion of background radiation and how to minimize it applies to all the infrared wavelengths. However, in the 5-μm photometric band the radiation from a 300 K blackbody is about a factor of 20 less than in the 10-μm band and at 3.6 μm the background is about another factor of 10 less. As a result, the background limit is seldom reached in astronomical applications at wavelengths short of 5 μm.

Telescopes designed for minimum background emission in the infrared suffer from a number of shortcomings for use at other wavelengths. For example, the elimination of sky baffles makes the instrument unsuitable for use in the visual region and in particular makes it very difficult to find faint objects for visual guiding of the telescope. This problem can be overcome by building baffles which can be automatically moved into place for viewing and hidden behind the secondary or in the central hole of the primary for infrared observations. Another problem is the poor durability of silver mirror coatings, which must be protected against both tarnishing and attack by water. This protection can be provided by placing the mirrors in airtight enclosures when the telescope is not in use and by not using the telescope on excessively humid nights when there is a risk of dew forming on the mirror surfaces.

Although large, ground-based Cassegrain telescopes can be built with effective emissivities comparable to that of the sky in the 10-μm photometric band, even lower emissivities are desirable in certain applications. For example, the mean emission of the sky from 3 km and between 10.3

and 12.3 μm is more than a factor of 2 lower than the mean over the whole 10-μm band; from higher sites or airborne telescopes, even lower emission levels could be expected. In space, the background level is much lower still and has an effective temperature of only a few degrees Kelvin.

An off-axis, or Herschelian, telescope, with no central obstruction and no central hole in the primary, could be built with a substantially lower emissivity than an equivalent Cassegrain instrument. It would be necessary to surround the primary mirror of such a telescope with a low-emissivity, cylindrical skirt in order that the extraneous rays not strike highly emitting surfaces, such as the mirror cell or the telescope tube. There appears to be no fundamental reason why a telescope of this design with an effective emissivity of about 0.01 could not be built. Modulation of the signal with a Herschelian instrument would require chopping the primary mirror; as a result, this design could only be used with relatively small apertures.

In space, where the formation of dew on exposed cold surfaces is not a problem, the telescope emission can be reduced dramatically by cooling the instrument. An instrument cooled with solid nitrogen could be operated at a temperature of about 50 K. An $f/45$, low-background, 150-cm telescope at this temperature with a field of view slightly larger than its diffraction limit could achieve a background-limited detector sensitivity more than 100 times greater than a similar ground-based instrument at 10 μm. Even at 50 μm, near the peak of a 50 K blackbody, the background emission of the cooled telescope would be more than 100 times lower than that from a similar instrument at 280 K; thus, the sensitivity would be increased by a factor of 10 or more. Our choice of an $f/45$, 150-cm telescope was made to facilitate the comparison with the discussion on ground-based telescopes and does not imply that such a telescope would be optimum for space applications. The space-borne infrared telescopes that are currently contemplated have an aperture of about 100 cm (40 in.) and a focal ratio considerably less than $f/45$.

In this section, we have discussed the sensitivity of infrared telescopes when the limiting noise is photon noise from the background radiation incident on the detector. In this case, it was shown that the signal-to-noise is proportional to the telescope area for small telescopes, where the minimum field of view is determined by diffraction, and is proportional to the aperture for large telescopes, where the minimum field is determined by seeing and guiding errors. Two other sources of noise can limit the sensitivity under certain conditions: detector noise and sky noise. Under these conditions, the dependence of the signal-to-noise on the telescope size is different from the case we have discussed.

If the limiting noise originates in the detector, the signal-to-noise is proportional to the telescope area up to a maximum size determined by the optimum size for the detector. This situation holds at the shorter infrared wavelengths, where the background is very low. For nearly all infrared detectors, the performance in principle improves as the sensitive element is reduced in size. However, because of practical difficulties, reducing the detector diameter below about 0.25 mm seldom results in any gain. The energy collected by the telescope must be coupled onto the detector by the field optics; this coupling is inefficient if the field optics suffer from serious aberrations. If we assume that the optimum detector size is 0.25 mm, that the minimum useable field of view is 3 arcsec, and that field optics faster than $f/1.5$ will suffer from serious aberrations, the proportionality between signal-to-noise and telescope area will hold until apertures of about 12 m are reached. Thus, this proportionality holds even with the largest existing telescopes.

The sensitivity can also be limited by sky noise; this situation can be expected, for example, when a large field of view is used to observe an extended source at 10 or 20 μm. Sky noise is discussed in more detail in Sections 9.5 and 9.6; for a given field of view, it increases approximately in proportion to the telescope area. Thus, under these conditions, the signal-to-noise will be approximately independent of telescope size.

9.5. Modulation and Space Filtering Techniques

Observing at 10 μm with a ground-based telescope has been likened to observing visually through a telescope lined with luminescent panels and surrounded by flickering lights as though the telescope dome were on fire. Quantitatively, the problem can be stated as follows: the background power falling on the detector is on the order of 10^{-7} W originating partly from within the telescope and partly from the atmosphere, whereas, the signals to be measured may be as small as 10^{-14} W. The sky emission has both temporal and spatial fluctuations and the telescope emission also varies because of mechanical flexure and constantly changing thermal gradients. Thus, the performance of such a system is determined by its ability to reject background fluctuations.

In the simplest arrangement, a single infrared detector is direct-coupled to a recorder through a high-gain amplifier. The telescope is alternately moved from sky to sky plus star and the difference is taken between the two deflections. There are two serious problems with this method: (1) all of the low-frequency noise of the detector will be sensed, (2) changes in the telescope and sky background level during the sample period will not be cancelled. The second problem can be greatly alleviated by adding a second carefully matched detector adjacent to the first detector and sensing only the difference between the two outputs as they alternately sample star plus sky and sky alone. This produces twice the deflection, increases the noise by only root two, and greatly suppresses drift; however, the excess noise at low frequencies is still present.

We will now enumerate several alternative arrangements before proceeding to what we believe is an optimum approach for ground-based astronomy. In the above example, the low-frequency excess noise would be eliminated if the telescope could be wobbled between the two sky positions rapidly enough, perhaps 5 or 10 times/sec, to permit capacitive coupling of the detector to the amplifier. This, of course, is not feasible with large telescopes. Within the optical train there are at least three alternatives: (1) wobble the primary mirror in its cell, (2) wobble the secondary mirror, or (3) move a small mirror near the focal plane to

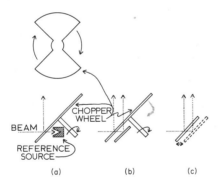

FIG. 8. Three types of focal plane modulators (see text for discussion).

simulate moving the detector back and forth between the two positions. All three alternatives are roughly equivalent in that the beam of the telescope is switched rapidly between two adjacent positions on the sky, modulating only the difference signal. In reality, only (2) and (3) are feasible. Historically, the focal plane modulators were the first to be used, so we will describe several ways this can be done.

Figure 8 shows schematically three arrangements that have been employed. Note that in Part (a) the reference beam is not on the sky, but looks into a black cavity whose temperature is regulated to simulate the sky-telescope background. There is always a finite difference between the two beams; this output deflection which arises from imperfect cancellation of the background is called *offset*. This arrangement is the infrared equivalent to the so-called "Dicke switch" used in radio astronomy and suffers from drift and fluctuation of the offset; only very bright sources can be detected using this method. The arrangement shown in part (b) represents an improvement in that both beams are returned to the sky; however, it is difficult to adjust the two mirrors so that the two beams are equivalent, resulting in large offsets. The advantage of this arrangement is that large beam separations can be generated at high modulation rates. The mirrors must be large to simulate true square wave modulation. The third alternative shown in part (c) produces the same action but with only one mirror. Experience proves that better balance of the two beams can be achieved. This type of focal plane modulator, when carefully

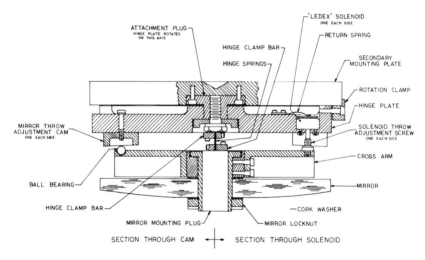

FIG. 9. Chopping secondary mount of the 155-cm (61-in.) telescope.

adjusted and diaphragmed to reduce edge effects, produces the lowest offset and the smallest modulation noise. There are several methods of supporting and driving the mirror to approximate square wave modulation.

Finally, we come to modulation by means of wobbling the secondary mirror, a technique which is feasible with many large Cassegrain telescopes. The earlier discussion on telescope background shows that in a high performance infrared telescope it is necessary to reduce the central obscuration as much as possible; it is also desirable to make the secondary slightly undersized so that it serves as a cold stop. Figure 9 shows how the 15-cm (6-in.) diameter secondary of the 155-cm telescope is mounted so that it can be wobbled to produce nearly perfect square wave modulation at 10 and 20 Hz for beam separations up to about 1 arcmin. When the telescope is in good collimation, extremely small stable offsets can be achieved and modulation noise remains below true sky noise for beam separations less than 30 arcsec. This modulation technique produces the highest performance yet achieved with ground-based telescopes. It has also proven extremely useful in far infrared airborne telescopes.

Infrared telescopes in space will be cooled to reduce their background; however, modulation of the signal and rejection of background from the earth, spacecraft, sun, moon and space debris will still be important. Therefore, experience with ground-based telescopes will influence the design of space telescopes.

In concluding this discussion, it should be noted that our objective was to maximize the signal/noise for measurements of discrete sources smaller than the beam size. When extended sources are to be mapped or large areas of the sky are to be scanned for unknown sources, the problem is more complex and involves factors not considered here.

9.6. Atmospheric Limitations

Atmospheric effects result in three fundamental limitations in ground-based infrared photometry:

(1) Observations can only be made in spectral regions where the atmosphere is transparent.

(2) The atmospheric emission places the ultimate lower limit on the amount of background radiation falling on the detector. In addition, nonuniform emission combined with turbulence can result in excess noise, commonly called *sky noise*.

(3) On large telescopes and particularly for the shorter wavelengths, the "seeing" (i.e., the increase in the angular spread of the light from the source due to optical inhomogeneities in the atmosphere) determines the minimum field of view which can be used for quantitative photometry.

The atmospheric transmission as a function of wavelength is shown in Fig. 10. The photometric designations of the infrared windows are

$H(\lambda_{\mathrm{eff}} = 1.6$ µm$)$, $K(2.2$ µm$)$, $L(3.6$ µm$)$, $M(5.0$ µm$)$, $N(10.6$ µm$)$, and $Q(21$ µm$)$. In addition, small but usable windows exist at 350 and 460 µm. The atmospheric constituents responsible for the dominant absorptions are $CO_2(2.0$ µm, 2.8 µm, 4.1–4.5 µm, and 13–17 µm$)$, $N_2O(4.5$ µm, and 7.4–7.8 µm$)$, $O_3(9.3$–10.0 µm$)$ and H_2O (most other wavelengths).

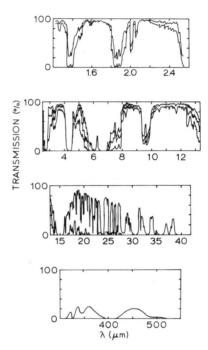

FIG. 10. Atmospheric transmission as a function of wavelength. The transmission is shown for a variety of atmospheric conditions.

The concentration of these atmospheric constituents, particularly H_2O, varies from site to site and from time to time at a given site. Therefore, curves corresponding to different levels of extinction are shown in the figure. A more detailed discussion of the photometric bands can be found in Section 9.7.

Of course, Fig. 10 applies only to the atmospheric extinction as seen from the ground. The extinction for airborne experiments can be estimated through the data of Murcray et al.,[5] Furashov,[6] Kundratiev et

[5] D. G. Murcray, F. H. Murcray, and W. J. Williams, *J. Opt. Soc. Amer.* **54**, 23 (1964).
[6] N. I. Furashov, *Opt. Spectrosc.* **20**, 234 (1965).

al.,[7] Hall,[8] Kyle *et al.*,[9] Murcray *et al.*,[10] and Murcray *et al.*[11] Figure 10 is based on the data in the above references and on that in Farmer and Key,[12] Johnson *et al.*,[13] Finke,[14] and Gillett.[15]

In addition to absorbing the radiation from extraterrestrial objects, the atmosphere emits. The atmospheric emission determines the minimum background level under which any ground-based detector must operate. The temperature of the emitting regions is of the order of 260 K; therefore, atmospheric emission is not a serious problem for observations at wavelengths short of 5 μm. Of course, the level of emission is strongly dependent on meteorological conditions; nonetheless, under reasonably good conditions and at an altitude of 3 km, the average emissivity of the atmosphere between 8.0 and 12.7 μm is of the order of 0.08.[16] Most of the emission in this region occurs between 9.2 and 10.1 μm and 12.3 and 12.7 μm and is caused, respectively, by ozone and CO_2. Since ozone is distributed to very high altitudes, the emission due to it is virtually independent of the altitude of observation up to 30 km; however, the emission from the other atmospheric constituents decreases rapidly with increasing altitude. Between 10.3 and 12.3 μm, emissivities of the order of 0.04 are observed at 3 km. Ground-based telescopes are usually at an ambient temperature slightly higher than the effective temperature of the sky. Assuming that the telescope is at 280 K, its effective emissivity would have to be reduced to less than 0.06 to permit operation in a sky-background-limited condition in the *N* photometric band. Between 10.3 and 12.3 μm, a telescope emissivity less than 0.03 would be required to achieve this condition. The design of telescopes

[7] K. Ya. Kundratiev, G. A. Nicolsky, I. Ya. Badinov, and S. D. Andreev, *Appl. Opt.* **6**, 197 (1967).

[8] J. T. Hall, *Appl. Opt.* **6**, 1391 (1967).

[9] T. G. Kyle, J. N. Brooks, D. G. Murcray, and W. J. Williams, *Appl. Opt.* **8**, 1926 (1969).

[10] D. G. Murcray, F. H. Murcray, W. J. Williams, T. G. Kyle, and A. Goldman, *Appl. Opt.* **8**, 2519 (1969).

[11] D. G. Murcray, J. N. Brooks, J. J. Kosters, and W. J. Williams, AFCRL-71-0359 (1971).

[12] C. B. Farmer and P. J. Key, *Appl. Opt.* **4**, 1051 (1965).

[13] H. L. Johnson, I. Coleman, R. I. Mitchell, and D. L. Steinmetz, Comm. Lunar and Plan. Lab. #113 (1968).

[14] U. Finke, private communication (1972).

[15] F. C. Gillett, private communication (1972).

[16] D. G. Murcray, J. N. Brooks, J. J. Kosters, and W. J. Williams, Data Rep., Balloon Flight 18 Nov. 1970, AFCRL Contract F 19628-68-C-0233 (1971).

for optimum performance in the infrared has already been discussed; emissivities of the order of 0.06 should be achievable.

The preceding discussion has assumed that the sky emission is perfectly well-behaved, i.e., that the only noise associated with it is photon noise. In fact, the nonuniformity of the sky emission can constitute the dominant source of noise at 10 and 20 μm. Very little systematic study of this source of noise has been carried out, although a number of its properties can be determined from observations taken for other purposes. An extensive study of the sky noise at various sites is being carried out by J. Westphal; the results of this work are not yet available.

In the following, the term "sky noise" will be understood to include two phenomena. The first is a fluctuating sky emission associated with nonuniform sky emissivity and atmospheric turbulence. This component of sky noise can be detected by comparing the noise observed (in the absence of any modulation) when the detector views the sky through the telescope with that observed from a nonvarying object at a similar background level. The second phenomenon is the presence of large-scale spatial gradients in the sky emission. Since the sky emission is much less than that from the telescope, such gradients are difficult to detect directly. Because of the sky noise, particularly the spatial gradients in the sky emission, it is necessary to make infrared observations with a beam switching technique in which the source and reference beams lie very close to each other on the sky, have equal weight, and are alternated at a frequency of a few Hertz or more. With such a modulator, the two beams pass through very nearly the same air path. As a result, any spatial gradients in the sky emission are strongly cancelled. With a good modulator, the spatial gradients do not limit the sensitivity of the system. A poorly designed beam switching system which does not give the two beams equal weight can result in "modulator noise." A system in which the noise with the modulator in operation exceeds that when the modulator is off suffers from modulator noise.

With a chopping-secondary modulator operating at 10 Hz, a 150-cm telescope, a 6-arcsec field of view, and a detector with an NEP of 10^{-14} W/Hz$^{1/2}$, the sky noise at N under good conditions is less than the system and background noise. More complete measurements with a detector NEP of 3×10^{-14} W/Hz$^{1/2}$ (at 20 Hz) indicate that the sky noise from 5 to 50 Hz at both N and Q is less than the system noise under good conditions. However, the sky noise increases rapidly with increasing field of view and is usually dominant in good systems with fields larger than 15 arcsec on a 150-cm telescope. Sky noise also appears to decrease rapidly with increasing frequency and can be minimized by beam switching rapidly.

High sky noise can occur under a variety of conditions. Since sky noise is associated with turbulence, bad seeing is frequently associated with high noise. Atmospheric turbulence and emission are increased in the daytime, and the daytime sky noise is typically a factor of 1.5 to 2 times greater than the nighttime value. Since water vapor emits strongly in the infrared, even very thin clouds that may be invisible under other observing conditions greatly increase the level of sky noise.

The final atmospheric effect to be considered here is "seeing." The diameter of the seeing disk decreases slightly as one goes farther into the infrared; under marginal conditions, it is possible to carry out reasonably quantitative photometry at N when only qualitative results can be obtained at K. No detailed studies of the infrared seeing have been carried out; however, from experience at the telescope, it appears that the seeing disk at 10 μm is at least half as large as it is in the visual. When additional allowance is made for setting and guiding errors, it seems unlikely that quantitative 10-μm photometry can be carried out with fields of view smaller than 3–4 arcsec. The effect of this limit on the maximum sensitivity level achievable with a given telescope has already been discussed.

9.7. The Infrared Photometric System

For convenience in comparing observations, a system of photometric bands has been defined in the infrared. These bands are analogous to the U, B, V bands employed in photoelectric photometry. The observed flux levels are often described on a scale of magnitudes which is also analogous to the practice in photoelectric photometry; a difference of one magnitude in flux corresponds to a ratio of $10^{0.4} = 2.51$ and the magnitude values increase as the flux level decreases. Absolute measurements of astronomical objects are particularly difficult in the infrared because of the high level of background radiation and poorly determined instrument efficiency. As a result, all photometric measurements are made relative to a system of standard stars. In addition to the brightness of different objects at a given wavelength, the ratios of the fluxes at two different wavelengths are often compared for two different objects. These ratios, often called the "colors" or "color indices" of the object, are usually expressed as the difference of the magnitudes in the two photometric bands. For example, the $K - L$ color of an object is the difference between the K magnitude and the L magnitude, which is equivalent to the ratio of the K and L fluxes times a factor that is dependent on the absolute calibration of the magnitude scales. This infrared photometric system, particularly when combined with even a crude absolute calibration of the fluxes from the standard stars, has proved to be a powerful tool for understanding the general properties of the astronomical objects which emit strongly in the infrared.

9.7.1. Infrared Photometric Bands

The characteristics of the infrared photometric bands are summarized in Table VII. Relatively narrow bands centered at approximately 11.5 and 13 μm are often designated O and P. A useful parameter in describing

TABLE VII. Characteristics of the Infrared Photometric Bands

Photometric band	λ_1 (cuton; μm)	λ_2 (cutoff; μm)	λ_0 (effective wavelength; μm)	$(\delta m\ 4000)$ (mag)
H	1.45	1.8	1.63	0.02
K	1.9	2.5	2.22	0.03
L	a		3.5	
L'	3.05	4.1	3.6	0.06
M	4.5	5.5	5.0	0.02
N	7.9	13.2	10.6	0.23
Q	17	28	21	0.19

[a] Defined in H. L. Johnson, *Astrophys. J.* **141**, 923 (1965).

a photometric band is its effective wavelength

$$\lambda_0 = \int_0^\infty \lambda\phi(\lambda)\,d\lambda \Big/ \int_0^\infty \phi(\lambda)\,d\lambda, \tag{9.7.1}$$

where $\phi(\lambda)$ is the relative sensitivity of the measuring instrument. To a first-order approximation, the observed flux levels behave like monochromatic flux levels at wavelength λ_0.

Table VIII contains a list of suggested standard stars, selected because they are reasonably bright and are distributed fairly uniformly over the

TABLE VIII. Magnitudes of Standard Stars

Name	B.S.	K	L	M	N	Q
β And	337	−1.85	−2.10	−1.97	−2.05	−2.20
α Ari	617	−0.65	−0.75	−0.80	−0.75	−0.82
α Tau	1457	−2.89	−3.00	−2.89	−2.96	−3.09
α Aur	1708	−1.78	−1.86	−1.92	−1.85	−1.90
α CMi	2943	−0.64	−0.67		−0.45	−0.54
α Hya	3748	−1.16	−1.36	−1.25	−1.27	
α Boo	5340	−2.99	−3.14	−2.98	−3.02	−3.27
γ Dra	6705	−1.29	−1.50		−1.33	−1.49
γ Aql	7525	−0.59	−0.80			

sky. A number of bright stars, such as α Ori, α Her, and β Peg, have been rejected because of possible variability, particularly at the shorter infrared wavelengths. However, these stars may be useful for calibration at the longer wavelengths, particularly 20 μm, where the fluxes from many of the adopted standards are rather weak. Particularly with PbS detectors, operating at the shorter infrared wavelengths, an additional means of calibration is sometimes provided in the form of a standard source of well regulated intensity which can be brought into the field of view of the detector.

The photometric bands were selected to match the regions of maximum atmospheric transmission, as can be seen from Fig. 10. Nonetheless, all of the bands contain wavelengths where there is atmospheric absorption. Observations at different zenith angles are made through atmospheric slant paths of different lengths and therefore are subject to different amounts of absorption. As a result, it is necessary to correct the apparent fluxes for the atmospheric extinction. Observations of standard stars at widely different zenith angles can be used to estimate the extinction, which is usually assumed to be proportional to the secant of the zenith angle. A measurement of the extinction is essential whenever observations are being made in a spectral region where it is high, such as M or especially Q; in the regions where the extinction is low, such as H, K, L, or N, average values can be used if a direct measurement for the night of the observations cannot be obtained.

Johnson[17] has computed λ_0 for the photometric bands. He did not consider the atmospheric absorption features in determining $\phi(\lambda)$. Basing $\phi(\lambda)$ only on instrumental parameters has the advantage that λ_0 is determined by readily measurable factors and is independent of atmospheric conditions. However, where there are strong absorptions within the photometric band, the value of λ_0 computed in this way may not correspond very closely to the situation actually realized on the telescope. The values of λ_0 in Table VII have been computed with allowance for the atmospheric absorptions. Unfortunately, this procedure makes λ_0 dependent on atmospheric conditions; the extent of this dependence is shown in Table IX. Three cases have been considered, corresponding to the three levels of extinction shown in Fig. 10. The range of extinctions actually observed at the Catalina Observing Station, Lunar and Planetary Laboratory, generally falls within the range computed from Fig. 10, as shown in the table. Significant dependence of λ_0 on extinction is predicted for M (2.2%) and Q (8%). The large dependence of λ_0 on extinction at Q undoubtedly contributes to the difficulties in achieving high accuracy in this band.

In the wider photometric bands, higher-order corrections should be applied to the monochromatic flux levels assigned at λ_0. The most exact procedure would involve integrating the assumed flux distribution over the instrument response function $\phi(\lambda)$. The result of such a calculation is shown in Table VII; $\delta m(4000)$ is the amount (in magnitudes) by which

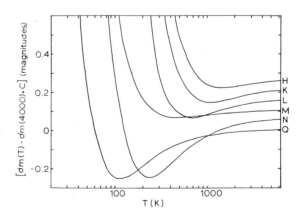

FIG. 11. $\delta m(T)$, the correction to the monochromatic levels, assigned at λ_0. The values of C are 0 (Q), 0.05 (N), 0.10 (M), 0.15 (L), 0.20 (K), and 0.25 (H).

[17] H. L. Johnson, Comm. Lunar and Plan. Lab. #53 (1965).

TABLE IX. Dependence of Effective Wavelength on Atmospheric Extinction

Photometric band	Observed extinction range (mag/air mass)	Low extinction		Intermediate extinction		High extinction	
		mag/air mass	λ_0	mag/air mass	λ_0	mag/air mass	λ_0
H		0.10	1.63			0.15	1.63
K	0.07–0.18	0.14	2.22			0.23	2.22
L	0.12–0.28	0.13	3.56	0.24	3.57	0.43	3.61
M	0.27–0.55	0.28	5.03	0.45	4.98	0.82	4.92
N	0.14–0.32	0.14	10.64	0.18	10.62	0.26	10.56
Q	0.30–1.4	0.39	21.49	0.54[a]	20.87	2.1	19.80

[a] Transmission curve not shown in Fig. 10.

the assigned flux level for a 4000 K blackbody must be reduced in order that the total flux radiated into the photometric band not be overestimated.[18] $[\delta m(T) - m(4000)]$ is shown in Fig. 11 for all of the photometric bands.

We can also estimate $\delta m(T)$ using a second-order approximation developed by King.[19] Even in the N band, where the largest corrections must be applied, this approximation is within 4% of the exact value for $T > 80$ K. However, for very cold objects or objects with complex spectra, the exact procedure should be used.

9.7.2. Absolute Calibration

In order to interpret photometric measurements physically, it is usually necessary to employ a calibration of the measurements in terms of absolute physical units. Originally, the photometric system was calibrated by Johnson.[17] His method was based on the observation of a number of stars similar to the sun in order to obtain their colors in the photometric system. Under the assumption that the colors of the sun are the same as the average colors of the comparison stars, it is possible to calibrate the photometric system through reference to absolute measurements of the flux from the sun. Unfortunately, Johnson had no direct measurements of suitable comparison stars at M or N; the colors he assumed at these wavelengths were based on interpolation and extrapolation from data on other types of stars. Johnson determined the zero points of the magnitude scale so that, for an average $A0$ V star, all magnitudes are the same, i.e., all color indices are zero. Enough $A0$ V stars have not been observed at M, N, and Q to set the zero points at these wavelengths; therefore, the values in Table X are based on a 10,000 K blackbody curve fitted to the zero point at 3.6 μm. These values differ slightly from those obtained by Johnson.

In the following, we shall revise Johnson's calibration on the basis of new data. The most important aspect of this revision is the presentation of observations of solar comparison stars at M and N which permit a more accurate application of his method at these wavelengths. The pertinent data are summarized in Table XI; all the $V - I$ and many of the $V - K$ and $V - L$ measurements are from Johnson et al.[20] The

---

[18] Similar calculations have been carried out by R. I. Mitchell (unpublished).

[19] I. King, Astronom. J. **57**, 253 (1952).

[20] H. L. Johnson, R. I. Mitchell, B. Iriarte, and W. Z. Wisniewski, Comm. Lunar and Plan. Lab. #63 (1966).

TABLE X. Flux Level for 0.0 Magnitude

Filter	λ_0 (μm)	$F(\lambda)$ (W/cm² μm)	$F(\nu)$ (W/m² Hz)
K	2.22	4.14×10^{-14}	6.80×10^{-24}
L	3.6	6.38×10^{-15}	2.76×10^{-24}
M	5.0	1.82×10^{-15}	1.52×10^{-24}
N	10.6	$9.7 \ \times 10^{-17}$	3.63×10^{-25}
Q	21	$6.5 \ \times 10^{-18}$	9.56×10^{-26}

TABLE XI. Observations of Solar Comparison Stars

B.S.	Sp	$(V - I)'$	$(V - K)'$	$(V - L)'$	$(V - M)'$	$(V - N)'$
219	G0 V	0.91	1.50		1.58	1.52
458	F8 V	0.85	1.42	1.48		
483	G2 V	0.86	1.42	1.54		1.47
937	G0 V	0.87	1.43			
1729	G0 V	0.90	1.51			1.56
2047	G0 V	0.87	1.48			1.50
3881	G1 V	0.89	1.42			
4496	G8 V	0.87	1.52			
4785	G0 V	0.90	1.55	1.60	1.58	1.58
4983[a]	G0 V	0.84	1.44	1.55	1.36	1.35
5072	G5 V	0.95			1.47	
5868	G0 V	0.88	1.47	1.50		
7503	G2 V	0.78	1.44	1.54		
7504	G5 V	0.73	1.48	1.55		
8729	G4 V	0.85	1.47	1.67		
Mean		0.86	1.47	1.55	1.54	1.53

[a] Not included in computation of mean colors.

tabulated numbers represent the observed colors adjusted to those of spectral type $G2$ V (that of the sun) by means of the mean colors as a function of spectral type.[21] Accurate values of the solar flux at the effective wavelengths of the photometric bands can be obtained from the calculations of Labs and Neckel,[22] which closely fit the absolute measurements. The resulting calibration is incorporated in the magnitudes of the standard stars listed in Table VIII.

It is not possible to extend Johnson's method to obtain an absolute calibration at Q because the fluxes from the comparison stars are too low. However, an approximate calibration can be obtained by extrapolation from the shorter wavelengths. The Q calibration in Table VIII is based on fitting blackbody curves at the effective temperature of the star to the fluxes of a number of stars at N and averaging the results. The major uncertainty with this procedure is the possibility that the stars show some systematic departure from the blackbody curves in the far infrared.

The data in Table XI for $(V - K)'$ have a standard deviation of only 4%; presumably, the overall accuracy of the calibration at K is of the order of 10% or better. However, at M and N the colors of the comparison stars show discrepancies which exceed the errors of measurement. The stars showing large deviations from the mean colors at M and N show smaller deviations in the same directions in $(V - I)'$ and $(V - K)'$; there appears to be a scatter in the intrinsic infrared colors of the comparison stars which is comparable with the measurement errors at the shorter wavelengths, but which becomes fairly large at M and N. Therefore, the true far infrared colors of the Sun cannot be determined very accurately by this method; the calibration at L, M, and N is probably uncertain by 15 or 20%.

Stebbins and Kron[23] compared the colors of the Sun directly with a number of stars out to $V - I$. The $V - I$ color of the Sun appeared to differ from that of the comparison stars. They attributed this difference to instrumental problems, but they could not exclude the possibility that the Sun has peculiar infrared colors. The work of Labs and Neckel[22] can be used to obtain a value of 0.89 for the $V - I$ of the sun; this result is in excellent agreement with the mean value obtained in Table XI and indicates that the sun behaves normally at least in the near infrared.

[21] H. L. Johnson, *Ann. Rev. Astron. Astrophys.* **4**, 173 (1966).
[22] D. Labs and H. Neckel, *Solar Phys.* **15**, 79 (1970).
[23] J. Stebbins and G. E. Kron, *Astrophys. J.* **126**, 266 (1957).

Because of the scatter in the intrinsic far infrared colors of the comparison stars and the resulting possibility that the colors of the sun differ substantially from the mean colors of these stars, Johnson's calibration method cannot yield high accuracy in this region of the spectrum. For this reason, we have carried out an experiment in which the standard stars are compared directly at 11.5 μm with a standard blackbody source. The results of this experiment are still being analyzed, but show that the Johnson method as we have applied it is probably accurate to within ±15%.

The revision in Johnson's absolute calibration is fairly minor. The extent of the revision is obscured by the changes made above in the effective wavelengths and zero points of the photometric bands. To convert to the new calibration, fluxes estimated under the old one should be multiplied by factors of 1.10, 0.98, 0.82, 0.74, and 0.79, respectively, for the K, L, M, N, and Q bands. These revisions are comparable in size to the second-order corrections to the monochromatic flux levels assigned at λ_0. For example, a 300 K blackbody might be measured at N by reference to a 4000 K standard star. From Fig. 12, the flux level

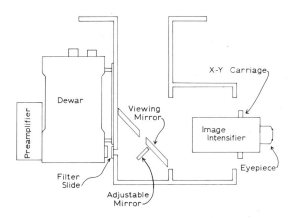

FIG. 12. An infrared photometer for use with a chopping secondary modulator.

inferred without the second-order corrections would be 0.27 mag or a factor of 1.28 too low. On the other hand, the old absolute calibration is a factor of 1.35 higher than the new one; thus, the assigned flux level with the old calibration would have agreed to within 6% with the corrected equivalent monochromatic flux level under the new calibration.

Beyond 20 μm the fluxes from the standard stars are too weak and the instruments too insensitive to permit calibrating the observations by means of standard stars. Instead, the measurements are usually referred to the bright planets, whose fluxes can be estimated on the basis of their temperatures: Venus (255 K), Mars (235 K), Jupiter (145 K), and Saturn (85 K).

9.8. Observing Procedure

The discussion of the equipment used for infrared photometry and of the photometric system is now complete; in this final section, we describe the procedures followed in actually carrying out ground-based observations in the infrared. After the cryostat has been filled and is ready for use, three additional steps must be taken before photometry can begin: the optical elements in the photometer must be adjusted to provide optimum performance, the lock-in amplifier has to be adjusted to match the phase of the incoming signal, and a short series of performance checks should be carried out.

The optical adjustment procedures can be understood by reference to the drawing of an infrared photometer shown in Fig. 12. In one position of the viewing mirror, the energy from an astronomical object passes through a hole and is reflected off the small adjustable mirror into the instrument dewar. In the other position, the energy is reflected off the viewing mirror into the image intensifier where an image can be seen by the observer. After acquiring an object that is compact and bright in the infrared, such as one of the standard stars, the tilting mirror is adjusted. When this mirror is properly adjusted, the cooled field lens projects an image of the detector which falls precisely on the secondary mirror. This maximizes the signal from the star and, if the telescope is well collimated, minimizes the sky offset. This procedure is easiest if carried out at one of the shorter infrared wavelengths, where the signal from a star is greater and the sky offset is negligible. After the maximum signal has been obtained, and it is verified that the offset at 10 μm is small, the image intensifier-eyepiece assembly is moved so the cross hairs are centered on the position of the object. Next, the infrared focus of the telescope is checked by recording the signal as the object drifts through the field of view. The focus is adjusted until a beam pattern with maximum signal, steep sides, and a flat top is obtained. The location of the position of maximum signal relative to the cross hairs is checked again. This position may differ slightly for the different filters if the two faces of the filters depart significantly from plane parallelism.

After the optical adjustments have been carried out, the phase of the lock-in amplifier must be shifted to coincide with that of the modulated signal from the object. This adjustment can be made by observing the phase-detected signal with an oscilloscope or by maximizing the signal.

Finally, a series of performance checks are carried out on the system. It is invaluable to have records of the performance achieved previously in order that substantial changes in system characteristics can be readily identified and any necessary repairs or adjustments made. First, the signal strength from a standard star is measured. Then a region of sky is observed. The noise is measured both with the modulator on and with it off. These measurements determine respectively the modulator noise, sky noise and system noise. If the system noise is high, repairs are necessary. Modulator noise can usually be diminished by reducing the distance in the sky between the two comparison beams. Figure 13 shows a series of noise measurements on a night with high sky noise.

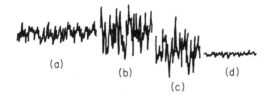

FIG. 13. A series of noise measurements on a night with high sky noise: (a) detector noise; (b) sky noise, modulator on; (c) sky noise, modulator off; (d) preamplifier noise.

If the system is performing satisfactorily, photometric observations can begin. The modulator produces two images of the object which differ in phase by 180°. Therefore, the phase-detected signal from one of these images contains the sky and telescope contributions (offset) plus the object signal, while the phase-detected signal from the other image contains the sky and telescope minus the object. The signal from the object is determined by subtracting measurements of one image from those of the other. It is desirable to adjust the sky offset electronically so that the deflection, when no astronomical object is observed, lies near the center of the dynamic range of the measuring equipment. Observations are then carried out by centering one image and integrating the deflection a set period of time (usually 15 sec), then centering the other image and integrating again. It is convenient to have equipment which automatically moves the telescope to shift the two images on and off the detector as the integrations are carried out. Since a single pair of observations does not

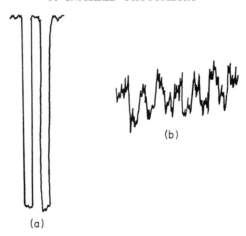

(b)

(a)

FIG. 14. Typical records of infrared photometry: (a) α Boo; (b) NGC 4151. The gain has been increased by a factor of 100 for NGC 4151.

allow slow drifts of the sky offset to be separated from the source contribution, the minimum measurement consists of an observation with the positive beam, one with the negative beam, and then another one with the positive beam. Drifts of the offset can be cancelled in the first order by averaging the first and third observations and subtracting the second to obtain the signal from the object. On faint objects, many deflections are usually made, but always an odd number of deflections are taken in order to permit the cancellation of drifts. A typical observation is shown in Fig. 14.

10. DIFFRACTION GRATING INSTRUMENTS*

10.1. General Spectrometer Considerations

Much of the information which forms the basis for the detailed study of celestial objects has been obtained through the use of diffraction grating instruments. This article is an introduction to the general properties of gratings and grating spectrometers, and a consideration of a number of grating instrument designs, as they are used in astronomy. The emphasis is on the instruments themselves, with only brief mention given where necessary to the detectors with which they are used.

10.1.1. Definition of Terms

Prior to a discussion of specific instruments it is essential to consider the properties of spectrometers generally. The term *spectrometer* is used here in a general way for either a spectrograph or a monochromator. The former is a spectrometer which detects many spectral elements simultaneously, while the latter is a spectrometer which scans spectral elements sequentially.

The most used configuration of a spectrometer is that in which the spectrometer entrance slit is located in the focal plane of a telescope. The general arrangement for this situation is shown in Fig. 1. A telescope of diameter D and focal length f feeds a spectrometer with rectangular entrance slit of width w and height h, collimator of diameter d_1 and focal length f_1, and camera of diameter d_2 and focal length f_2. The projected slit width and height in the camera focal plane are $w' = w(f_2/f_1)$ and $h' = h(f_2/f_1)$, respectively. The dispersing element has angular dispersion $d\beta/d\lambda$, with the dispersion parallel to the slit width, and a spectral element of width $\delta\lambda$ is isolated on the projected slit width. The dispersing element accepts all of the light from the collimator. The angles subtended by the slit on the sky are $\varphi = w/f$ and $\varphi' = h/f$, and the angles subtended by the slit at the collimator are $\theta = w/f_1$ and $\theta' = h/f_1$. In terms of the angular

---

* Part 10 is by Daniel J. Schroeder.

dispersion, $\delta\lambda = \theta\, d\lambda/d\beta = w'/(f_2\, d\beta/d\lambda) = w'P$, where P is the reciprocal linear dispersion expressed in angstroms per millimeter. For the telescope, collimator, and camera, respectively, the focal ratios are: $\mathscr{F} = f/D$, $\mathscr{F}_1 = f_1/d_1$, $\mathscr{F}_2 = f_2/d_1$. The focal ratio of the camera is defined in terms of the beam diameter of a monochromatic beam at the camera. In the following discussion it is assumed that the spectrometer is matched to the telescope, that is, $\mathscr{F} = \mathscr{F}_1$. Only in this way is optimum use made of the spectrometer.

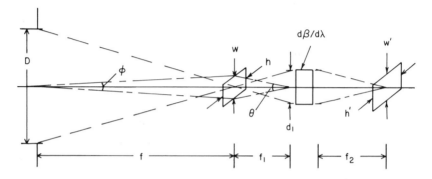

FIG. 1. General layout of a telescope and spectrometer. See text for definitions of symbols.

In these definitions it is assumed that the dimensions of a monochromatic collimated beam are unchanged by the dispersing element. This is not strictly true for diffraction gratings and, although the effect is often small, it is worthwhile noting its effects. In terms of the grating angles as defined in Eq. (10.2.1) let $r = \cos\alpha/\cos\beta$. The effects of this factor r are as follows: (1) the ratio of the dispersed beam width to the incident beam width at the grating is $1/r$, (2) the projected slit width in the camera focal plane $w' = rw(f_2/f_1)$, (3) the angle subtended by the projected slit width w' at the camera is $r\theta$. Because r affects the beam and projected slit widths, but not the heights, it is often called the anamorphic magnification of the grating. The relation of the anamorphic magnification to the spectrometer parameters is discussed in Section 10.1.2.7.

10.1.2. Spectrometer Parameters

The important parameters of any spectrometer are (1) spectral purity, (2) spectral resolution, (3) throughput (sometimes called *luminosité*),

(4) throughput-resolution product, (5) efficiency, and (6) spectrographic speed. Their values determine the suitability of any particular spectrometer for a given observational program.

10.1.2.1. Spectral Purity. Assume that the entrance slit is illuminated by two monochromatic wavelengths separated by $\Delta\lambda$. The two images in the camera focal plane will just be separated when the angle between corresponding sides of the images is slightly greater than θ. When $\Delta\lambda = \delta\lambda$ the images are on the verge of being resolved, and this spectral width is called the *spectral purity*. Making use of the definitions above, we find that $w' = f_2\theta = f_2\,\delta\lambda \cdot d\beta/d\lambda = w(f_2/f_1)$. Therefore,

$$\delta\lambda = \frac{w}{f_1\,d\beta/d\lambda} = \frac{D\varphi}{d_1} \cdot \frac{1}{d\beta/d\lambda}. \tag{10.1.1}$$

It is clear from Eq. (10.1.1) that better spectral purity (smaller $\delta\lambda$), for a given telescope and angle on the sky, is achieved by increasing either the diameter or angular dispersion of the dispersing element.

10.1.2.2. Spectral Resolution. By definition the spectral resolution $\mathscr{R} = \lambda/\delta\lambda$. Using Eq. (10.1.1), we find for the spectrometer of Fig. 1 that

$$\mathscr{R} = d_1/D\varphi \cdot \lambda\,d\beta/d\lambda. \tag{10.1.2}$$

In most astronomical application this resolution is considerably smaller than the maximum resolution $\mathscr{R}_0$ of the spectrometer. Only in situations where there is sufficient light can the actual resolution approach the maximum resolution.

10.1.2.3. Throughput. Because most astronomical sources are faint, the amount of light transmitted by the telescope–spectrometer combination is an important parameter. It was shown by Jacquinot[1] that the flux transmitted F by any optical system is the product of the luminance L at the entrance aperture and the throughput $\mathscr{L}$ of the system. Considering only the spectrometer, the luminance is the intensity per unit area at the entrance slit (see Longhurst[2] for a complete discussion of the fundamental photometric quantities). The throughput, as defined by Jacquinot,[1] is the product of the solid angle of the entrance slit subtended at the collimator, the collimator area, and the spectrometer transmittance

[1] P. Jacquinot, *J. Opt. Soc. Amer.* **44**, 761 (1954).

[2] R. S. Longhurst, "Geometrical and Physical Optics," 2nd ed., Chapter 18. Wiley, New York, 1967.

τ. Assuming for the moment that $\tau = 1$, it follows from the sine relation (Longhurst[2]) that the throughput is an invariant for a given instrument with a particular entrance aperture. For our general spectrometer we find

$$\mathscr{L} = \tau(\pi d_1^2/4)\theta\theta' = \tau(\pi D^2/4)\varphi\varphi', \qquad (10.1.3)$$

where the telescope characteristics have been incorporated into the second part of Eq. (10.1.3).

10.1.2.4. Throughput-Resolution Product. An important result is found by forming the product of Eqs. (10.1.2) and (10.1.3). The resulting product is

$$\mathscr{L}\mathscr{R} = \tau(\pi d_1/4) \cdot (D\varphi') \cdot \lambda \, d\beta/d\lambda, \qquad (10.1.4)$$

which can be considered a measure of the merit of a given spectrometer. If we assume that φ' is the angular diameter due to atmospheric "seeing" for a stellar source, then for a given telescope the throughput-resolution product is a constant. Increasing this product for a given telescope requires either a larger collimator or larger angular dispersion. Meaburn[3] has examined this product for a wide variety of astronomical spectrometers including prism, grating, and Fabry–Perot instruments.

The relation in Eq. (10.1.4) is most useful when comparing different spectrometers for use on the same telescope under similar conditions of seeing. When considering different seeing conditions, or extended sources, it is also necessary to take into account the luminance of the source at the spectrometer entrance slit. From the definition of luminance we see that for a point source $L \propto (1/\varphi')^2$, while for an extended source of uniform surface brightness $L = $ constant when the source's angular diameter is large compared to the seeing limit. If these results are incorporated into Eq. (10.1.4), we find the following expressions for the products of the transmitted flux and resolution:

$$F\mathscr{R} = \begin{cases} K\tau(d_1 D/\varphi') \cdot \lambda \, d\beta/d\lambda & \text{for point sources,} & (10.1.5a) \\ K\tau(d_1 D\varphi') \cdot \lambda \, d\beta/d\lambda & \text{for extended sources,} & (10.1.5b) \end{cases}$$

where K is a constant.

From Eq. (10.1.5a) we see that better seeing (smaller φ') results in a larger transmitted flux-resolution product, as expected. If the seeing disk is larger than the angular width of the entrance slit ($\varphi' > \varphi$), then

[3] J. Meaburn, *Astrophys. Space Sci.* **9**, 206 (1970).

increased resolution can be obtained only at the expense of transmitted flux. In Eq. (10.1.5b) the factor φ' is not related to the seeing; it is simply the angular height of the entrance slit projected on the sky. In both expressions it is assumed that the boundary of the area through which the light passes is rectangular, and that within this area the luminance is constant. These assumptions are not strictly true, but for the purpose of illustrating the general relationship between throughput, transmitted flux, and resolution it is useful not to unduly complicate the discussion.[4]

10.1.2.5. Efficiency. In Eqs. (10.1.3)–(10.1.5) the factor τ has an important role in determining the final transmitted flux. Clearly it is essential to have τ as large as possible. Most grating spectrometers are designed for use over wide spectral ranges, and only limited use has been made of high-efficiency multilayer dielectric coatings for which $\tau > 0.95$ per optical element over limited spectral ranges. In most cases spectrometer mirrors are coated with bare or overcoated aluminum films for which $\tau \sim 0.85$ per surface. It is not possible to consider in detail here the question of efficiency, but anyone planning to use a spectrometer should be aware of the importance of this factor. Grating efficiencies are discussed in Section 10.2.1.5.

10.1.2.6. Spectrographic "Speed." The results summarized in Eqs. (10.1.4) and (10.1.5) are true in general, but in the case of a spectrometer used as a photographic spectrograph it is essential to also consider the speed of the instrument. Ignoring such complicating factors as reciprocity failure of photographic emulsions we can say that the speed of a spectrograph is proportional to the flux per unit area on the detector. Following the results of Bowen[4] we will take the width of this area as that distance which covers a wavelength range of 1 Å. If the reciprocal linear dispersion (hereafter called the *plate factor*) is P, in units of angstroms per millimeter, then this distance is simply $1/P$ mm. For an effective entrance slit height h (which may be the diameter of a seeing-broadened point source) the height of the area on the detector is h'. The detector area is then h'/P. More generally the spectrum is broadened to height H so that better photometric accuracy is possible from the recorded spectrum. In this case the pertinent area is $H/P = (h'/P) \cdot (H/h')$. Using the results for the luminance given in Section 10.1.2.4, and combining these with

[4] I. S. Bowen, "Stars and Stellar Systems" (G. P. Kuiper and B. M. Middlehurst, eds.), Vol. 2, p. 34. Univ. of Chicago Press, Chicago, Illinois, 1962.

Eq. (10.1.3), we obtain the following expressions for the transmitted flux:

$$F = \begin{cases} K\tau D^2\varphi/\varphi' & \text{for point sources,} & (10.1.6a) \\ K\tau D^2\varphi\varphi' & \text{for extended sources,} & (10.1.6b) \end{cases}$$

where K is a constant. Denoting the speed by S we find for a seeing-broadened point source that

$$S = K\tau(D^2\varphi/\varphi') \cdot (P/h') \cdot (h'/H), \qquad (10.1.7)$$

where $h'/H = 1$ for an unbroadened spectrum. For extended sources the spectrum is not broadened and

$$S = K\tau(D^2\varphi\varphi') \cdot (P/h'). \qquad (10.1.8)$$

Following the results of Bowen[4] there are various cases of Eq. (10.1.7) to be considered.

Case A: All of the light passes the entrance slit, in which case $\varphi = \varphi'$, $h' = w'$. Equation (10.1.7) then becomes

$$S = K\tau D^2 P/H \; (= K\tau D^2 P/w') \qquad (10.1.9)$$

for broadened and unbroadened spectra, respectively. For photographic emulsions w' is normally taken as the limit of resolution of the emulsion.

Case B: In this case the entrance slit is narrower than the stellar image so that $\varphi' > \varphi$, and the spectrum is broadened. For this situation the result is

$$S = K\tau Dw'P/(H\mathscr{F}_2\varphi') = K\tau Dw'P^2 d_1 \, d\beta/d\lambda/(H\varphi'). \qquad (10.1.10)$$

For a given plate factor, Eq. (10.1.10) shows clearly how the speed depends on such factors as collimator diameter, angular dispersion, and camera focal ratio.

Case C: In this case the entrance slit is narrower than the stellar image ($\varphi' > \varphi$), but the spectrum is not broadened. The resulting speed is

$$S = K\tau Dw'P/(h'\mathscr{F}_2\varphi') = K\tau w'P^3 d_1{}^2(d\beta/d\lambda)^2/(\varphi')^2. \qquad (10.1.11)$$

For extended sources Eq. (10.1.8) applies and we obtain

$$S = K\tau w'P/\mathscr{F}_2{}^2 = K\tau w'P^3 d_1{}^2(d\beta/d\lambda)^2. \qquad (10.1.12)$$

In this case seeing does not play a part. For a given spectral purity $\delta\lambda = w'P$, we see that the speed is determined solely by the camera focal ratio.

Equations (10.1.9)–(10.1.12) show once again the importance of the collimator diameter and angular dispersion. These should clearly be as large as possible for maximum speed. The other factors in these equations are set by the telescope, emulsion, and desired spectral resolution.

10.1.2.7. Anamorphic Magnification. All of the relations above are derived assuming that the anamorphic magnification $r = 1$. This assumption is valid for prism instruments and for some types of grating instruments. If $r \neq 1$ then it necessarily affects the spectral purity, spectral resolution, and spectrographic speed because the projected slit width depends on r. The effects are as follows: (1) the spectral purity is directly proportional to r, (2) the spectral resolution and spectrographic speed are inversely proportional to r. Consideration of these effects makes it clear that $r < 1$ is the desired choice, hence $\cos\alpha < \cos\beta$, and $\alpha > \beta$.

10.2. Diffraction Gratings

In this section a brief summary of some of the pertinent properties of diffraction gratings is given. It is assumed that the reader is familiar with the basic features of gratings, such as given by Longhurst.[5]

10.2.1. General Grating Properties

A cross section of a small part of a diffraction grating is shown in Fig. 2, where the grating is a reflection type. Assuming collimated light is incident, each incident ray makes angle α with the grating normal. The reflected and diffracted rays make angle β with the grating normal. Constructive interference occurs when the optical path difference between rays from adjacent grooves is an integral number of wavelengths. The result is the well-known grating relation

$$m\lambda = \sigma(\sin\alpha + \sin\beta), \tag{10.2.1}$$

where α and β are the angles of incidence and diffraction, respectively, σ is the grating constant, and m is the order number for wavelength λ.

[5] R. S. Longhurst, "Geometrical and Physical Optics," 2nd ed., Chapter 12. Wiley, New York, 1967.

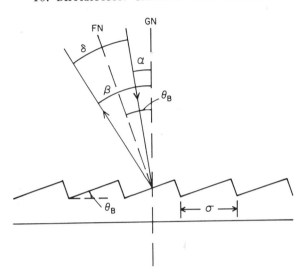

FIG. 2. Small cross section of a reflecting diffraction grating with blaze angle θ_B. The groove length is perpendicular to the plane of the diagram, GN denotes the grating normal; FN is the normal to an individual facet (see text for definitions of remaining angles).

If α and β are on the same sides of the grating normal, their signs are the same; if they are on opposite sides of the normal, their signs are also opposite. Equation (10.2.1) holds for transmission gratings as well.

10.2.1.1. Angular Dispersion. For a fixed angle of incidence we can find the angular dispersion by differentiation of Eq. (10.2.1),

$$d\beta/d\lambda = m/(\sigma \cos \beta) = (\sin \alpha + \sin \beta)/(\lambda \cos \beta). \qquad (10.2.2)$$

If a grating is used in low order, $m = 1$ or 2, then high angular dispersion is obtained only with a finely ruled grating, hence σ small. On the other hand, it is also possible to obtain large angular dispersion with a coarsely ruled grating, if the order number is sufficiently high. Gratings specifically designed for use at large values of m, and with large angles α and β, are called *echelles*.

In many grating spectrometers the angles α and β are not too different. The difference δ between these angles is usually set by mechanical considerations so that the collimated and diffracted beams just clear one another at the locations of the spectrometer optics. The angular dispersion takes on a particularly simple form when $\alpha = \beta$, the so-called *Littrow condition,* and its value is close to that of many actual spectrometers for which $\alpha \neq \beta$. For this special case $d\beta/d\lambda = (2/\lambda)\tan\beta$; hence at a particular wavelength the angular dispersion depends only on the angle at which the grating is used. Most gratings are designed for use at small β, while echelles are generally designed for use at large β with $\tan\beta$ of 2 or more. Further details on these two types of gratings are given in Sections 10.2.2 and 10.2.3. Angular dispersions for various assumed grating constants, order numbers, and values of δ have been shown by Schroeder.[6]

10.2.1.2. Plate Factor. The linear dispersion in the camera focal plane is $f_2 \cdot d\beta/d\lambda$; its units are generally millimeters per angstrom. Most generally it is the reciprocal linear dispersion or plate factor which is specified for a spectrometer. For the Littrow condition $P = \lambda/(2f_2\tan\beta)$. More generally $P = \sigma\cos\beta/mf_2$.

10.2.1.3. Spectral Resolution. We can now combine the grating angular dispersion according to Eq. (10.2.2) with Eq. (10.1.2) to obtain the spectral resolution of a grating spectrometer with slit of angular width φ. For $\alpha = \beta$ we have

$$\mathscr{R}\varphi = (d_1/D) \cdot 2\tan\beta. \qquad (10.2.3)$$

Figure 3 shows the product $\mathscr{R}\varphi$ for various ratios of the collimator to telescope diameter, as a function of $\tan\beta$. For several typical choices of m and σ the corresponding values of $\tan\beta$ are shown for $\lambda = 5000$ Å. The results in Fig. 3 can be used directly to determine what is needed in the way of collimator diameter and angular dispersion for a particular desired resolution and angular slit width.

10.2.1.4. Free Spectral Range. The free spectral range $\Delta\lambda$ is the wavelength difference between two wavelengths in successive orders at the same angle β. This difference is found by setting $m(\lambda + \Delta\lambda) = (m + 1)\lambda$, from which we obtain

$$\Delta\lambda = \lambda/m. \qquad (10.2.4)$$

When m is small the free spectral range is large and unwanted orders are easily suppressed with colored glass filters. When m is large, as it is for an echelle, it is usually necessary to separate successive orders by using an auxiliary dispersing device. The details of how this is done is considered in Section 10.4.

---

[6] D. J. Schroeder, *Publ. Astron. Soc. Pacific* **82**, 1253 (1970).

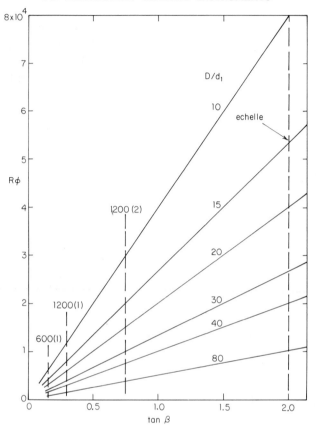

Fig. 3. Product of spectral resolution and angular slit width as a function of grating diffraction angle. Results are shown for various ratios of telescope to collimator diameter. The units of φ are arcseconds. Each dashed line denotes a typical grating used in the Littrow arrangement; the designation is grating grooves per millimeter (order number) (see text).

10.2.1.5. Grating Efficiency.

The earliest ruled gratings were wasteful of light because the diffracted light went into many different orders, including the zero order. It is now possible, however, to shape the individual grooves so that most of the light of a specified wavelength is sent into a single order. Such a grating is said to be *blazed*. The effect of the blaze for a reflection grating can be seen by reference to Fig. 2. If the individual groove facets are inclined at angle θ_B, the so-called blaze angle, with respect to the back of the grating, then the incident ray makes angle $\theta_B - \alpha$ with respect to the normal to the facet. Since the groove acts as a narrow mirror the diffracted light will have its maximum intensity when the diffracted ray makes the same angle with respect to the facet normal, that is when $\beta - \theta_B = \theta_B - \alpha$. The first-order wavelength for which this relation is satisfied is the blaze wavelength λ_B. In terms of δ of Fig. 2 we find $\lambda_B = 2\sigma \sin \theta_B \cos(\delta/2)$. The blaze wavelengths in other orders are given by $\lambda_m = \lambda_B/m$. Transmission gratings are also blazed, in which case each of the grating grooves acts as a narrow prism.

The quoted efficiencies of reflection gratings at the blaze peak are generally in the range of 60 to 75%. It has been claimed that some transmission gratings put up to 90% of the light of a particular wavelength into a single order, though this is apparently the exception rather than the rule.

The intensities for wavelengths other than the blaze wavelengths are determined by the single slit diffraction pattern of the individual groove. The angular width of this pattern, called the *blaze function*, is inversely proportional to σ, hence a fine-toothed grating will have a broad blaze function. As shown by Davis[7], the wavelength interval over which the intensity in order m is appreciable is approximately λ_m. This interval, equal to one free spectral range, is centered on λ_m, and at the ends of the interval the intensity is approximately one-half that at the diffraction peak.

10.2.2. Grating Types

In the sections which follow the term "grating" denotes ruled diffraction gratings intended for use in low orders and at small angles.

10.2.2.1. Reflection Type.

Of the various types of gratings available, the most widely used in astronomical spectrometers are plane reflection gratings. These are available in a wide and ever-increasing variety of groove spacings, blaze angles, and sizes. Ruled gratings as large as 300 $\times 400$ mm have been made, and larger ones are likely to be made soon. Because of the quality of these gratings, and the many choices available, most grating spectrometers built in recent years for astronomical use have used them as the dispersing element.

Concave reflection gratings are also used by astronomers, though to a much lesser extent than plane gratings. Concave gratings are generally more difficult to rule than plane gratings, and for a particular radius of

[7] S. P. Davis, "Diffraction Grating Spectrographs," Chapter 1. Holt, New York, 1970.

curvature the selection of groove spacing, blaze, and size is much more restricted. The main advantage of concave gratings is that no additional optical elements are needed for collimator or camera, and hence the overall spectrometer efficiency can be quite high. Their use in astronomy has been confined primarily to ultraviolet observations from space.

Liller[8] has evaluated a number of concave grating designs and has concluded that some are suitable for ground-based astronomical spectrometers. In this article the wide variety of concave grating systems is not considered, primarily because they have been little used in astronomical instruments. The interested reader is asked to consult Ref. 8. For a more detailed treatment of concave grating mountings and their aberrations the reader is referred to the review article by Namioka.[9]

10.2.2.2. Transmission Type. As with concave gratings, transmission gratings are little used in astronomical spectrometers. In most situations they have disadvantages as compared with reflection gratings. They are limited to spectral regions where the material of which the grating is made is transparent; the spectrum cannot be scanned by rotating the transmission grating; and the basically straight-through optical path may require an inconvenient mechanical arrangement for the overall spectrometer.

One application for which the blazed transmission grating does show considerable promise is that of objective spectroscopy. For this situation the straight-through optical path is an essential feature. Some of the details of one such application are discussed in Section 10.3.2.

10.2.3. Echelles

An echelle is basically a diffraction grating designed for use at a high angle of incidence and diffraction.[10] As shown in Section 10.2.1.1 the angular dispersion is directly proportional to $\tan \beta$, so, in comparison to a grating as normally used, the angular dispersion of an echelle may be 5–10 times larger. In view of the dependence of resolution and spectrographic speed on angular dispersion, this potential gain is significant. Echelles are rather coarse gratings, typically 30–300 grooves/mm, used in a high order of interference, typically 10–100. Most commercially

---

[8] W. Liller, *Appl. Opt.* **2**, 187 (1963).

[9] T. Namioka, "Space Astrophysics" (W. Liller, ed.), p. 228. McGraw-Hill, New York, 1961.

[10] G. R. Harrison, *J. Opt. Soc. Amer.* **39**, 522 (1949).

available echelles have blaze angles of 63.5° and are intended for use near the Littrow condition, hence $\tan \beta = 2$. The angle at which the echelle is used is sufficiently large so that the short side of the groove facet is used rather than the wide side as in the case of a grating. Echelles of high quality have been ruled in sizes up to 200×400 mm, with sizes up to 300×600 mm considered feasible.[11] In view of these sizes it is clear that the $d_1\, d\beta/d\lambda$ product can be several times larger for large echelles than for the largest gratings.

Because of the high orders in which an echelle is used, it is necessary to use an auxiliary disperser "crossed" with the echelle. If, in the camera focal plane, the direction of the echelle dispersion is vertical, then the cross-disperser will separate the various orders horizontally. Both prisms and gratings are used to provide cross-dispersion.

One consequence of working in high orders because of the large grating spacing is the narrow width of the blaze function, typically a few degrees. At the same time, however, all wavelengths are located in some order within the blaze function. As noted in Section 10.2.1.5, free spectral ranges centered on the blaze peak have efficiencies at the ends of the ranges about one-half of the peak efficiencies. Schroeder[6] has given a detailed analysis of the relationship between free spectral range and blaze function, as a function of the angular separation of the incident and diffracted beams.

Although the potential of echelles has been recognized for many years, it is only within the last few years that ruling engines using interferometric control have succeeded in producing large echelles of high quality. They are now used in a few astronomical spectrometers, and it is anticipated that their use will increase greatly in the near future. A discussion of echelle spectrometers is given in Section 10.4.

10.3. Grating Spectrometers

A large variety of plane grating spectrometers have been used for astronomical studies. Some are designed primarily for use as spectrographs, with a single entrance slit and either photographic film or an image tube as the detector. Other designs are intended primarily as monochromators, with either single or multiple exit slits, and with one detector per slit. When these various designs are analyzed it is found that

[11] G. R. Harrison and S. W. Thompson, *J. Opt. Soc. Amer.* **60**, 591 (1970).

there are only a few basic optical arrangements which are widely used. In this section a description of each of these arrangements is given with stress placed on their aberrations if any, the limitations put by the aberrations on the performance, and the mode in which each design can be satisfactorily used. Specific examples of actual instruments of each type are cited.

10.3.1. Plane Gratings—Collimated Light

With few exceptions plane gratings are used in collimated light. When so used the grating itself does not contribute directly to the aberrations of the spectrometer, except insofar as it changes the dimensions of the diffracted beam with respect to the collimated beam. The following three sections describe the characteristics of spectrometers in which the grating is so used.

10.3.1.1. Czerny–Turner Mount. A schematic diagram of the general arrangement of the Czerny–Turner mount is shown in Fig. 4. The main features of this mount are as follows: (1) spherical collimator M_1 and

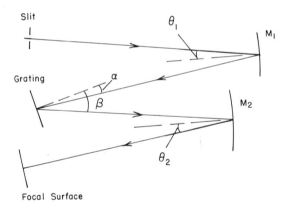

Fig. 4. Schematic arrangement of Czerny-Turner grating mount (see text for definitions of symbols).

camera M_2 mirrors with angles θ_1 and θ_2 between their respective optical axes and the chief ray through the system. The mirror radii of curvature are R_1 and R_2, respectively; (2) an entrance slit located at the tangential focal point of the collimator. This ensures that the tangential image surface is independent of the grating orientation and grating-camera separation[12]; (3) a grating with rulings perpendicular to the plane of the diagram, and arranged to rotate about an axis perpendicular to the diagram plane at the point where the chief ray strikes the grating; (4) a focal surface through the tangential astigmatic images, those images perpendicular to the dispersion direction.

When spherical mirrors are used as shown, the aberrations of astigmatism, coma, and spherical aberration are present. However, as first shown by Rosendahl,[13] the mirrors can be arranged to give coma compensation, exact at one wavelength and nearly so over a wide range of wavelengths. This compensation is essential for obtaining high resolution. Astigmatism and spherical aberration, on the other hand, are additive and set the limit on the speed of the instrument. To keep the effect of the spherical aberration at or below the limit of detection requires a camera focal ratio of about $f/10$ or slower. Analyses of the third-order aberrations of this arrangement lead to the following results[13,14]:

$$AST = 2f_2 d_1 \cdot (\theta_1^2/R_1 + \theta_2^2/R_2), \qquad (10.3.1)$$

$$TC = \tfrac{3}{4} f_2 d_1^2 (\theta_1/R_1^2 - A^3 \cdot \theta_2/R_2^2), \qquad (10.3.2)$$

$$TSA = \tfrac{1}{8} f_2 d_1^3 (1/R_1^3 + A^4 \cdot 1/R_2^3), \qquad (10.3.3)$$

where AST is the length of the astigmatic image at the tangential focus, TC is the tangential coma, and TSA is the transverse spherical aberration. AST is measured perpendicular to the direction of dispersion; TC and TSA are measured parallel to the dispersion direction. The quantity A is the ratio of the camera and collimator beam widths, where $A = (\cos \beta / \cos \alpha) \cdot (\cos \theta_1/\cos \theta_2)$. Except for the most demanding situations it can be assumed that $\cos \theta_1 = \cos \theta_2$. Equation (10.3.2) sets the condition for zero coma at the desired wavelength. As an example of the magnitude of the aberrations assume $A = 1$, $\theta_1 = \theta_2 = 0.05$ rad, $R_1 = R_2 = 150$ cm, $d_1 = 5$ cm. For this case we find $TSA = 7$ μm and $AST = 0.25$ mm. In practice of course A is not unity though usually within a few percent of it.

Equations (10.3.1)–(10.3.3) apply specifically when the Czerny–Turner mount is used in the monochromator mode. This system can also be used effectively as a spectrograph, in which case coma and astigmatism vary along the focal surface. The aberration expressions for the

[12] P. E. Rouse, B. Brixner, and J. V. Kline, *J. Opt. Soc. Amer.* **59**, 955 (1969).

[13] G. R. Rosendahl, *J. Opt. Soc. Amer.* **52**, 412 (1962).

[14] A. B. Shafer, L. R. Megill, and L. Droppleman, *J. Opt. Soc. Amer.* **54**, 879 (1964).

general case are rather complicated, but if we assume that $A = 1$ and that the angles of diffraction are small enough so that $\sin \beta \approx \beta$, then the following relatively simple expressions can be derived:

$$AST = 2f_2 d_1 \left[\frac{\theta_1{}^2}{R_1} + \frac{1}{R_2} \left[\theta_2 - \left(1 - \frac{W}{R_2} \right) \cdot \frac{m}{\sigma} (\lambda_0 - \lambda) \right]^2 \right], \quad (10.3.4)$$

$$TC = \frac{3}{4} f_2 d_1{}^2 \cdot \frac{1}{R_1{}^2} \left(1 - \frac{W}{R_2} \right) \cdot \frac{m}{\sigma} (\lambda_0 - \lambda), \quad (10.3.5)$$

where W is the separation of the grating and camera mirror, measured along the line joining the center of curvature of the camera and the center of the grating.[14] The wavelength λ_0, located at the plate center, is that wavelength for which the coma is zero. The distance W enters into the aberration expressions for the spectrograph mode because the grating is the effective stop for the camera mirror. Note that if $W = R_2$, then $TC = 0$ for all wavelengths and AST is independent of wavelength. Willstrop[15] has described a spectrograph of this particular type. In his design θ_2 is made sufficiently large so that the plateholder clears the diffracted beam from the grating, and although the coma is essentially zero the astigmatism is appreciable.

For any spectrograph it is essential that the film lie along the camera focal surface, in this case the surface on which the tangential images are located. According to third-order aberration theory the curvature k_t of the tangential image surface of a spherical mirror is

$$k_t = \frac{1}{R_2} \left[4 - 6 \frac{W}{R_2} \left(2 - \frac{W}{R_2} \right) \right], \quad (10.3.6)$$

where k_t is the reciprocal of the radius of curvature of the surface. When k_t and R_2 have opposite signs, the surface is convex as seen from the camera mirror. Setting $k_t = 0$, the condition for a flat field, we find that $W/R_2 = 1 \pm 1/3^{1/2}$ is required. The choice $W = 0.423R_2$ is the more convenient one as it places the focal plane slightly farther from the camera than is the grating. In the flat-field condition, as derived from third-order theory, the focal plane is perpendicular to the line joining the camera mirror center of curvature and the grating center. From more exact calculations Reader[16] has shown that there is a "superflat-field" condition in which W is slightly less than $0.423R_2$ and with the focal

[15] R. V. Willstrop, *Mon. Notices Roy. Astron. Soc.* **130**, 233 (1965).
[16] J. Reader, *J. Opt. Soc. Amer.* **59**, 1189 (1969).

plane slightly tipped. Except for situations where the highest possible resolution is required the condition according to Eq. (10.3.6) is sufficient.

In some cases it is necessary to violate the flat-field condition in order to provide adequate clearance between the grating and the focal surface. This is especially the case for spectrometers small in size. If, however, the depth of focus is sufficiently large the effect on spectral resolution will not be noticeable. Such is the case for a spectrometer designed by the author for use on the 91-cm (36-in.), $f/13.5$ telescope at Kitt Peak National Observatory. For this instrument $R_2 = 55$ cm, $W = 0.33R_2$, and the detector is an image dissector tube. With a 600-groove/mm grating used in first order, the plate factor of this spectrometer is about 60 Å/mm.

Because of the flexibility of the Czerny–Turner mount, in that it can be used either as a monochromator or as a flat-field spectrograph, it is widely used. One of the most interesting spectrometers based on the Czerny–Turner design is the polychromator designed by Oke.[17] His instrument has 33 exit slits, each feeding a separate photomultiplier. The camera mirror is considerably larger than the collimator so that wavelengths in both the first and second orders can be detected. In Oke's design $W = 0.50R_2$ and the camera mirror is slightly tipped, putting the focal surface above the grating in Fig. 4.

One disadvantage of the Czerny–Turner mount for telescope use is that an additional mirror must be used so that the final beam to the detector is not heading back toward the telescope. This requirement for convenience means a somewhat reduced spectrometer efficiency. For most situations, however, the desirable features of this mount outweigh this disadvantage.

10.3.1.2. Ebert–Fastie Mount.

This mount is a special case of the Czerny–Turner arrangement in which the collimator and camera mirrors are each part of a single spherical mirror. The features of this arrangement were first fully examined by Fastie.[18] All of the results given in the previous section apply to the Ebert–Fastie mount if we set $R_1 = R_2$ and make the centers of curvature of the two mirrors coincident. This type of spectrometer has been used primarily in the monochromator mode, though it can also be used as a spectrograph. There are no further features of this system, as compared with the Czerny–Turner mount, and further discussion is unnecessary.

[17] J. B. Oke, *Publ. Astron. Soc. Pacific* **81**, 11 (1969).
[18] W. G. Fastie, *J. Opt. Soc. Amer.* **42**, 641 (1952).

10.3.1.3. Schmidt-type Mounts. All of the arrangements in this class are characterized by the nature of the camera, a spherical mirror for which the effective aperture stop is located at the center of curvature of the mirror. A spherical mirror so used has no preferred optical axis, hence there is no coma or astigmatism. Apart from field curvature, spherical aberration is the only remaining image defect, and it can be eliminated by means of an aspheric corrector located at the center of curvature. The advantage of a camera system of this type is that the restriction on camera speed is now largely eliminated. As noted in Section 10.1.2.6 the camera focal ratio plays an important part in determining spectrographic speed, with smaller focal ratios giving greater spectrographic speed at a given plate factor. Systems of the Schmidt-type are generally used in the spectrograph mode.

Some of the basic features of this type of spectrometer can be deduced from the relations given in Section 10.3.1.1. In Eqs. (10.3.1)–(10.3.4) the terms containing θ_1 refer to the collimator while those containing θ_2 refer to the camera. It is clear from these equations that the camera coma and astigmatism are zero only if $\theta_2 = 0$, and $W = R_2$. If, in addition, we require that the total coma of the spectrometer be zero, then we must also choose $\theta_1 = 0$. From this it follows that the total astigmatism is also zero. These choices of angles of course do not affect the spherical aberration, and this defect is most effectively eliminated, if necessary, by an aspheric corrector. From Eq. (10.3.6) we find that the image surface curvature $k = -2/R_2$.

Basically then, we find that this type of spectrometer is one in which the collimator and camera are each essentially aberration free, at least for a point source at the collimator focal point. Various choices of the collimator can be made. If the beam to the collimator is $f/10$ or slower, as is the case at the Coudé focus, a simple spherical mirror is adequate. For faster collimator beams, as is often the case at the Cassegrain focus of a telescope, it is necessary to use either an on-axis paraboloid with a flat mirror to fold the telescope beam or an off-axis paraboloid. Both of these arrangements are used in Cassegrain spectrometers. Various camera choices are also possible. For camera beams about $f/10$ or slower a spherical mirror with center of curvature at the grating is satisfactory. Because of the large focal ratio a corrector is not required. Faster cameras are almost always of the classical Schmidt-type with either a twice-through corrector plate located just in front of the grating or a once-through corrector in the dispersed beam from the grating. When focal ratios smaller than about $f/1.2$ are desired, it is necessary to use one of the several varieties of semisolid Schmidt designs. If a flat field is required, it is obtained by means of a suitable lens located close to the focal surface. For further details of various kinds of Schmidt cameras see the articles by Bowen[4] and Schulte.[19]

Most Coudé spectrographs employ the Schmidt arrangement. One example is the spectrograph of the 2.13-m (84-in.) telescope at Kitt Peak National Observatory. The $f/31.2$ spherical collimator has a focal length of 6.86 m and a point source beam diameter of 22 cm. Following a grating with 600 grooves/mm, used in either first or second order, are cameras ranging from $f/16$ to $f/2.7$ with plate factors of 2.2 to 13.5 Å/mm in the second order, respectively.

This telescope is also equipped with a Cassegrain spectrograph of the Schmidt-type. Its $f/7.54$ collimator is an off-axis paraboloid with a point source beam diameter of 10.2 cm. Various gratings and fast cameras provide a variety of plate factors in the range of 40 to 400 Å/mm.

10.3.2. Plane Gratings—Convergent Light

For high resolution the use of a plane grating in collimated light is essential. For studies of faint celestial sources at moderate or low resolution the efficiency of the spectrometer is of prime importance, and it may be desirable to abandon the collimated light condition if greater efficiency can be thereby achieved. When a plane grating is used in noncollimated light, aberrations heretofore absent are introduced. These aberrations may either be compensated by other optical elements within the spectrometer, or may be negligible as compared with limitations imposed by the seeing conditions. The following paragraphs consider two arrangements, one for each of the conditions noted above.

In Section 10.3.1.1 it was noted that a minimum of four optical elements is required for a Czerny–Turner mount in order to have convenient access to the focal surface. Higher efficiency can be achieved if a plane reflection grating is combined with a single spherical mirror, in which case only two optical elements are needed. The mirror in this design sends a convergent light beam to the grating, which, after diffraction, forms a real spectrum. Both the grating and mirror show spherical aberration, coma, and astigmatism, but calculations[20,21] show that it is

[19] D. H. Schulte, *Appl. Opt.* **2**, 141 (1963).

[20] T. Kaneko, T. Namioka, and M. Seya, *Appl. Opt.* **10**, 367 (1971).

[21] D. J. Schroeder, *J. Opt. Soc. Amer.* **60**, 1022 (1970).

possible to arrange the two elements so as to eliminate both coma and astigmatism at one wavelength. At wavelengths other than the corrected one there are residual aberrations which are negligible at moderate resolution. For more detailed discussion of this arrangement the reader is referred to the references cited.

In another application of a plane grating in convergent light, effective use has been demonstrated for photography of low-dispersion spectra over a moderate field. The arrangement is simply a blazed transmission grating ahead of the plateholder in the converging beam of a telescope. Unlike the case of a grating or prism over the entire objective of a telescope, the transmission grating so used has aberrations, but for low resolution the spectral purity is determined entirely by the seeing, and not by the aberrations. In tests made by Hoag and Schroeder[22] they used a grating with 150 grooves/mm, blaze wavelength about 3550 Å, located about 51 mm from the focal plane of a 91-cm (36-in.) $f/7.5$, telescope. The plate factor was about 1260 Å/mm. With a grating diameter of 80 mm the field covered was about 30 arcmin in diameter. The results clearly demonstrated the efficiency of this method as the spectra of 18th magnitude stellar objects were obtained in one hour on IIa–O plates. Features of this technique are high efficiency, low cost, ease of adaptation to any telescope, and flexibility in choice of plate factor.

10.4. Echelle Spectrometers

Spectrometers which use echelles as the primary dispersing elements are sufficiently different from grating spectrometers that they merit separate attention. The difference is largely the result of the high order of interference in which the echelle is used, and consequently order separation is required. When the necessary cross dispersion is introduced the spectrum on the focal surface is two-dimensional in nature, a format which is well suited for image tubes.

In this section the general features of echelle spectrometers are considered, with attention given to the methods of order separation and the nature of the focal surface. A comparison of the spectral resolution and throughput of typical echelle and grating instruments is given, as well as comments on specific echelle instruments.

[22] A. A. Hoag and D. J. Schroeder, *Publ. Astron. Soc. Pacific* **82**, 1141 (1970).

10.4.1. Methods of Order Sorting

A schematic diagram of a general echelle spectrometer is shown in Fig. 5. The basic elements of the system are a collimator mirror M_1, an echelle whose dispersion is perpendicular to the plane of the diagram E, a cross disperser whose dispersion is in the plane of the diagram CD, and a camera mirror M_2. The arrangement in Fig. 5 is intended only for illustration of a general layout and not as an actual design.

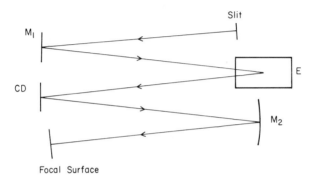

FIG. 5. Schematic arrangement of echelle spectrometer with internally mounted cross-disperser (see text for definitions of symbols).

When the cross disperser is located as shown between the collimator and camera optics the mounting is of the internal type. This arrangement is preferred when broad spectral coverage is required. An alternate scheme has the cross disperser located external to the echelle spectrometer, as when the spectrum of the latter is focused on the entrance slit of a prism or grating spectrograph which provides the cross dispersion. This type of external mounting arrangement is wasteful of light because two separate collimator–camera systems are needed, and for this reason is little used. Another external arrangement has the cross-dispersion system ahead of the echelle instrument in the optical train. The predisperser, which may be a narrow-pass filter, is chosen to isolate a single echelle order. Possible internal mountings which can be used are (1) a prism located just in front of the echelle and used double-pass, (2) a transmission grating in the same location, (3) a concave grating which serves as both collimator and cross disperser, (4) a concave grating which serves as both camera and cross disperser, and (5) a plane grating, either reflection or transmission, between the echelle and camera and used single-pass.

An example of an instrument using a prism is the echelle spectrometer designed by Liller[23] for use at the Coudé focus of Harvard's 1.55-m (61-in.) Wyeth telescope. With an echelle having 73 grooves/mm, and with camera focal lengths of 120 and 360 cm, adequate order separation is achieved. One disadvantage of a prism cross disperser arises when short focus cameras are used. If, for example, a 30° quartz prism is used double-pass with a camera of 70-cm focal length and with the above echelle, the order separation at 4200 Å is about 0.7 mm. This amount of space is marginal for a stellar spectrum and accompanying comparison spectra. For shorter cameras still less room is available. For situations where the prism does provide adequate order separation, as in Liller's instrument, it has the advantage of providing broad spectral coverage with constant efficiency. Another example of an instrument using a prism cross disperser is that reported by Tousey, Purcell, and Garrett.[24] This rocket-borne echelle spectrograph was used to obtain near-ultraviolet spectra of the sun.

A transmission grating located at the echelle could provide higher cross dispersion than a prism, but with the disadvantage of being markedly less efficient when used double-pass. This is particularly true for wavelengths off the blaze peak, and for this reason this mode of cross dispersion is not practical.

A concave grating cross disperser also serving as the spectrometer collimator should be more efficient than the transmission grating arrangement, and can be selected to give adequate order separation. A given grating, because of its blaze characteristics, is limited to a particular spectral range in a given order. A disadvantage of having the cross-disperser precede the echelle is that diffracted beams of different wavelength strike different parts of the echelle with an angle which is a function of the wavelength. In Fig. 2 this amounts to the chief ray of different wavelengths making different angles with respect to the plane of the diagram. For a stellar image trailed along the entrance slit to give broadened spectra, the result is varying line tilt of the spectral lines on the focal surface. If we call the angle between the chief ray and the diagram plane γ, then the slope of the spectral lines is $2 \tan \gamma \tan \beta$, where $\tan \beta = 2$ for the most used echelles.[6] Over a reasonable camera field the slope will vary by several degrees, making spectrum analysis more complicated. This effect is not serious if the spectrometer is intended for use as a

[23] W. Liller, *Appl. Opt.* **9**, 2332 (1970).
[24] R. Tousey, J. D. Purcell, and D. L. Garrett, *Appl. Opt.* **6**, 365 (1967).

high-resolution monochromator, in which case γ is essentially constant. An echelle spectrometer for which this is the case is currently being built at the University of Wisconsin Space Astronomy Laboratory. Its intended use is for studies of selected spectral features of early stars in the far ultraviolet.

One of the first regularly employed echelle spectrographs used a concave grating doubling as camera and cross-disperser.[25] In this case the arrangement of the grating is essentially that of a Wadsworth mounting.[9] Measured spectral resolution exceeding 400,000 was obtained with this setup. Aberrations of the concave grating limit the speed of this arrangement to about $f/15$ or slower.

The final method of order-sorting involves a single-pass plane grating between the echelle and camera. A design using a transmission grating has been reported by Bayle *et al.*[26] This type of setup has the disadvantage that the wavelength at the center of the camera field can be changed only by rotating the entire camera around an axis on the grating face. If, instead, a reflection grating is used, any desired wavelength can be placed at the center of the camera. Such a design is used for order-sorting in the Cassegrain echelle spectrograph on the 91-cm (36-in.) telescope at the Pine Bluff station of Washburn Observatory. Results obtained with this instrument have recently been reported by Schroeder and Anderson.[27] Further details regarding this spectrograph are given in Section 10.4.4.

In summary, for spectrographic use either a single-pass reflection grating or a double-pass prism provides effective cross-dispersion. The former has the defect of limited spectral coverage with reasonable efficiency but can give adequate order separation with any camera, while the latter can provide broad spectral coverage but is limited in the amount of order separation it gives. For echelle monochromators or relatively slow spectrographs, concave gratings are satisfactory.

10.4.2. Nature of Focal Surface

The display of the spectrum of the focal surface is most easily visualized by considering a continuous flat spectrum broken up into a series of segments, each having a length of one free spectral range. This spectral range is given by Eq. (10.2.4). The spectral display is obtained by placing these segments side by side. The echelle dispersion is along the segments

[25] G. R. Harrison, J. E. Archer, and J. Camus, *J. Opt. Soc. Amer.* **42**, 706 (1952).
[26] A. Bayle, J. Espiard, C. Breton, M. Capet, and L. Herman, *Revue Opt.* **41**, 585 (1962).

while the cross-dispersion provides the separation of the strips. Each strip corresponds to one echelle order, with the intensity a maximum at the center and decreasing to about one-half of the peak intensity at each end. Examples of echelle stellar spectra are shown by Liller[23] and by Schroeder and Anderson.[27]

Using the equations in Section 10.2.1 we can determine the characteristics of the spectral display. As seen from the echelle the angular spread of one echelle order $\delta\beta = \Delta\lambda \cdot d\beta/d\lambda$. Using Eqs. (10.2.2) and (10.2.4) we find

$$\delta\beta = \lambda/\sigma \cos\beta, \qquad (10.4.1)$$

where $\cos\beta$ is taken at the blaze peak. At the focal surface the length of one echelle order is $f_2\,\delta\beta$. Assuming the Littrow condition we also find that

$$\Delta\lambda = \lambda^2/2\sigma \sin\beta. \qquad (10.4.2)$$

Consideration of the above relations shows that the echelle order length increases directly with the wavelength, while the number of angstroms per order goes as the square of the wavelength. This is as expected because according to Section 10.2.1.2 the plate factor also increases directly with the wavelength.

If we denote the angular dispersion of the cross-disperser by $d\beta'/d\lambda$, then the separation of successive echelle orders $\Delta x = f_2\,\Delta\lambda\,d\beta'/d\lambda$ $= \Delta\lambda/P'$, where P' is the plate factor of the cross-disperser.[6] If, for example, $\Delta\lambda = 60$ Å and $P' = 30$ Å/mm, the order separation is 2 mm. The variation of order separation with wavelength depends on the type of cross-disperser. For a grating $d\beta'/d\lambda$ is independent of wavelength and using Eq. (10.4.2) we find $\Delta x \propto \lambda^2$. For prisms $d\beta'/d\lambda \propto \lambda^{-3}$ and hence $\Delta x \propto \lambda^{-1}$. Thus gratings provide greater order separation at longer wavelengths, while for prisms the reverse is true.

The total spectral coverage which can be achieved with a given detector size depends on the parameters of both the echelle and the cross-disperser. Typically the spectral coverage on a detector 40 mm square is 1500–2000 Å when the average echelle plate factor is 3 Å/mm and the average order separation is 2.0 mm.[6] The two-dimensional spectral format is advantageous from the point of view of image tubes as detectors, though there are difficult design problems in combining an image tube with a fast camera without excessive beam obscuration.

[27] D. J. Schroeder and C. M. Anderson, *Publ. Astron. Soc. Pacific* **83**, 438 (1971).

10.4.3. Comparison of Echelle and Grating Instruments

In making comparisons between echelle spectrometers and grating instruments it must be remembered that such comparisons are of meaning only when the two instruments considered are intended for making the same kinds of observations. It does little good to check the performance of an echelle system with plate factor of 5 Å/mm against that of a grating spectrometer whose plate factor is 100 Å/mm. The discussion in this section is therefore restricted to a comparison of echelle spectrometers whose plate factors are a few angstroms per millimeter to grating spectrometers with similar dispersions.

Grating instruments of this dispersion are almost always used at the Coudé focus because of the large cameras required. Because of the higher angular dispersion of an echelle, spectrometers which use them can be placed at the Cassegrain focus of a medium- or large-diameter telescope. For example, an echelle camera focal length of 50 cm gives a plate factor of about 2.5 Å/mm at 4200 Å. The advantage of using the Cassegrain focus is that more light reaches the spectrometer entrance slit as compared with the same instrument at the Coudé focus.

Further comparison of these two types of instruments shows that at the same resolution the throughput of an echelle spectrometer can be significantly higher than that of the grating system, a point already noted in Section 10.2.3. Conversely, for the same throughput the spectral resolution of the echelle instrument can be higher. These results are easily obtained from Eq. (10.1.4) from which it follows that

$$\left(\frac{\mathscr{L}_e}{\mathscr{L}_g}\right)_{\mathscr{R}_e=\mathscr{R}_g} = \left(\frac{\mathscr{R}_e}{\mathscr{R}_g}\right)_{\mathscr{L}_e=\mathscr{L}_g} = \frac{\tau_e \cdot d_{1e} \cdot (d\beta/d\lambda)_e}{\tau_g \cdot d_{1g} \cdot (d\beta/d\lambda)_g}, \qquad (10.4.3)$$

where e and g denote the echelle and grating, respectively. If the comparison is made between Cassegrain echelle and Coudé grating instruments, then τ_g must include the effect of the additional mirrors used to reach the Coudé focus. If, for example, we compare an echelle with tan β = 2 with a 600-groove/mm grating used in second order at 5000 Å, and assume $\tau_e = \tau_g$ and $d_{1g} = 2d_{1e}$, then the ratio in Eq. (10.4.3) is about 3.1. Note that the grating collimator beam diameter is twice that of the echelle collimator beam. If we compare the spectrographic speeds at the same resolution for Case B of Section 10.1.2.6, the situation which most often applies in photographic stellar spectroscopy, the ratio given in Eq. (10.4.3) is again obtained. Clearly the higher angular dispersion of the echelle makes possible significant potential gains in speed or resolution.

10.4.4. Echelle Spectrometer Designs

As in the case of plane grating spectrometers, there is a wide variety of possible echelle spectrometer designs. It is also true again, however, that there are only a few basic optical arrangements upon which a design is normally based. Because an echelle system with internally mounted cross-disperser is basically nothing more than two dispersion elements between a collimator and camera, most of the comments made in Section 10.3.1 about plane grating instruments also hold true for echelle instruments. The two-dimensional nature of the spectral output does, however, make the design of echelle systems somewhat more complicated, particularly of the camera. In this section only a few brief comments are made about actual and proposed echelle spectrometers, with the reader referred to the literature for further details. Only Cassegrain echelle instruments are considered.

Characteristics of one type of echelle spectrometer have been reported by Schroeder and Anderson.[27] This design has the echelle and single-pass reflection grating located between the collimator and camera mirrors of a modified Czerny–Turner mount. With a camera focal length of 75 cm the plate factor at 5000 Å is approximately 1.9 Å/mm. The spectrum may be recorded with an image tube, direct photographic plates, or an exit slit–photomultiplier combination. Though relatively slow at $f/13.6$, this design does provide a focal surface which is easily accessible to bulky image tubes. Resolution of about 100,000 has been demonstrated with this spectrometer. To date over 1000 spectrograms have been obtained with this instrument.

The design of another type of echelle instrument[6] has the echelle and reflection grating located between a parabolic collimator and a Schmidt camera. Possibilities for the latter include a classical photographic type or a folded type for use with image tubes. The sizes of image tubes, and the consequent difficulty of getting the photocathode to the image surface, place serious constraints on the design of folded Schmidt cameras. A version of this type of arrangement, currently being built at Yerkes Observatory, employs an echelle with 154×306-mm ruled area and a folded camera with focal length of 30 cm for use with an electrostatically focused image tube.

One of the instruments being built for the 3.8-m (150-in.) Anglo–

Australian telescope is an image tube echelle spectrograph with prism cross-dispersion.[28] This instrument has a camera focal length of 50 cm and is designed specifically to give broad spectral coverage at high resolution. Correction for the spherical aberration of the camera mirror is achieved by means of an aspheric figure on one face of the prism.

10.5. Concluding Comments

The discussion of the properties of echelle and grating spectrometers in the preceding sections is general in nature and concentrates on the features of each type of instrument, rather than presenting numerous details on specific instruments. It is hoped that this information, in conjunction with the discussion of spectrometer parameters, will provide a useful guide to the intelligent choice of the type of spectrometer best suited for a particular observational program. In view of the premium placed on telescope observing time, it is essential that the observer strive to use the most appropriate instrument on a given problem so as to maximize the amount of useful information obtained per unit observing time.

[28] R. C. M. Learner, *Observatory* **91**, 93 (1971).

11. FOURIER SPECTROMETERS*

11.0. Introduction

The ready access to high-speed computing facilities, the convenience of commercially produced spectrometers, and the acceptance of the advantages of Fourier transform spectroscopy in the infrared wavelength region have contributed to the successful application of Fourier spectrometers to infrared astronomy. In this chapter we present a summary of the history, theory, and practice of Fourier spectroscopy. The instrument used most frequently in astronomical investigations is the rapid scanning Michelson interferometer. Therefore, this discussion of Fourier spectroscopy is based on its properties.

We will not deal extensively with the technical aspects of the mechanical and optical systems. The convenience of high-quality commercial systems and the existence of a vast literature guarantees the user an adequate supply of information on these subjects. For similar reasons we will avoid a detailed discussion of the detection system. We have chosen to treat the interferometer as a kind of mathematical machine. In doing this we will focus our discussion on the answer to the following question: What is the minimum detectable signal that a given combination of telescope, observation time, detector sensitivity, amplifier system, and instrument resolution will detect?

Spectroscopists have never made a practice of making scientific interpretations directly from interferograms. The usable results of the experiment are based on Fourier analysis; the computer is as much a part of the experiment as the interferometer. Thus, since a good deal of the experiment really takes place in the computer, we tend toward a mathematical presentation.

---

* Part 11 is by Herbert W. Schnopper and Rodger I. Thompson.

11.1. Historical Background

11.1.1. Introduction

Spectrometers, using prisms, gratings, crystals, or even magnetic fields in the case of charged particles, are dispersive in nature. Each spectral component is spatially separated and examined by an appropriate detector. The observed spectrum is broken down into many components and only a small percentage of the observing time is spent on any single component. Some spectrometers attempt to overcome this disadvantage by using more than one detector; however, when many components are observed, the number of detectors and the problems of normalizing the output become prohibitive.

Multiplex spectroscopy operates more efficiently by observing simultaneously all spectral components with a single detector for the total observing time. Each spectral component is coded in such a way as to allow it to be separated from the others by numerical, electronic, or other means. In Fourier transform spectroscopy, each spectral component is coded by being sinusoidally modulated at a different frequency, and the individual spectral components are then separated by the well-known methods of Fourier analysis.

11.1.2. Early History

A Michelson interferometer produces a different modulation frequency for each spectral component, and our study of Fourier transform infrared spectroscopy will be based on this instrument. Michelson developed the interferometer in order to study the properties of light, the most famous being the effect of an ether in the propagation of light.[1] In other experiments using the interferometer, Michelson was able to define the meter in terms of the wavelength of the red cadmium line and to formulate first the principles of Fourier transform spectroscopy.[2,3] His inability to compute efficiently long Fourier transforms prevented Michelson from performing detailed analysis of complicated spectra. He was able, however, to construct an 80-channel frequency analyzer by using a combination of springs and levers,[4] and became the first person to use an interferometer in conjunction with a frequency analyzer to study a spectrum.

[1] A. A. Michelson and E. W. Morley, *Silliman J.* **34**, 33, 427 (1887).
[2] A. A. Michelson, *Phil. Mag.* **31**, 256 (1891).
[3] A. A. Michelson, *Phil. Mag.* **34**, 280 (1892).
[4] E. V. Lowenstein, *Appl. Opt.* **5**, 845 (1966).

In more recent times, although electronic computers capable of performing the calculations involved in the Fourier transform had come into existence, researchers had no compelling reason to develop Fourier spectroscopy since prism and grating spectrometers were already capable of providing directly high-resolution spectra. The necessity of performing a complicated and lengthy Fourier transform to regain the spectrum inhibited the use of the Michelson interferometer until 1951 when Peter Fellgett, in his thesis at Cambridge University in England,[5] provided the reason for extra computation.

11.1.3. Fellgett Advantage

Fellgett was the first to point out that under *appropriate conditions*, an interferometer operating for a fixed time and at a fixed resolution enjoyed a superiority in the signal-to-noise ratio over a scanning grating or prism spectrometer operating for the same time and at the same resolution. If the spectrum being observed is divided into N spectral components, the signal-to-noise gain is found to be a factor of $N^{1/2}$. The gain in signal-to-noise is known as "Fellgett's advantage" in honor of his pioneering work. Fellgett's advantage is fully realized only when the dominant noise in the recorded signal is independent of signal level. This type of noise is characteristic of background and electronic noise. When the noise is proportional to the square root of the signal as in photon counting experiments, Fellgett's advantage is exactly cancelled. Full realization of Fellgett's advantage is, however, achieved in the infrared region. At present all available detectors are limited by electronic noise since they are not sensitive enough to detect individual photons.

A Michelson interferometer is thus ideally suited for work in high-resolution infrared spectroscopy of astronomical objects. In this application the available signal-to-noise ratio limits the number of observable objects; therefore, any gain in signal-to-noise is of tremendous importance. Fellgett proposed that interferometry be applied to this field.

There are other advantages of the Michelson interferometer which make it desirable for astronomical observations even when Fellgett's advantage is not fully realized. The absence of slits in the instrument increases its optical throughput or luminosity over grating or prism instruments, thus allowing more of the incident radiation to be utilized.[6] For a given resolution, an interferometer can be made lighter and more

---

[5] P. Fellgett, Thesis, Cambridge Univ. (1951).

[6] P. Jacquinot, *J. Opt. Soc. Amer.* **44**, 761 (1954).

compact than dispersive instruments, and this is a distinct advantage when an instrument must be mounted on a telescope. Also, errors which arise from changes in sky conditions during the time between the observation of two spectral elements are also eliminated since all spectral elements are observed simultaneously.

11.1.4. Recent History

Most of the work in interferometry up to 1956 was presented at the C.N.R.S. Belevue Colloquium in 1957. The papers for this conference are published in *Le Journal de Physique et le Radium*, Volume 19, 1958. Stellar spectra were presented at this conference by Fellgett as were some interferograms by Lawrence Mertz. The Fellgett spectra are probably the first stellar spectral observations made with a Michelson interferometer.

The mathematical basis for most of the later work in Fourier spectroscopy appeared in 1961 when Connes published her thesis.[7] Her work covers basic areas such as apodization, convolution, sampling, resolution, and signal-to-noise.

Serious use of the Michelson interferometer in astronomical spectral observations began in the 1960s. A medium-resolution interferometer constructed by Mertz was used to study stellar spectra in the near infrared from 1 to 2.5 μm.[8] The emphasis in this instrument was placed on its compactness and portability. As attached to a telescope the interferometer had a weight of about 3 kg. The Connes team worked on developing a high-resolution Michelson interferometer. They were able to develop an interferometer capable of 0.1-cm^{-1} resolution laboratory spectra and 1.0-cm^{-1} planetary spectra.[9] This high-resolution instrument is, however, cumbersome and must be used at the Coudé focus of the telescope.

One of the common criticisms of interferometry during the 1960s was that it was particularly susceptible to atmospheric scintillations.[10] Since an interferometer codes each wavelength with a different frequency, modulations in intensity caused by atmospheric scintillation are not distinguishable from true signal modulations. However, the frequency spectrum of the atmospheric scintillation is significant only below 250 Hz and the problem is handled easily by scanning the interferometer rapidly

[7] J. Connes, *Rev. Opt.* **40**, 45, 116, 171, 231 (1961).
[8] L. Mertz, *Astronom. J.* **70**, 548 (1965).
[9] J. Connes, and P. Connes, *J. Opt. Soc. Amer.* **56**, 896 (1966).
[10] P. Kuiper, *Com. Lunar Planet Lab.* **23**, 179 (1963).

so that the lowest modulation frequency, for a spectral element of interest, will lie above 250 Hz.[11]

The final ingredient which made it practical to perform high-resolution interferometric spectral measurements was the algorithm developed by Cooley and Tukey for performing fast Fourier transforms.[12]

A second conference of Fourier transform spectroscopy was held in 1967 at Liege, Belgium. Many of the papers from that conference are published in *Journal de Physique Colloque, C2, supplement au no. 3–4, Volume 28*, March–April 1967. Spectra of Jupiter at 0.3-cm⁻¹ resolution and Venus at 0.07-cm⁻¹ resolution obtained with an interferometer designed by P. Connes were presented.[13] This resolution is much higher than has been obtained using grating or prism instruments. The most recent and comprehensive conference on Fourier spectroscopy was held at Aspen, Colorado, in April of 1970. The proceedings of this conference have been published.[†]

The first balloon-borne interferometer was launched in August of 1965 to study the solar spectrum in the far-infrared region.[14] Minimal results were obtained, however. The most serious problem facing balloon-borne astronomical observations is not the performance of the interferometer, but rather the pointing accuracy obtainable from balloon-borne platforms. An alternative to balloon flights is the use of airplane operating at high altitudes. Stellar and planetary studies from airplanes are currently being carried out at the Lunear and Planetary Laboratory and Steward Observatory of the University of Arizona.

11.2. Theory of Fourier Transform Spectroscopy

11.2.1. The Michelson Interferometer

11.2.1.1. Basic Properties. The Michelson interferometer is a device for interfering two beams of light. Light entering the interferometer is divided into two separate beams by a beam splitter and is recombined after a controllable phase delay has been applied to one of the beams

---

[11] L. Mertz, *Infrared Phys.* **7**, 17 (1967).
[12] J. W. Cooley and J. W. Tukey, *Math. Comp.* **19**, 297 (1965).
[13] J. Connes, P. Connes, and J. P. Maillard, *J. Phys. Radium* **28**, c2-120 (1967).
[14] R. Beer, *J. Phys. Radium* **28**, c2-113 (1967).

[†] AFCRL-71-0019, January 5, 1971, Special Report No. 114, Aspen International Conference on Fourier Spectroscopy, 1970.

(Fig. 1). The incoming beam is partially reflected along path A and partially transmitted along path B. After traversing the instrument, the partial beams either proceed back through the input or out through the output to the lens-detector assembly. It is important to note that the beam splitter has its wave function divided equally (ignoring inhomogenities in the beam splitter) between path A and path B. This means that *individual* photon wave functions interfere upon recombination at the beam splitter and the phase of the wave function in either path is a meaningful concept.

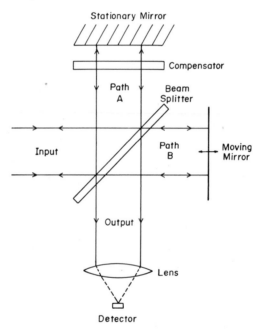

FIG. 1. A functional diagram of the Michelson interferometer.

11.2.1.2. Interference Equations. Interference is produced by the difference in phase caused by the difference in optical path lengths A and B. If the optical path length difference is x, then the phase difference $\Phi(k)$ is

$$\Phi(k) = 2\pi k x \qquad (11.2.1)$$

for light of wave number k upon recombination at the beam splitter.

If the amplitude of the incoming light is A, the amplitude in path A and path B will be $A/(2)^{1/2}$ for a perfect beam splitter. After the second transmission or reflection at the beam splitter, an amplitude $A/2$ from path A and $A/2$ from path B contribute to the beam which exits through the output. A similar beam exits through the original input. The amplitude of the radiation received at the detector is

$$A_D = (A/2) + (A/2)e^{2\pi i k x} \tag{11.2.2}$$

since the light beams from paths A and B are out-of-phase by the angle given in Eq. (11.2.1). The intensity of radiation at the detector $I_D = |A_D|^2$ is

$$I_D = I/2[1 + \cos(2\pi k x)], \tag{11.2.3}$$

where I is equal to $|A|^2$.

It has been confirmed by experience that for dielectric beam splitters, the light which emerges from the input is complementary,[13] i.e.,

$$I_I = I/2[1 - \cos(2\pi k x)]. \tag{11.2.4}$$

Thus, with no absorption loss occurring in the beam splitter, the total power into the interferometer equals the total power out as required by energy conservation. For metallic beam splitters, however, it has been shown experimentally that I_D and I_I are in-phase.[15] This is due to absorption in the metallic beam splitter. Since the absorption is on the order of 50% it is still true for metallic beam splitters, as it must be, that the total power in equals the total power out plus absorbed power.

Equation (11.2.3) shows that the output intensity has a term independent of the optical path length difference x and a term sinusoidally dependent on x. Since x may be varied with the moving mirror shown in Fig. 1, the output signal may be modulated sinusoidally in time by displacing the mirror at a constant velocity. The frequency of the modulation is determined by the wave number k of the incident light thus each wave number is coded with a different frequency. If the incident light is made up of different wave numbers, each of intensity $I(k)$, the modulated output signal $f(x)$ is

$$f(x) = \tfrac{1}{2} \int_{-\infty}^{\infty} I(k)e^{2\pi i k x} \, dk, \tag{11.2.5}$$

where $I(-k) = I(k)$ since $\cos(2\pi k)$ is symmetric. The constant term has been dropped from Eq. (11.2.5) as it only contributes to the zero-frequency (dc) term when the signal is analyzed. Most infrared detectors

[15] L. Mertz, "Transformations in Optics," p. 34. Wiley, New York, 1965.

have negligible dc response, therefore, the constant term is usually not detected.

Dispersion in the beam splitter will introduce a wave number dependent phase delay $\theta(k)$ in the transmitted beam, and the output of the interferometer is

$$f(x) = \tfrac{1}{2} \int_{-\infty}^{\infty} I(k)e^{i\theta(k)}e^{2\pi ikx}\, dk \tag{11.2.6}$$

or

$$f(x) = \int_{-\infty}^{\infty} g(k)e^{2\pi ikx}\, dk, \tag{11.2.7}$$

where $g(k)$ equals $\tfrac{1}{2}I(k)e^{i\theta(k)}$ The spectrum $g(k)$ thus has both cosine and sine components or equivalently real and imaginary components in the exponential formulation. An exponential form has been used as a mathematical convenience with the real part corresponding to the cosine terms and the imaginary part corresponding to the sine terms.

Only the basic properties of the Michelson interferometer have been given in this section. More detailed accounts of the theory of interferometry are given in several books.[15–17]

11.2.2. Mathematics of Fourier Transform Spectroscopy

11.2.2.1. Fourier Transform.
Many different definitions of the Fourier transform exist which may differ by factors of 2π and signs of exponential; however, all are mathematically equivalent with a simple change of variables. For the purposes of this paper, the following definition is appropriate.

$$f'(k) = \int_{-\infty}^{\infty} f(x)e^{-2\pi ikx}\, dx. \tag{11.2.8}$$

A prime is used to denote the Fourier transform of a function. It is also possible to define an inverse transform denoted by a double prime as

$$f''(x) = \int_{-\infty}^{\infty} f'(k)e^{2\pi ikx}\, dk. \tag{11.2.9}$$

If Eq. (11.2.9) is compared with Eq. (11.2.7), it is evident that the output of a Michelson interferometer is the inverse transform of the spectrum $g(k)$.

[16] W. H. Steel, "Interferometry." Cambridge Univ. Press, New York and London, 1967.

[17] M. Francon, "Optical Interferometry." Academic Press, New York, 1966.

The function $f'(k)$ can be recovered by Fourier transforming $f''(x)$. This can be shown by substituting $f''(x)$ for $f(x)$ in Eq. (11.2.8)

$$f'(k) = \int_{-\infty}^{\infty} f'(\xi)e^{2\pi i \xi x} \, d\xi \, e^{-2\pi i k x} \, dx. \qquad (11.2.10)$$

Integration over x gives

$$f'(k) = \int_{-\infty}^{\infty} f'(\xi) \, \delta(\xi - k) \, d\xi. \qquad (11.2.11)$$

Finally, integration over ξ gives

$$f'(k) = f'(k). \qquad (11.2.12)$$

This result can be summarized by the following two equations which relate the Fourier transform and its inverse

$$f''(x) = \int_{-\infty}^{\infty} f'(k)e^{2\pi i k x} \, dk, \qquad (11.2.13)$$

$$f'(k) = \int_{-\infty}^{\infty} f''(x)e^{-2\pi i k x} \, dx. \qquad (11.2.14)$$

For a given input spectrum $f'(k)$, the output of a Michelson interferometer is the inverse transform $f''(x)$. The original spectrum $f'(k)$ is then recovered by Fourier transforming the output $f''(x)$.

Often sine and cosine Fourier transforms are used instead of the exponential transform of Eq. (11.2.8). In this case, the real part of the exponential transform corresponds to the cosine transform and the negative of the imaginary part of the exponential transform corresponds to the sine transform. If the function to be Fourier analyzed is broken up into symmetric (cosine) and antisymmetric (sine) terms, it is evident that the Fourier transform of symmetric terms will be nonzero only for the cosine transform and the Fourier transform of the antisymmetric term will be nonzero only for the sine transform. In other words, the real part of the exponential Fourier transform will be the transform of the symmetric part of that function and the imaginary part of the transform will be the transform of the antisymmetric part of the function. In many applications, it is useful to break up a function into the sum of its symmetric and antisymmetric parts before the Fourier transform is taken.

11.2.2.2. Convolution Integral. One of the most recurrent problems in Fourier transform spectroscopy is the Fourier transform of the product of two functions, $f(x)$ and $g(x)$. Morse and Feshbach[18] give the solution as

$$\int_{-\infty}^{\infty} g(x)f(x)e^{-2\pi i k x}\, dk = \int_{-\infty}^{\infty} g'(y)f'(k-y)\, dy. \qquad (11.2.15)$$

The right-hand side of (11.2.15) is called the *faltung* (German for folding) or *convolution integral* of g' and f'. A symbol $*$ will be used to denote the convolution of two functions

$$g' * f' = \int_{-\infty}^{\infty} g'(y)f'(k-y)\, dy. \qquad (11.2.16)$$

Equation (11.2.15) can be generalized to include the Fourier transform of more than two equations

$$(f \cdot h \cdot g \cdot \cdots)' = f' * h' * g' * \cdots. \qquad (11.2.17)$$

Very simply, Eq. (11.2.17) states that the Fourier transform of the product of several functions is the convolution of the Fourier transform of the individual functions. An important property of the convolution integral is its symmetry under the exchange of variables.[16] This symmetry is expressed as

$$\int_{-\infty}^{\infty} g(y)f(k-y)\, dy = \int_{-\infty}^{\infty} g(k-y)f(y)\, dy. \qquad (11.2.18)$$

11.2.2.3. Shift and Scale. Let a function $f(x)$ be shifted by an amount c and scaled by a factor a. The result of this operation is

$$\int_{-\infty}^{\infty} f\left(\frac{x-c}{a}\right)e^{-2\pi i k x}\, dx = af'(ak)e^{-2\pi i k c}, \qquad (11.2.19)$$

where $f'(k)$ is the Fourier transform of $f(x)$. Appendix A has a table of useful Fourier transforms.

Only a few of the many mathematical techniques involved in Fourier transform spectroscopy have been presented here. More will be presented in the context of the following sections.

11.2.3. Resolution and Apodization

11.2.3.1. Apparatus Function. For a Michelson interferometer, the spectral resolution is determined by the range of optical path length for which the interferogram is known. If the moving mirror in Fig. 1 moves

---

[18] P. M. Morse and H. Feshbach, "Methods of Theoretical Physics," p. 464. McGraw-Hill, New York, 1953.

a distance a, the range of the optical path length is $2a$. This limitation on the optical path length difference can be represented by multiplying the interferogram $f(x)$ by a function $R(x, a)$.

$$R(x, a) = \begin{cases} 1, & |x| < a, \\ 0, & |x| > a. \end{cases} \qquad (11.2.20)$$

When the function $f(x)R(x, a)$ is Fourier transformed, the result is, according to Eq. (11.2.17), the convolution of the spectrum $f'(k)$ with $R'(k, a)$, where

$$R'(x, a) = \frac{\sin(2\pi ak)}{\pi k}. \qquad (11.2.21)$$

The function $(\sin x)/x$ is known as the *sinc function* and therefore $R'(k, a)$ is equal to $2a \, \text{sinc}(2\pi ak)$. The Fourier transform of $f(x)R(x, a)$ is

$$g(k) = 2a \int_{-\infty}^{\infty} f'(k) \frac{\sin[2\pi a(k - y)]}{2\pi a(k - y)} \, dy \qquad (11.2.22)$$

For any given wave number k_0, the output is no longer the exact spectrum $f'(k_0)$ but the spectrum at all k weighted by the function $\text{sinc}[2\pi a(k_0 - k)]$. The spectrum is thus "smeared out" by the sinc function. In spectroscopy it is common practice to call the function which is convolved with the exact spectrum the *apparatus function*. Figure 2 gives a plot of the apparatus function, in this case $\text{sinc}(2\pi ak)$ plotted in units of $1/a$.

A resolution element of $k = 1/a$ is defined by the width of the sinc function's central maximum. This is not an exact resolution element as the individual elements are not completely independent of one another. Each resolution element overlaps the other as is shown by comparing the dashed and solid line plots in Fig. 2. Since some overlap occurs for

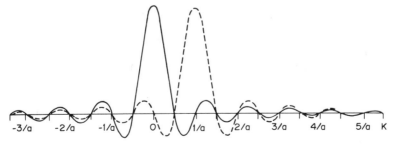

FIG. 2. The apparatus function for a Michelson interferometer whose maximum optical path difference is $2a$. The response to two different frequencies spaced apart by one "resolution element" ($\Delta k \approx 1/a$) is shown.

widely separated resolution elements, the definition of resolution is somewhat arbitrary.

In regions of sharp spectral discontinuities (e.g., molecular band heads), the sinc function introduces ripples in the spectrum on either side of the discontinuity. This phenomena is known as "ringing" in electrical engineering, or mathematically as the *Gibbs phenomenon*.[16] These ripples could possibly either be confused with emission and absorption lines or mask smaller features in the spectrum.

It is therefore desirable to change the apparatus function in a way which will reduce the "ringing." For high resolution, the apparatus function $A(k - k_0)$ should have a sharp maximum for $k = k_0$ and rapidly decrease to zero for values of k other than k_0. The best apparatus function is obviously the Dirac delta function $\delta(k - k_0)$. This apparatus function is achieved only if the interferogram is known for all optical path length differences, but this requires an infinite mirror motion.

11.2.3.2. Apodization. Apodization is the method used to reduce the ringing or introduction of "false feet" discussed above. Since the sinc apparatus function was the result of the function $R(x, a)$ which multiplied the interferogram $f(x)$, a new apparatus function can be obtained by replacing $R(x, a)$ with a new function $A(x, a)$. The only restrictions on $A(x, a)$ are that it must have a Fourier transform and be zero for the region $|x| > a$. Since $A(x, a)R(x, a)$ is simply equal to $A(x, a)$, the new apparatus function is simply $A'(k, a)$.

One of the most common apodization functions is the triangle function $T(x, a)$

$$T(x, a) = \begin{cases} 1 - |x|/a, & |x| < a, \\ 0, & |x| > a. \end{cases} \qquad (11.2.23)$$

The apparatus function which corresponds to $T(x, a)$ is

$$T'(k, a) = a \frac{\sin^2(\pi k a)}{\pi^2 k^2 a^2}. \qquad (11.2.24)$$

Since the function $T'(k, a)$ is always positive, there are no negative side lobes as with $R'(k, a)$. In addition, the amplitude of the side lobes now falls off as $(\pi k a)^{-2}$, instead of $(\pi k a)^{-1}$. For a triangular apodization, the effect of spectral elements which are several resolution widths away from the spectral element of interest is reduced. This reduction is offset by a loss of resolution since the central peak of $T'(k, a)$ is twice as wide as that of $R'(k, a)$. The choice between the two functions will depend on the nature of the spectrum being analyzed.

Other apodization functions have been proposed by various authors. J. Connes proposes the function[7]

$$A(x, a) = \begin{cases} 0, & x < -a, \\ (1 - x^2/a^2)^2, & -a < x < a, \\ 0, & x > a, \end{cases} \qquad (11.2.25)$$

with the apparatus function

$$A'(k, a) = (2\pi ka)^{-5/2} J_{5/2}(2\pi ka), \qquad (11.2.26)$$

where $J_{5/2}(2\pi ka)$ is the half integer Bessel function of order 5/2. The width of the central peak is the same order as that of $T'(k, a)$, but the higher-order side lobes decrease faster after an initial negative lobe. Other functions have been proposed by von Hann[19] and Hamming.[19]

The transforms presented above are for on-center or symmetric apodization functions. Off-center transforms can be obtained by applying the scale and shift theorem, Eq. (11.2.19).

11.2.4. Sampling

In the previous discussion the interferogram $f(x)$ has been treated as a continuous function of the optical path length difference x. The use of digital computers in performing the Fourier transform of the interferogram, however, implies that the data be supplied in digital form. This section discusses methods for sampling or digitizing the continuous function $f(x)$. The consequences of digitizing the interferogram have been discussed by Connes[7] and Steel.[16]

11.2.4.1. Dirac Comb. Digitization can be described mathematically by multiplying the interferogram by a Dirac comb $III(x/d)$

$$III(x/d) = \sum_{n=-\infty}^{\infty} \delta(x/d - n). \qquad (11.2.27)$$

A Dirac comb is a series of Dirac delta functions spaced a distance d apart much like the teeth of a comb. The effect of this function is to sample or digitize the interferogram at intervals of x equal to d. A Fourier transform of the interferogram then becomes

$$g(k) = \int_{-\infty}^{\infty} f(x) A(x, a) III(x/d) e^{-2\pi ikx} \, dx \qquad (11.2.28)$$

[19] R. B. Blackman and J. W. Tukey, "The Measurement of Power Spectra." Dover, New York, 1958.

or

$$g(k) = \sum_{n=-\infty}^{\infty} f(nd)A(nd, a)e^{-2\pi iknd}, \qquad (11.2.29)$$

where $A(x, a)$ is the function used for apodization.

Equation (11.2.29) defines the Fourier series of a function. This is the process which is performed by the computer in determining the amplitude for a wave number k in the spectrum.

11.2.4.2. Consequences of Digitization. The effect of the digitization can be found by performing the integral in Eq. (11.2.28) according to the convolution theorem (11.2.17)

$$g(k) = \int_{-\infty}^{\infty} f'(k - y)A'(y - z)III'(z/d) \, dy \, dz, \qquad (11.2.30)$$

where the Fourier transform $III'(z/d)$ is given by

$$III'(z/d) = dIII(zd). \qquad (11.2.31)$$

Integration over z in (11.2.30) gives

$$g(k) = d \sum_{n=-\infty}^{\infty} \int_{-\infty}^{\infty} f'(k - y)A'[y - (n/d)] \, dy. \qquad (11.2.32)$$

The result of the digitization is that at $k = 0$ the spectrum is convolved with not one but with an infinite number of apparatus functions spaced at distances of $1/d$ from each other. As k, i.e., n, increases, the apparatus functions corresponding to positive frequencies moves to the right and those corresponding to negative frequencies move to the left.

Digitization with a Dirac comb and its consequences are illustrated in Fig. 3 where the case of an on-center interferogram $f(x)$ which extends a distance a on either side of $x = 0$ is considered. If the interferogram is sampled N times, the spacing between samples is $d = 2a/N$; therefore, the Dirac comb for this function is given by

$$III[Nx/(2a)] = \sum_{n=-\infty}^{\infty} \delta[Nx/(2a) - n]. \qquad (11.2.33)$$

The Fourier transform of the interferogram is

$$g(k) = \int_{-\infty}^{\infty} f(x)A(x, a)III[Nx/(2a)]e^{-2\pi ikx} \, dx, \qquad (11.2.34)$$

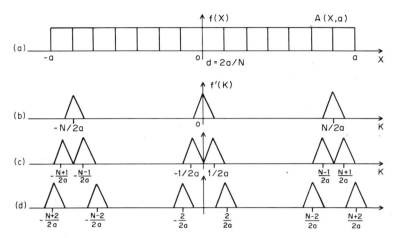

FIG. 3. (a) The data are defined and weighted by $A(x, a)$ and are sampled in N equal intervals. The optical path difference between intervals is given by $d = 2a/N$. (b)–(d) A digital Fourier transform output is obtained and is evaluated at values of $k = j/(2a)$ where j is an integer. Aliased frequencies are introduced at values of $k = N/(2a) - j$. This effect is shown for $j = 0$ (b), 1 (c), and 2 (d). The triangular shape is an abbreviation of the true apparatus function shown in Fig. 2. The highest frequency present in the data must be restricted to the range less than $k = N/4a$ otherwise the presence of high-frequency components will cause aliasing at low frequencies.

where $A(x, a)$ is the apodization function. Since a digital computer is being used, digital output as well as input is obtained. The discrete values of k are taken at $k = j/(2a)$ where j is an integer.

11.2.4.3. Sampling Theorem. Since $A(2an/N, a)$ is nonzero, only for $-N/2 < n < N/2$, integration over the Dirac comb in Eq. (11.2.34) produces the finite sum

$$g[j/(2a)] = \sum_{n=-N/2}^{n=N/2} f(2an/N)A(2an/N, a)e^{-2\pi inj/N}. \qquad (11.2.35)$$

Equation (11.2.34) can also be written as

$$g[j/(2a)] = \sum_{n=-\infty}^{\infty} \int_{-\infty}^{\infty} f'[j/(2a) - y]A[y - nN/(2a)] \, dy. \qquad (11.2.36)$$

As was discussed above, the consequence of digitization is that the spectrum is explored not by just one apparatus function, but by an infinite number of apparatus functions which make the computed spectrum symmetric about $k = N/(4a)$, and periodic with a period of $N/(2a)$. This result is known as the *sampling theorem*.

The consequences of the sampling theorem make it clear that the spectrum being observed must be contained within a range $\Delta k = N/(4a)$ to prevent overlapping of spectral features. This spectral range is usually defined by the sensitivity of the detectors and the response of the various optical and electronic filtering systems used. In turn, the spectral range of interest sets the limit on the interferogram sampling interval. For a spectral region with a maximum wave number k_m and an interferogram of total optical path length $2a$, the number of samples must be at least equal to $4ak_m$. This number of samples corresponds to a sampling interval in optical path length equal to $1/(2k_m)$ or one half the wavelength of the highest-frequency radiation. If the spectral range of interest does not include $k = 0$, but has a lower bound k_0, the sampling interval may be increased to $1/2(k_m - k_0)$.[7]

The noise spectrum often cannot be contained between certain limits. For every computed spectral element, the noise from all n in Eq. (11.2.36) is added into the spectrum. The consequences of this noise addition will be explored further in Section 11.2.7.

11.2.5. Off-Center Interferograms

In the previous sections it has been assumed that the white light fringe (the fringe which corresponds to zero path difference) of the interferogram occurs at the center of the recorded section of the interferogram. In many cases, however, the center of the recorded section of the interferogram may not correspond to the white light fringe. The range of the interferogram is no longer centered at $x = 0$ but is shifted to a position $x = c$. Unless this shift is included in the analysis, it will cause distortions in the spectrum.

11.2.5.1. Introduction of Imaginary Components.

In the absence of an apodization function, the off-center apparatus function is due only to the shifted rectangle function expressed as

$$R'(k, a, c) = 2a[\sin(2\pi ka)/(2\pi ka)]e^{-2\pi ikc}. \qquad (11.2.37)$$

For $c = a$, Eq. (11.2.37) becomes

$$R'(k, a, a) = 2a[\sin(4\pi ka)/(4\pi ka)] - ia[\sin^2(2\pi ka)/(2\pi ka)]. \qquad (11.2.38)$$

The real part of the off-center apparatus function has a resolution which is twice that for the case of a centered interferogram. A large imaginary term is also present in the apparatus function of Eq. (11.2.38) and, therefore, all imaginary terms in the spectrum *must* be eliminated if the extra resolution in the real term is to be realized. If there are no phase errors in the interferogram, this is accomplished by retaining only the real terms of the Fourier transform. When phase errors are present as is usually the case, phase correction procedures such as described in the next section are needed.

Figure 4 shows the real and imaginary parts of Eq. (11.2.37) for a case where c is less than a. As in (11.2.38), the real part of the apparatus function is symmetric and the imaginary part, antisymmetric. An important feature which occurs when a is not equal to c is the effect which modulates the real part and makes it more positive around the origin.

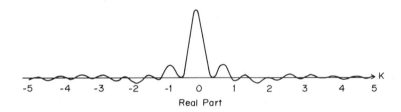

Real Part

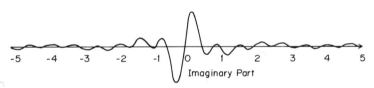

Imaginary Part

FIG. 4. The apparatus function for an off-centered unapodized interferogram.

The effect of the slow-beat frequency is to make the contributions from spectral elements near the calculated element essentially positive rather than averaging to zero. As an example, consider a single absorption line in a flat continuum. Let the intensity of the spectrum in the continuum be C and the intensity at the absorption line be A. The true percentage of absorption is then $1 - (A/C)$. If the spectrum is convolved with the apparatus function (11.2.38), a certain fraction F of the surrounding spectrum is added to each spectral element. If the absorption line is sharp (less than one resolution element), its intensity will be increased by FC because of the apparatus function. The continuum, however, will also be increased by the amount FC again assuming a sharp absorption line. The measured percentage of absorption then becomes $1 - [(A + FC)/(C + FC)]$ giving a smaller absorption than is actually the case. As an example if we let C equal 1, A equal 0.7, and F equal 0.3, the true absorption is 30% but the measured absorption is 24%.

11.2.5.2. Mertz Apodization Function. An apodization function has been proposed which eliminates the beating effects described above for off-center interferograms with $a \neq c$.[11] Equation (11.2.39) gives the apodization function $M(x, b, d)$ which consists of a sloped section about the origin from $-b$ to b and a rectangle function from b to $b + 2d$

$$M(x, b, d) = \begin{cases} 0, & x < -b, \\ (x + b)/2b, & x < |b|, \\ 1, & b < x < b + 2d, \\ 0, & x > b + 2d. \end{cases} \quad (11.2.39)$$

Figure 5 shows $M(x, b, d)$. Its corresponding apparatus function $M'(k, b, d)$ is

$$M'(k, b, d) = \frac{\sin[2\pi k(b + 2d)]}{2\pi k} + \frac{i \cos[2\pi k(b + 2d)]}{2\pi k}$$

$$- \frac{i \sin(2b)}{4\pi^2 k^2 b}. \quad (11.2.40)$$

As Eq. (11.2.40) shows, the real part of $M'(k, b, d)$ is the sinc function of frequency $b + 2d$. The imaginary part of the apparatus function is an antisymmetric function, which for $d \gg b$, consists mainly of the term $\cos[2\pi k(d + 2d)]/(2\pi k)$. This term can be eliminated with accurate phase correction.

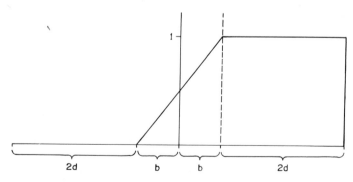

FIG. 5. The Mertz apodization function.

11.2.6. Phase Correction

If phase errors are present in the interferogram, then it is clear that the real part of the Fourier transform is not sufficient to specify the spectrum. These phase errors can be introduced by various phenomena among which are dispersion in the beam splitter, displacement of the sample point for $x = 0$ by an amount Δ, and random noise.

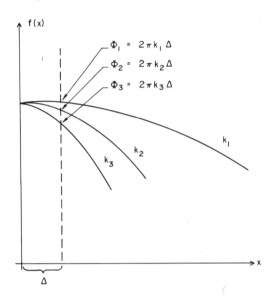

FIG. 6. A phase error is introduced when the first sample point is at $x = \Delta$ and there is no sample point at $x = 0$. The effect is shown for three different frequencies k_1, k_2, and k_3 which are all in phase at $x = 0$ but have phase errors Φ_1, Φ_2, and Φ_3 at $x = \Delta$. Similar errors are introduced by incomplete beam splitter compensation.

11.2.6.1. Phase Errors. Figure 6 shows for three different wave numbers the phase errors which are due to off-center sampling. The sampling Dirac comb is no longer given by $III(x/d)$, but by $III[(x - \Delta)/d]$ which shifts the phase of each wave number k by an amount $2\pi k\Delta$, i.e., a phase error which is linear in k.

The dispersive properties of the beam splitter can cause wave number-dependent phase shifts. Most of this error is removed when a compensator of the same material and thickness as the beam splitter is placed in the beam as shown in Fig. 1. The residual phase shifts result in a different optical path length for each wave number. This shift in zero-phase point has the same effect as the shift in sampling position described above. The phase error becomes $2\pi ky(k)$, where $y(k)$ is the shift in zero-phase point for wave number k. Both phase shifts are continuous functions of k and, for beam splitter compensation and accurate sampling, are usually slowly varying.

11.2.6.2. Methods of Phase Correction. The following method of phase correction is due to Mertz.[11] Let the phase error which is due to off-center sampling or beam splitter dispersion be the slowly varying function $\Phi(k)$. If $f'(k)e^{-2\pi ikx} \, dk$ is the input before phase shift, where f' is real, then the phase-shifted input is $f'(k)e^{i\Phi(k)}e^{-2\pi ikx} \, dk$. The Fourier transform of this input with an apodization function $A(x)$ is

$$g(k) = [f'(k)e^{i\Phi(k)}] * [A_R'(k) + A_I'(k)] \qquad (11.2.41)$$

where A_R' and A_I' are the real and imaginary parts, respectively, of the apparatus function. The phase error $\Phi(k)$ is slowly varying, and if it is assumed to be nearly constant over the range of the apparatus function, it can be removed from the convolution function. Equation (11.2.41) may then be written as

$$G(k) = e^{i\Phi(k)}(F_R + F_I) \qquad (11.2.42)$$

and F_R and F_I are $f' * A_R$ and $f' * A_I$, respectively. The object of the phase correction procedure is to determine F_R, i.e., the convolution of the input spectrum with the real part of the apparatus function.

$\Phi(k_0)$ is determined from a Fourier transform of a small part of the interferogram about the white light fringe. Since $\Phi(k_0)$ is slowly varying, a low-resolution transform will give a good approximation with a high signal-to-noise ratio. A symmetric apodization function is used in the transform to ensure that the apparatus function is real.

Let F and β be defined by

$$F = (F_R^2 + F_I^2)^{1/2} = |G| \qquad (11.2.43)$$

and

$$\beta = \tan^{-1}(F_I/F_R). \qquad (11.2.44)$$

With these definitions, $G(k) = F \exp i[\Phi(k) + \beta]$ and $\Phi(k) + \beta = \tan^{-1}(G_R/G_I)$ from which β may be obtained. The real part of the apparatus function is then

$$F_R = |G| \cos \beta = F \cos \beta. \qquad (11.2.45)$$

An equivalent method of phase correction has been proposed by Forman, Steel, and Vanasse in which the correction is performed on the interferogram rather than on the spectrum.[20,21] The phase function $\Phi(k)$ is
determined in the same way with a small part of the interferogram, but
the correction is convolved with the interferogram rather than multiplied
by the spectrum as in (11.2.45). Equation (11.2.8) shows the mathematical
equivalence of the two methods. Both methods of phase correction
introduce a subjective estimation of how much of the interferogram to
include in the transform to obtain $\Phi(k)$. This estimate is a function of
the instrument being used and of the signal-to-noise ratio of the interferogram.

11.2.7. Noise

The properties of noise under Fourier transform will play an important
role in the data analysis. Noise and true signal are recorded simultaneously and then Fourier transformed to achieve the final output. In order
to proceed, a quantitative measure of the signal-to-noise ratio must be
established. The signal-to-noise ratio $R(\nu)$ is defined as the amplitude
of the true signal at frequency ν divided by the integrated root mean
square noise at the same frequency. There are two problems which must
be solved in order to determine $R(\nu)$. First the auto-correlation function
of the noise at frequency ν must be found, and second, the auto-correlation function must be integrated over the observation time T to find the
integrated root mean square error as a function of T.

Let $S(t)$ be the true signal and $n(t)$ be the noise. For this discussion,
time and frequency are more convenient variables than x and k. The
results may now be expressed in terms of the observation time and the
noise frequency spectrum of the detector. It is a simple matter to connect
with the previous discussion as it is assumed that the interferometer
mirror moves with a constant velocity v so that $t = x/2v$ and $\nu = 2kv$.
A factor of 2 is present because the optical path length velocity dx/dt
is twice the mirror velocity v.

11.2.7.1. Autocorrelation Function.
The derivation of the noise auto-
correlation function follows a method derived by Connes.[7] A difference
exists in the method used for recording the data. Connes treats the case
in which the interferometer mirror is moved very slowly and an optical

[20] W. H. Steel and M. L. Forman, *J. Opt. Soc. Amer.* **56**, 982 (1966).
[21] M. L. Forman, W. H. Steel, and G. A. Vanasse, *J. Opt. Soc. Amer.* **56**, 59 (1966).

chopper is used to modulate the signal. This discussion assumes that the mirror is moved and that the wave numbers of interest are modulated at frequencies from approximately 250 to 1000 Hz. In this case it is usual to coadd to produce the total interferogram.

The recorded interferogram $I(t)$ is the sum of the true signal $S(t)$ and the noise $n(t)$. The linearity of the Fourier transform permits separate transformations of noise and signal. The noise part of the interferogram is

$$n(v_0) = \int_{-\infty}^{\infty} n(t)A(t)\mathrm{III}(t/\tau) \exp(-2\pi i v_0 t)\, dt, \qquad (11.2.46)$$

where $A(t)$ is the apodization function and $\mathrm{III}(t/\tau)$ is the sampling function and both are expressed as functions of time.

11.2.7.2. Variance. True noise has zero mean, i.e., $\overline{n(t)} = 0$. In the frequency domain the mean square value of the noise $\overline{\eta(v)^2}$ equals the variance $\sigma^2(v_0)$. However, $n(t)$ is the random variable and the mean square of the noise $\overline{\eta(v_0)^2}$ must be written as

$$\sigma^2(v_0) = \int_{-\infty}^{\infty} \int_{-\infty}^{\infty} n(t)n^*(t')A(t)A^*(t')\mathrm{III}(t/\tau)\mathrm{III}^*(t'/\tau)$$
$$\times \exp[-2\pi i v_0(t - t')]\, dt\, dt'. \qquad (11.2.47)$$

Equation (11.2.47) is equivalent to the autocorrelation function of $n(t)$ for zero delay.

The value of $\overline{n(t)n^*(t')}$ is given by the autocorrelation function of $n(t)$

$$n(t)n^*(t') = \lim_{T' \to \infty} (1/T') \int_{-T'/2}^{T'/2} n(t)n^*(t + \Delta)\, dt \qquad (11.2.48)$$

at $\Delta = 0$. A theorem due to Wiener allows $n(t)n^*(t')$ to be expressed in terms of frequency. The theorem states that the inverse Fourier transform of the power density spectrum of a function $f(v)$ is the autocorrelation of the inverse transform $f''(t)$. Equation (11.2.48) then becomes

$$n(t)n^*(t') = \int_{-\infty}^{\infty} P_n(v) \exp[-2\pi i v(t - t')]\, dv, \qquad (11.2.49)$$

where $P_n(v)$ is the power density spectrum given by

$$P_n(v) = \lim_{T' \to \infty} (1/T) \left| \int_{-T^{1/2}}^{T^{1/2}} n(t)e^{-2\pi i v t}\, dt \right|. \qquad (11.2.50)$$

Substitution of (11.2.49) into (11.2.47) gives

$$\sigma^2(\nu_0) = \int_{-\infty}^{\infty} P_n(\nu) \int_{-\infty}^{\infty} \int_{-\infty}^{\infty} A(t)A^*(t')\mathrm{III}(t/\tau)\mathrm{III}^*(t'/\tau)$$
$$\times \exp[-2\pi i \nu_0(t - t')] \exp[2\pi i \nu(t - t')]\, dt\, dt'\, d\nu. \quad (11.2.51)$$

Terms in ν and t may be regrouped to give

$$\sigma^2(\nu_0) = \int_{-\infty}^{\infty} P_n(\nu) \int_{-\infty}^{\infty} \int_{-\infty}^{\infty} A(t)A^*(t')\mathrm{III}(t/\tau)\mathrm{III}(t'/\tau)$$
$$\times \exp[-2\pi i t(\nu_0 - \nu)] \exp[-2\pi i t'(\nu_0 - \nu)]\, dt\, dt'\, d\nu. \quad (11.2.52)$$

The integrations over t and t' can now be performed with the help of (11.2.31)

$$\sigma^2(\nu_0) = \tau^2 \int_{-\infty}^{\infty} P_n(\nu) \left[\sum_{n=-\infty}^{\infty} A'(\nu_0 - \nu - n/\tau) \right]$$
$$\times \left[\sum_{n=-\infty}^{\infty} A'(\nu_0 - \nu - n/\tau) \right]^* d\nu. \quad (11.2.53)$$

For any reasonable apodization function, the apparatus function $A'(\nu)$ is sufficiently narrow that no significant overlap occurs for different values of n in (11.2.53). All of the cross terms in (11.2.53) therefore go to zero upon integration and the equation may be written as

$$\sigma^2(\nu_0) = \sum_{n=-\infty}^{\infty} \tau^2 \int_{-\infty}^{\infty} P_n(\nu) \, |\, A'(\nu_0 - \nu - n/\tau)\, |^2\, d\nu. \quad (11.2.54)$$

11.2.7.3. Necessity for Filtering. Equation (11.2.54) indicates the need for proper filtering of the signal. As a result of the digital sampling function $\mathrm{III}(t/\tau)$, the noise spectrum is sampled by an infinite number of apodization functions spaced at intervals of $1/\tau$. If the frequency range of the noise power density spectrum of $P_n(\nu)$ is not limited by appropriate filtering, the noise contribution will become infinite. The signal must also be limited to the same range, i.e., $\Delta\nu = 1/\tau$ to avoid aliasing. Thus the same filtering process is appropriate for both the signal and the noise. The filtered noise is given by

$$P_n^{\,0}(\nu) = P_n(\nu) \, |\, G(\nu)\, |^2, \quad (11.2.55)$$

where $G(\nu)$ is the frequency response of the filters. If an RC filter of time constant τ' is used, $|\, G(\nu)\, |^2$ is given by

$$|\, G(\nu)\, |^2 = 1/[1 + (2\pi\nu\tau')^2]. \quad (11.2.56)$$

Then the variance of the filtered noise is

$$\sigma^2(\nu_0) = \tau^2 \int_{-\infty}^{\infty} P_n{}^0(\nu) \mid A'(\nu_0 - \nu) \mid^2 d\nu. \qquad (11.2.57)$$

Equation (11.2.57) states that the noise power density spectrum is viewed through the spectral window $\mid A'(\nu) \mid^2$ in the same way the spectrum is viewed through the window $A'(\nu)$. An intuitive argument leads to the same conclusion.

11.2.7.4. Noise Time Dependence. Equation (11.2.57) is the auto-correlation function of the noise frequency ν_0 for zero time delay. For a time delay $\varDelta$ other than zero and a filter function such as the one in (11.2.56), the autocorrelation function $\Phi(\nu_0, \varDelta)$ is

$$\Phi(\nu_0, \varDelta) = \sigma^2(\nu_0)e^{-|\varDelta|/\tau}. \qquad (11.2.58)$$

The filter time constant τ' is used to set the limits of the spectral interval. The value of τ' is approximately the same as the sampling interval τ and both may be set equal to τ.

The noise must be integrated over a time T equal to the observation time in order to determine the integrated variance. If the noise is integrated for a time $T = 2T'$, the variance is[22]

$$\sigma^2(\nu_0, T') = \int_{-T'}^{T'} \frac{T - \mid \varDelta \mid}{T'^2} \, \Phi(\nu_0, \varDelta) \, d\varDelta. \qquad (11.2.59)$$

Substitution of (11.2.58) into (11.2.59) gives

$$\sigma^2(\nu_0, T') = \sigma^2(\nu_0) \int_{-T'}^{T'} \frac{T' - \mid \varDelta \mid}{T'^2} \, e^{-|\varDelta|/\tau} \, d\varDelta \qquad (11.2.60)$$

$$\sigma^2(\nu_0, T') = 2\sigma^2(\nu_0)[(T'/\tau) - 1 + e^{-T'/\tau}](\tau/T')^2. \qquad (11.2.61)$$

For delay times longer than T' and with the substitution of $T = 2T'$, Eq. (11.2.61) reduces to

$$\sigma^2(\nu_0, T) = 4\sigma^2(\nu_0)\tau/T. \qquad (11.2.62)$$

Equation (11.2.51) may be substituted into (11.2.62) to give the full expression for the root mean square of the noise at frequency ν_0 after

---

[22] Y. W. Lee, "Statistical Theory of Communications," p. 286. Wiley, New York, 1967.

integration for the time T

$$\sigma(\nu_0, T) = 2\tau \left[\int_{-\infty}^{\infty} P_n'(\nu) \, | \, A'(\nu_0 - \nu) \, |^2 \, d\nu \right]^{1/2} (\tau/T)^{1/2}. \quad (11.2.63)$$

Thus, $\sigma(\nu_0, T) \to 0$ as $T^{-1/2}$.

The value of the true (time-independent) signal at frequency ν_0 is

$$\xi(\nu_0) = \int_{-\infty}^{\infty} S'(\nu) A'(\nu - \nu_0) \, d\nu. \quad (11.2.64)$$

After an integration time T, the signal-to-noise ratio $R(\nu_0, T) = \xi(\nu_0)/\sigma(\nu_0, T)$ is given by

$$R(\nu_0, T) = \tfrac{1}{2}(T/\tau)^{1/2}$$

$$\times \left[\int_{-\infty}^{\infty} S'(\nu) A'(\nu_0 - \nu) \, d\nu \right] \Big/ \left[\int_{-\infty}^{\infty} P_n(\nu) \, | \, A'(\nu_0 - \nu) \, |^2 \, d\nu \right]^{1/2}.$$

$$(11.2.65)$$

Equation (11.2.65) yields a result which is typical of most measurement processes, namely, that the signal-to-noise ratio increases as $T^{1/2}$. If $S'(\nu)$ and $P_n(\nu)$ are approximately constant over the frequency interval of the apparatus function $A'(\nu)$, Eq. (11.2.65) may be written as

$$R(\nu_0, T) = \tfrac{1}{2}(T/\tau)^{1/2}[S'(\nu_0)/P_n^{1/2}(\nu_0)]Q/Q', \quad (11.2.66)$$

where

$$Q = \int_{-\infty}^{\infty} A'(\nu - \nu_0) \, d\nu \quad (11.2.67)$$

and

$$Q' = \left[\int_{-\infty}^{\infty} | \, A'(\nu_0 - \nu) \, |^2 \, d\nu \right]^{1/2}. \quad (11.2.68)$$

The width of $A'(\nu)$ is the width of the spectral resolution element $\delta\nu$. Thus Q and Q' are of the order of $\delta\nu$ and $(\delta\nu)^{1/2}$, respectively.

11.2.7.5. Noise Equivalent Power.
It is more useful to express Eq. (11.2.66) in terms of the noise equivalent power (NEP) for infrared detectors. The photovoltage detectors typically used in the short wavelength region are characterized by an output voltage which is proportional to the power in watts of the incident photon flux on the detector. NEP is defined as the incident power in watts of radiation wave number k, in a band width $\delta\nu$ about a modulation frequency ν_0, needed to produce a signal-to-noise ratio of one. Noise equivalent power is therefore a function of the wave number k and the modulation frequency and has units

of watts per (hertz)$^{1/2}$. If $S'(\nu_0)$ is expressed in watts per (hertz) in (11.2.66), the term $P_n^{1/2}(\nu_0)$ may be replaced with $\text{NEP}(k_0, \nu_0)$. Equation (11.2.66) may be rewritten as

$$R(\nu_0, T) = \frac{1}{2} \left(\frac{T}{\tau} \right)^{1/2} \frac{S'(k_0)\delta\nu}{\text{NEP}(k_0, \nu_0)(\delta\nu)^{1/2}}, \qquad (11.2.69)$$

where k_0 is the wave number corresponding to the modulation frequency ν_0.

11.2.7.6. Comparison of an Interferometer and a Scanning Spectrometer. The efficiency of a scanning spectrometer and an interferometer can be compared directly in terms of their respective signal-to-noise ratios. Let both of the instruments use the same detector, have the same resolution, and operate for a time T. If the resolution is such that the spectral interval is divided into M elements, a single detector scanning spectrometer spends a time T/M observing each element.

The apodization function for the interferometer is functionally equivalent to the slit function of the spectrometer; thus the numerator in (11.2.69) is the same for both instruments. In the denominator $(\delta\nu)^{1/2}$ should be approximately the same for both instruments, however, τ may be different. τ for the scanning spectrometer will be approximately $1/\nu_0$, where ν_0 is the chopping frequency. Since ν_0 is usually chosen to minimize the NEP, it will probably lie in the middle of the range of frequencies used by the interferometer and a difference of a factor of two may exist in the sampling times. Fellgett's advantage for this case is then proportional to $[M\tau_S/\tau_I]^{1/2}$, where S and I refer to the scanning spectrometer and interferometer, respectively.

11.3. Fourier Spectroscopy in Practice

11.3.1. The Interferometer

The Michelson interferometer is the Fourier spectrometer most often used for astronomical observations. It will be used as the basis for the following discussion, but many of the topics discussed here are equally applicable to other Fourier transform spectrometers.

11.3.1.1. Scanning Drive. The Michelson interferometer is, in principle, a very simple device with only one active element, the mirror drive or scanning system (Fig. 1). It is not surprising therefore that a large portion of the development work on Michelson interferometers has been devoted to improving methods for accurately positioning and driving the moving mirror. Two systems, the continuous and the stepped drive system, have been developed.

11.3.1.1.1. CONTINUOUS DRIVE. In the United States, most commercially available interferometers are continuous drive instruments, i.e., the moving mirror is driven continuously at a constant velocity over the total scan distance. At the end of the scan the mirror may either flyback to start another scan or perform a scan in the reverse direction.

Movement of the mirror is commonly accomplished with a magnetic drive system similar to a loudspeaker drive. The mirror may be attached directly to a metallic slug, or to a precision slide which is in turn connected to the slug. The objective is to drive the mirror at a constant velocity throughout the scan distance in a manner which is independent of the orientation or vibration environment of the interferometer. This can only be accomplished with a very stiff feedback system. The position and velocity of the mirror are fed back to the drive mechanism which then adjusts the acceleration of the mirror by varying the current to the coils of the drive magnet. Position and velocity of the mirror can be determined by suitable electronic transducers or by an optical means described below.

Most precision instruments use a slide-mounted mirror. The critical mechanical component of the drive system is the bearing which supports the mirror assembly and guides its linear motion. The bearing must allow the mirror to move freely along the desired direction of travel but must not allow any tilt of the mirror over the length of travel. (Tilt errors are discussed extensively by Steel.[16]) Ways equipped with precision ball bearing slides and air bearings are used for support of the mirror system. Both systems are sufficiently free of tilt jitter to be used in wavelength regions from the far infrared through the ultraviolet.

11.3.1.1.2. STEPPED DRIVES. An alternative to the continuous drive system is offered by the stepped drive system,[9] where, after each incremental step, the signal is integrated for a fixed time. Motion of the mirror is again accomplished by a magnetic drive system; however, a very stiff system of positional feedback is required. This mode of motion does relax almost completely the requirements on the velocity feedback system, since the constancy of the motion need only be sufficient to ensure proper operation of the positional transducer and to minimize the dead time of the system while the mirror is moved. Bearings for this system may be the same as for the continuous drives described above.

11.3.1.2. Reference Signals. With either scanning system, the position of the mirror must be monitored at all times. Positional information is used to update the mirror drive feedback system and to determine accurately where the interferogram should be sampled in order to provide uniformly spaced digital data output. The most accurate and elegant way of producing this information is to introduce a beam of monochromatic radiation into the interferometer and monitor the output. The output from the monochromatic beam will be a sinusoid, and each zero crossing of the sinusoid indicates that the optical path length of the interferometer has changed by one half of the wavelength of the monochromatic beam, which in turn indicates that the mirror has moved one quarter of a wavelength. Velocity information may also be obtained from the monochromatic signal by integrating the signal under the sinusoid between zero crossings. For purposes of comparison, a standard clock can be used to provide a reference signal. Values of the integrated monochromatic signal larger than the reference indicate the velocity is too low, lesser values indicate that the velocity is too high.

It is most common now to use a laser to produce the monochromatic beam. A laser has the desirable properties of being highly monochromatic and highly collimated, and, in particular, a He–Ne laser has the advantages of a convenient wavelength and low cost.

The mechanical system which establishes the optical path of the monochromatic reference beam can be one of two designs. One method is to introduce the beam parallel to the radiation being observed but displaced from it so as not to interfere with the primary beam. The mirrors of the interferometer should be made large enough to accommodate this beam. The reference beam uses the same beam splitter and the same surface of the reflecting mirrors as the signal beam. The phase of the reference signal is subject to mirror tilt errors, and in some wavelength regions, scattered light from the reference beam can become a source of noise in the primary signal.

Another method is to construct a second interferometer which uses the other face of the moving mirror of the primary interferometer as the moving mirror for the reference interferometer (Fig. 7). Thermal expansion and contraction of the instrument can lead to drift between the two interferometers since they have different stationary mirrors and beam splitters. Thus, thermal effects can change the optical path lengths in each interferometer by different amounts and introduce a variable phase shift between them. Thermal cycling often occurs during a night's observation time at a telescope and, if no corrections are made, the effects of unknown phase shifts may limit the number of successive data scans which may be coadded.

The monochromatic reference signal discussed above provides only a relative reference, i.e., it determines how far the mirror has moved from some arbitrary position but not the absolute position of the mirror with respect to zero path difference. An absolute position fiducial mark is provided by a second polychromatic or white light reference beam. Since the radiation is polychromatic, a sharp white light interferogram

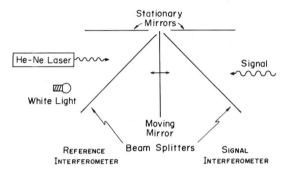

FIG. 7. A "two-cube" Michelson interferometer which features a common moving mirror but separate signal and reference optics.

occurs and a sharp output voltage spike is produced when the optical path lengths of the two arms of the interferometer are equal. If the white light beam is parallel to the signal beam, the white light pulse can be used as the fiducial mark for the signal beam. If the reference beam is used in a secondary interferometer as described above, then the phase of the white light pulse is determined by the position of the stationary mirror in the second interferometer. For suitable positions of this mirror the sharp pulse can then be used as a signal to begin recording data, but this method is subject to the thermal drifts described above. In either case, the position of the interferometer at any time is determined by counting the number of laser beam zero crossings from the occurrence of the white-light-pulse fiducial mark.

11.3.2. Data Systems

The data recording and processing systems are as important as the optical and mechanical systems of the interferometer. When used for astronomical observations, continuous scan instruments usually have very high data sampling rates and require data processing systems which are more complex than those for stepped drive interferometers. Because of its widespread usage, most of the following discussion will be directed toward a continuous scan data system.

11.3.2.1. Filtering.   A schematic diagram of a typical continuous scan interferometer data system is shown in Fig. 8. After the signal from the interferometer has been detected and amplified by an appropriate detector–preamplifier system for the wavelengths being observed, the first component of the system that the signal encounters is a set of electrical filters. Section 11.2.4.3 has discussed the need for limiting the signal

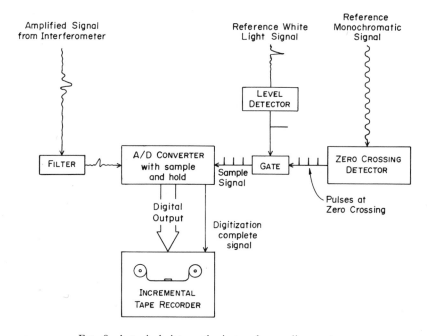

FIG. 8. A typical data gathering and recording system.

frequencies to a definite range in order to avoid aliasing in the final spectrum. Electrical noise from the detectors is also limited by the filters which makes electrical filtering preferable to optical filtering which would only limit the spectral noise. A sharp frequency cutoff is very desirable and a typical filter system can have 48 dB/octave frequency cutoffs.

11.3.2.2. Sampling.   After the filters have limited the signal frequencies to the proper range the signal must be sampled at equally spaced intervals of optical path length difference.

11.3.2.2.1. ZERO CROSSINGS. The zero crossings of the sinusoidal reference signal mark off equally spaced intervals of optical path length difference, and can be used to mark the sample points. A zero-crossing detector is used to produce a pulse which then signals an analog-to-digital (A/D) converter to sample the interferometer signal, digitize it, and display the result on its output wires.

11.3.2.2.2. A/D SYSTEMS. The A/D converter must have a sample and hold feature with a small aperture time. This is important because the point at which the signal is sampled should be independent of the reference signal voltage height. The signal should be sampled at the instant of the monochromatic reference zero-crossing, not at some variable time after the pulse. An aperture time of less than a microsecond is generally used. A digitization complete pulse is sent by the A/D converter which signals the recording device that the data may be read from the wires. It is of course essential that the time to complete this operation, the digitization time, be less than the time between successive samples. Most modern A/D converters are capable of handling the typical date rates produced by an interferometer.

11.3.2.2.3. SIGNAL INTEGRATION. It is important to recognize that signal is not lost in the sampling process. At first it seems that only the signal seen during the aperture time of the A/D converter is being recorded. It must be remembered, however, that the filters which come before the A/D converter act as integraters. Now, the time response of a filter system is the Fourier transform of its frequency response and if the frequency bandpass is of the order $\Delta \nu$, then the integration time τ of the filter is of the order $1/\Delta \nu$. The time for the mirror to travel between sample points is adjusted to be of the order τ since the sampling frequency is twice that of the highest frequency to be recorded, thus practically all of the signal between the two sample points is included in the output of the A/D converter.

Sakai[23] has pointed out that if the velocity of the moving mirror is not constant throughout the scan, either a loss of efficiency or an excess noise in the final spectrum will occur. If the mirror velocity is not constant, the time between successive sample points varies even though the integration time τ of the filter does not. Unequally spaced sample points produce ghosts in the spectrum which contribute to the spectral noise.

[23] H. Sakai, in AFCRL-71-0019, Spec. Rep. No. 114, *Aspen Int. Conf. Fourier Spectrosc. 1970*, January 5, 1971.

(These are similar to the ghosts produced by unavoidable variations in the line spacings of a diffraction grating.) In order to reduce this noise the integration time must be shortened to reduce the variation in sample spacing. This in turn reduces the efficiency of the system since the integration time is now less than the time between samples. Sakai's calculations indicate that for a 10% variation in the speed of the moving mirror the efficiency drops by a factor of six. This result is a compelling reason to guard against fluctuations in the mirror speed.

An alternative is to have a circuit which integrates the signal between sample points and then divides the result by the time of travel between these points. This circuit then acts as a filter with a variable bandpass equal to $1/t$, where t is the time of travel between the two sample points.

11.3.2.3. Recording. Recording the digital output of the A/D converter presents some practical problems which are due to the relatively high rate of the output data. As an example, consider a system which is used to investigate the 1–4-μm spectral region. In order to avoid noise from atmospheric fluctuations the lowest frequency of modulation must be above 250 Hz. If the 4-μm signal is modulated at 250 Hz the 1-μm signal will be modulated at 1000 Hz. Since the sampling occurs at twice that frequency, the sample rate is 2000 samples/sec. At least two tape words (signal and reference) must be written for every sample, thus the word rate for the tape recorder is on the order of 4000 words/sec. Continuous mode digital tape recorders such as are used on high-speed computers cannot be used. Unfortunately, the constancy of the data rate required by these recorders cannot be guaranteed because of the small variations in mirror scan speed. An incremental tape recorder must, therefore, be used for recording data directly. Although most of these tape recorders have maximum data rates less than 1000 words/sec, a few newer models have data rates up to 10,000 words/sec. These use a small buffer memory to allow for variations in data rate. However, they still require the data rate to be constant within about 10%.

Alternative solutions to the problem of high data rate are available. One is to record the interferometer signal and reference signals on separate tracks of an analog tape recorder. This tape can be played back at a lower speed in the laboratory thereby achieving a lower data rate at the expense of adding tape noise and requiring greater than real time data analysis. Care must also be taken that the frequency response of the tape recorder is flat over the entire frequency range covered by both the fast record and slow playback signals. This method, however, is often the most viable alternative, especially when cost is an important consideration.

A second alternative is to use a small computer to provide a constant data rate to the digital tape recorder. The computer acts as a large memory buffer which accepts data at an irregular rate from the interferometer and supplies data to the tape recorder at a constant rate. In this system the tape recorder acts essentially in a continuous mode with a greatly increased data rate capability. Other data format functions such as blocking the data into small records on the tape can be performed by the computer at the same time. This is essential when reading the output tape on some large computers which have small input buffers in their tape drives. The computer may also be used to control the interferometer through interaction with the laser and white light reference signals and the drive system.

11.3.2.4. Dynamic Range in the Signal.   The presence in the signal of a high-amplitude white light fringe at the center of the interferogram and the relatively small amplitude in the wings imposes the special problem of dealing with the limited dynamic range over which the signal can be recorded accurately. Most good amplifiers are linear throughout the dynamic range of approximately 10^4 present in the output signal, but a 12-bit A/D converter has a dynamic range of only 4096. If the peak of the white light fringe is full scale on the A/D converter, then the accuracy of the measurement is 1/4096 or approximately 0.025%. In the wings of the interferogram, however, where the average amplitude is often 0.1% of that at the white light peak the accuracy of the A/D conversion is only $\frac{1}{4}$ or 25%. The problem is that the signal in the wings of the interferogram needs only 1 or 2 bits in the A/D converter. Unless the sensitivity to the signal in the wings is increased, this effect adds noise to the spectrum and increases the integration time needed to achieve a given signal-to-noise ratio.

One solution to this problem is to increase the number of bits converted. A/D converters are available with 16-bit accuracy, but the digitization times are often long and costs are high. An alternative solution is to use a variable gain amplifier or nonlinear amplifier which has a much higher gain for low voltages than for high thereby reducing the dynamic range present in the signal. The amplifier must be gain stable and carefully calibrated in order to remove the nonlinearity from the interferogram *before* the spectrum is extracted.

Another similar solution is to use two amplifiers each with a different gain. A high-gain amplifier records the wings and a low-gain amplifier records the region around the white light peak. The low-gain amplifier may be switched on and off at predetermined points in the interferogram and one of the bits of the A/D may be used to flag the gain state.

11.3.2.5. Coadding. A rapid scanning interferometer will scan a 1-cm distance in about 20 sec. Many scans are therefore produced in a typical observing period of about 2-hr. If the system is stable, these scans can be coadded to produce a single interferogram which represents the total 2-hr observing time. Digitized individual scans recorded on magnetic tape may be read into a computer and added together. In many computers the input buffer of the tape drive is too small to handle a total run of over 60,000 samples and the data must be broken into smaller records when recorded.

The most difficult problem of coadding is the proper alignment of data. If there is no thermal drift in the interferometer, the use of reference signals and recording procedure discussed in Section 11.3 will assure that the first sample in each scan is from the same mirror position and that all subsequent samples in each scan are taken from the same positions.

If thermal drift is a critical problem, other techniques can be employed. It is usually better in these cases to record the data in an analog rather than digital form. This allows corrections to be made on the sample points when the tape is played back in the laboratory. If the signal-to-noise ratio is large enough to easily identify the position of the white light peak in each scan, individual scans may be aligned by superposing white light peaks as seen on an oscilloscope trace. A variable time delay in the monochromatic reference signal playback channel is used to accomplish this. The region of the interferogram about the white light fringe is added, stored, and displayed in a pulse height analyzer. An operator can then adjust the time delay in the reference signal so that all subsequent white light peaks are aligned.

If the white light fringe cannot be identified in each scan, it may be necessary to limit the range of observing time over which scans can be coadded. Data from long observing runs will require many separate transformations and thus be expensive to process.

11.3.3. Optics

A detailed account of the optical theory of interferometers is given in several books,[16,17] therefore it is not discussed rigorously here. Instead the problem of the proper fore optics for the interferometer is discussed in some detail.

For standard laboratory usage, such as for chemical analysis absorption spectra, the optics may be quite simple. A collimated light beam is passed through the sample to be analyzed and enters the interferometer through one of the inputs. Focusing optics are placed at the complementary output and the emerging beam is focused on a detector. Astronomical for optics are usually more complicated often involving tilted and diverging or converging beams.

11.3.3.1. Uncollimated Light. The light which is collected by the telescope is neither collimated nor plane parallel when it enters the interferometer system. In order to examine the effect of converging or diverging light, consider the converging beam in Fig. 9. Let the total optical path length difference for a full sweep of the mirror be d which is equal to twice the mirror sweep distance. The central ray A of the converging beam will have a total optical path length difference of just d. An off-axis ray B, however, will have a larger optical path length difference l, where $l = d/\cos \theta$. The difference is $\Delta d = l - d(1 - \sec \theta)$. The angle θ may be expressed in terms of the f number of the beam as $\theta = \frac{1}{2} \cot^{-1}(f)$. If the shortest wavelength being considered is λ, the phase shift between the central and off-axis ray for this wavelength is $\Phi = 2\pi(\Delta d/\lambda)$. By setting a limit of $\lambda/10$ on Φ the smallest useful (limiting) f number may be calculated. The same geometry applies to diverging beams. For beams of smaller than the limiting f number calculated above, the shorter wavelengths become "washed out" at the ends of the mirror sweep. This effect degrades the resolution of the instrument for these wavelengths since the effective total optical path

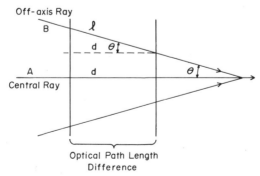

FIG. 9. The difference in path length for on- and off-axis rays limits the resolution of the interferometer.

length difference is less than d. This result is similar to the observations which occur for off-axis rays in the case of grating spectrometers and this result is not surprising.

11.3.3.2. Dual Beam Fore Optics

11.3.3.2.1. TWO-DETECTOR SYSTEMS. In the example given above only one of the outputs of the interferometer was used. Since there is usually no problem with intensity in laboratory experiments this is of no great importance. In astronomical experiments, however, many objects are at the limit of detectability and it is desirable to use the signal from both outputs. Since the second output is also the input, the beams must be separated without vignetting the beam excessively. One possibility is to allow the input beam to diverge as it enters the interferometer as shown in Fig. 10. Upon return from the interferometer the beam diameter has grown to the point where the input reflector removes only a small amount of the radiation. A focusing system located beyond the reflector focuses the radiation on the detector.

The limiting divergence or f number is set by the discussion in Section 11.3.3.1. For a sufficiently high resolution (large d) the limiting f number can be too large to allow the system to be of reasonable dimensions. An alternative geometry uses a collimated beam which is tilted so that it enters off-axis as in Fig. 11. A simple ray tracing shows that for all mirror displacements if the focusing optics have a large enough aperture, then all rays in the beam interfere and fall on the detectors. The large aperture required may cause problems in systems which are limited by background radiation.

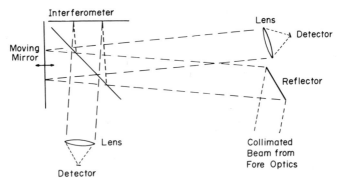

FIG. 10. Divergent beam fore optics which allow the use of both outputs. This arrangement is most useful when the interferometer is used at the Cassagrain focus of a telescope, or in cases where the f number is small.

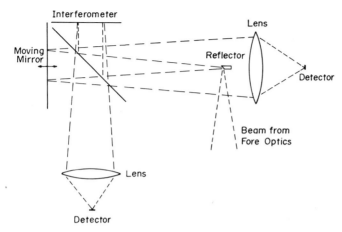

FIG. 11. Tilted fore optics which allow the use of both outputs. This arrangement is most useful when the interferometer is used at the Coudé focus of a telescope, or in cases where the *f* number is large.

The complementary aspect of the two outputs for dielectric beam splitters discussed in Section 11.2.1.2 means that the two detectors should be connected in a difference circuit since the outputs are 180° out of phase. The detectors should be matched since it is better to have one very good detector alone than to have a very good detector and a noisy detector coupled in a difference circuit.

11.3.3.2.2. SKY SUBTRACTION. An automatic sky background subtraction is also possible because of the complementary aspect of the interferometer inputs and outputs. If a sky background signal is placed in one input and the star plus background signal is placed in the other input, the result at each output will be star signal only. Compensation for any minor asymmetry in the transmission properties of the interferometer can be accomplished by switching the beams between the two inputs. This requires additional logic in the scanning system and provides an excellent opportunity to make use of the control capabilities of a small computer.

11.3.4. Limiting Magnitude

In planning observations, the magnitude of the object in the wavelength band of interest is the parameter upon which the feasibility of the measurement rests. With a given detection system which includes a telescope, an interferometer, and a detector, there is a limiting magnitude above which adequate signal-to-noise ratio cannot be obtained within a reasonable observing time. These are highly subjective quantities and what is adequate and reasonable depends upon the ultimate use of the data.

The discussion in Section 11.2.7 lead to Eq. (11.2.69) in which the signal-to-noise ratio was defined as

$$R(k_0) = \frac{1}{2} \left(\frac{T}{\tau} \right)^{1/2} \frac{S'(k_0)\delta\nu}{\mathrm{NEP}(k_0,\nu_0)(\delta\nu)^{1/2}}. \tag{11.3.1}$$

The power at wave number k_0 incident on the detector in the modulation frequency interval $\delta\nu$ is $S'(k_0)\delta\nu$. The modulation frequency ν is related to the frequency of the incoming radiation f by

$$\nu = 2k\nu = 2f\nu/c = \nu_0 f/k_0 c, \tag{11.3.2}$$

where ν_0 is the modulation frequency for the wave number k_0. If $P(k_0)\delta f$ is the power per unit area incident on the telescope in the frequency range df, then

$$S'(k_0)\delta\nu = \varepsilon AP(k_0)(df/d\nu)\delta\nu = \varepsilon AP(k_0)(k_0 c/\nu_0)\delta\nu, \tag{11.3.3}$$

where A is the collecting area of the telescope and ε is an efficiency factor which allows for losses in the optical system between the telescope and the detector. The signal-to-noise ratio now becomes

$$R(k_0) = \frac{1}{2} \left(\frac{T}{\tau} \right)^{1/2} \frac{\varepsilon AP(k_0)(k_0 c/\nu_0)\delta\nu}{\mathrm{NEP}(k_0,\nu_0)(\delta\nu)^{1/2}}. \tag{11.3.4}$$

The desired resolution element δk fixes $\delta\nu$, i.e., $\delta\nu = (\nu_0/k_0)\delta k$ and finally

$$R(k_0) = [\varepsilon(1/2)(Tk_0/\tau\nu_0)\varepsilon AP(k_0)(\delta k)^{1/2}]/\mathrm{NEP}(k_0,\nu_0). \tag{11.3.5}$$

An adequate signal-to-noise ratio might be chosen to be 100 while a convenient observing time might be 1 hr. The following are typical parameters for an experiment in the 1–4-μm wavelength range where lead sulfide is a good choice of detector material:

telescope diameter $= 150$ cm $= 60$ in.,

$\mathrm{NEP} = 10^{-14}$ W Hz$^{-1/2}$, $\quad \tau = 1$ sec, $\quad T = 1$ hr, $\quad \delta k = 1$ cm^{-1},

$R(k_0) = 100$, $\quad \varepsilon \approx 0.10$, $\quad k_0 = 10^4$ cm^{-1}, $\quad \nu_0 = 10^4$ Hz.

With these values, $P(k_0) \approx 10^{-28}$ W cm^{-2} Hz^{-1}. This compares with 6.3×10^{-28} W cm^{-2} Hz^{-1} which is the standard assumed for a zero-magnitude star in the K band. Thus it should be possible to observe a first or second magnitude star with no difficulty.

The literature survey and preparation of this article was completed in September 1971.

ACKNOWLEDGMENTS

The authors wish especially to thank Dr. Harold L. Johnson for his countless discussions on the practice of astronomical Fourier spectroscopy. We also appreciate the help and advice given us so freely by Dr. James Pritchard. One of us (R.I.T.) would also like to thank Dr. L. Mertz for the initial opportunity to enter the field of Fourier spectroscopy.

Appendix A

Table of Fourier Transforms

Function	Fourier transform
$f(x)$	$f'(k)$

Rectangle

$$R(x) = \begin{cases} 1, & |x| < 1 \\ 0, & |x| > 1 \end{cases}$$

$$R'(k) = \frac{\sin(2\pi k)}{\pi k}$$

Slope up

$$SU(x) = \begin{cases} (x+1)/2, & |x| < 1 \\ 0, & |x| > 1 \end{cases}$$

$$SU'(k) = \frac{\sin(2\pi k)}{2\pi k} + \frac{1}{2} i \left[\frac{\cos(2\pi k)}{\pi k} - \frac{\sin(2\pi k)}{2\pi^2 k^2} \right]$$

Slope down

$$SD(x) = \begin{cases} (1-x)/2, & |x| < 1 \\ 0, & |x| > 1 \end{cases}$$

$$SD'(k) = \frac{\sin(2\pi k)}{2\pi k} + \frac{1}{2} i \left[\frac{\sin(2\pi k)}{2\pi^2 k^2} - \frac{\cos(2\pi k)}{\pi k} \right]$$

Triangle

$$T(x) = \begin{cases} 1-x, & |x| < 1 \\ 0, & |x| > 1 \end{cases}$$

$$T'(k) = \frac{\sin^2(\pi k)}{\pi^2 k^2}$$

Dirac delta function $\delta(x)$ $\delta'(k) = 1$

Dirac comb $III(x)$ $III'(k) = III(k)$

Gaussian $G(x) = e^{-\pi x^2}$ $G'(k) = e^{-\pi k^2}$

Lorentzian $L(x) = (1 + x^2)^{-1}$ $L'(k) = e^{-(2\pi k)}$

12. FABRY–PEROT INSTRUMENTS FOR ASTRONOMY*

12.0. Introduction

The object in writing this chapter is to set down, without great detail, the basic considerations which are relevant to the use and design of Fabry–Perot spectrometers for problems in astronomy. This article is intended as a beginning, and not as a review. The underlying thought has been that if the broad basic principles are well understood and put in proper perspective, detailed solutions can be worked out readily by most experimentalists. Often the most crucial part of instrument design is isolating the important things about which to think. Hopefully the references, which are by no means exhaustive, will guide the interested reader to detailed answers to many specific questions. In particular, there are several good reviews treating at least in part Fabry–Perot instruments for astronomical applications,[1-3] as well as treatments of the general field of Fabry–Perot spectroscopy,[4-6] with at least one attempt at a comprehensive formal treatment.[7]

[1] A. D. Code and W. C. Liller, in "Stars and Stellar Systems," Vol. 2, "Astronomical Techniques," p. 281. Chicago Univ. Press, Chicago, 1962.

[2] A. H. Vaughn, Ann. Rev. Astron. Astrophys. 5, 139 (1967).

[3] John Meaburn, Astrophys. Space Sci. 9, 206 (1970).

[4] P. Jacquinot, Rep. Progr. Phys. 23, 267 (1960).

[5] A. Girard and P. Jacquinot, in "Advanced Optical Techniques," p. 73. North Holland Publ., Amsterdam, 1967.

[6] Methodes Nouvelles de Spectroscopie Instrumental, 25–29 April 1966. Many Papers on Fabry–Perot Techniques in J. Phys. Suppl. C2 28 (1967).

[7] R. Chabbal, J. Rech. Centre Nat. Rech. Sci. Lab. Bellevue (Paris) No. 6, 91 (1948). English translation by R. B. Jacobi, UKAEA Res. Group, Harwell 1958/JMR Hx 4128.

---

* Part 12 is by **F. L. Roesler**.

12.1. The Ideal Fabry–Perot Interferometer

12.1.1. Basic Properties and Definitions

The ideal Fabry–Perot interferometer (or etalon) consists of a pair of transparent perfect optical flats having semitransparent mirrors assembled a small distance apart with the mirrored surfaces facing one another and accurately parallel. The transmission of light through the interferometer is given by the Airy function $\mathscr{A}$

$$\mathscr{A} = [1 + 4R(1 - R)^{-2} \sin^2(2\pi n l \sigma \cos \theta)]^{-1}, \qquad (12.1.1)$$

where R is the reflectance of the coatings, l the spacing between the plates, n the refractive index of the material (usually air) between the plates, θ the angle of incidence, and $\sigma = 1/\lambda$ the wave number of the incident light. The transmittance peaks of $\mathscr{A}$ occur whenever

$$2nl\sigma \cos \theta = m, \qquad (12.1.2)$$

where m is an integer called the *order number*. For fixed n, l, and θ, the transmitted light consists of a series of narrow peaks uniformly spaced in wave number by the wave number *free spectral range* Q of the interferometer. For small angles θ and $n = 1$,

$$Q = 1/2l \quad \text{cm}^{-1}. \qquad (12.1.3)$$

The wavelength free spectral range is $Q_\lambda = (\lambda^2/2l)10^{-8}$ Å for λ measured in angstroms and l measured in centimeters. The full width at half maximum of a peak, $\delta_{1/2}\sigma$, which may loosely be called the *resolving limit*, depends on the reflectance, and with sufficient accuracy for R greater than about 0.85 is given by

$$\delta_{1/2}\sigma = (1 - R)/(2l\pi R^{1/2}) = Q/N_R, \qquad (12.1.4)$$

where N_R is the reflective finesse given by

$$N_R = \pi R^{1/2}/(1 - R). \qquad (12.1.5)$$

Similarly $\delta_{1/2}\lambda = Q_\lambda/N_R$. The theoretical resolution is

$$\mathscr{R}_t = \sigma/\delta_{1/2}\sigma = \lambda/\delta_{1/2}\lambda = N_R\sigma/Q = N_R/\lambda Q_\lambda = 2N_R\sigma l. \qquad (12.1.6)$$

The contrast C of the interferometer is the ratio of the maximum transmittance (unity for an ideal interferometer) to the minimum transmittance, and likewise depends only on the reflectance:

$$C = (1 + R)^2/(1 - R)^2. \qquad (12.1.7)$$

12.1.2. The Monochromatic Ring System of the Fabry–Perot

When a uniform monochromatic field is viewed through a Fabry–Perot interferometer, the field is seen as a concentric narrow-ring pattern at infinity centered at normal incidence to the interferometer, with ring positions given by Eq. (12.1.2). For illumination with light of wave number σ, the angular diameters of the jth and kth rings from the center are related by

$$2(\cos \theta_j - \cos \theta_k) = (k - j)/l\sigma = 2(k - j)Q/\sigma \approx \theta_k^2 - \theta_j^2; \quad (12.1.8)$$

the difference of the squares of adjacent ring diameters is a constant. The angular width of a ring defined by its half-intensity points, is

$$\delta_{1/2}\theta = Q/N_R\sigma\theta. \qquad (12.1.9)$$

The solid angle subtended by such a ring is

$$\Omega = 2\pi\theta\,\delta_{1/2}\theta = 2\pi Q/N_R\sigma = 2\pi/\mathscr{R}_t; \qquad (12.1.10)$$

hence all rings of a given Fabry–Perot subtend equal solid angles and have equal spectroscopic resolution.

The transmitted wave number increases with increasing angle of incidence according to the expression

$$\sigma_m = \sigma_{om}/\cos \theta \approx \sigma_{om}(1 + \theta^2/2), \qquad (12.1.11)$$

where $\sigma_{om} = 1/\lambda_{om}$ is the normal incidence wave number in order m. The corresponding expression for wavelength is

$$\lambda_m = \lambda_{om} \cos \theta \approx \lambda_{om}(1 - \theta^2/2). \qquad (12.1.12)$$

Thus if a ring of σ_m' occurs at $\theta = \theta'$, the ring of $\sigma_m' + Q$ will occur at θ'' determined from $\theta''^2 - \theta'^2 = 2Q/\sigma_{om}$. This is the same as the separation between adjacent monochromatic rings determined from Eq. (12.1.8); hence the rings spectroscopically calibrate the field for the measurement of line widths or shifts smaller than a free spectral range. For a line shifted by an amount εQ, one has $\theta_k{}^2 - \theta_{k+\varepsilon}^2 = 2\varepsilon Q/\sigma$. From measurements of the radii of the rings ($r = f\theta$, where f is the camera focal length, not necessarily well known)

$$\varepsilon = (r_{k+\varepsilon}^2 - r_k{}^2)/(r_{k+1}^2 - r_k{}^2). \tag{12.1.13}$$

12.2. Application of the Fabry–Perot Interferometer as a Spectrometer

The basic elements of a Fabry–Perot spectrometer are shown in Fig. 1. Light from the source A or its image generally passes first through a premonochromator B which restricts the spectral extent of light incident

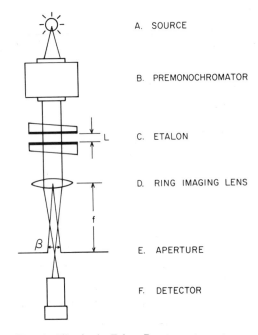

FIG. 1. The basic Fabry–Perot spectrometer.

on the Fabry–Perot C. The ring-imaging lens D focuses the ring pattern in the plane of the exit aperture E. Light passing through the aperture is detected by the detector F. The fundamental design details of this type of Fabry–Perot spectrometer are discussed in the following sections.

12.2.1. Premonochromatization Requirement

Because at any angle θ the Fabry–Perot passes a multiplicity of spectral elements $\sigma_m = m/2nl \cos \theta$, the spectral extent of the light incident on the Fabry–Perot must be restricted to a range less than Q. When the nature of the source is such that only one or a few lines are present, this restriction is naturally satisfied. In general a premonochromator (element B, Fig. 1) must be used along with the Fabry–Perot to isolate a line or a spectral region less than Q. This premonochromator may be one or a combination of dye filters, an interference filter, a grating or prism monochromator, or one or more additional Fabry–Perot interferometers. Some details of methods of premonochromatization using Fabry–Perot interferometers are discussed below (Section 12.3). Examples of instruments employing other methods of premonochromatization are found elsewhere.[8,9] In the material immediately following the light is assumed sufficiently monochromatic.

12.2.2. Isolation of a Spectral Element

A monochromator isolating a single spectral element, that is, a spectral region of width $\delta\sigma = \sigma/\mathscr{R}_t$, can be achieved by placing an annular aperture (element E, Fig. 1) concentric with the Fabry–Perot ring images in the focal plane of a lens (element D, Fig. 1) placed after the Fabry–Perot. In the usual axial-fringe instrument this annulus is the one having zero internal radius, i.e., a hole on axis.

The dimension of an annulus for isolating a spectral element is found following Eq. (12.1.9): $\theta\delta_{1/2}\theta = Q/N_R\sigma = \delta_{1/2}\sigma/\sigma = 1/\mathscr{R}_t$. If the focal length of the lens forming the Fabry–Perot ring images is f, then the dimensions of the annulus must be related by $r\delta_{1/2}r = f^2/\mathscr{R}_t$. More exactly $r_2^2 - r_1^2 = 2f^2/\mathscr{R}_t$; hence any annulus used with appropriate f is suitable to isolate a spectral element $\delta_{1/2}\sigma = \sigma/\mathscr{R}_t$. In the case of the axial fringe spectrometer, the angular diameter $\beta = 2\delta_{1/2}\theta$ of the hole is related to the theoretical resolution $\mathscr{R}_t$ through the expression

$$\beta = (8/\mathscr{R}_t)^{1/2}. \qquad (12.2.1)$$

When an annulus of the type just described is used to isolate a spectral element, the band-pass of the instrument is increased by the combined

[8] A. A. Wyller and T. Fay, *Appl. Opt.* **11**, 1152 (1972).

[9] James Kaplan, *Appl. Opt.* **11**, 1978 (1972).

effects of the Airy function and the annulus. The band-pass of the annulus is a rectangular function F [since equal spectral intervals correspond to equal solid angles, Eq. (12.1.10), for example] of width $\delta_{1/2}\sigma$, and the resulting bandpass is the convolution

$$W(\sigma_0) = \int_{\sigma_0 - \delta_{1/2}\sigma}^{\sigma_0 + \delta_{1/2}\sigma} \mathscr{A}(\sigma - \sigma_0)\, d\sigma = F * \mathscr{A}.$$

This reduces the resolution to $R = 0.7R_t$. Moreover, the transmission is not constant across the annulus, resulting in a peak transmittance of $W(\sigma_0)$ of $\pi/4 \approx 0.79$.

12.2.3. Luminosity and Luminosity-Resolution Product of Fabry–Perot Spectrometers

The luminosity[5,10] L of a spectroscopic instrument is the ratio of the flux $\Phi(\sigma)$ arriving at the detector to the monochromatic luminance $B(\sigma)$ of the source for the spectral element considered: $L = \Phi(\sigma)/B(\sigma)$. However, $\Phi(\sigma) = \tau B(\sigma)S\Omega$, where S is the area of the dispersing element of the spectrometer, Ω is the solid angle of light on the dispersing element, and τ is a transmittance factor which is the lumped effect of the transmittance of the various optical elements and the shape of the instrumental profile; hence $L = \tau S\Omega$. The quantity $S\Omega$ is the etendue U of the instrument

$$U = S\Omega. \tag{12.2.2}$$

Hence

$$L = \tau S\Omega = \tau U. \tag{12.2.3}$$

At a specified theoretical resolution $\mathscr{R}_t$, the solid angle of a resolution element of a Fabry–Perot is given by Eqn. (12.1.10). Hence

$$U_{FP} = 2\pi S/\mathscr{R}_t, \tag{12.2.4}$$

$$L_{FP} = 2\pi\tau S/\mathscr{R}_t. \tag{12.2.5}$$

Equation (12.2.5) illustrates the constancy of the luminosity-resolution product: $L\mathscr{R}_t = 2\pi\tau S$. When the effects of the annulus discussed in Chapter 12.2 are included in Eq. (12.2.5), one obtains $L_{FP} = \tau U_{FP} = 3.5S/\mathscr{R}$, where τ is now the transmittance of the optical elements and $\mathscr{R}$ the resolution achieved.

[10] P. Jacquinot, *J. Opt. Soc. Amer.* **44**, 761 (1954).

Jacquinot[10] has compared the luminosity-resolution products of prism, grating, and Fabry–Perot spectrometers, showing that gratings are much superior to prisms, and that Fabry–Perots are 70–350 times superior to gratings having the same area. The gain arises because at a given resolution the Fabry–Perot spectrometer accepts light from a much larger solid angle than a grating spectrometer. This high luminosity-resolution product of Fabry–Perot spectrometers (a feature it shares with the Michelson interferometer as used in Fourier transform spectroscopy[†]) is largely responsible for its importance, allowing in many cases large reductions in observing time needed to obtain adequate signal-to-noise ratios (see Section 12.7).

12.2.4. Scanning Fabry–Perot Spectrometers

A scanning spectrometer is achieved by providing means for changing the wavelength of the light passing through the annulus and means for measuring and recording the light intensity as a function of wavelength, usually a photomultiplier (element F, Fig. 1) and associated electronics. The variables which may be altered to effect the scan are [Eq. (12.1.1)] n, l, and θ.

One of the easiest and most precise methods of scanning is that of *pressure scanning* in which l and θ are held fixed and the refractive index n of the gas between the plates is changed by changing the gas pressure in a windowed gas-tight chamber containing the interferometer. The scanning range that can be achieved by this method is $\Delta\sigma = -\sigma\,\Delta n/n \approx \sigma\,\Delta n$ or $\Delta\lambda = +\lambda\,\Delta n/n \approx \lambda\,\Delta n$, where Δn is the maximum change in index. For air at standard conditions $n = 1.00029$ in the visible (the exact value is wavelength-dependent), and the scanning range per atmosphere is about $\Delta\sigma = 5.8$ cm^{-1} or $\Delta\lambda = 1.45$ Å at $\sigma = 20{,}000$ cm^{-1} ($\lambda = 5000$ Å). The range may be increased conveniently by scanning between vacuum and 2 atm. By using SF_6, for which the STP refractive index is roughly 1.00078 the scanning range in the visible may be extended to 15.6 cm^{-1} or 3.9 Å/atm scan. To the extent that the gases are ideal, the scan is linear in pressure, provided that the temperature of the gas does not change during the scan. The extent of the scan may exceed the free spectral range of the system, in which case, with increasing pressure, the spectrum scanned first in order m is repeated in order $m + 1$, and so on. A feature of wave number σ_m in order $m + k$

[†] See Part II of this volume.

falls at the position of a feature at $\sigma_m + kQ$; hence the repetition of the structure serves to calibrate the spectrum. Assuming the scan is linear the relative position of two features on the spectrum may be found by linear interpolation.

When the plates are mounted with a provision for carefully changing the plate spacing l in a uniform manner, *mechanical scanning* is achieved by virtue of the wavelength dependence on l at fixed n and θ. The scan is related to the change δl in l through the equation $\Delta\sigma = -\sigma \, \delta l/l$, $\Delta\lambda = +\lambda \, \delta l/l$. Whereas for pressure scanning the total spectral range scanned is independent of l, for mechanical scanning the range increases with decreasing l; an entire free spectral range is scanned whenever l changes by $\lambda/2$.

Angular scanning, that is, scanning by virtue of a change in the angle θ, can also be employed to scan Fabry–Perot spectrometers. When the angle θ is changed simply by tilting the etalon and accepting the light coming through a small hole of fixed angular diameter the bandwidth increases with increasing θ. This method is practical only when the resolution $\mathscr{R}_t$ is so low that the spread of incidence angles that can be accepted without loss of resolution is much larger than can be fed into the interferometer by a practical coupling device, or when the source is so bright that (or is by its nature such that) its angular spread can be restricted to less than a ring width at the angle of maximum tilt. This is generally the case for interference filters, for example, which may be conveniently tilt tuned; however, even for interference filters the restriction must be understood.

A nested set of annuli with radii $r_0, r_1, r_2, \ldots$ such that $2f/\mathscr{R}_t = r_1^2 = r_2^2 - r_1^2 = r_3^2 - r_2^2 \cdots$ (Section 12.2.3), avoids the problem of changing resolution and maintains optimum efficiency of the spectrometer. In this case the magnification of the source image must be adjusted to fill the largest diameter ring. For extended sources this is trivial, but for small sources may be difficult. The annuli may be stepped sequentially onto the axis of the fringe pattern,[11] or may be provided with means for sending light from each annular area onto a separate photomultiplier.[12,13] This method is useful for solid-gap etalons which may not be scanned conveniently by other means.

[11] G. G. Shepherd, C. W. Lake, J. R. Miller, and L. L. Cogger. *Appl. Opt.* **4**, 267 (1965).

[12] J. G. Hirschberg and P. Platz. *Appl. Opt.* **4**, 1375 (1965).

[13] Mark Daehler, *Appl. Opt.* **9**, 2529 (1970).

12.2.5. Limitations Imposed by Surface Nonflatness

The discussion to this point has been concerned with a perfect Fabry–Perot interferometer in which the fringe shape is determined only by the combined effects of the reflective finesse and the angular diameter of the entrance aperture (Section 12.2.2). In practice the interferometer mirrors will have irregularities due to nonflatness which will broaden a fringe and reduce its peak height. The effect is easily analyzed in terms of a mosaic of microetalons making up the total plate area.[7] Each microetalon has the spacing equal to the gap of the imperfect interferometer at its particular location. The effect on the transmitted wave number of the departure from a mean gap l_0 by an amount $\delta l = (l - l_0)$ follows from Eq. (12.1.2) and is $\delta\sigma = -\sigma\,\delta l/l_0$. Roughly, if a pair of interferometer plates is characterized as having a range of defects δl, the limit of resolution has a defect limit given by $(\delta\sigma)_D = -\sigma\,\delta l/l_0 = 2\sigma\,\delta l$. By analogy with Eq. (12.1.4) a defect finesse N_D may be defined such that $(\delta\sigma)_D = Q/N_D$, where $N_D = 1/2\sigma\,\delta l$. Since plate flatness tolerances are typically quoted as fractions $(1/k)$ of a wavelength, that is $\delta l = \lambda/k = 1/k\sigma$, one has the convenient approximation $N_D = k/2$. For example, for plates flat to $\lambda/100$, $N_D = 50$.

The exact shape of the passband will of course depend upon the exact distribution of defects. If the area dS of the plates having a particular spacing l is given by a distribution function $D'(l - l_0)$ such that $dS = D'(l - l_0)\,dl$, the corresponding expression for wave number is, using Eq. (12.1.2), $dS = (l/\sigma)D'[(l/2\sigma)(\sigma - \sigma_0)]\,d\sigma = D(\sigma)\,d\sigma$. The resultant effect on the transmittance function is the convolution product $\mathscr{A} * D = \int \mathscr{A}(\sigma - \sigma_0)D(\sigma)\,d\sigma$. The resultant instrumental function is found by convoluting this with the rectangular function describing the isolation aperture (Section 12.2.2). Thus $W = \mathscr{A} * F * D$.

If the reflective finesse N_R [Eq. (12.1.5)] of a pair of imperfect plates is greater than the defect finesse, then monochromatic light will be transmitted by only a fraction of the total area, resulting in a loss of etendue. If $N_R = N_D$, then the transmittance at each point of the etalon will be one half or more of its maximum possible value. Thus the reflectance of a pair of plates should be matched to the defects by choosing $N_R \leq N_D$ to minimize loss of efficiency due to plate defects.

It follows also that an interferometer must be adjusted for parallelism to within the same tolerance, i.e., $\delta l < Q/N_R$; however, the methods for adjusting plates (Section 12.6) easily achieve this accuracy naturally, although some judgement is required for achieving optimum adjustment for nonflat plates.

12.2.6. Instrumental Finesse

It is generally convenient to speak of an instrumental finesse N, which includes the effects of the finite entrance aperture and plate defects. From the discussion following Eq. (12.2.1) it follows that due to the finite entrance aperture $N = 0.7N_R$. This values may be achieved when defects can be neglected. If defects are such that $N_D = N_R$, then $N \approx 0.6N_R$. A convenient approximation is $\delta\sigma_W{}^2 = \delta\sigma_{\mathscr{A}}{}^2 + \delta\sigma_F{}^2 + \delta\sigma_D{}^2$, where $\delta\sigma_W$, $\delta\sigma_{\mathscr{A}}$, $\delta\sigma_F$, and $\delta\sigma_D$ are the respective full half widths of the instrument, Airy function, aperture function, and defect function. Introducing an aperture-limited finesse $N_F = Q/\delta\sigma_F$ we can write $1/N^2 = 1/N_R{}^2 + 1/N_F{}^2 + 1/N_D{}^2$, approximately.

It is difficult to do better than approximate the instrumental finesse or the instrumental profile from the measured properties of the system, and whenever possible it is best to measure them directly for the instrument as finally used, and to use the measured final profile for decomposition of the recorded spectra.

12.3. Multiple Fabry–Perot Spectrometers

The application of a single Fabry–Perot interferometer as a spectrometer is restricted by the free spectral range Q of the interferometer. If the finesse of the interferometer is N, then the single interferometer can study at most N spectral elements of width Q/N. In the case of an emission source having a number M of elements less than about $N/2$, this may be sufficient. However, in general, the free spectral range must be extended to permit the unambiguous study of a spectrum.

The free spectral range may be extended without loss in etendue by putting in series two or more Fabry–Perot interferometers with different spacers.[7,14] We consider in this section the properties of multiple etalon spectrometers.

12.3.1. Method of Extending the Free Spectral Range

An etalon of spacing l_1 having $Q_1 = 1/2l_1$ and tuned to transmit wave number σ_0 at normal incidence will also transmit $\sigma_1 = \sigma_0 \pm k_1Q_1$ [Eq. (12.1.2)]. Similarly a second etalon of spacing l_2 having $Q_2 = 1/2l_2$

tuned to transmit σ_0, also transmits $\sigma_{k_2} = \sigma_0 \pm k_2 Q_2$. If $l_1 \neq l_2$, then in general $\sigma_{k_1} \neq \sigma_{k_2}$ except when $k_1 = k_2 = 0$. Thus the second (suppression) etalon blocks some of the undesired or parasitic orders of the first, giving an extension of the free spectral range as shown in Fig. 2.

The figure shows the extension of the free spectral range of the resolving etalon (the one with the largest spacer) by a factor of five for three different choices of suppression etalon spacers. Spacer ratios may be grouped roughly into three categories. Lines (a) and (b) for which

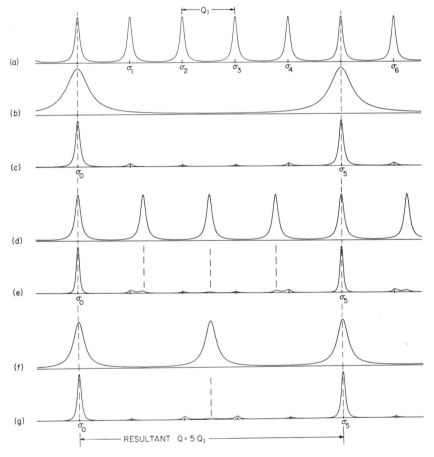

FIG. 2. Extension of free spectral range with a series of two Fabry–Perot etalons. The free spectral range of the etalon having spacer l_1 is extended by a factor of five by the addition of a second etalon in series. Three different choices of the spacer l_2 are shown. A very low finesse value was used to emphasize the ghosts for illustration. (a) l_1; (b) $l_2 = l_1/5$; (c) spacers (a) and (b); (d) $l_2 = 4l_1/5$; (e) spacers (a) and (d); (f) $l_2 = (2/5)l_1$; (g) spacers (a) and (f).

$l_1/l_2 = 5$ illustrate the group of *multiple ratios* in which the spacers have quite different lengths, that is, the ratio $l_1/l_2 > 2$. Lines (a) and (d) for which $l_1/l_2 = \frac{5}{4}$ represent the group of *vernier ratios* in which the spacers have about the same length, that is, $l_1/l_2 < 2$. Lines (a) and (f) for which $l_1/l_2 = \frac{5}{2}$ illustrate the group of half-multiple ratios in which $l_1/l_2 = n + \frac{1}{2}$, where n is an integer. This grouping is to some extent arbitrary, but it serves to illustrate the general consideration.

The essential object is the cancellation of parasitic peaks; however, this can be done only approximately because of the finite contrast of the etalon. Consequently there appear a number of ghosts in the resulting transmittance function. The magnitude of a ghost relative to the tuned peak may be estimated roughly by calculating the ordinate of the trans-mittance function of the suppression etalon at the position of the peak to be suppressed. This can be written as $G = (1 + 4k^2)^{-1}$, where k is the distance to the suppressed peak from the nearest suppression peak in units of the suppression etalon half-width Q/N. [This follows readily from Eq. (12.1.1) by approximating the sine of the phase by the phase, less an appropriate multiple of π, in the vicinity of a peak, and rewriting the variable in units of the half width, Eq. (12.1.4).] One has always $k \leq N/2$. If we desire the strongest ghost to be no more than 0.01, then $k \geq 5$. Thus with a two-etalon system we may cancel $(N/5 - 1)$ peaks to less than 1% of the tuned peak and achieve an extension of the free spectral range by a factor $N/5$. If we are willing to tolerate ghosts of about 0.06, then $k \geq 2$ and the free spectral range may be extended by a factor $N/2$.

It is apparent for the example in the figure that for each of the three groups of ratio pairs the ghost strengths are approximately the same, i.e., suppressed peaks fall at multiples of $Q_2/5$ from the suppression etalon peaks. The positions of the various strength ghosts may, however, fall in different places. This is an example of a general rule that any p-fold extension of the free-spectral range will result in the same set of ghosts with k-values which are multiples of N/p but possibly in different loca-tions.

The principle may be extended to three or more Fabry–Perot inter-ferometers in series. The three-etalon case has been treated in several papers describing the PEPSIOS spectrometer[14-16] and will not be treated

[14] J. E. Mack, D. P. McNutt, F. L. Roesler, and R. Chabbal, *Appl. Opt.* **2**, 873 (1963).
[15] F. L. Roesler and J. E. Mack, *J. Phys. Suppl.* C2 **28**, 313 (1967).
[16] D. P. McNutt, *J. Opt. Soc. Amer.* **55**, 288 (1965).

in detail here. In general, for each etalon in the series, the free spectral range may be extended by a factor $N/2$.

12.3.2. Method of Coupling Etalons in Series

Fabry–Perot etalons having the same useful aperture diameter may be stacked with their axes parallel in a simple series without any intermediate optics. The etendue is limited by the etalon having the largest spacer. There is no restriction on the spacers which may be used. Unless the resolution of the system is very low, it is generally possible to stack the etalons close enough that vignetting losses due to the angular spread β of the light through the series can be neglected.

In some cases it is desirable to couple the etalons with magnification between the etalons, as shown in Fig. 3, in order that plates of different diameter may be used without loss of etendue. This may be necessary for reasons of economy, for example, where the purchase of more than one very large diameter interferometer may be prohibitively expensive. The coupling may be accomplished with an afocal lens system, that is, a lens system which expands a beam of collimated light. Such a system consists of a combination of lenses spaced so that the second focal point of the first coincides with the first focal point of the second.

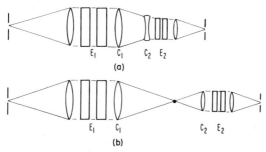

FIG. 3. Two methods of series-coupling etalons of different diameters: (a) positive–negative coupling; (b) positive–positive coupling.

When the coupling lenses C_1 and C_2 are a positive–negative combination as shown in Fig. 3a, the coupling is accomplished with a minimum spacing between E_1 and E_2. However, this system increases the vignetting problems, and losses may become important at low resolutions.

When both coupling lenses are positive, as shown in Fig. 3b, the combination may be spaced so that the system images E_1 on E_2 to minimize vignetting. Alternatively a field lens may be added at the common focal point to accomplish the imaging of E_1 on E_2 with minimized interetalon spacing.

In either case the beam diameters D_1 and D_2 and angular spreads β_1 and β_2 are related to the focal lengths f_1 and f_2 of the coupling lenses by the equation $f_1/f_2 = D_1/D_2 = \beta_2/\beta_1$. This also places a restriction on the spacing ratio of the coupled etalons, since the spacing is proportional to the resolution which is in turn inversely proportional to β^2 [Eq. (12.2.1)]. Consequently, for etalons of the same finesse,

$$\mathscr{R}_1/\mathscr{R}_2 = l_1/l_2 = \beta_2{}^2/\beta_1{}^2 = D_1{}^2/D_2{}^2. \qquad (12.3.1)$$

For example, if it is desired to couple an etalon having a diameter of 150 mm to an etalon having a diameter of 50 mm without loss of etendue, the focal length ratio of the coupling lenses must be 3:1 and the spacer ratios must be 9:1 or greater. Obviously it is not practical to use vernier ratios in this case.

12.3.3. Comparison of Single- and Multiple-Etalon Properties

The transmittance $\mathscr{A}'(\sigma - \sigma_0)$ of a simple series of etalons tuned to pass wave number σ_0 and having spacings $l_1, l_2, l_3, \ldots, l_m$ is the product of the transmittances of the individual etalons

$$\mathscr{A}'(\sigma - \sigma_0) = \mathscr{A}_1(\sigma - \sigma_0) \cdot \mathscr{A}_2(\sigma - \sigma_0) \cdot \cdots \cdot \mathscr{A}_m(\sigma - \sigma_0), \quad (12.3.2)$$

where $\mathscr{A}_j(\sigma - \sigma_0)$ is the transmittance of an etalon [Eq. (12.1.1)] having spacing l_j. (Reflections between etalons have been neglected. The case with reflections has been treated in the literature.[14] In practice they may be avoided by putting a slight amount of absorption between the etalons or by tilting their axes slightly with respect to each other.) For a series with interetalon magnification this equation holds only at normal incidence. For the case with magnification the incidence angle θ_j on the jth etalon must be multiplied by the magnification between etalons j and $j - 1$.

We have already discussed in Section 12.3.1 the free spectral range of a multiple-etalon system, indicating that with suitable choice of spacers it may be extended by a large factor.

The resolution of a multiple etalon is approximately the resolution of the widest spaced etalon in series; however, because of the multiplication of passband ordinates, the resolution is increased by a factor which depends on the spacer ratios and the number of etalons in the series.[14,17]

For two etalons of equal spacing, the resolution is increased by a factor of 1.5. For three equally spaced etalons the resolution is increased by a factor of two, very nearly.

The contrast of the multiple etalon increases very greatly. For a single etalon C is typically 10^2. For double and triple etalons contrast values of 10^4 and 10^6, respectively, have been readily achieved.[17]

The shape of the instrumental profile for a multiple etalon given by Eq. (12.3.1) is appreciably improved over that of a single etalon in the vernier spacer ratio case. The single-etalon function has a very broad wing characteristic of the resonance function which it approximates. This wing is greatly suppressed in the multiple-etalon case,[15,17] increasing the purity of spectral isolation.

Since the angular variation of the transmitted wave number for any etalon [Eq. (12.1.11)] does not depend on the spacer, the wave number dependence on angle for a multiple etalon is the same as for a single etalon when the etalons are coupled without magnification between etalons. Consequently apertures for isolating a spectral element are chosen in the same way for a simple series polyetalon as for a single etalon. Furthermore the etendue for a multiple etalon at a given resolution is the same as for a single etalon acting at the resolution of the multiple etalon. Adjustment must of course be made for the increased resolution of the polyetalon in the vernier ratio case.

When etalons are coupled with magnification, the variation in σ across the intermediate aperture position is different for the two etalons because the angle subtended at each etalon by a point in the intermediate aperture is different for each. Consequently only a single annulus—a hole in the central fringe case—can be used. Since the fringe image width is the same for both etalons under the coupling condition of Eq. (12.3.1), the transmission of the combination decreases faster with angle off the fringe peak than in the case of a single etalon. Thus strict adherance to Eq. (12.3.1) results in a loss in etendue (about 0.67) with only a negligible increase in resolution occasioned by the effective reduction in entrance aperture diameter. Consequently the $L \cdot \mathscr{R}$ product is reduced by a factor about 0.7 due to this effect. While the trimming of fringe image width occurs also for the simple coupling case, it is accompanied by an increase in resolution which matches the fringe image width reduction, and leaves the $L \cdot \mathscr{R}$ product nearly constant. The loss in $L \cdot \mathscr{R}$ product for coupling with magnification can be corrected by decreasing the spacing

[17] M. Daehler and F. L. Roesler, *Appl. Opt.* **7**, 1240 (1968).

of the small etalon by a factor about 0.7. One might also reduce the reflective finesse of the small etalon by a factor of 0.7, but caution must be exercised to avoid introducing excessive ghost intensities.

The scanning of multiple Fabry–Perot spectrometers is the same as for single etalons in the case of pressure and, for a simple series, angular scanning. The law for pressure scanning, $\Delta\sigma = \sigma \Delta n$ is independent of spacer, hence all etalons tuned to pass a common peak will scan the same way. The techniques of angular scanning, which depend upon the angular variation in transmittance, apply to the simple series polyetalon as well as to the single etalon since, as noted above, the angular transmittance variation is independent of spacing.[13] It is assumed the etalons are tuned to pass the same wavelength at normal incidence. The technique of tuning will be considered later (Chapter 12.6). The angular scanning technique is not applicable in the case of etalons coupled with magnification because the angles are changed by the coupling.

In the case of mechanical scanning the scan for a change δl in spacing is inversely proportional to the spacing l. Consequently the mechanical displacements for the same spectral scan $\Delta\sigma$ of coupled etalons, regardless of the coupling method, must be in the ratio $\Delta l_1/\Delta l_2 = l_2/l_1 = Q_1/Q_2$. Any technique of mechanical scanning multiple etalons must therefore provide separate adjustable scanning mechanisms for the etalons, preferably with some optical method of verifying the synchronism of the scan.[18]

Plate defect tolerances are somewhat tighter for a series of etalons as compared to an etalon used singly.[19] The effect is similar to the fringe width narrowing effect which increases the resolution of the polyetalon instrument. The permissible defect range is smaller by a factor of about 0.6 for two etalons, and about 0.5 for three etalons.

12.3.4. Required Number of Etalons

The object of this section is to provide guidelines for selecting the number of Fabry–Perot etalons needed to achieve an integral spectrometer, that is, a spectrometer capable of isolating a single spectral element. A spectral element is a band of spectral width $\delta\sigma = \sigma/\mathcal{R}$, where $\mathcal{R}$ is the required instrumental resolution. The number M of spectral elements in a source may be small in the case of a source consisting of a few emission lines. For continuous sources the number of elements in the visible is

[18] J. V. Ramsay, *Appl. Opt.* **5**, 1291 (1966).
[19] F. L. Roesler, *Appl. Opt.* **8**, 829 (1969).

roughly $\mathscr{R}/2$, where the factor $\frac{1}{2}$ arises because the visible range is approximately 10,000 cm^{-1} centered at $\sigma = 20,000$ cm^{-1}.

In general, the Fabry–Perot interferometer of finesse N is capable of isolating N elements; however, to assure a high purity of isolation as discussed in Section 12.3.1, the number isolated is reduced to approximately $N/2$.

In an all-interference spectrometer the number of spectral elements reaching the Fabry–Perot system is first reduced by an interference filter. The interference filter may be characterized by a resolution $\mathscr{R}_F = \lambda/\Delta\lambda$,

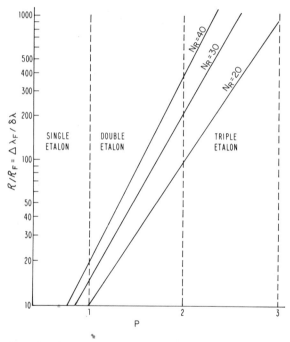

FIG. 4. Number of etalons of equal finesse, for various values of the finesse, needed in series with an interference filter of resolution $\mathscr{R}_F$ to achieve a resolution $\mathscr{R}$. Immediately to the left of the indicated boundaries the strongest ghosts from suppressed peaks are about 0.06 times the tuned peak. When greater purity of isolation is required an additional etalon or, if quality permits, a higher finesse should be used.

where $\Delta\lambda$ is the filter bandwidth. The filter reduces the number of elements to $(\mathscr{R}/2)/(\mathscr{R}_F/2) = \mathscr{R}/\mathscr{R}_F$ in the case of a continuous source. In the case of sources having only a few emission lines an interference filter generally is sufficient to isolate a single line since $\mathscr{R}_F$ is typically greater than 100. In this case the single line isolated by the interference filter may be studied with a single Fabry–Perot spectrometer at nearly any desired resolution.

Since $\mathscr{R}/\mathscr{R}_\mathrm{F}$ may be much larger than $N/2$ it is frequently necessary to use two or more etalons in series. To achieve isolation of a single element requires a number p of etalons such that $(N/2)^p \geq \mathscr{R}/\mathscr{R}_\mathrm{F} = \Delta\lambda/\delta\lambda$, where $\Delta\lambda$ is the filter width and $\delta\lambda$ is the required limit of resolution. Figure 4 shows $(\mathscr{R}/\mathscr{R}_\mathrm{F})$ vs. p for various values of the finesse, and indicates regions where one, two, or three etalons are required. For example, with $\mathscr{R}_\mathrm{F} = 300$ (20 Å at 6000 Å), a single etalon with $N_R = 30$ is sufficient for problems requiring a resolution less than about 4500. Using the same $\mathscr{R}_\mathrm{F}$ and N_R two etalons are sufficient for resolutions up to about 60,000, and three etalons are required for higher resolution up to about 10^6. Figure 4 is, of course, a guide only, and other requirements of the spectrometer or the peculiarities of the source characteristics may dictate an instrumental design different from that predicted by the figure.

12.4. Observation of Astronomical Sources with Fabry–Perot Spectrometers

12.4.1. Relevant Characteristics of Astronomical Sources

There are basically two source characteristics to be considered in the design of Fabry–Perot spectrometers for astronomical observations: angular extent and spectral character. The source as seen by the observer subtends a solid angle ω which may vary from nearly zero to nearly 2π depending on the source to be observed. Stars are point sources, broadened in practical cases by atmospheric seeing for a ground-based observer, and by diffraction and guiding error limits for a space telescope. Planets, nebulae, and diffuse galactic emission fill the gap to the other extreme presented by the night sky which covers a complete hemisphere. The spectral character includes the source brightness, characteristic resolution, and line density. Except for the sun all sources are faint. Line density ranges from a few emission lines to continuous, and the characteristic resolution required to achieve the object of a study may range from a few hundred to a few hundreds of thousands.

The object of the following sections will be to outline the basic considerations prerequisite to designing instrumentation for the range of problems presented.

12.4.2. Geometrical Restrictions in Coupling Sources to Fabry–Perot Spectrometers.

The source–telescope system has an etendue [Eq. (12.2.2)] $U_T = A\omega$, where A is the area of the telescope. To accept all of the light from a perfect telescope, the spectrometer must have an equal or larger etendue. If there are optical imperfections (or guiding errors) the etendue U_S of the spectrometer must be at least $A\omega'$, where $\omega' > \omega$ is the effective solid angle of the source in the focal plane of the telescope. The matching condition for a perfect telescope is $U_S \geq U_T$ or

$$4.4S/\mathscr{R}A\omega \geq 1. \qquad (12.4.1)$$

The actual coupling is achieved generally by imaging the source on the entrance aperture of the spectrometer and the telescope objective on the interferometer plates. An additional lens may be necessary to match the f number of the telescope and spectrometer collimator.

If ψ is the angular diameter of the object (assumed circular), Eq. (12.4.1) can be written $D/d \leq 2.4/\psi\mathscr{R}^{1/2}$, where D is the telescope diameter and d is the interferometer diameter. Figure 5 shows the ratio D/d vs. ψ for various resolving powers. For a given object and resolution, any D/d below a line of constant $\mathscr{R}$ results in a telescope limited system. Above the line the system is spectrometer limited. For example, at $\mathscr{R} = 10^5$, the ratio D/d for stellar sources is typically about 700. Thus for even the largest telescope (510 cm) an interferometer less than 1 cm in diameter is sufficient to accept all the light from the telescope. Nevertheless it is advisable to use larger plates to minimize the effects of plate defects (discussed in Section 12.2.5) by selecting the best area, and to reduce the noise caused by wandering of the star image in the entrance aperture. The latter effect is a result of the inevitable transmission change with angle [a factor of two when β is chosen according to Eq. (12.2.1)], and can be reduced by confining the stellar image more nearly to the peak of the Airy function, thus reducing Ω in Eq. (12.2.2), and requiring a larger area of the interferometer to be used to preserve the match with the telescope. This also reduces the possibility of apparent wavelength shifts due to a noncentral mean image position [Eq. (12.1.11)]. While it is possible to change the imaging to avoid transmission changes and wavelength shifts associated with wandering of the star image on the entrance aperture by focusing the telescope objective on the entrance aperture and the star image on the interferometer plates, one must then face the equally troublesome effects caused by localized plate defects.

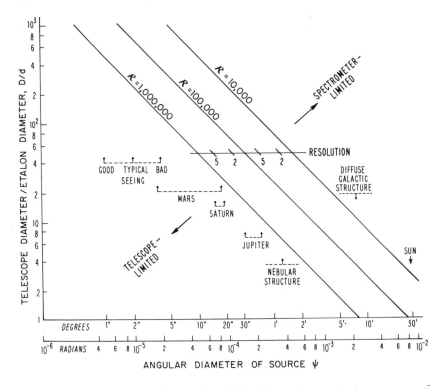

FIG. 5. Ratio of telescope diameter D to Fabry–Perot diameter d when a source of angular diameter ψ is studied with spectral resolution $\mathscr{R}$. Above lines of constant $\mathscr{R}$ the etendue is limited by the spectrometer, and below lines of constant $\mathscr{R}$ the etendue is limited by the telescope. The angular diameters of various sources are indicated. The lines apply to the case of negligible defects. When $N_D = N_R$, read values for a resolution 1.22 times the desired resolution (see Section 12.2.6).

For an object as large as Jupiter, $D/d = 35$ at $\mathscr{R} = 10^5$, requiring 17-cm interferometers for a proper match to a 510-cm telescope. However, at this resolution in the particular case of Jupiter, it would be necessary to use the doppler compensating adjustment[20] to cancel the effect of Jupiter's rapid rotation.

This discussion illustrates an aspect of interference spectroscopy that has not been fully exploited, although it has been frequently pointed out. It is that the telescope image quality tolerance may be considerably relaxed without loss of spectroscopic efficiency for the study of stellar objects. For example, a 5-m telescope operating at a resolution of 200,000

[20] J. T. Trauger and F. L. Roesler, *Appl. Opt.* **11**, 1964 (1972).

used with a 10-cm aperture interferometer at the Coudé focus would need to resolve only about 10^{-4} rad or 18 arcsec. The saving in telescope cost would more than pay for the large aperture interferometers. There would, of course, be limitations on the problems one might study, but there is a sufficient selection of suitable problems to realize a great saving in time and cost compared to using unnecessarily high-resolution telescopes.

12.4.3. Source Spectral Characteristics and Instrumental Resolution

The resolution of a Fabry–Perot spectrometer may be varied over a wide range according to Eq. (12.1.6) by varying the plate spacing l and, to a lesser extent by varying the reflective finesse N_R. The spectral characteristics of the source to be studied set a characteristic resolution $\mathscr{R}_c$ determined by the spectral width $\delta\sigma_c$ of the features to be studied: $\mathscr{R}_c = \sigma/\delta\sigma_c$. We consider below an idealized spectrometer observing an idealized spectrum with the object of establishing guidelines for determining the optimum instrumental resolution for a particular problem.

The total light flux reaching the detector when the instrument is tuned to σ_0 is proportional to

$$F(\sigma_0) = \iint B(\Omega, \sigma)W(\Omega, \sigma)\, d\sigma\, d\Omega.$$

The problem is idealized for simplicity as shown in Fig. 6 by assuming that the spectral feature to be detected has a rectangular profile of spectral width w corresponding to a characteristic resolution $\mathscr{R}_c = \sigma/w$, and that the passband profile of the spectrometer is likewise rectangular of width W corresponding to an instrument resolution $\mathscr{R} = \sigma/W$. The feature

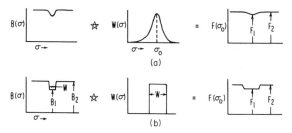

FIG. 6. The source function $B(\sigma)$ is scanned by the instrumental function $W(\sigma)$ to produce an output function $F(\sigma)$: (a) real case; and (b) the idealized case used in the analysis given in the text.

will be assumed most easily detected when the ratio of the difference between the flux F_1, measured with the spectrometer passband centered off the feature and the flux F_2 measured when the spectrometer is centered on the feature to the noise in the spectrum is a maximum. The noise will be assumed proportional to the square root of the sum of the flux F_2 and the dark response D of the detector system. In other words, we seek the maximum of the function $E = | (F_2 - F_1) |/(F_2 + D)^{1/2}$ as a function of the ratio $\mathcal{R}/\mathcal{R}_c$ of the instrumental resolution to the characteristic resolution.

There are basically two cases to consider: (1) continuous point source, and (2) continuous extended source. Line emission sources generally have a continuous background, so they can be considered in the above cases. For the purposes of this discussion, a point source is one for which D/d and ψ on Fig. 5 correspond to a point below a line of constant $\mathcal{R}$.

For case (1), changing the instrumental resolution does not change the etendue of the system, so the changes in F_1 and F_2 are a result only of changing the spectral width accepted. Hence,

$$E \approx \begin{cases} w(B_2 - B_1)/(WB_2 + D)^{1/2} & \text{if } w/W = \mathcal{R}/\mathcal{R}_c < 1, \\ W(B_2 - B_1)/(WB_2 + D)^{1/2} & \text{if } w/W = \mathcal{R}/\mathcal{R}_c > 1. \end{cases}$$

For case (2), changes in F_2 and F_1 arise from changing the system etendue as well as from changing the spectral width accepted. Since the etendue is proportional to W, that is, to $1/\mathcal{R}$, there results

$$E \approx \begin{cases} Ww(B_2 - B_1)/(W^2B_2 + D)^{1/2} & \text{if } w/W = \mathcal{R}/\mathcal{R}_c < 1, \\ W^2(B_2 - B_1)/(W^2B_2 + D)^{1/2} & \text{if } w/W = \mathcal{R}/\mathcal{R}_c > 1. \end{cases}$$

The results are summarized in Table I which gives the dependence of the ease of detecting a spectral feature on the ratio of the instrumental resolution to the source characteristic resolution for the idealized source and instrument for the photon limited case and the detector limited case. The photon limited case assumes D is negligible, and the detector-limited case assumes D dominates.

The most obvious, and probably the most important feature illustrated is that it is never advantageous to use higher resolution than required by the nature of the problem to be solved. This is particularly true in the case of faint extended sources when, for example, photomultiplier dark count is appreciable. When the simple detection of a line from a faint extended source is all that is required, it may be advantageous to use an instrumental resolution much less than the characteristic resolution (for example, using interference filters to measure night sky emission lines); however in practical cases a limit is set by other lines which begin to fall within the instrumental passband as it is increased. In most cases the instrumental resolution should be set equal to the characteristic resolution.

TABLE I. Dependence of the Ease of Detecting a Spectral Feature on the Ratio $\mathscr{R}/\mathscr{R}_c$ of the Spectrometer Resolution to the Source Characteristic Resolution

	Photon limited	Detector limited
Point source		
$\mathscr{R}/\mathscr{R}_c < 1$	$(\mathscr{R}/\mathscr{R}_c)^{1/2}$	$(\mathscr{R}/\mathscr{R}_c)^0$
$\mathscr{R}/\mathscr{R}_c > 1$	$(\mathscr{R}/\mathscr{R}_c)^{-1/2}$	$(\mathscr{R}/\mathscr{R}_c)^{-1}$
Extended source		
$\mathscr{R}/\mathscr{R}_c < 1$	$(\mathscr{R}/\mathscr{R}_c)^0$	$(\mathscr{R}/\mathscr{R}_c)^{-1}$
$\mathscr{R}/\mathscr{R}_c > 1$	$(\mathscr{R}/\mathscr{R}_c)^{-1}$	$(\mathscr{R}/\mathscr{R}_c)^{-2}$

To go beyond the simplified treatment expands the complication enormously with little gain in principle. In practice the question can become very important as, for example, in the case where detailed profiles of planetary absorption lines are required for model fitting, and the problem should be expanded in some detail according to particular problems.

12.5. Examples of Basic Fabry–Perot Spectrometer Design for Astronomical Observations

We consider the basic optics of only a few examples to illustrate the application of the principles discussed. A wide variety of examples is readily found in the literature.

12.5.1. A Single-Etalon Nebular Spectrometer

The spectra of nebulae are characterized by a few relatively bright emission lines such as the 5007-Å forbidden line of O III which in general may be satisfactorily isolated by an interference filter. Conditions are typically such that electron temperatures and velocity microstructure produce lines about 0.5 Å (2.0 cm^{-1}) in width.

Let us say we wish to design a spectrometer for a 1-m telescope using the 5007-Å line to study the temperature and velocity field of a particular nebula with a detail of about 1.5 arcmin. If only the detection of the line were of interest, the discussion in Section 12.4.3 would suggest a design resolution limit of 2.0 cm^{-1}; however, since profile is assumed to be of interest, it is well to use a design resolution limit somewhat less, say about 1.0 cm^{-1} corresponding to an instrumental resolution $\mathscr{R} = 20{,}000$, so that the broadening of the recorded spectrum is about as much from the source as from the instrument.

[The problem of fitting the recorded data with a model source profile is a general one in spectroscopy, and is not considered further in this discussion of the characteristics of Fabry–Perot spectrometers. The reader should be aware, however, that the recorded spectrum differs from the source profile, and that computational fitting methods are commonly used to reconstruct the source spectrum. For this purpose, the instrumental profile is generally measured by scanning a laboratory line known to be much narrower than the instrumental profile, but illuminating the spectrometer as nearly as possible identically to the illumination by the astronomical source to avoid systematic shift or profile distortion.]

Once the requirements imposed by the source and study objectives have been determined, the properties of the interferometer are required to determine the design parameters. Assume the available interferometer plates have a flatness tolerance of $\lambda/100$, a reasonable figure for the current state of the art, and that they can be coated without appreciably degrading the surface figure. The discussion of Section 12.2.5 implies that a reflective finesse $N_R = N_D = 50$ is reasonable. From Eq. (12.1.5), a reflectance of 0.94 is suggested. It is somewhat safer to allow for possible mounting and thermal induced distortions, for example, which typically occur in the field, by choosing a reflectance somewhat less than this, say 0.90. This helps keep the transmittance of the instrument high both by the reduced loss due to plate defects (Section 12.2.5) and by the lower absorption in the lower reflectance coatings. Thus $N_R = 30$ which implies (Section 12.2.6) an instrumental finesse $N \approx 0.6 N_R = 18$. Combined with the required resolution limit of 1.0 cm^{-1}, this implies a free spectral range $Q = N\delta\sigma = (18)(1.0)$ cm^{-1}, and a spacer length $l = 1/2Q = 0.28$ mm. The theoretical resolution $R_t = 2N_R\sigma l = (2)(30)(20000)(0.033)$

$= 3.36 \times 10^4$, and the acceptance angle of the Fabry–Perot in this case is $\beta = (8/R_t)^{1/2} = 1.55 \times 10^{-2}$. Since the original field given was 1.5 arcmin $(4.5 \times 10^{-4}\,\text{rad})$, the sky magnification (ratio of the effective telescope focal length to the spectrometer collimator focal length) must be 33, so the useful interferometer aperture is, for the 1-m telescope, 30 mm. Equivalently using the etendue matching equation (12.4.1) as represented in Fig. 5 for $\mathscr{R} = (1.2)(2 \times 10^4) = 24{,}000$ gives $D/d \approx 33$ and hence $d = 30$ mm. (The factor 1.2 is introduced because defects are not included in Fig. 5.)

Even though Fig. 4 suggests a double-etalon solution since $\mathscr{R}_c/\mathscr{R}_F \approx 50$, the line characteristic of the source makes the single-etalon solution acceptable. Actually Meaburn,[21] for example, has shown that a double-etalon spectrometer can give more satisfactory results because of the ever-present continuous background.

12.5.2. A Large-Aperture, Double-Etalon Spectrometer

The following instrument design problem[22] relates to the study of Hα emission not associated with classical H II regions, and differs from the preceding in that the source is fainter and the required resolution is slightly higher. For example, the intensity of the Hα line away from classical H II regions is about 10^{-2} times that from typical H II regions, and the expected line width is such that a resolution limit of about 0.15 Å $(0.35\ \text{cm}^{-1})$ is appropriate. The faintness of these lines requires a large-aperture Fabry–Perot, and the addition of a second Fabry–Perot is required to reduce the no longer negligible contribution from the continuous night sky background and the several other faint night sky lines in the vicinity. According to Section 12.3.4 with $\mathscr{R}_c/\mathscr{R}_F = 44{,}000/500 = 88$, the double-etalon instrument will provide sufficient purity of isolation.

To achieve the desired resolution with an instrumental finesse of about 15, the reflectance is chosen (following reasoning similar to Section 12.5.1) to be $R \approx 0.90$, with a spacer $l = 1.0$ mm. The primary etalon has a diameter of 150 mm, and for reasons of economy the secondary, or suppression etalon is 50 mm in diameter. The coupling is accomplished as shown in Fig. 2b with $f_1/f_2 = D_1/D_2 = 3$. Thus, following Eq. (12.3.1), $l_2 = l_1/9 = 0.11$ mm. Since $l_2 \ll l_1$, the resolution is determined almost

[21] J. Meaburn, *Astron. Astrophys.* **17**, 106 (1972).

[22] R. J. Reynolds, F. L. Roesler, and F. Scherb, *Astrophys. J.* **179**, 651 (1973).

entirely by the large etalon. The coupling arrangement of Fig. 2b was chosen rather than the arrangement of Fig. 2a because at the low resolution required, especially for the suppression etalon, vignetting losses would be appreciable. Assuming the instrument is to be coupled to a 90-cm (36-in.) diameter telescope, one has (Section 12.4.2) for D/d = 90/15 = 6 and $\mathscr{R}$ = 44,000, a field of view $\psi = 1.6 \times 10^{-3}$ rad = 5.5 arcmin.

12.5.3. A High-Resolution Triple-Etalon Spectrometer for Continuous Sources—PEPSIOS

For studies of continuous sources requiring resolutions in excess of 10^5, such as in studies of planetary atmospheres,[20,23] interstellar lines,[24] or solar lines,[25,26] the PEPSIOS spectrometer[14,15] with three interferometers and an interference filter provides a satisfactory solution. For a problem requiring a resolution of 2.5×10^5 at 5000 Å, for example, suitable purity of isolation is provided by a combination of three etalons with spacers in the ratios 1.000:0.8831:0.7244, as suggested by McNutt.[16] The instrumental resolution is higher than that of the largest spaced etalon by a factor about 1.5 for reasons discussed in Section 12.3.3. Thus the resolution of the largest etalon to achieve the desired resolution should be 1.67×10^5. Using the parameters of the first example which gave a single-etalon instrumental finesse of about 15 suggests a largest spacer of nominal magnitude $l_1 = \mathscr{R}/2\sigma N = 1.67 \times 10^5/(2)(2 \times 10^4)(15)$ = 2.8 mm with the others nominally $l_2 = 2.47$ mm and $l_3 = 2.02$ mm. The ratios l_2/l_1 and l_3/l_1, however, must be precisely maintained. It has already been pointed out that for stellar sources even under poor seeing conditions $D/d \approx 200$, so that even for the largest telescopes only a small diameter (about 25 mm) of the etalon is used.

In order to avoid the effect of reflections between etalons which tends to reduce the contrast of the instrument, it is best to put the etalon with the smallest spacer in the center, and to tilt it slightly off axis. Since the smaller-spaced etalon has the greatest angular acceptance [Eq. (12.2.1)], this can be done with only negligible loss in transmittance.

[23] N. P. Carleton, A. Sharma, R. M. Goody, W. L. Liller, and F. L. Roesler, *Astrophys. J.* **155**, 323 (1969).

[24] L. M. Hobbs, *Astrophys. J.* **142**, 160 (1965); **157**, 135 (1969).

[25] Mark Daehler, *Astrophys. J.* **150**, 667 (1967).

[26] W. Traub and F. L. Roesler, *Astrophys. J.* **163**, 629 (1971).

Plate defect tolerances are somewhat stricter for a polyetalon than for a single etalon,[19] and extra care must be exercised in the selection, coating, and mounting of plates used in a PEPSIOS spectrometer.

12.6. Adjustment and Calibration of Fabry–Perot Spectrometers

12.6.1. Adjustment of Single-Etalon Spectrometers

There are basically three adjustments necessary for the operation of a single-etalon spectrometer: (1) adjustment of plate parallelism, (2) adjustment of the interferometer axis parallel to the spectrometer axis, and (3) fine adjustment of the spacing to bring the line of interest within the scanning range of the interferometer.

The test for plate parallelism is conveniently made by looking through the Fabry–Perot at a large area monochromatic source and observing the ring pattern directly with the eye. Since the pupil of the eye is a few millimeters in diameter, the ring pattern seen is that associated with the average spacing of a very small region of the aperture normal to the line between the eye and the center of the bull's-eye ring pattern. When the eye is moved to a different region of the plate, the spacing of the new region will be in general different if the plates are nonparallel. The ring pattern will be seen expanded if the new region has larger spacing, and contracted if the spacing is smaller. When the ring observed is nearly closed on axis, this test for parallelism is easily sensitive to $\lambda/4N_R$, or typically $\lambda/400$. Since plates are rarely produced and mounted to this degree of precision, one sees a combination of plate defects and parallelism error, and some judgement is required to achieve an optimum adjustment. When the aperture is large it is advisable to mask off regions observed to have large defects; if the aperture is small, as in the observation of stellar sources, it may be advantageous to have means for selecting the plate region used.

In view of the extreme sensitivity to small spacing charges, particular care must be taken with the mounting of the plates and with the means for adjusting them with the required precision. Generally the plates are mounted with a spacer ring between the plates, with provision for applying force to achieve fine adjustment by slight compression of the spacers. This configuration can provide the greatest stability. The only essential points are the following: The spacer should be made of a stable low-expansion material such as Invar or fused silica, and have three areas for contacting the plate faces near their edges. The plate-contacting regions of opposing land areas of the spacer must be accurately opposite each other. The mean planes determined by the upper and lower land areas, respectively, should be lapped parallel to within about ½ fringe. The contacting areas (observed by the dark zero-thickness fringe in the light reflected from the spacer land and plate contact region) should be roughly equal and at least 2-mm square, but not so large as to require excessive force for adjustment of about ± 1 fringe. (A maximum force of about 2 kg has been found satisfactory.) The planes of the respective land areas must be parallel to the mean plane of the three which contact a plate face to within about $\lambda/4$ over the contacting area. The contact areas of each land must be accurately centered on the line of the adjusting force. If edge contact with the plates is required, for example, when it is necessary to hold the plates in place when the instrument is to be swung from the back of the telescope, extreme care must be exercised to avoid small torques which will twist the plates. (Note that changes in gravitational force may cause changes in etalon adjustment depending on the spacer contact area.)

It is possible to achieve a mounting of unusual stability by optically contacting separate spacer buttons to the plates to produce essentially a monolithic interferometer. Tolerances must be extremely high since the plate faces conform to the spacer faces. Furthermore the assembly and disassembly of the contacted interferometer requires great skill.

The optical axis of the interferometer is put parallel to the spectrometer axis by autocollimation. Light sent through the entrance aperture is reflected from the interferometer. When the spectrometer axis and interferometer axes are parallel, the light returns through the entrance aperture. If the interference filter is used between the etalon and entrance aperture, it may be removed to make this adjustment easier; however interference filters might produce a slight beam deviation and should be checked. Since the separate plates of the interferometer are almost always wedged to avoid interference from the unwanted reflected light from the outside faces of the interferometer, it is usually necessary to adjust the exit aperture after the interferometer is in place.

When the spacing of the interferometer plates is so small that, in the case of a pressure scanned instrument, a full free spectral range can barely be scanned within a convenient pressure range, the forces on the three spacer feet must be changed to change the gap while checking its uniformity using the ring patterns. Increasing the force compressing the spacer increases the gas pressure at which a given transmittance peak occurs, since the refractive index must be increased to keep the optical spacing constant.

12.6.2. Adjustment of Polyetalon Spectrometers

When two or more interferometers are used in series one has in addition to the requirements described above for the single etalon, the requirement that the optical spacing of the additional etalons must be carefully adjusted to give a transmittance peak coincident with the desired transmittance of the first. There are several ways of accomplishing this.

The gap of the second etalon may be fine tuned mechanically by adjusting the magnitude of the gap. This is possible even for fixed gap etalons since the contact areas and adjusting force range are properly designed to provide a full fringe adjustment by compressing the spacers. With the first etalon adjusted to pass a convenient laboratory source line at or near the wavelength to be studied, the second etalon is mechanically fine tuned, constantly verifying parallelism, until maximum transmittance of the combination is achieved. The principle is readily extended to three etalons.

While considerable skill is required to produce a good alignment, this method has the advantage of very high stability (PEPSIOS instruments adjusted by this method have operated months at a time without need of adjustment), and permits synchronous scanning of all the etalons in a common pressure chamber. It has the disadvantage that the wavelength adjustment of the instrument cannot be shifted readily.

When the etalons are housed in separate pressure chambers, the first may be pressure scanned until the desired wavelength is transmitted. With the pressure in the first etalon chamber held fixed, the second etalon is pressure scanned until it too passes the desired wavelength. The two etalons may be scanned synchronously provided the *pressure difference* between the two is held constant during the scan.[14] This requires a precise pressure difference control system. While the pressure difference method requires a rather complex pressure difference control system, it offers the greatest ease in tuning. Moreover, opposed to the above technique which required a line at or very near the wavelength range to be studied, tuning at an arbitrary wavelength is readily accomplished by the pressure stepping technique.

Figure 7 shows an example of the pressure-stepping technique used to tune a PEPSIOS spectrometer for the study of planetary CO_2 band lines near 10493 Å (9527.55 cm^{-1}). The nearest convenient laboratory line is the Ar I line at 10470.05 Å (9548.44 cm^{-1}). The passbands of the three etalons in the McNutt ratios with largest spacer $l_1 = 4$ mm are shown tuned on the Ar line. With this tune the required scanning range of nearly 21 cm^{-1} is outside practical limits. However, as shown in the figure, the free spectral ranges of the interferometers are accurately calculable, both in terms of spectral interval and pressure interval. Therefore the pressure steps required to bring into coincidence the 16th, 14th, and 12th peaks of the three etalons, thereby effecting an alignment

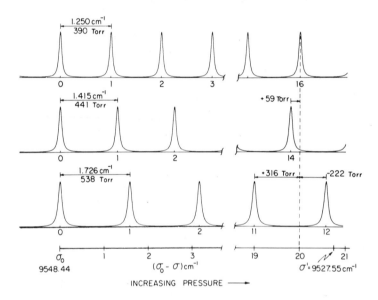

Fig. 7. Illustration of the pressure-stepping technique used for tuning polyetalon spectrometers. A series of three etalons tuned to pass the Ar I 9548.44 cm^{-1} line from a convenient laboratory source can be adjusted to the region of the CO_2 band line at 9527.55 cm^{-1} by adjusting the pressure by the differences indicated in the figure.

near the CO_2 band line, are accurately calculable. The tune at the Ar I line will in general give a first set of pressure differences. The required steps for alignment near the CO_2 line are added to these by readjustment of the (calibrated) pressure difference sensors to give a new set of differences. Fine adjustment of the new pressure differences can be achieved by the technique of white light tuning in which the second and third etalons are tuned to the light transmitted by the desired peak of the first. Unless the pressure steps are grossly in error there will be no other near coincidence of peaks which might pull the tune to an erroneous position.

A technique somewhat different from that used to adjust the axis of the single-etalon instrument is usually required for adjusting the axes of additional etalons. This is simply because it is often difficult to see the return reflection from the second etalon through the first. The first etalon is adjusted as for a single-etalon instrument. One technique for adjusting the remaining etalons is to observe the image of the entrance aperture at the exit aperture. If the axes are not parallel, multiple reflections between the etalons will produce a series of images which trail off away from the direct image. As the axes become more parallel, the images come together and coincide when the axes are accurately parallel. It is usually possible to observe this separately for each of the additional etalons. Actually, as pointed out in Section 12.5.3, it is best to have a slight axis tilt. The observation of the flare spot that is seen with the axis slightly inclined serves as a convenient way to set the desired tilt.

An alternative way of adjusting the axes is to observe the ring patterns of all etalons in the series by looking through the series illuminated by a spatially broad monochromatic source. It is easy to judge when the ring patterns are concentric, indicating that the axes are parallel. However, in an actual instrument this method may be less readily applicable than the flare spot method.

12.6.3. Calibration of Fabry–Perot Spectrometers

The Fabry–Perot spectrometer is capable of precise relative and absolute wavelength measurements. In a scanning Fabry–Perot spectrometer some means must be provided for calibrating the scan. In the case of a single Fabry–Perot used to study one or a few closely spaced wavelengths, the system may be self-calibrating for relative measurements as discussed in Section 12.2.4. Absolute calibration is provided by scanning a nearby laboratory line of known wavelength. Even if the known line lies some distance from the wavelength region being studied, absolute calibration may be achieved in the region of interest by relating the known line to the region to be calibrated through the accurately measurable free spectral range of the interferometer. The method is readily inferred from Fig. 7. The interferometer spacing must be known, and the region being studied must be known to within a free-spectral range.

For example, referring to Fig. 7, assume that it is desired to study a hypothetical emission line expected at $\sigma = 9528.3$ cm^{-1} using a single pressure scanned Fabry–Perot system. The Fabry–Perot has a spacing of 2.897 mm giving a free spectral range of 1.726 cm^{-1} as illustrated in the bottom line of Fig. 7. A transmittance peak of the Ar I line at $\sigma = 9548.44$ cm^{-1} ($\lambda = 10,470$ Å) which is to be used for calibration is found when $P_0 = 321$ Torr and again at $P_0 + \Delta P = 859$ Torr. The pressure P_0 also gives transmission peaks at wave numbers differing from the calibration line by multiples of the free spectral range. In the present example, the transmittance peak 12 orders to the red is at $\sigma = 9548.44 - (12) \times (1.726) = 9527.73$ cm^{-1}, or 0.57 cm^{-1} toward the red from the expected emission line. The expected emission line may be reached by scanning the interferometer towards the blue (larger wave numbers) by a fraction of an order $f = 0.57/1.726$ corresponding to a pressure decrease of $(0.330)(538) = 177.5$ Torr. Thus the emission line will be expected at $P = 143.5$ Torr and at $P = 681.5$ Torr. If the line appeared at, say, exactly 154.3 Torr, the implied wave number would be $\sigma = 9527.73 + (321.0 - 154.3)(1.726)/538 = 9528.26$ cm^{-1}. Note carefully that vacuum wave numbers have been used. It is obviously very important to know whether the value of wavelength or wave number used for calibration is the vacuum or standard air value, and to be consistent in the use of either. Wavelength lists are usually for standard air, and conversion to vacuum wavelength or vacuum wave number must be made using the refractive index of air at the wavelength of interest. The NBS wave number tables[27] are convenient for this purpose. Note also that this calibration method assumes that the refractive index of the scanning gas is the same for the calibration and the unknown line. This is generally a sufficiently precise assumption so long as the calibration and unknown line are within several angstroms. If necessary, correction can be made readily if the dispersion of the scanning gas is accurately known. It should also be pointed out that the pressure of the scanning gas is temperature-dependent. In the above example, compensation for the temperature was accomplished automatically by measuring the pressure difference corresponding to a free spectral range by scanning two adjacent peaks of the calibration lines at the time of the experiment and using that pressure difference in the calibration calculations. This method of calibration is extended in a more or less obvious manner to multiple Fabry–Perot spectrometers.

The exact functional relationship between the parameter effecting the scan and the wave number must be accurately known. In the case of mechanical scanning by changing the interferometer spacing by the

[27] C. D. Coleman, W. R. Bozman, and W. F. Meggers, Tables of Wavenumbers. Nat. Bur. of Std. Monograph 3. U.S. Govt. Printing Office, Washington, D.C., 1960.

piezoelectric effect, for example, the relationship between displacement and voltage can be determined by measuring several orders of a known wavelength. If more frequent points are needed on the displacement vs. voltage curve, the experiment may be repeated using different wavelengths. Alternatively, ring diameters may be measured as a function of voltage and related to the spacing through Eq. (12.1.2), for example.

In the case of pressure scanning an auxiliary interferometer may be used to obtain more frequent calibration points. This interferometer may be either a large gap Fabry–Perot interferometer housed in a chamber having the same pressure as the primary etalon of the spectrometer, or a Michelson interferometer with a chamber in one arm connected to the primary etalon chamber. If the length of the Fabry–Perot gap or chamber in the Michelson is l, then when the calibrating interferometer and spectrometer are observing the same wavelength, the calibrator provides fringes at spectral intervals $\Delta\sigma_r = \frac{1}{2}l$. Typically for a calibration Fabry–Perot $l = 5$ cm and $\Delta\sigma_r = 0.1$ cm^{-1}. For the Michelson type calibrator l may easily be 25 cm giving calibration marks at intervals of $\Delta\sigma_r = 0.02$ cm^{-1}. When the calibrator and spectrometer are used at different wavelengths, as is usually the case, the spectral interval $\Delta\sigma_s$ at the spectrometer wave number σ_s may be related to the spectral interval $\Delta\sigma_r$ at the reference wave number σ_r by the equation

$$\Delta\sigma_s = \Delta\sigma_r(\sigma_s/\sigma_r)(n_r/n_s)(1 - n_r)/(1 - n_s),$$

where n_r and n_s are the refractive indices of the scanning gas at standard conditions for the reference and spectrometer wavelengths, respectively.

Optically, the basic differences in the two calibration schemes arise from the different orders of interference [Eq. (12.1.2)] in the two cases. For the Fabry–Perot the order of interference m_F is high; typically $m_F = 10^5$ for a practical case. For the Michelson, the order of interference is roughly $m_M = 2\sigma(l_1 - nl_2) \approx 10^3$, where l_1 and l_2 are the arm lengths and n is the maximum refractive index of the scanning gas. This has at least two important consequences. First $\Delta\sigma_F/\Delta\sigma_M = m_M/m_F \approx 10^{-2}$, where $\Delta\sigma_F$ and $\Delta\sigma_M$ are the spectral ranges between intensity maxima for the light illuminating the Fabry–Perot and Michelson calibration interferometers. (Note that whereas $\Delta\sigma_F$ is equal to the $\Delta\sigma_r$ referred to above, $\Delta\sigma_M = \frac{1}{2}(l_1 - nl_2) \neq \Delta\sigma_r = \frac{1}{2}l_2$). Secondly, the acceptance angles, which determine the etendue, or light gathering power, are related by

$\Delta\theta_F/\Delta\theta_M = m_M/m_F \approx 10^{-1}$. The practical consequence of this is that a bright and spectrally narrow light source is needed for Fabry–Perot calibration, while a much broader and fainter source is adequate for the Michelson. Typically, an electrodeless Hg[198] tube is needed for Fabry–Perot calibration, while natural mercury germicidal lamps are adequate for Michelson calibration. These considerations indicate that the Michelson calibrator is a more practical device. While generally the reference fringes provided by the Fabry–Perot device have been found to be more stable than those provided by the usual Michelson interferometer, the Michelson device has been found entirely adequate for astronomical purposes.

When the exact values to be used in calculating the correct spectral equivalent of the calibration fringes at the wavelength being studied are not known or not readily measured, a known spectral line may be scanned using a Fabry–Perot in the spectrometer with a spacer large enough to give at least one full free spectral range. The free spectral range of the interferometer divided by the number of calibration marks gives the spectral interval per calibration mark.

Several of the references cited give examples of work in which calibration was done with a Fabry–Perot[24–26] and a Michelson interferometer.[20,23] In the case of the Michelson, the calibration fringes can be closely enough spaced to be used to trigger the data readout system. This is especially useful when a long series of scans is to be added, since the data points are recorded into precise spectral channels.

12.7. Comparison with Other Instruments

12.7.1. Comparison of Fabry–Perot and Grating Spectrometers

The basic question for any spectrometer is whether, at a given spectral resolution, it can accept all the light per spectral element delivered by the telescope, that is, whether the spectrometer etendue (Section 12.2.3) is greater or less than the etendue of the source–telescope system. Consider a telescope of objective diameter D observing a circular source of angular diameter ψ coupled to a spectrometer with effective disperser diameter d. If the telescope aperture is imaged on the dispersing element, the aperture magnification is d/D and the source angle magnification is D/d. The dispersing element is thus illuminated with a light bundle of angular divergence $\eta = \psi D/d$ unless the angular divergence is limited by the entrance aperture of the spectrometer. The latter case is the spectrometer-limited case, and the former the telescope-limited case.

For the Fabry–Perot spectrometer the onset of the spectrometer-limited situation for a given d, D and ψ depends upon the resolution. The limiting resolution is reached when $\eta > \beta = (8/\mathscr{R}_{L})^{1/2}$, or $\mathscr{R}_{L} = 8/\eta^{2}$. Thus when $\mathscr{R} > 8/\eta^{2}$ the system is spectrometer limited, and the etendue (or optical acceptance) decreases proportional to $\mathscr{R}^{-1}$ [Eq. (12.2.4)] from the limit $U_{T} = A\omega = \pi^{2}D^{2}\psi^{2}/16$ (Section 12.4.2).

To establish the equivalent criterion for a grating, the geometry of a grating spectrometer must first be briefly treated. Consider an echelle grating spectrometer with a blaze angle of $63°26'$, typical of several in current use in astronomical applications.[28] The grating equation is $n\lambda = s(\sin \theta_{1} + \sin \theta_{2})$, where s is the grating spacing and θ_{1} and θ_{2} are, respectively, the angles of incidence and diffraction. When the grating is illuminated with a range of angles ε about θ_{1}, then $\delta\lambda = (s/n)\varepsilon \cos \theta$. In the usual manner of illuminating an echelle $\theta_{1} \approx \theta_{2} \approx \theta_{B}$, so $\mathscr{R} = \lambda/\delta\lambda = 2 \tan \theta_{B}/\varepsilon = 4/\varepsilon$ for $\theta_{B} = 63°26'$. This holds so long as $\mathscr{R} \ll \mathscr{R}_{t}$, where $\mathscr{R}_{t}$ is the theoretical resolution limit of the grating given by $\mathscr{R}_{t} = (2W \sin \theta)/\lambda \approx 2d \tan \theta/\lambda = 4d/\lambda$, where W is the grating width here taken as approximately equal to $d \sec \theta$. For a 15-cm effective aperture echelle $\mathscr{R}_{t} = 1.2 \times 10^{6}$, so the above limit is the usual case in astronomy where almost always $\mathscr{R} < 3 \times 10^{5}$.

When the resolution required is such that $\mathscr{R} = 4/\varepsilon > 4/\eta$, the spectrometer slits begin to trim off the edges of the image of the circular source. The etendue decreases gradually until roughly $\mathscr{R} > 8/\eta$, and then falls off nearly proportional to $\mathscr{R}^{-1}$ since $\mathscr{R}$ is inversely proportional to the slit width. We take as the limiting resolution for the grating $\mathscr{R}_{L} = 8/\eta$.

In Fig. 8 the limiting resolutions for the grating and Fabry–Perot are plotted as a function of $\eta = \psi D/d$. Several general conclusions appear. First, for stellar sources under usual seeing conditions observed with telescopes having primary mirrors in the 1-m range there is little or no advantage to the Fabry–Perot since either instrument accepts all the light that is available at most practical resolutions. For a 100-cm telescope equipped with 15-cm aperture echelle, for example, the limiting resolution for stellar sources under typical seeing conditions ($\psi = 10^{-5}$) is $\mathscr{R}_{L} = 120{,}000$, or adequate for most stellar problems.

Secondly, for the largest telescopes the Fabry–Perot can offer a large

[28] D. J. Schroeder, *Appl. Opt.* **6**, 1976 (1967).

advantage, especially at the higher resolutions required for the investigation of interstellar lines, for example. A 4-m telescope observing a source with effective diameter $\psi = 10^{-5}$ gives for the echelle a limiting resolution $\mathscr{R}_L = 30,000$. If the required resolution is 3×10^5 for an interstellar line problem, then the grating accepts only 1/10 the available light. For a planetary source requiring the same spectral resolution the fraction of the available light accepted may be five to ten times smaller still. For a 5-cm diameter Fabry–Perot, on the other hand, the conditions give a

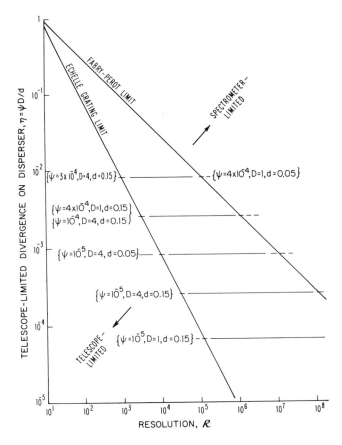

FIG. 8. Resolution at which a Fabry–Perot or echelle grating spectrometer limits the etendue of the source–telescope–spectrometer system characterized by the parameter $\eta = \psi D/d$. η is also the telescope-limited divergence of light on the disperser. Several combinations referred to in the text are indicated on the graph. Below a line, all the light accepted by the telescope is accepted by the spectrometer. Above a line the spectrometer fails to accept the available light by a factor proportional to $\mathscr{R}_L/\mathscr{R}$.

limiting resolution $\mathscr{R}_L = 12 \times 10^6$, indicating that the Fabry–Perot is still far from the spectrometer limited case.

Thirdly, for nebular work a Fabry–Perot is highly advantageous. For even a 1-m telescope and an angular field of view of 1.5′ one has for the 15-cm grating aperture $\mathscr{R}_L \approx 2600$, which is a factor of ten lower than typically desirable, indicating that only about 1/10 of the available light can be accepted by the spectrometer. A 5-cm Fabry–Perot, on the other hand, still has excess capacity at a desirable resolution for nebular problems. On a larger telescope or with larger fields of view the Fabry–Perot system could easily be a hundred times more efficient. The various examples are indicated in Fig. 8.

The treatment above needs to be moderate somewhat. The relative transmittance of the grating and Fabry–Perot system has not been mentioned. In practice they can be comparable at wavelengths greater than about 4300 Å if sufficient care is taken with the optical quality of the interferometer. For shorter wavelengths, increasing absorption and increasing difficulty in achieving the required optical quality begins to reduce the Fabry–Perot gain. However, for special applications in astronomy, uv Fabry–Perots have been used.[29,30] Also, some of the light lost when the grating slit must be made smaller than the source image can be recovered by an image slicer which reimages the cut off portion through the slit above and below the central portion of the image. Furthermore, the comparison is directed towards scanning spectrometers. Current progress in the development of image sensors having the high quantum efficiency and photon-limited performance previously realized only with photomultiplier tubes indicates that high-efficiency grating spectrographs recording large spectral ranges may soon become widely used. This will increase the efficiency of grating instruments for most astronomical applications.

The question of efficiency as treated above should not be considered independently of the objective of a program. For example, if a spectrum over a very wide spectral range is needed, then one can accept considerable loss in spectrometer etendue to make up for the difficulty of step-scanning a Fabry–Perot spectrometer.

[29] D. J. Bradley, B. Bates, C.O.L. Juulman, and T. Kohno. *J. Phys. Suppl. C2* 280 (1967).

[30] B. Bates, C. D. McKeith, G. R. Courts, and J. K. Conway, *Proc. Congr. Int. Commission Opt., 9th, 1972* (in preparation).

12.7.2. Comparison of Fabry–Perot and Michelson Fourier Transform Spectrometers

In the usual case the Fabry–Perot interferometer and Michelson interferometer as used in Fourier transform spectrometry are equivalent from the standpoint of etendue. With the Michelson interferometer all spectral elements are observed simultaneously, and each is coded by the ratio of the mirror displacement to the wavelength of the spectral element to produce a different frequency in the interferogram, which is the output intensity as a function of mirror displacement. The interferogram is then frequency analyzed (Fourier transformed) to recover the original spectrum. This multiplexing technique has a great advantage in the detector-noise-limited case where it can be shown that one achieves an additional advantage equal to the number of spectral elements analyzed. This may be many thousands in addition to the etendue gain compared to grating spectrometers, and accounts for the wide use of Fourier transform interferometers for wavelengths greater than about 1.0 μm. For shorter wavelengths, detectors are quantum-noise-limited and the multiplex gain largely vanishes. In some cases multiplexing may be a disadvantage when for example, in the quantum-noise-limited case one is obliged to spend time observing more spectral elements than he really wishes. In general, practice seems to indicate that for wavelengths shorter than 1.0 μm, the Fabry–Perot spectrometer is most practical for detailed studies of small spectral ranges at higher resolutions, or for the study of isolated emission lines in nebulae, for example. This is especially true when looking for short time variations, as in some planetary problems. The Michelson interferometer is useful for obtaining broad-range visible spectra especially at low and moderate resolutions.

As detectors in the infrared approach the quantum limit, as seems to be indicated on the horizon of detector development, much of the multiplex advantage may disappear. In the case of far-infrared spectra observed from ground-based or airplane-borne telescopes, noise associated with the thermal radiation from the telescope and atmosphere prevents the acquisition of spectra of faint astronomical sources by Fourier transform spectroscopy. In that case, the filtering potentially provided by a liquid helium cooled Fabry–Perot spectrometer can be expected to offer a large advantage.

Since the Michelson interferometer in principle has only one transmitting element, the beam splitter, and requires an optical tolerance of only about $\lambda/10$ as contrasted with the Fabry–Perot tolerance of $\lambda/100$, it could potentially achieve with useful efficiency the advantage characteristic of interference spectrometers in at least the near ultraviolet where the transmittance of Fabry–Perot instruments is excessively low for most applications. Recently an all-reflection interferometer useful for Fourier transform spectroscopy has been described.[31] Such an interferometer may prove useful in spectral regions where transmitting materials are not available.

While it may be possible to define quantitative analytical criteria for the comparison of spectrometers, the choice of a particular spectrometer for a particular problem by a particular observer is correctly dictated in part by factors which cannot be readily described. The reader is invited to read a few pages by Jacquinot[32] which, although not particularly related to Fabry–Perot spectrometers, are noteworthy for their reasonable approach toward instrument selection.

[31] R. A. Kruger, L. W. Anderson, and F. L. Roesler, *J. Opt. Soc. Amer.* **62**, 938 (1972).

[32] P. Jacquinot, *Appl. Opt.* **8**, 497 (1969).

AUTHOR INDEX

Numbers in parentheses are footnote numbers. They are inserted to indicate that the reference to an author's work is cited with a footnote number and his name does not appear on that page.

SUBJECT INDEX

A

Aberrations, in spectrometers, 477–479
ADP, *see* Ammonium dihydrogen phosphate
Ag–O–Cs, 24
Air showers, 317, 356
  distinction between proton and gamma-ray induction, 356
Airy function, 532
Ammonium dihydrogen phosphate (ADP), 378
Analyzers for linearly polarized light, 363
Anamorphic magnification, 464, 469
Apparatus function, 502
Apodization, 404, 502, 508
Atmosphere, 416
  absorption of infrared by, 416, 440, 447–448, 453–454
  emission of infrared radiation by, 416, 435, 438, 442, 444, 447–451
  transmission through infrared windows, 447–448
Atmospheric fluorescence, 316–343
  absorption of X-rays in, 320–324
  arrival direction of exciting radiation, 329
  background sources of light, 338–343
  decay time of, 321
  efficiency of, 319
  ground-based detection of, 324
  molecular transitions involved, 318
  sensitivity of, 329
  time duration of a pulse, 327

B

Background, infrared, 422–426, 429, 435–444, 447–449, 451

Background, light-dependent, in image tubes, 242
Baking of photographic emulsions, 233
Band-gap energy, 8
Beam pattern, 466
Beam splitters, 428–429
Birefringent interferometers, 401–406
  (*see also* Fourier spectrometers)
Blackening, 421
Blaze, of diffraction gratings, 472
Bouguer plots, 142
Brightness, 193

C

Cadmium sulfide, 379
Calcite, 368, 374
Calibration, infrared, 440, 451–460
Calibration, photometric, 112–120
Cements, for optical components, 370
Cerenkov light, 40, 317, 343–350
Cesium iodide, 379
Channel plate, 239
Chopping secondary mirror, 446–447, 450
Circular dichroism, 397
Circular polarization, 395
Coadding, 524
Color indices, infrared, 451, 456–460
Compensator, 401, 404
Conduction band, 8
Conical scanning, 215–219
Convolution integral, 499
Crab-nebula pulsar, 332
Cross-disperser, 475, 483
Cryogens, 429–432
Cryostats, 420, 429–432, 438, 460
$CsNa_2KSb$, 22
Cs_3Sb, 19–21
Szerny–Turner spectrometer, 476–479

A 4
B 5
C 6
D 7
E 8
F 9
G 0
H 1
I 2
J 3

DORDT COLLEGE LIBRARY
Sioux Center, Iowa 51250